AF342092

Perfumes, Soaps, Detergents & Cosmetics

Volume II

(Perfumes & Cosmetics)

Perfumes, Soaps, Detergents & Cosmetics

Volume II
(Perfumes & Cosmetics)

S.C.Bhatia
B.E.(Chemical), M.B.A.

CBS

CBS PUBLISHERS & DISTRIBUTORS
4596/1A, 11 Darya Ganj, New Delhi - 110 002 (India)

ISBN : 81-239-0726-5

First Edition : 2001
     Reprint : 2009

Published by :
Satish Kumar Jain for CBS Publishers and Distributors,
4596/1-A, 11 Darya Ganj, New Delhi - 110 002 (India)

Printed at :
Asia Printograph,
Shahdara, Delhi - 110 032

Preface

This compact but comprehensive source book on perfumes and cosmetics is the second part of an informative textbook on the perfume, soap, detergent and cosmetic industries. It is concerned primarily with the manufacturing processes, raw materials and equipments related to these industries but also considers ancillary topics needed for a full understanding of the subject. This reference textbook is intended for chemists, postgraduates in chemistry, specialists in oil, fats, soap, detergents, perfumes and cosmetics, chemical engineers and research students.

Section I deals with the perfume industry. The agents that cause olfactory sensations have been researched and understood over the years and odours play a very important role in a wide range of household, agricultural, industrial, food and personal products. The creation of a good perfume oil for use in a consumer product requires an intimate knowledge of all fragrance compounds. As the manufacture of essential oils or perfumes has been carried out in India for centuries, this topic has been highlighted in Chapter 1, an overview of the perfume industry in India. Chapter 2 focuses on natural raw materials used in making perfumes. These raw materials are of plant or animal origin and were the base for perfumes upto the end of the nineteenth century until about 90 per cent of all ingredients used in the perfume industry were replaced by synthetics in the twentieth century. These synthetic raw materials are described in Chapter 3, featuring synthetic odorants and isolates, corresponding natural odourants and modifying as well as accessory odorants, including fixatives. Classification of odours into groups and subgroups is necessary in view of the great number of materials with varied characteristic odours, odour intensities and different properties. Fragrance being a subjective phenomenon, odour classifications are often subject to controversy but are usually of two general types – one for odorants based on odour persistence and the other for odours based on their type. The classification of odours and odorants is the topic of Chapter 4 which focuses especially on the 'semi-quantitative evaluation of odours' which accurately divides odours into four principal categories – fragrant (sweet), acid (sour), burnt (empyreumatic) and caprylic. Chapter 5 describes the process of creating a perfume, a process that requires perseverance, in-depth knowledge of perfumery chemicals, imagination and a trained mind. Drawing heavily from his knowledge of the chemistry and odour properties of all fragrance materials, the perfumer starts with base notes, the middle notes to enhance the blend and finally adds the top-notes, then each blend tested and judged carefully. Chapter 6 describes flower perfumes and their formulations, approaching the true note of each flower and listing the constituents used in its reproduction. Complete formulae are given by way of illustrations.

Sophisticated or fantasy perfumes are described in Chapter 7, roughly dividing them into categories such as floral perfumes, aldehydic perfumes, chypre types, oriental and green types and perfumes with a distinguishing dominant note. Chapter 8 focuses on colognes or perfumes for men, highlighting the raw materials used and the formulae followed for some famous colognes. A working knowledge of olfaction and gustation, the sense of smell and taste, is essential for a good practitioner of perfumery. This area is explored in Chapter 9, particularly the complexity of odour-stimulus and the chemical, physical and physiological factors involved. The astonishing range of perfumery usage is compiled into categories in Chapter 10 while Chapter 11 describes the presentation, packaging and marketing of perfumes. Finally, Chapter 12 lucidly explains the sampling, analysis, physical and chemical methods of testing perfumes.

Section II of this book deals with cosmetics. Part I describes product ingredients or raw materials of cosmetics and form the subject of Chapter 13. Chapter 14 concentrates on surface-active agents which have the property of altering surface energy. The use of surfactants in toilet products and cosmetics is explained. Chapter 15 is devoted to humectants & antiseptics. It is imperative that a micro-biologist is involved in the formulation of toilet preparations since the materials used in the manufacture are susceptible to biological degradation by micro-organisms. Therefore, Chapter 16 is devoted to the description of preservatives used to prolong the shelf-life of a product and to protect consumers from infection. It also explains the functions of anti-oxidants. Emulsions form a very important part of the cosmetic market so the fundamentals of emulsion technology used in the manufacture of creams and lotions is the topic of Chapter 17. An overview of the broad range of processes used in the manufacture of cosmetics is presented in Chapter 18 to help the reader understand the basic principles of cosmetic production technology. Manufacture of a quality cosmetic requires many procedures for quality control and this is elaborated on in Chapter 19 which explains how control of microbial contamination is carried out. The principles and procedures of packaging, an area of prime importance in marketing cosmetics, are detailed in Chapter 20.

Part II is devoted to cosmetics for the skin and begins with Chapter 21 which covers skin creams, astringents and skin tonics which cater to the epidermis, the skin's most important layer where cosmetics are concerned. Bath preparations form the subject of Chapter 22, describing the vast of bath products such as bath oils, shower gels, after-bath body lotions and hydro-alcoholic products or bath satins. Chapter 23 concentrates on products especially made for babies and young people, keeping in mind the need to protect soft, smooth young skin from the hostile environment, to clean the skin thoroughly from sebum, grime and excreta and to keep the skin surface dry. The mechanism, properties and manufacturer of deodorants designed to reduce axillary odour, and anti-perspirants, designed to reduce axillary wetness are explained in Chapter 24. Preparations for the removal of unwanted hair or depilatories are described in Chapter 25, while Chapter 26 covers the criteria and manufacture of non-irritating shaving preparations which should assist in removing shaving debris from the face, remainst able in a range of temperatures, be resistant to rapid drying out collapse and be non-corrosive to the razor blade. Chapter 27 discusses preparations for the foot that should counteract a sweat accumulation in shoes.

Exposure to sunlight can have both beneficial and harmful effects on the human body and Chapter 28 deals with the products that protects the skin from exposure to sunlight such as sunscreen, suntan and anti-sunburn preparations. Chapters 29 and 30 describes the characteristics and production of colour make-up preparations such as lipsticks, eye make-up, face powders and other make-up preparations.

Part III focuses on nail and manicure products which cover nail lacquers providing colour and polish to the nails, nail lacquer removers, nail creams and lotions and cuticle removers. The constituents and production of these preparations are described in detail in Chapter 31.

Part IV concentrates on hair and hair products. Chapter 32 explains the characteristics of hair while Chapter 33 highlights the constituents and production of shampoos or suitable detergents for washing of hair, the primary function of which is to cleanse the hair of sebum, scalp debris and residues of hair-grooming preparations. The aim of hair-dressing is to improve the manageability of hair and to impart some lustre to it. These hair-setting lotions and conditioners constitute Chapter 34, while hair colourants form the topic of Chapter 35. The process of permanent waving of hair and the constituents of hair strengthener and straighteners is dealt within Chapter 36 which ends the section on hair and hair-products.

Part V begins with Chapter 37, explaining the role of oral hygiene which is to keep the teeth free of stains, to freshen the breath and also in the modern day world, to prevent dental caries. The two main areas of preventive self applied dental formulations is reduction of dental caries and prevention of gum diseases, both of which are adversely affecting the industrial population. These are dentifrices such as toothpaste which are primarily needed to remove adherent soiling matter from a hard surface (enamel) with minimal damage to that surface. The basic requirements and structure of dentifrices are detailed in Chapter 38 while mouth-washes of the antibacterial type are dealt with in Chapter 39.

Part VI goes on to miscellaneous cosmetic items. Chapter 40 is devoted to the process of imparting a pleasant and suitable odour to a product during the process of perfuming cosmetic preparations. Keeping in view the modern new-line way of thinking that 'nature is the best alternative', Chapter 41 highlights herbal cosmetics. These cosmetics are getting a good place in the cosmetic markets and are rapidly growing in popularity.

Chapter 42 explains the physiology and biochemistry of skin colour, particularly the pigmentation process and then goes on to describe skin lightening agents and formulations as well as bleaches.

Finally, Part VII deals with testing of products and their quality control as well as environmental considerations. Chapter 43 emphasises the fact that quality control means constant striving for excellence and its various facets, especially ensuring of safety, is discussed. Chapter 44 provides specific procedures as well as general methods of testing cosmetics, detailing the aspects of analytical chemistry used to identify inorganic and organic components in extremely complex mixtures. The environmental factors other than processes and equipments, which directly influence the microbiological quality of a finished product, are discussed in Chapter 45. These include the design of building and processes, the choice of construction materials and surface finishes, the provision of ancillary services such as lighting, ventilation and power, and the exclusion of vermin.

This reference textbook also includes a glossary of technical terms that will help processors, consumers and others connected with the trade, in correct interpretation of the terms. Useful information, references, an index and appendices listing plant and machinery and raw materials manufacturers add to the value of the book. The text is throughout supplemented by Flow diagrams, Figures and Tables, wherever needed.

The author, S C Bhatia, is a chemical engineer with management qualifications who has written several books on chemical and allied subjects. At present, he is a renowned consultant in the field of environment, waste heat recovery/energy conservation and petrochemicals.

Acknowledgements are due to Mr. Santosh Kumar Shrivastava, the computer operator who composed the matter and drew & labelled the flow diagrams. Proof-reading and editing has been done by Ms. Namita Mohanty Das, M.Sc. (Chemistry) and. Mr. Madhuresh Kumar. The author also wishes to express the appreciation for the work of Mr. Harinder Singh Negi who assisted in various other matters related to this book.

S C Bhatia

Author

CONTENTS AT A GLANCE

CONTENTS IN DETAIL

CHAPTER 1

CHAPTER 2

RAW MATERIALS OF PERFUMES (NATURAL ORIGIN)

CHAPTER 3

RAW MATERIALS OF PERFUMES (SYNTHETIC ORIGIN) **64–97**

CHAPTER 4

CLASSIFICATION OF ODOURS AND ODOURANTS 98–106

CHAPTER 6
FLOWER PERFUMES AND THEIR FORMULATION 126–159

CHAPTER 7

SOPHISTICATED/FANTASY PERFUMES AND THEIR FORMULATION

CHAPTER 8

COLOGNES : PERFUMES FOR MEN

CHAPTER 9

OLFACTION AND GUSTATION : THE SENSE OF SMELL AND TASTE

CHAPTER 10

APPLICATIONS OF PERFUMES 198–217

CHAPTER 11

CHAPTER 12

CHAPTER 13

RAW MATERIALS FOR COSMETICS 270–282

CHAPTER 14

CHAPTER 15

HUMECTANTS AND ANTISEPTICS

CHAPTER 16

PRESERVATIVES AND ANTIOXIDANTS

CHAPTER 17

CHAPTER 18

EQUIPMENT FOR MANUFACTURE OF COSMETICS 324–351

CHAPTER 19
CONTROL OF MICROBIAL CONTAMINATION 352–367

CHAPTER 22

CHAPTER 23

SKIN PRODUCTS FOR BABIES AND YOUNG PEOPLE 416–420

CHAPTER 24

ANTIPERSPIRANTS AND DEODOURANTS 421–426

CHAPTER 25

DEPILATORIES 427–432

CHAPTER 26

SHAVING PREPARATIONS 434–444

CHAPTER 29

FACE POWDER AND MAKE-UP **468–489**

CHAPTER 31

NAIL AND MANICURE PRODUCTS **521–535**

CHAPTER 32

CHAPTER 33

CHAPTER 34
HAIR SETTING LOTIONS AND CONDITIONERS 550–562

CHAPTER 41

CHAPTER 42

SKIN LIGHTENERS OR BLEACHES 652-656

CHAPTER 43

QUALITY CONTROL OF COSMETICS 657-664

CHAPTER 44

TESTING OF COSMETICS 665-731

CHAPTER 45

ENVIRONMENTAL ASPECTS **732-746**

<h1 align="center">Chapter 1</h1>

The Perfume Industry in India: An Overview

INTRODUCTION

The manufacture of essential oils or perfumes and *attars* has been carried out in India since times immemorial and Indian products have been popular for their quality and aroma throughout the world. With the development of the perfume industry in the western countries on modern scientific lines, the Indian industry received a set-back due to higher cost of production and lower yield of the oils, because Indian perfumers did not adopt modern technique and equipment.

The introduction of synthetic essential oils and aromatic chemicals, which are cheaper and have an aroma somewhat similar to those of natural oils, have revolutionalised the Indian perfumes.

ESSENTIAL OILS MANUFACTURED IN INDIA

The Indian essential oil industry manufactures rose, kewda, lemon grass oil, peppermint oil, oil of champa, citrus oil, hina, khas, cyperiol, ajowain, mayur pankhi, palmarosa and rose water. The major cities which have essential oil plants in India are Mumbai, Mysore and Chennai. The major producer of palmarosa oil is Maharashtra, Andhra Pradesh, Madhya Pradesh and Uttar Pradesh. Kerala contributes major share of lemon grass oil. Karnataka is known for sandalwood oil. Vetiver or Khas oil is produced in Rajasthan, Uttar Pradesh and Kerala. Orange, linaloe, and eucalyptus and cinnamon oil are produced in Madhya Pradesh, Karnataka, and Chennai. 'Attars' which is popular name of essential oils in India are manufactured in various parts of India, specifically Ghazipur, Jaunpur, Kannauj, Hassayan, and Barmana, all from Uttar Pradesh. Orissa produces essential oil from

1

Kewra and Champa flowers. The discussion follows on each type of oils produced in India.

Citrus Oils

Cultivation of citrus fruits on a commercial scale has been taken up during the last fifty years. India produces 13.5 million maunds of citrus fruits grown in an area of 1,30,011 acres which is 6.5 per cent of the total area of 2 million acres under all fruits in the country. The principal citrus fruit growing states in India are Chennai, Uttar Pradesh, Punjab, Maharashtra and Assam. The average yield of citrus fruits is 98–100 maunds per acre.

Malta oranges are cultivated at Gujranwala, Renalakhurd and a few other places in Punjab.

Karna Khatta fruits grow wildly in the hilly regions of Uttar Pradesh, i.e., Almora and Garhwal which are not utilised commercially. Citrus oils are extensively used in perfumery trade for flavouring purposes. They are of three types :

(1) Oils prepared from the peels of the fruits are named after the fruits. They are generally made by the expression methods and rarely by distillation.

(2) Oils prepared from the flowers are called neroli oils.

(3) Oils prepared from the leaves and twigs are called petitgrain oils.

Flower Oils

Otto of Kewda or Pandanus odoratissimus L

The *Pandanus odoratissimus L.*, is the only aromatic member of the N.O. Pandanacae. It is called in Sanskrit—Ketaki, Dhuli Puspika; English—Fragrant screw pine, Hindi—Keora, Gagandhul, Kewda, Bengali—Kewra, Ketuki; Bomb.—Kevda; Tel—Mogili, Gajangi, Ketaki, Tamil—Talamchedi, Tazhai Kedagai and Guj.—Kewoda.

It is a shrub with fragrant flowers found in India, Arabia, Persia, Burma and Andaman islands. Kolapalli, Meghna, Agraram in the Ganjam District in Orissa, Kotabamali in Madras State, Jaunpur, Ghazipur and Sahaswan in Uttar Pradesh and Alwar in Rajasthan are the centres of production.

The superior quality of flowers yielding an excellent odour are met in Ganjam district, which supplies more than 90 per cent of the total production of flowers in India. The average height of the plant is 12 feet to 18 feet and in exceptional cases goes up to 25 feet.

In India the flowers are used mainly for the manufacture of Kewda water and Kewda Attar. Kewda Attar is very popular in the country. The otto of Kewda, as is known in South Orissa is not the real otto or oil as it is based on Palmarosa oil.

Rose oil or otto of rose

In India there are mainly two types of roses which are grown on large scale for commercial purposes: (i) *Rosa damascena* mill called "Fasli" which is used for making rose water, otto of rose and rose attar. (ii) Bourbon hybrid perpetual or Edward called "Cheenia" used for making gulkand and rose attar but not for otto of rose.

The chief centres of cultivation for rose flowers are Barmana and its neighbouring villages in Aligarh District, Kannauj, Ghazipur, Kanpur and Sikandarpur (District Ballia) all in Uttar Pradesh.

The flowers should be plucked early in the morning before sunrise to obtain the maximum yield of oil.

Jasmine Oils

The jasmine flowers have been valued since long in perfumery on account of their peculiar, mild and pleasing odour. Every high class perfume has a touch of jasmine perfume. There are nearly one hundred varieties of jasmine mostly natives of India, Arabia, China and Egypt, etc. In India, the jasmine flowers are used for making hair oils and attars and for making garlands for personal decoration and also offering to deities.

Species and varieties—The following varieties of jasmine are commonly found in India :

(1) *Jasminum arborescens*

(2) *Jasminum grandiflorum*

(3) *Jasminum auriculatum*

(4) *Jasminum humile*

(5) *Jasminum ajonicum*

(6) *Jasminum pubescens*

(7) *Jasminum sambac*

(8) *Jasminum undulatum*

(9) *Jasminum revolutum*

The propagation of jasmine plants is done by branching which are transplanted at a distance of 3–4 feet apart in the rainy season in the fields

In one acre there are about 2,000 plants. A jasmine field lasts for 10–15 years but on well drained soil the plantations may last much longer. The bushes should be trimmed every year to facilitate the harvesting of the flowers.

The flowers of *J. auriculatum* are very light, 26,000 flowers weigh one kgm, while 3,000–4,000 flowers of *J. sambac* and 10,000–12,000 flowers of *J. grandiflorum* weigh the same quantity. The perfumers in India prepare attars and not ottos or absolutes from them. In Ghazipur and Jaunpur, etc., hair oils are made from chameli and bela flowers. The quantity of juhi flowers available is very low and therefore they are used mainly for making attar. The best method for the extraction of otto from jasmine flowers is the enfleurage method.

Absolute or ottos are yellowish red liquids soluble in 90 per cent alcohol and possess characteristic flowery notes. On ageing they darken in colour.

Oil of Champaca

The oil of champaca is obtained from the flowers of *Michelia champaca L*; which belongs to N.O. Magnoliaceae. It is a tall evergreen tree native of India. It is cultivated as well as it grows wild. Assam, Nilgiris, Chennai and S. Orissa are the main champaca producing regions in India. The tree is 15–25 ft. high, 2 ft. in diameter and bears golden yellow flowers.

The tree is known as Champaca or Kusum in Sanskrit, Golden Yellow Champa in English; Champac in French, Champa in Bengali and Hindi; Sonchampa in Marathi; Sampang in Tamil, Gulia Champa in Nepali, Rae Champa in Gujarati and Sampige in Canarese.

Besides possessing pharmaceutical properties, the flowers contain an essential oil. In S. India the flowers are used as offerings to deities and also for personal decoration. The tree is frequently planted near the temples and is considered sacred.

The tree is propagated either from the seed or cuttings on sandy soils and planted at a distance of 15–20 ft. apart. The tree flowers after 4–5 years. There are two flowering seasons per year (i) April, and (ii) September-October.

The average yield of flowers per tree is 50–100 flowers per day during the season. Owing to the presence of an oxidising ferment in the flowers, they become brown in a few hours after being picked, hence to prevent its odour being impaired. the oil must be prepared soon after picking. The flowers have eight golden yellow rolled petals. In S. India Pertappur, Parlakimedi and Jaspur (S. Orissa) are the production centres of flowers.

About 2 million flowers weighing nearly 120 mds. are produced every year and 375–425 flowers weigh one kilogram.

Indian perfumers make champa attar. Experiments were conducted to prepare champa oil. Solvent extraction method was used to prepare the oil using benzene as a solvent.

The concrete obtained has yield of 0.26 per cent which gave 26.3 per cent of steam volatile oil. It has the following physical properties given as under :

Properties	Values
M.P.	29°C
Congealing point	28°C
Acid value	29.1
Sap. value	177.2

To determine the composition of champa oil it was fractionated under vacuum. The oil was found to contain cineole, isoeugenol, phenyl-ethyl-alcohol, benzaldehyde and methyl anthranilate and no resinification of the oil took place during fractionation.

Marigold Flower Oils

There are several varieties commonly known as marigolds (Tagetes) bearing flowers from golden yellow to dark red. They belong to the family Compositae and are mostly annuals. The important species of tagetes are *T. patula, T. erecta, T. grandulifera, T. signata, T. minuta and T. lucida,* etc.

It is called Genda in Hindi and Bengali, Gendu in Oriya, Tangla metok in Punjabi, Makhmal Gule Jafri in Bombay, Roji Chapful in Marathi, Banti in Telugu and Marigold in English.

Tagetes can be broadly divided into two groups for garden purposes according to the nature of the growth.

(1) *T. erecta* and *T. lucida* which grow upright and are of somewhat open growth.

(2) *T. patula* and *T. signata* which grow in the form of bushes, the lower branches being very close to the ground.

T. erecta plants are not suitable for bedding purposes on account of their open growth. They should be grown with plenty of space for roots to spread, air and rich soil to get flowers of double variety and bigger in size.

This plant is a hardy annual growing 3-4 ft. in height with leaves much bigger than those of *T. patula*. The colour of the flowers varies from light sulphur yellow to reddish yellow. The flowering season is from last week of November to January.

T. patula flowers are dark red in colour and are late blooming. The flowering season lasts from January to March. The plant is two and a half feet in height and forms a compact bush. The yield of flowers per acre varies from 80–100 maunds.

All the parts of the above plants contain essential oil. The essential oil from flowers has been found to be a fly repellent. In India, these flowers are used for making garlands to decorate the idols in the temples, etc. The juice of the flowers is sometimes used as a blood purifier and as a remedy for piles.

The yield of oil from the flowers varies from 0.015–0.02 per cent. The oils are reddish yellow in colour and possess characteristic marigold odour. They have a tendency to polymerise readily in presence of air. The oils from leaves and stems are greenish yellow in colour. The properties of the oils are given in Table 1.1

Table 1.1. Properties of essential oil of Marigold.

Particulars	Essential oil from *T. erecta* using			Essential oil from *T. patula* using		
	Flowers	*Leaves*	*Stems*	*Flowers*	*Leaves*	*Stems*
1. Percentage yield of oil	0.017–0.018%	0.05–0.07%	0.009%	0.018–0.02	0.09–0.1%	0.03%
2. Sp. gr. at 30°C	0.9360	0.9526	0.9093	0.9326	0.9416	0.9034
3. Refractive index at 30°C	1.5025	1.5027	1.4975	1.5069	1.5054	1.4981
4. Opt. rot.	1.2	4.2	–6.6	–1.2	+2.4	
5. Acid value	5.4	2.3	12.7	5.6	9.01	9.08
6. Ester value	33.5	50.2	30.8	45.09	31.13	26.0
7. E.V. after acetylation	119.4	113.57	124.1	106.09	98.70	88.53
8. Carbonyl compounds as $C_{10}H_{16}O$ by Stillman and Read method.	23.5%	45.9%	24.5%	39.2%	33.67%	29.29%
9. Solubility in alcohol all the oils are soluble in equal vols. of 90% alcohol.						

The oils from flowers and leaves were examined in detail and found to contain :

	Flower oil	Leaf oil
Ocimene %	8.5	6.0
Limonene %	14.03	21.8
Linalool %	21.14	16.57
Linalylacetate %	13.75	11.00
Tagetone %	40.38	42.53

Essential Oils from Kadamb, Moulsari, Harsinghar and Night Queen Flowers

India abounds in a number of perfume bearing flowers, some of which have not yet been worked out for their essential oils. A few of the flowers, e.g. Kadamb (*Anthocephalus cadamba*), Harsinghar (*Nyctanthes abortristes*), Moulsari (*Mimusops elengi*) and queen of night (*Cestrum nocturnum*) have been distilled in places renowned for them and their oils prepared and analysed.

Anthocephalus cadamba Miq. or Kadamb

It belongs to N.O. Rubiaceae family. It is called Kadamba, Halpriya, Sisupala or Nipa in Sanskrit, Kadamb in Gujarati, Bengali and Hindi; Vellai Cadamba in Tamil, Ruduraksh Kamba or Kadambamu in Telugu, Kadam in Marathi, Raghu in Assam, Man-lettan-she in Burmese and Wild Cinchona in English.

The tree is found all over India and grows wild in Northern and Eastern Bengal, Pegu and the western Ghats. It is cultivated in North India, especially in Nandgaon, Bersana, Kosi Kalan and Brindaban—all in Mathura District of Uttar Pradesh. It thrives well in drained alluvial soil. It is considered to be a sacred tree. It is a large deciduous tree about 30–40 ft. high and 5–7 ft. in girth with horizontal branches rather drooping at the ends.

The flowers are numerous, conspicuous round in shape, 1–2" in diameter, terminal and scented at night. They are yellow or orange coloured with white stigmas.

The Kadamb blossoms at the end of summer and produces very beautiful flowers. The blossoming period is June-July and lasts for about a month. The flowers are generally used as offerings to the Gods and not put to any commercial utility. The flowers emit a fascinating perfume.

Mimusops elengi or moulsari

It belongs to N.O. Sabataceae. It is called Sinhakesara, Bakula in Sanskrit, Moulsari. Molsari in Hindi, Bakul in Bengali, Ranjansal in Marathi, Pogadamanu in Telugu, Vakulam, Mogadam in Tamil, Khraja in Burmese, Bholsari, Gholsari in Gujarati and Affengesict in German.

It is a large evergreen ornamental tree about 50 ft. in height. It is found wild in the forests of South India, Burma and Malaya Peninsula. It is cultivated in North India. Its flowers are very fragrant. The fragrant smell of the corolla persists long after drying. The flower is white and fragrant. The main season of the flowers lasts from May to June in South India and July to September in North India. The flowers are very light and 14,000–15,000 flowers weigh one kg. The Moulsari attar is prepared from these flowers. Even the essential oil also has been obtained.

Nyctanthes arbortristes or harsinghar

It belongs to the N.O. Oleaceae. It is called Parijata, Siphalika, Rajanikasa in Sanskrit, Night Jasmine, Weeping Nyctanthes in English, Harsinghar in Hindi, Seoli or Shiuli Singhar in Bengali, Kuri, Laduri in Punjabi, Partaka, Khursali in Marathi, Shwetasurasa Payala-mully in Telugu and Manjapu, Pavala, Pavala Malligai in Tamil.

It is a large shrub 15–25 ft. high and is cultivated throughout India. It is found wild in the forests of Madhya Bharat and Sub-Himalayan region, Burma and Ceylon.

The flowers possess a cylindrical tube orange red in colour with white spreading limbs. The flowers are very light in weight and 3,000–4,000 flowers weigh one kg. The blossoming period begins in the month of September and lasts till October. They contain an essential oil similar to jasmine and are used for making attar.

Cestrum nocturnum

It belongs to Solanaceae family. It is called Ratki Rani in Hindi, Queen of Night or Night Jasmine in English and Hasnu Haria in Bengali. It is an evergreen shrub 3–5 ft. high. Flowers bloom at intervals throughout the year and emit very fragrant odour at night. It thrives best in tropical climate of America, India and West Indies.

The flowers are greenish white in colour. The plant gives abundant flowers in the months of April, July, September and November. They are propagated by cuttings in February or early in March and transplanted in July. The trees are pruned in the months of December or January.

OILS FROM LEAVES

Indian Peppermint

Indian mint *Mentha arvensis* grows wild in the northern and western Himalayan ranges in Kashmir. An excellent quality of oil was obtained from the mint grown at Nilgiri Hills several years ago.

Dimethyl sulphide was found to be present in traces in both the Indian variety oils, other samples did not give positive test for dimethyl sulphide.

Oil of Spearmint

Spearmint comprises several varieties of *Mentha spicata* and *Mentha virides*. It is called pudina in Hindi, Bengali, Marathi, Gujarati, Sindhi and Telugu. The plant is cultivated throughout India. It is used as a carminative, stomachic and stimulent. It is given in hiccough, bilious vomitting, flatulence, colic pains and cholera, etc. The oil is a powerful antiseptic and relieves toothache. The leaves are used in chutneys to impart flavour.

The plant is indigenous to India and grows throughout the country. The plants require frequent irrigation. The accumulation of much water in the plants is injurious for their growth. This is why spearmint plants are spoiled in the rainy season.

The green leaves are distilled. The yield of oil depends on several factors, e.g. soil, variety of spearmint, manure and locality. The yield of oil varies from 0.20–0.25 per cent. From an acre 25–30 pounds of oil can be obtained.

Spearmint oil improves quality on storage. One year old oil being finer and more characteristic in odour and flavour than a freshly prepared oil. The oil should be kept in well stoppered amber coloured bottles in a cool place protected from light.

The properties of spearmint oil are as follows :

Properties	*Values*
Sp. gravity	0.925–0.940
Opt. rot.	–30° to –50°
Ref. Index	1.4800 to 1.4930
Carvone content	42–60 %

Palmarosa and Gingergrass Oils

Palmarosa and gingergrass oils i.e. (oils from Cymbopogon Martini Var Motia and Cymbopogon Martini Var Sofia) also known as Rusha oils or

Motia and Sofia oils, are valued on account of the presence of geraniol which is one of the main constituents of the otto of rose. Geraniol being stable in contact with alkali can be used with advantage in all kinds of soaps. Palmarosa oil containing higher percentage of geraniol (85–95 per cent) is exported for the isolation of geraniol.

Both the varieties occur wild in Madhya Pradesh, Maharashtra, Berar, Karnataka, Hyderabad (Deccan), and few parts of Southern India. Though both the varieties are closely related they do not grow side by side except in a few places. Motia variety grows in open forests on dry sunny slopes and the sofia in lower altitudes, valleys and in dense moist areas with poor drainage.

A loamy soil is suitable for their cultivation. A well drained clayey soil can also serve the purpose. The grasses can be raised from seeds or from nursery plants. Farm yard compost can be used as a manure. Care should be taken to irrigate the plants regularly so that the plants may not dry up.

The plants should be harvested in October at the time of flowering when the leaves, stems and flowers contain the maximum quantity of oil. Two cuttings per year can be made in July and in October, if the soil is very rich but in general only one cutting is done in October.

The average yield of each grass is about 12,000 lbs. per acre per annum. The flowering tops contain the maximum quantity of oil than leaves. The whole plant of Motia variety in fresh state yields 0.15–0.17 per cent oil and that of sofia 0.4–0.42 per cent on distillation.

The best grades of Palmarosa (Motia) oil containing 90–95 per cent geraniol are produced in Berar while those prepared in Maharashtra contain 80–85 per cent geraniol. Sofia oil contains much less geraniol, generally about 60 per cent.

Lemongrass Oil

This grass grows wild and cultivated abundantly in Travancore-Cochin and Tinnevely District in South India. The oil extracted from it is extensively used for perfuming soap and cheap scents. The oil is exported for the isolation of citral used in the preparation of ionone. Recently a process has been developed to prepare Vitamin A from ionone. The oil has also been shown to be a mosquito repellent.

It is a hardy plant and grows in almost any kind of soil. The more fertile the soil, the lower is the citral content, although the yield is high. The plants require little attention. The plants once sown remain in the field for 8–10 years. The roots are left in the field and only the grass is cut off. It is propagated by clumps of fully matured plants and planted in the field at a distance of 2–3" apart. It requires a warm typical climate and intermittent rainfall or water. The grass is ready for harvesting after 2–3 months depending on weather conditions. When the plant is about 4 ft. high and has 4–5 leaves, it is ready for harvesting. There can be 3–4 cuttings per year. The average yield of oil varies from 30 to 50 lbs. per acre per annum.

The green grass is distilled yielding 0.2–0.25 per cent oil on the weight of fresh grass. The samples of oil obtained by distilling the grass grown in the fields were analysed. It had been observed that the yield of oil was only 0.14 per cent in the rainy season and 0.22 per cent in the summer.

THE ESSENTIAL OIL FROM THE LEAVES AND FLOWERS OF OCIMUM BASILICUM

It is a variety of ocimum called "Marua Dauna" in Hindi. It grows wild throughout the country. All parts of the plant excepting the root contain oil especially the leaves and flowers, which emit a very pleasing fragrance. The oil possesses some insecticidal action against house flies, blue beetle flies and especially mosquitoes. The plants were cultivated systematically in the fields at a distance of 2–2.5' apart. Ordinary loamy or clayey soil is suitable. Farm yard manure can be used. They require frequent irrigation. There can be two harvests per year. The oils were distilled from leaves and flowers separately at different intervals. The flowers can yield 18 lbs. and the leaves 12 lbs. oil per acre per year. The yield of flowers or leaves obtained per acre per annum was 6,000 lbs. each. The harvesting is done when the plants flower profusely.

The flower oil is more pleasant and fragrant than the leaf oil. The oils can be used for flavouring, confectionary, baked goods and condiments like chillisauces, catchups, tomato paste and pickles, etc. It can also be employed in certain dental preparations. The oil can also be used in soaps.

Oil from Mango Leaves (Mangifera Indica)

Mango tree has been well known on account of its delicious fruits since many centuries. It is called Amra, Chuta in Sanskrit, Am in Hindi, Gujarati and Bengali, Amba in Marathi, Thayet in Mumbai, Mamidi in Telugu, Mamaram in Tamil, Mampalain in Malaya.

The tree is indigenous and is cultivated in many varieties throughout the tropical countries. Its fruit is diaphoretic, astringent and refrigerant; the ripe one is slightly laxative, diuretic, nourshing and invigorating. The bark is astringent and tonic. In cases of asthma, diarrhoea, chronic dysentery, hematemesis and menorrhagia, leucorrhoea, bleeding piles, round worms, etc., powdered seed or kernel is given. A decoction of leaves is useful in aphonia and the tender leaves after drying and powdering and useful in diabetes. The ash of the leaves is a popular remedy for burns and scalds. Practically all parts of the plant, i.e., fruits, flowers and leaves contain some essential oil.

It is a thick and viscous liquid having dark red colour. Its sp. gravity at 32° is 1.002 and refractive Index at 32°C is 1.505. The oil is soluble with turbidity in 1.5 vols. of 80 per cent alcohol.

Blackberry Leaves Oil

Eugenia jambolana belongs to the family Myrtaceae. It is called Nilaphala, Rajphala or Jambul in Sanskrit, Jamun in Hindi, Kalajam in Bengali and Blackberry in English. The tree is found throughout the plains of India. It is valued on account of its fruits which are said to possess stomachic, carminative and diuretic properties.

The leaves of the tree possess an agreeable odour due to the presence of an essential oil. The oil was, extracted by steam distillation with the help of solvents and examined. It was observed that the percentage of oil in leaves in winter is more than that in rainy reason varying from 0.013 to 0.02 per cent.

The oil is dark brown in colour and possesses an agreeable odour. It can be used in soap perfumes and cosmetics.

The oil is fractionated under reduced pressure and the fractions examined in detail. The oil is found to contain 20–30 per cent terpenes (chiefly limonene and dipentene), sesquiterpenes of cadenenic type about 40 per cent and small quantity of an azulenic sesquiterpene.

OILS FROM ROOTS AND TUBERS

Oil of Cyperiol (Cyperus Scariosus)

It belongs to N.O. Cyperacea and is known in Sanskrit as Nagar Musfaka and in Hindi as Nagar motha. It occurs in damp places of Bengal, Pegu, U.P., Southern parts of India and common in Sunderban. The plant produces deep brown tubers with aromatic odour, which are used for medicinal purposes. The oil extracted from it is used as a hair tonic.

The oil contains 36–40 per cent of bicyclic sesquiterpene ketones, 30 per cent of sesquiterpene alcohols most of which are tertiary and 20–25 per cent of tricyclic sesquiterpene hydrocarbons which are most probably azulenic in nature. The oil also contains small quantities of a solid sesquiterpene ester.

Vetiver Oil

Vetiver (*Vetiveria zizanioides*) belongs to N.O. Gramineae and is known in India as Khas-Khas or Khas and in Java it is called Akar wangi. It grows wild on the plains and slopes of mountains in N. India, Ceylon and Malaya while it is cultivated in Java, Reunion, Haiti, Guatemala, Mexico and Brazil.

It grows wild particularly on the banks of rivers, canals and marshy lands in Uttar Pradesh, Punjab, Bharatpur, Baroda, Hyderabad, Mysore, Assam, Bihar, Orissa etc., and is cultivated in Malabar, Tuticorin, South Tranvancore and some parts of East Godawari.

The roots contain an essential oil having an agreeable sweet odour. They are also used for making aromatic mats and screens during summer season, because these when sprinkled with cold water, impart a fine refreshing aroma and cool the surrounding atmosphere.

Khas grows in any soil, though loamy cum sandy soil is most suitable. In North India, the harvesting of roots commences from November to February. The length of the roots grown in India varies from 4" to 14" depending on the age of the plants. Thicker roots produce more oil than thinner ones. The thickness of the roots varies from 1.0 to 3.0 mm. The digging of roots commences from early morning till dark and one labourer can dig 15–20 seers of roots per day.

The roots vary in colour from light yellow or yellowish brown to reddish-yellow. If the roots are very light in weight or almost white in colour they contain very little oil. The following are the characteristics of good quality khas roots :

(1) It should expose a hard surface when the skin is peeled off.

(2) It should not break on bending.

(3) It should be long, thick and hairy.

(4) It should be bitter in taste when chewed and more bitter the roots, the higher the oil content.

The distillation of khas roots is beset with considerable difficulties due to the viscid nature of the oil, low volatility and high boiling constituents. The separation of the oil from water is difficult owing to its specific gravity approximating water. It may be observed that North Indian vetiver roots yield less quantity of oil than those from South India.

Essential Oil from the Rhizomes of Acorus Calamus

Acorus calamus linn (Family Araceae) is a semi-aquatic perennial growing wild in damp marshy places. It is indigenous to India and Burma and is known as Bach in Hindi. The dried rhizomes are used in medicinal preparations. The rhizomes possess strong germicidal properties. The oil is also employed in perfumery as it has a peculiar spicy odour, which blends well with the compositions of heavy oriental type. The oil obtained from the powdered rhizomes in an yield of 5.1 per cent had the following physico-chemical properties. Specific gravity at 30°C–1.0784; Refractive index at 30°C–1.5489; Opt. rot— +4.2°; Acid value—2.08; Ester value—5.9; Ester value after acetylation—15.45; Methoxy content–35.8 per cent; Solubility—clearly soluble in 80 per cent alcohol.

The oil was fractionated at a reduced pressure of 16 m.m. after the removal of free fatty acids. The individual fractions were analysed for their physico-chemical properties and the constituents present in them were identified. The oil was found to contain asarone 80.1 per cent; free alcohols as Calamene-nol 3.7 per cent; esters as Ethyl Butyrate—1.2 per cent and free acids as Palmitic acid—0.95 per cent.

OILS FROM SOME ESSENTIAL OIL-BEARING SEEDS

Oil from Dill Seeds (Anethum Sowa)

It is called Satpushpi or Misariya in Sanskrit, Sowa or Soya in Hindi, Bengali and Punjabi, Satakuppi in Tamil, Surva in Gujarati, Shepu in Marathi. The seeds are often used as a carminative, stomachic, stimulent and diuretic. The distilled water of fruits is much used in flatulence, hiccough, colic and abdominal pains of children and adults.

The leaves are widely used as a flavouring agent in vegetables, meat and curries. Dill herb oil is employed for flavouring of soups and sauces, etc. Dill vinegar forms household condiment.

The seeds are sown in winter and the crop is harvested in April. One acre of land yields 500–700 lb. of dill seeds. To extract the essential oil, the seeds are crushed and then distilled by steam for a period of 10–12 hours. The yield of oil varies from 2.3 to 3.0 per cent. The oil is pale yellow in colour.

The dill herb oil is prepared from the fresh green herb by distillation with steam for 8–10 hours. The yield of oil is from 0.4–0.6 per cent. The oil is brownish yellow in colour. It may be noted that the herb oil contains less carvone than that from seeds.

As Indian dill seed oil contains less carvone and does not satisfy the B.P. specifications, according to which the oil should contain not less than 43 per cent and not more than 63 per cent carvone, hence it is not used in pharmaceutical preparations. To meet the specifications, the Indian dill seed oil was fractionated such that 10 per cent fraction in the beginning and 20 per cent fraction at the end was rejected, the remaining oil contained about 50 per cent carvone, which lies within B.P. specifications.

The seed oil contains limonene, carvone and dillapiol as the main constituent while the herb oil contains d L phellandrene and carvone.

Ajowain (Phycotis Ajowan) Oil

Ajowain is largely cultivated in U.P., Bihar, Madhya Bharat, Punjab and Bengal. It is called Ajmada yavanika in Sanskrit, Jowan in Bengali, Ajowain in Punjabi, Jawind in Kashmir, Vova in Mumbai, Yavan in Gujarati, Omum in Tamil, Omainu in Telugu and Jamhun in Burmese.

The seeds are useful in flatulence, colic, atonic dyspepsia, diarrhoea; cholera, hysteria and spasmodic affections of the bowels. Externally it is applied to relieve rheumatic and neuralgic pains. The hot seeds are made used as a dry fomentation to the chest in asthma and to the hands and feet in cholera and fainting. It checks chronic discharges from bronchitis. The leaves are used as a vermicide.

The seeds contain an essential oil which is rich in thymol. The seeds are sown in November and plants are harvested in May. The seeds on distillation yield 2.5–4.0 per cent oil. The exhausted seeds contain 10–25 per cent fixed oil.

The oil was commercially manufactured at Rao in Indore up to the second World War, but now it is not in operation. Small quantities of oil are now being made at Kannauj. The production of the oil has decreased due to the synthetic preparation of thymol.

The seeds cultivated in our fields yielded 2.7 per cent oil. The oil had the following properties. Ajowain herb oil was also prepared. The yield of oil was 1.52 per cent. The seed oil contains gamma terpinene, dipentene and paracymene along with thymol and carvacrol.

Cumin Seed Oil

Cumin or *Cuminum cyminum L.* belongs to the family Umbelliferae and is known as Safed Jira in Hindi, Bengali and Persian. The plant is a native of Syria, Egypt and is grown extensively in India, China, South-Eastern Europe and North Africa bordering Mediterranean Sea. Cumin seeds have an aromatic odour and a spicy bitter taste. Hence they are essential ingredient in all mixed spices and curry powders for flavouring soups, pickles and for seasoning breads and cakes. The seeds are also used in medicinal preparations.

The yield of the seeds varies from .250–300 lb. per acre per year and they contain 2.5–3.0 per cent essential oil.

The oil obtained from the cumin seeds (Yield 2.5 per cent) has the following physico-chemical properties.

Constituents	Values
Sp. gr. at 15°C	0.9144
Optical rotation	4°20'
Refractive index at 20°C	1.5001
Acid value	2.4
Aldehydes as cumin aldehyde	41.35%
Solubility in 80 per cent alcohol	1.5 Vols.

The oil is analysed after systematic fractionation under reduced pressure and the properties are given as under :

Constituents	Values
Terpenes (para cymene and dipentene)	33.73%
Aldehydes as cumin aldehyde	41.35%
Alcohols as cuminic alcohol	7.23%

Essential Oil from Tomar (Zanthoxylum Acanthopodium) Seeds

The seeds are extensively used in the preparation of toothpowders and other medicinal preparations. The seeds contain 1.5–2.0 per cent of a pleasant smelling volatile oil. The oil has been previously analysed at the Indian Institute of Science, Bangalore. The physico-chemical properties of the oil are as follows :

Constituents	Values
Sp. gr. at 20°C	0.8637
Ref. index at 20°C	1.4678
Optical rot. at 20°C	6.4
Acid value	0.868
Ester value	21.37
Ester value after acetylation	143.42

The oil is fractionated under reduced pressure and the individual fractions analysed and the constituents are as under :

Constituents	Values
Terpenes, dipentene and dl. phellandrene	47.00%
Alcohols as linalool	37.75%
Aldehydes	6%
Esters as methyl cinnamate	6.0%
Free acids in traces	

Carrot Seed Oil

Carrot or *Daucus carota L* belongs to the family Umbelliferae. It is a native of India, Africa and Europe and is cultivated throughout the world for the sake of its edible tap roots which are universally relished as an article of food. Carrots are exceptionally rich in iron, purify the blood and have beneficial influence on the kidneys. It is a rich source of alpha, beta and gamma carotenes.

The seeds contain both fixed and essential oils. The seeds are used for the preparation of alcoholic tinctures, which are used in the flavouring of certain liquors. The crushed seeds on steam distillation yield 1–1.6 per cent of a thick yellowish oil having persistent odour resembling orris.

Essential Oil from Laurel Berries (Laurus Nobilis)

The plant *Laurus nobilis* belongs to the family Lauraceae and its berries or fruits are known as Sugandh Kokila in Hindi. The berries possess medicinal properties and are often used in diarrhoea and dropsy. The berries (both the shell and the kernel) contain essential oil.

The essential oil obtained by the steam distillation of the shell of the Laurel berries has yield of 3.9–4.1. per cent. It has been analysed for its physic -chemical properties which are given as under :

Properties	Values	Properties	Values
Specific gravity at 30°C	0.9218°	Ref. Index at 20°C	1.4898
Opt. rot.	18.9°	Acid value	5.9
Sap. value	67.9°	Sap. value after acetylation	99.8
Percentage of carbonyls		Alkali solubles	9.0%
By neutral sulphite method	12.0%		

The pale yellow oil has a characteristic cineolic odour is fractionated under reduced pressure after the removal of free acids and phenols using packed column. The individual fractions are analysed and their constituents are as under :

Constituents	Values
Cineole	12.8 %
Free alcohols (mostly tertiary)	10.87 %
Esters chiefly methyl cinnamate	17.91 %
Free cinnamic acid	1.56 %
Free phenols	2.0 %
Terpene hydrocarbons (chiefly alpha pinene) and beta pinene	15.4 %

It also contained carbonyl compounds and some sesquiterpenes, whose identification and characterisation is in progress.

Cinnamon Oil

It is prepared from the barks and leaves of *Cinnamomum zeylanicum*. The largest producer of cinnamon oil is Ceylon. The oil is distilled in Malabar, South Kanara, Bangalore, Augarakandy estate near Tellicherry. Only pungent and bitter leaves are distilled for about 16 hours. The oil is heavier and therefore it sinks to the bottom. The yield of oil is 0.5 per cent to 0.75 per cent.

Cinnamon leaf oil is a brown pungent liquid with a camphoraceous odour. It is used in flavouring confectionary and is a common adulterant for cinnamon bark oil. Eugenol is easily isolated from the oil is used for the preparation of vanillin. It is used in rheumatism. Cinnamon bark oil is a valuable flavouring ingredient used widely in all kinds of confectionaries, baked goods, candies, soft drinks and table sauces.

Eucalyptus Oil

The genus Eucalyptus includes over 500 species, a major of which is used for essential oils. Three types of oils are generally used. (i) Medicinal oil or cineole oil; (ii) Industrial oil or Phellandrene oil; (iii) Perfumery oil.

In India, Eucalyptus oil is obtained by the distillation of leaves of E. Globulous which grows on Nilgiris hills and Annamalai and Palni hills. The leaves of trees 15 years of age and above specifically matured are preferred. They are dried for three days in shade and then distilled. The crude oil is refined by redistillation after adding caustic potash. The yield varies from 0.9 to 1.2 per cent.

The oil is a colourless to pale yellow liquid with a characteristic odour. The oil is used mainly for medicinal purposes. It is used as an antiseptic and deodorant in the treatment of chronic bronchitis, asthma, catarrhal colds and hookworm.

Geranium Oil

The only true geranium species from which essential oil is derived is *G. macrorrhizum L.*, which grows in Bulgaria and Balkan countries. The oil distilled from it is employed for the adulteration of rose oil. The commercial oil of geranium is derived from several species, varieties and strains of Pelargonium Spp. and is grown in Reunion, Algeria, Morocco, France, Spain, Belgian-Congo and East Africa.

Owing to its agreeable and pronounced strong rose-like odour, oil of geranium is one of the most important items in perfumery. It blends well with all kinds of scents and is widely used in scenting of soaps.

Pelargonium graveolens grows wild in Nilgiri. The cultivation of the plant on a large scale has been taken up at Yercaud hills near Salem from cuttings of *P. odoratissima* from Reunion islands. The oil is being produced there on a small scale. The yield is reported to be 8 pounds of oil per acre.

Linaloe Oil

Bursera delpichiana or linaloe tree is a native of Mexico. In India its plantation was started in Tatguni estate 11 miles from Bangalore about 30 years ago, the area of cultivation is being expanded every year. Unlike the Mexican practice, the oil is obtained exclusively from the outer husks of berries. The yield of oil is 1.8 per cent as compared to 2.5 to 3.0 per cent obtained from the wood in Mexico. The oil is called Mysore Linaloe oil. The ripe berries are collected when they fall to the ground. After drying and removing the shell, the berries are distilled by steam.

Mysore Linaloe oil is a light coloured mobile liquid. It resembles Bois de Rose. It can be used as fixative in perfuming lily, lavender, cananga and sweat-pea soaps. It can also be employed in transparent soaps.

The main constituent of Mexican Linaloe wood oil is Linalool (60–75 per cent) along with geraniol, terpineol, methyl-heptenone, nerol and sesquiterpenes. Mysore oil contains mainly linalyl acetate (35–44 per cent) along with small quantities of methyl heptenyl, Linalol (47.7 per cent) and sesquiterpenes, etc. The average annual production of oil from Tatguni estate is 6,000–7,000 lb.

The leaves of linaloe trees yield an oil (0.15–0.25 per cent). It possesses a sweet odour and is very rich in linalyl acetate (67.67 per cent), hence preferred by perfumers.

Patchouli Oil

It is obtained from *Pogostemon patchouli* which is cultivated in Malaya, Sumatra, Java, Madagaskar and Reunion, etc. About 90 per cent of the world's output is supplied from North Sumatra. Efforts have been made to grow these plants in India for the last 3–4 years at Bangalore. The seeds were imported from Singapore and Johore.

The leaves after cuttings are dried and matured without permitting fermentation to occur. The distillation of leaves free from foreign matter is carried out by steam. The yield as well as quality of the oil is improved by prolonging the distillation. The more valuable components of the oil are present in high boiling fractions. The aged oils have finer aroma than fresh ones. The yield of oil varies from 2.5–3.0 per cent on the weight of the weight of dried leaves.

Patchouli oil is one of the best fixatives for heavy type of perfumes. It blends well with palmarosa, clove, khus, cassia and other oils. It is used in a number of blends of soap and other cosmetics, etc.

Sandalwood Oil

Indian Sandalwood oil commonly known as East Indian Sandalwood is obtained from the roots and heartwood of *Santalum album*. This tree occurs in Coorg, Chennai and Mysore.

The wood is chipped and disintegrated into coarse powder and distilled. The average yield of oil varies from 105–110 lb. per tonne of wood or 4.5–6.5 per cent on the weight of wood. The roots yield oil up to 10 per cent, while hard and soft woods yield only 2.0 per cent oil.

Besides the Government Sandal Oil Factory at Mysore, there are four small distilleries in Kuppam in Chennai State; one at Mumbai, one at Mettur and two at Kannauj (U.P. one at Faizabad, one at Kanpur and one at Calicut). The distillation is conducted for 72–120 hours with low pressure steam (20–40 lb. per sq. inch).

The oil is yellowish in colour, with a characteristic heavy sweet lasting odour. It contains at least 90 per cent santalol. It is used in medicine and in perfumery. It has excellent fixative properties and almost all compositions contain sandalwood oil. About 90 per cent oil is consumed in soap, perfumery and cosmetics.

Attar

The queen of all attars is Rooh Gulab, that was once discovered by Noorjahan, the wife of Mughal emperor, Jahangir once while she went for her morning bath, she found an oily layer over the water kept to cool overnight and when it was distilled later on her order, it turned out to be the costliest attar as it is priced at Rs. 3,500 per ten grams.

For Rooh Gulab, the attar of roses, the attar maters visit flower gardens of Kannauj, Sikatra near Aligarh, Ghazipur and Jaunpur. The condition is that the flowers must be plucked at dawn and used before the sun rises for after that, the fragrance begins to leave the blooms. For Chameli and Motia, they have to go to Sikandarpur and some villages in Varanasi. In the Ganjam district of Orissa's Kellua Palli, they get Keora, Molsri and Champa.

Multi-purpose potency of the attars is incredibly vast indeed. Passages in Indian literature are replete with examples of attar being quite an aid to romance. Young men and gods dreamt of maidens walking in a cloud of jasmine, roses and marigold scents.

Cleopatra used it, the heady art behind seduction, a symbol of prosperity and culture. In Aain-e-Akbari, Abul Fazal, has mentioned about Akbar using the fragrant attar along with the burning of incense sticks daily burnt in gold and silver censers. A princess's toilette was incomplete without incense and attar. But today, it appears that the attar is losing its market in competition to the synthetic perfumes. But there is a new ray of hope as the youngsters in certain sections of the Indian society are craving for the attars. In fact the daily turnover of the attar market in India is more than 50 crores according to a market survey report. According to Blossom Kochher, aromatherapy, a very important branch of Indian systems of medicines has long been subjected to blatant neglect. Unfortunately, it hasn't been played up and people are unaware about it. The therapeutic aspect of Indian perfumes must be elaborated by some big corporate sponsors. Since Indian perfumery system hasn't any big name to pat its back, it appears to be dying.

The oils that are used in making of attars too are known for their medicinal effects. For diabetes, asthma, boils and varicose veins, lemon oil is effective. Migraine and hangovers can be cured by taking three drops of marjorem with jaggery. The simplest example of aromatherapy is after Gill (*sondhi mitti*) that has the fresh aroma that emanates from the earth after the first summer showers. On smelling it cures blood pressures and flow of blood through nose owing to intense heat. Sherbet of attar Khus is relished in summer as it provides a cooling effect to the digestive system.

As it is quite unaffordable to buy pure attars these days for dearth of animal products like musk, amber and myrrh which are rare, synthetic attars too are manufactured that are quite cheaper in comparison to the pure ones.

Unfortunately, the flourishing position which India had attained in olden times and medieval periods dwindled down due to the lack of scientific knowledge and the inability of Indian manufacturers to adopt themselves to modern scientific advancements and equipments with the result that the perfume industry in India is in a deplorable condition. The introduction of cheap aromatic chemicals and synthetic perfumes, having odours almost similar to natural essential oils, further decreased Indian exports. The application of cheap synthetic perfumes in the manufacture of soaps, cosmetics and confectionary, etc., induced Indian perfumers to import them in large quantities. Moreover, the development of essential oil-bearing plants in other tropical countries under their Government's patronage took away from India the monopoly enjoyed by her in ancient times. The old fashioned equipments and methods of production could not compete with the modern scientific equipments and processes backed by mechanised agriculture and mass production etc. Therefore, improvements in the existing equipment and process are to be introduced early to develop the present state of perfume industry in the country.

Conclusion

India has a variety of climatic conditions and soils suitable for the growth of aromatic plants. It may be mentioned that a few varieties of Lavender have been grown in the lower altitudes of Nilgiris and Sheveroy hills. Clove trees have been grown in Travancore-Cochin and the possibility of their cultivation in other places has also been surveyed by the Indian Council of Agricultural Research, New Delhi.

Under the auspices of the Essential Oils Research Committee, the cultivation of patchouli plants has been taken up at Bangalore. The U.P. Government has been trying to cultivate patchouli, ylang ylang, peppermint, lavender plants, etc., in the State on commercial scale.

The Forest Research Institute, Dehra Dun has contributed a good deal towards the cultivation of aromatic plants and the analysis of their essential oils in U.P.

The Essential Oils Research Committee have proposed to spend large funds to devise a machine for the extraction of orange oil from their peels in Madhya Pradesh. A special variety of eucalyptus (*E. citrodora*) containing a high percentage of citronellol in its oil is being cultivated on a small scale in Calicut.

As regards aromatic chemicals sufficient work has been done in various laboratories in India. Mention may be made of the National Chemical Laboratories, Pune, Indian Institute of Science, Bangalore, Delhi, Calcutta

Universities and H.B. Technological Institute, Kanpur. The Government of India is imposing restrictions on the import of synthetic essential oils so that their production in the country may be encouraged.

The standardisation of important essential oils produced on large scale in India has also been started and standards for Indian Palmarosa, Lemongrass, Sandalwood, Eucalyptus oils, etc. have been drawn by the Bureau of Indian Standard, Delhi and the standard specifications for other oils are being drawn. This standardisation will improve the export trade of the country and the consumers will now get the genuine oils for industrial purposes.

Considering the progress in various States and by different agencies towards the cultivation of aromatic plants in the country and the preparation of aromatic chemicals and synthetic essential oils it is expected that soon the country may become self sufficient.

Chapter 2

Raw Materials of Perfumes
(Natural Origin)

INTRODUCTION

The diversified natural products are used in perfumery. Advances in synthetic chemistry had major repercussions on perfumery and marked the beginning of a new epoch. The enthusiasm resulting from obtaining odours similar to those of certain flowers led at one time to the belief that synthetic products could entirely take the place of natural products.

The quality of natural products is very important. It can be ensured through physical and chemical standards, however, olfactory analysis provides final decision. It is necessary that one must be adequately trained and exercised for judging olfactory purity and quality.

Points which have to be taken into consideration while examining natural perfumery raw materials are :

(1) The custom, more or less permissible, of making mixtures which physical and chemical analysis may or may not detect, and which olfactory analysis can readily detect.

(2) For a determined natural product, olfactory analysis recognises the origin of the product. For instance, a Grasse Jasmine as distinct from an Italian jasmine, or the jasmine of one supplier from that of another firm.

(3) The tastes and preferences of a perfumer play a part here. Thus one perfumer will look in a jasmine for the fresh part, the sharp top note; while another will seek rather the warm, rather 'jam-like' and sometimes indoloid note.

Thus a young perfumer who wishes to know the raw materials olfactorily must have available for his education the largest possible number of products.

NATURAL ODOURANTS EMPLOYED IN PERFUMERY

The natural odourants that are employed in perfumery can be assigned to seven categories, viz :

(1) Concrete oils (concretes).

(2) Absolute oils (absolutes).

(3) Essential oils derived from the distillation of flowers, leaves, roots and fruits.

(4) Essential oils obtained by expressing fruits.

(5) Aromatic odourants obtained by the fractional distillation etc. of certain essential oils.

(6) Odourants (such as tinctures) of vegetable or animal origin.

(7) Products in the form of resins and balsams.

Concrete Oils

Concretes and absolutes are sometimes loosely designated 'natural flower oils' in order to distinguish them from the distilled and expressed 'essential oils'. They are obtained by different processes. For example, the flowers, leaves, roots etc. are subjected to a kind of extraction by hydrocarbon solvents, which dissolve the waxes containing the odourous principles from the flowers. The hydrocarbons are then eliminated by evaporation under reduced pressure so that the heat does not affect the product so obtained, which is called a concrete oil or simply a 'concrete'.

The concrete of the flowers so treated has the appearance of a more or less solid wax. It is insoluble in water and virtually insoluble in alcohol. However, it is possible to make tinctures with it, i.e. to dissolve by mixing in the cold state the odourous principles contained in the concrete, by using enclosed blade mixers and 95° alcohol for a considerable number of days (about a month). An odourous product of great fineness is obtained, displaying marked fidelity to the perfume of the flower. This is quite sensational, particularly with jasmine, and certain great perfumes owe part of their fame to this practice, which is moreover of very great antiquity.

A variation on the process is based on ultrasonics and the energy which the latter liberate by their intense vibratory movements, producing a series of pressures and decompressions ringing about the 'bursting of the cells' which liberates the essential oils contained in those cells. The oils are then recovered, but not apparently without some difficulty. It seems quite possible, *a priori*, that an adequate medium may be found for each type of product, and that this will facilitate the recovery of the essential oils suspended in the supporting liquid. This process has at least the potential advantage of doing away with maloduorous solvents and of obtaining concretes or absolute oils that have not been overheated.

Absolute Oils

These are generally obtained by extracting the concretes with alcohol, then eliminating the alcohol at reduced pressure. The product so obtained is entirely soluble in alcohol and often has the consistency of honey.

The concrete is subjected to extraction by alcohol by heating to the lowest possible temperature in mixers. The product is then concentrated and the alcohol evaporated by heating to a low temperature and at low pressure. A so-called 'absolute' essential oil is obtained which will be termed commercially: 'ether type'.

A few years ago a process was invented which is a low temperature one and consists of passing a stream of butane gas (boiling point $-0.5°C$) through a kind of tube in which the flowers are placed whose perfume is to be extracted. The absolute oil so obtained tends to be relatively light-coloured and true to type, but in some cases it may not be entirely soluble in alcohol. This process first made it possible to obtain commercially interesting absolutes of lilac and lily-of-the-valley.

A very old process is that of the 'pommades'. There are two ways of carrying it out :

(1) The process of maceration in the hot state, which consisted of melting an odourless fat and allowing the flowers to be macerated in it, the fat or oil being kept at a medium temperature for the fat to be liquid, without excessive heating. The fat absorbs the perfume. In the past, forms of alcoholate were prepared, by extracting these pommades with alcohol in the cold state. Pommade washings (*lavages de pommade*) were obtained, which showed their concentration according to the number they bore.

Then extractions of these pommades with alcohol are carried out to obtained concentrated solutions, and the alcohol is evaporated, in order to produce the 'absolute essential oils ex pommade'.

Not all flowers can be treated by this process, but only those which are not too delicate: the flowers of *Cassie farnesiana*, rose, orange blossom, mimosa, hyacinth, carnation etc.

(2) The process of maceration in the cold state, or *enfleurage*.

In this process, fat or oil is also used. It involves coating the glass plates of glazed wooden frames with a fat or oil. Inside these frames (termed *cadre* or *chassis*), on the part coated with fat, are placed the flowers. The frames are stacked on one another so as to obtain a hermetic seal and they are allowed to stand for 24 hours, so that the effluvia liberated dissolve in the fat or oil.

The flowers that have given up their perfume are removed and replaced by fresh ones. This operation is repeated until the fat or grease is saturated with perfume. The perfumed fats are treated with alcohol, and by increasing these extraction operations, followed by chilling and filtration, an alcoholic solution is obtained, which is subsequently concentrated by evaporation to yield the 'absolute enfleurage essential oil'.

The flowers that have given up the greater part of their perfume in the frames still contain perfume. They are treated with volatile solvents (petroleum ether or benzene) so that products are obtained which bear the name of 'absolute ex chassis, ether type', or 'absolute ex chassis, benzene type'.

The flowers treated by this process are those which are the most delicate, but which in spite of all continue to 'live' and yield perfume after they have been cut. These include jasmine, tuberose, jonquil, narcissus and mignonette.

The fats used are lard or vegetable fat. Pork lard is best, and it is preferably deodorised. In these processes, a small amount of fat is always carried along by the alcohol. In the past this habitually gave, when used in perfumes, a liquid that was very slightly cloudy and which had to be chilled and filtered.

It is at present possible to obtain extremely odourless fats, some of them synthetic. On the other hand, there were once certain perfumers who were not opposed to the odour *sui generis* of these fats, which contributed a very slight suggestion of *rancio* (to use the special term adopted by perfumers in the early part of this century).

Essential Oils derived from Distillation

This process is applied to flowers, leaves, stalks, herbs, roots and even to certain fruits. There are three main ways of obtaining these products :

(1) Distillation in a still with naked flame. The essential oil of Bulgarian rose obtained by this process then bore the name of Bulgarian peasant rose

(*Rose Bulgare Paysanne*) to distinguish it from that obtained by the steam processes.

In naked flame distillation, water is placed in the still along with the ingredients to be distilled. It is heated to boiling point, the steam passed through a cooled coil and at the end of the coils the water and the essential oil are collected. The latter floats on the surface (except in the case of a very few oils of higher specific gravity) after some time and is separated from the water.

This process is still used for certain plants for which distillation on the spot is more advantageous. These include lavender, lavandin, rosemary, spike, thyme etc., and in this way the products treated have the advantage of being fresh and undried.

(2) The second process is to have large cylindrical stills heated by coils in which steam circulates. It makes it possible to heat more regularly and to avoid the heat surges inevitable with naked flame stills. Water is placed in the still with the raw materials from which it is desired to obtain the essential oil. It is heated and the steam is collected in more or less the same way as in the naked flame process.

(3) This process is the same as the previous one, but a stream of live steam is passed to the still, which carries the essential oils along more quickly. In certain cases a partial vacuum is produced which makes it possible to heat to a lower temperature and to obtain finer products.

In addition to leaves, stems and roots, citrus fruits are sometimes treated in this way, i.e. the peel of lemons and oranges. The essential oils obtained in this way are generally less fine and are above all used in the food and beverage industries.

Essential Oils obtained by Expression

The fruits that yield essential oils by expression are the citrus fruits or Hesperidaceae. These fruits include: lemons, sweet and bitter oranges, mandarins, bergamots, citrons, limes from Italy or Africa; the highly perfumed West Indian lime, and the shaddock or grapefruit.

All these fruits, whose essential oil is contained in the peel, can be simply squeezed out as they are. The whole of the juice of the fruit is thus collected, together with the floating essential oil. The drawback is that the juice is always acid and considerably impairs the essential oil. A number of mechanical processes have since been patented and used to obtain citrus peel type essential oils economically and of good quality.

Isolates from Essential Oils

These products are midway between the natural products and the pure synthetic products. An example will better show what is meant. From the natural essential oil of Brazilian bois de rose is extracted linalool, of which it contains up to 90 per cent. This natural organic product has a sweet odour, which is pleasant and recalls the odour of the parent oil, a slightly rosy and at the same time woody odour.

This is a very important ingredient in perfumery, which lacks the 'dryness' and 'harshness' of the products of pure chemical synthesis. That is why, in our opinion, it should not be assigned to the class of synthetic products, but to a kind of special synthetic-natural class.

Take the example of linalyl acetate obtained from bois de rose linalool and entirely synthetic linalyl acetate. It has to be admitted that for the perfumer, of the two linalyl acetates, that having bois de rose as its origin is far superior, while the other tends to suggest terpinyl acetate.

This will not prevent the use, in high-class perfumery, of isolates etc. obtained from natural products. Sometimes, even, exceptional products are found, e.g. a special bergamot linalyl acetate derived from bergamot oil and having a distinct but subtle odour variation all of its own.

However, there is a whole class of products derived for different reasons from essential oils and these are the 'deterpenated essential oils'. Over forty years ago, a perfumery chemist who caused a great stir at the time, conducted considerable publicity for deterpenated essential oils in the sole Fresh perfumery journal that existed in those days. According to him, all the essential oils should be done away with and replaced by the corresponding terpene-free essential oils. The latter are more soluble in alcohol than those which are not deterpenated, and it is therefore possible to concentrate more in a weakly alcoholic solution. But that is not the only criterion. The terpenes present in normal essential oils play a part, when incorporated in an alcoholic perfume, that is far from negligible and appears to 'exalt' the perfume as a whole. This is due in part at least to the fact that, in the scale of volatilities, terpenes are an important link between the alcohol itself and the low boiling esters etc.

The example has very often been cited of a perfume made with natural essential oils and the same perfume made with deterpenated essential oils. Taking into account differences in strength, that made with natural, non-deterpenated essential oils is often greatly superior. Moreover, what can a terpeneless oil of lemon mean to a perfumer, when the natural essential oil contains 90 per cent terpenes? On the other hand some well-produced terpeneless oils undoubtedly deserve a place on the perfumer's shelves.

It has been frequently stated that perfumes made with terpeneless products deteriorate less rapidly but perfumers are not in entire agreement on this point.

Deterpenation is carried out by fractional distillation at reduced pressure. It is also effected by appropriate solvents and other more recent methods.

Natural Odourants as Tinctures

Some natural products are frequently employed in the form of tinctures, based on their more or less prolonged maceration in 95° alcohol, which is initially used hot or cold. Commonly found in this category are animal musk, ambergris, civet and castoreum, but other materials may also be treated in this way, e.g. patchouli leaves, vetiver root, oakmoss. The corresponding essential oils and/or absolutes are in most cases also available. Infusions are similarly prepared, by means of such treatments as processing in 95° alcohol, in a heated closed container fitted with reflux cooling equipment.

Balsams and Resins

These products are as a rule types of resin occurring as exudations on trees or plants, e.g. Peru and Tolu balsams, benzoin and olibanum. These products were originally used in their raw state, as the bases of alcoholic tinctures. Firms have produced, however, a variety of more conveniently handled concentrates or absolutes, known collectively as resinoids. They are in fact the end-products of extracting these materials with different solvents, including hydrocarbons, petroleum ether, benzene, mineral oil, acetone, the glycols, trichlorethylene and butane. Other procedures include second extraction with alcohol, the inclusion of plasticising diluents, decolourisation etc. The finished, standardised and duly concentrated resinoid may often bear a special name, indicative of its origin, e.g. Gomodor, Clair, Verarome.

Using vacuum distillation, molecular distillation and combinations of extraction crude resins and balsams are also obtained. The various interesting extracts and components obtained are novel odourants.

But the principal part played by these resins and balsams, especially if their resinous texture is retained, is to 'hold back' the volatilisation of the perfumes in which they are used, probably by the physical or physico-chemical properties of the thin varnish-like film that they form. They do not flatten out or restrict the proper diffusion of the perfume. This is the most extra-ordinary feature of these products, in that they can 'fix' a perfume without interfering with its other characteristics, whereas materials such as the phthalates and benzoates slow down the evaporation but in doing so alter the perfume's behaviour and make it dull and flat.

We now take a closer look at these various categories and the principal substances contained in each of them.

Concrete oils or 'Concretes'

As the concrete, for each of the flowers to which it corresponds, is not normally used as such but is only an intermediate between the flower and the absolute, we shall not deal at any length with the concrete itself. It may occasionally be used as such, in the form of a tincture. Concretes are not fully soluble, even in 95° alcohol, but the soluble portion does in fact contain the fragrant principles. A typical concrete has the appearance of a more or less granular wax and its colour ranges from cream to yellow or almost brown. Sometimes it has a more or less greenish colour. All these shades differ, according to the flowers from which they are prepared.

The concrete, in brief, is the wax in the flower, accompanied by the essential oil which is responsible for the flower's perfume.

Absolute oils

Whatever the process of extraction, the absolute takes the form of a more or less consistent syrupy substance, ranging in colour from yellow to brown. Some are green or greenish. Others have been treated by decolourisation processes. The problem in removing colour is to ensure that a significant proportion of the odourous components is not removed or adversely affected in the process.

Floral Series

Rose notes

Absolute oil of rose de Grasse

This absolute essential oil comes either from the pommade through maceration of the petals, or through volatile solvents to obtain the concrete of *Rosa centifolia* in the first stage, and after that the absolute essential oil, which is generally denoted by the name of *Rose de Mai absolute*.

The odour of this oil of rose is characterised by a honey note, resembling very slightly a delicate peppery note shading off into a light tonality of natural carnation.

Sometimes it is possible to perceive a slightly green odour, depending on the ripeness of the flower and above all on the care taken in manufacture, so also to eliminate as much as possible the ends of stalks, calyxes etc. from the flowers.

Bulgarian rose absolute

This comes from the concrete of *Rosa damascena*, grown in Bulgaria, and in many cases treated in France to obtain the absolute.

Moroccan and Turkish rose absolutes

The absolute oil of Moroccan rose has a less peppery note than that of the Bulgarian species. The absolute of Turkish or Anatolian rose is produced from *Rosa damascena*, as in Bulgaria. So far, however, the absolute from these two countries is not as good as that of Bulgarian rose.

Absolute of rose geranium

The rose or *Rosat geranium* is produced from *Pelargonium roseum* or *Pelargonium odorantissimum*, or *Pelargonium graveolens*, *Pelargonium radula* etc. These different species are cultivated in Algeria, Morocco, Spain, the Bourbon islands, India etc. For a long time, a Grasse firm made from the concrete a colourless absolute oil of geranium, which was used with some sythetics to make a composition used for blending with Bulgarian rose, so as to have 'cut' or blended Bulgarian rose at reduced price.

It is regrettable that the Bulgarian government, which manages the rose industry, does not take the initiative of making concrete of Zdravetz (*Bulgarian geranium*) avaialble. One is convinced that it would be possible to make a colourless absolute oil of Zdravetz, as was done for *Rosat geranium*, and far better mixtures could thus be obtained with synthetics than the blends we have mentioned; and this without prejudice to the consumption of Bulgarian rose oil. On the contrary, it would favour the use of such Bulgarian oils in everyday products that must be produced at inexpensive levels.

Absolute Oil of Rose Leaves

In addition to the flowers that contain essential oils, it quite often happens that the stems, and more often the leaves, contain essential oils which impart perfume to the flower with very often a slight green note. This is what happens with Rose. The absolute oil or rose leaves performs a very valuable service in rose composition and in modern compositions with a green note.

Jasmine notes

Absolute oil of Jasmine

The white flower is that of *Jasminum grandiflorum* or *Jasminum officinale* or *Jasminum odorantissimum* etc., whose place of origin was India. It is grown in the south of France, in the region of Grasse.

The absolute essential oil of *Jasmine revolution* is very different from that of the jasmine from France, Italy etc. Its odour is warmer, more like that of orange-flower, very special, and if it were produced in sufficient quantity it would provide the perfumer's palette with a new and interesting note.

The absolute oil of jasmine from the species *Jasminum grandiflorum* is obtained by the pommade or enfleurage process - but above all from the concrete obtained by volatile solvent extraction or the other processes already described.

One of us (M.B.) has taken some jasmine absolute and carried out frantional distillation under vacuum. He found that a portion of this distillation had a very powerful odour that was exceptionally appealing, capable of giving rise, with two or three other original components, to a really fine perfume.

It is true that such a process involves a high-priced product, bearing in mind the price of jasmine absolute. But must this be taken into consideration for high-class perfumes.

In the same order of ideas, it is possible to take absolutes or ordinary essential oils and extract from them, by fractional distillation, components of the very greatest interest. It is in similar fashion that oil of *Virginia cedarwood* yields cedrol, which has a very agreeable and warm odour, recalling that of oil of carrot and also methyl ionone.

As previously referred to the *concrete de chassis* obtained from the exhausted flowers in the enfleurage process. This concrete is extracted with 95° alcohol, then the alcohol is evaporated off and a product is obtained termed absolute oil of chassis jasmine, ex ether or benzene according to the solvent used. This absolute is brown in colour, it has a very special jasmine odour, with a suggestion of caramel slightly recalling jam. It performs great services when used in a small percentage in synthetic jasmines, to soften their crudity. The type with ether is better than the benzene type. Unfortunately, the availability of these products continually declines, owing to the decrease in enfleurage.

Absolute of Ylang-Ylang

In some respects absolute oil of ylang-ylang bears a certain analogy to that of jasmine. The ylang-ylang is a tree of the Anonaceae family: *Cananga odorata*.

The absolute is obtained by the usual processes. The note of ylang-ylang is a very special one and the absolute oil (like the distilled oil) is one of the products that is indispensable in fine perfumery.

Hyacinth notes

The absolute oil obtained from *Hyacinthus orientalis* may be a product of the enfleurage process but usually it derives from volatile solvent extraction via the concrete. The best hyacinth absolutes come from Holland.

Lilac and Lily

Absolute oil of lilac

This was virtually unobtainable until the Butaflor process made it possible to produce a natural oil from lilac in commercial quantities. The shrub which yields the lilac flowers is *Syringa vulgaris*. Ordinary everyday lilac has an odour characterised by hydroxycitronellal and a rose odour. White lilac has a more jasmine-like odour. All the lilacs also have, in varying degrees, a mild odour of greenery.

Lily-of-the-valley

Lily-of-the-valley, *Convallaria majalis*, grows widely in the northern hemisphere. A Butaflor absolute oil is available.

Orange-blossom notes

Orangeflower absolute

For a long time perfumers denoted this essential oil by the name of absolute oil of orange an obviously incorrect name but one which is nevertheless found in the formulae of ancient books. It is obtained by the same processes as those used for jasmine (pommade, enfleurage, volatile solvents).

Absolute oil of orangeflower water

The distilled flowers of the bitter orange or bigarade yield oil of neroli, accompanied by a floral water which contains an important quantity of essential oil. By extracting this oil from the water by means of petroleum ether, the absolute oil of *Eaux de Fleurs d'Oranger* is obtained.

Absolute oil of Syringa

Its absolute, obtained by the usual process of volatile solvent extraction, followed by washing the concrete with alcohol, has a perfume that relates it to orange-blossom, but without indole and with a very fresh initial note of a slightly lemony type The sole drawback to this absolute is that its production in France is still very small. It could become greater and would certainly perform excellent service to perfumers.

Tuberose notes

This note includes those of tuberose and narcissus, which are flowers from which absolutes have been prepared. With the progress of modern technology, and in particular by the use of the Butaflor processes, it should be possible to obtain some other interesting absolute oils in this sector.

Absolute oil of Tuberose

This is a plant originating in India in the form of the tubers of Polyanthes tuberosa. The absolute oil is obtained by the process of pommade, enfleurage or by volatile solvents. Its odour is heady, and it blends very well, in small quantities, with carnation and lends it distinction.

Absolute oil of Narcissus

This absolute comes from the *Narcissus poeticus*. There are two types of absolute: the Narcissus cultivated on the plains of Grasse and the 'mountain' type, which is that of the Cevennes and the hills near Grasse. It is the volatile solvent process that yields the concrete from which the absolute is extracted. This is a rather dark greenish-brown or very dark yellow oil, according to its origin (mountain or plain). The type of odour is represented in the crude state by different esters of paracresol and methyl paracresol (acetate or phenylacetate etc.), with a middle note recalling isoeugenol, and finally a rather characteristic green note.

Absolute oil of Jonquil

The absolute oil of jonquil comes from *Narcissus jonquilla*. It is obtained mainly by enfleurage.

Absolute oil of Champaca

This oil is solvent-extracted from the flowers of two species, *Michelia champaca* (yellow champaca) and *Michelia longifolia* (white champaca), of the Magnoliaceae family.

Absolute oil of Honeysuckle

This is extracted by volatile solvents from the flowers of *Lonicera caprifolium*. Its odour somewhat recalls that of jasmine, as regards its fresh part, with a body suggesting tuberose, but rather more orangeflower-like and with a slightly fruity background suggesting apricot.

Absolute oil of Lily

This absolute oil, which is produced in France in only small quantities, comes from *Lilium candidum*. It has a heady and rather fatty floral note and could well be of sufficient interest for its production to be more intensively developed.

Violet notes

Different kinds of violet flowers have different types of odour. The *Viola odorata*, with a definite violet character, is rather earthy or humus-like. The double violet, also known as the Parma violet, has a very sweet, soft and slightly fruity odour.

Absolute oil of Violet

This oil is obtained from the concrete which is itself obtained by volatile solvents, in this case petroleum ether, from flowers of the *Victoria violet*, which is grown in the Grasse region. It has a definite odour of violet, with a green and earthy note.

Absolute oil of Parma Violet

Obtained from the solvent-extracted concrete, its odour is sweeter than that of the foregoing and can be related to that of beta ionone, gamma or delta methyl ionone. It is more flowery and less earthy in odour than the oil of the Victoria violet.

Absolute oil of Violet Leaves

This absolute oil is obtained by treating the leaves of the *Victoria violet* with petroleum ether, and subsequently processing the concrete in the usual way. It gives excellent service in violet compositions, particularly when used in association with other absolutes which also have a violet note, such as boronia, cassie, mimosa and iris.

Absolute oil of Boronia

The absolute oil of boronia comes from a herbaceous plant, the *Boronia polygalifolia*, of Australia. After remaining for a varying length of time in petroleum ether, the flowers undergo the ordinary treatment with volatile solvents and subsequently with 95° alcohol. The green-coloured syrupy flower oil contains beta ionone, and is much appreciated by perfumers, who have found that they can combine it with a wide variety of other notes. Its note *de depart* is very sharp and somewhat recalls the odour of a sorrel leaf crushed between the fingers. The other part of the boronia perfume is a mixture of the odours of violets, olibanum and ambergris.

Absolute oil of Cassie

Absolute oil of cassie is obtained from the flowers of *Acacia farnesiana*. The concrete is prepared with volatile solvents, and extraction by 95° alcohol is then carried out by the usual processes. The absolute is dark yellow in colour and contains a violet-smelling ketone together with farnesol. The absolute oil of cassie has a warm odour with a few points in common with violet, and some reminders also of mimosa. It imparts to a perfume, richness, body and depth and it can be associated with a large number of products.

Absolute oil of Mimosa

There are three kinds of concretes: the mimosa concrete made solely with flowers, the concrete of mimosa and leaves made from mimosa flowers and leaves, and finally the concrete of leaves made solely with mimosa leaves.

All these concretes obviously have a tonality of mimosa, but are accompanied by a more or less green note according to the individual case.

Mimosa flowers themselves possess a very sharp top note, almost lemony, while in the absolute this is quite absent. Mimosa absolute oil is a relatively inexpensive natural product, but sufficient advantage has not yet been taken of it.

Absolute oil of Iris

Absolute oil of iris (orris) is extremely costly, owing to the very low yield of iris concrete or butter from the root of *Iris germanica* or florentina Dykes, or from the root of *Iris pallida*, which is of the order of 0.2 to 0.3 per cent.

To obtain iris concrete, or butter, dried iris root is subjected to steam distillation. From this concrete, there is obtained by washing in 95° alcohol, according to the ordinary processes, the absolute oil of iris. Quite often, the concrete contains from 13 to 17 per cent iron.

Mignonette notes

Absolute oil of Mignonette (Reseda)

Absolute oil of reseda has an odour recalling that of violet leaves, but with a rather fine 'powdery' background and a suggestion of rose and a little basil. The synthetics representing its odour are first of all ethyl decin carbonate and also, to some extent, methyl hexyl ketone. The distillation of reseda root yields an oil with a piquant odour, like black radish, or rather horseradish.

Woody Series

Sandal notes

Solvent-extracted Sandalwood

These oils, the products of a technique midway between distillation and extraction, have a fine middle fraction santalol note but lack the character and bouquet (e.g. the 'early emergers' and mildly animal-empyreumatic notes) of the best quality distilled oils.

Peppery notes

Absolute oil of thyme

This is obtained from *Thymus vulgaris*, which is found in France and Spain. It has an herbaceous, woody, powerful and slightly biting odour.

Caryophyllaceous notes

Absolute oil of carnation

The flowers of *Dianthus caryophyllus* are extracted to give the absolute essential oil, which has the characteristic eugenol, isoeugenol odour, shaded with a honey and rose tone, suggestive of *Rose de Grasse absolute*.

Absolute oil of clovers

This is extracted from the buds of *Caryophyllus aromaticus*, which is a small tree of the Myrtaceae family. It has a spicy, hot, woody odour.

Absolute oil of tobacco flower

This is derived from various species of *Nicotiana*. It has a very sweet odour recalling methyl ionone, but is much sweeter, reminding one a little of clove pink.

Rural Series

Herbaceous notes

Absolute oil of flouve

Flouve or vernal grass is a field grass and among the various species that exist in France, it is the flowers, seeds and stalk ends of this species which serve as the raw material for this absolute.

New Mown Hay Absolute

In itself the expression cut or mown hay is very vague. It is represented as an odour by what is perceived in the air.

Absolute oil of Lavender

The plant yielding this oil is *Lavandula officinalis* or Lavandula vera. The odour of the absolute is sweet and herbaceous, with an almost fruity background.

Absolute oil of Lavandin

Lavandin is a hybrid, derived from *Lavandula vera* and *Lavandula spica* (aspic). The odour of lavandin absolute is less floral than that of lavender, and drier, slightly spicy.

Tea Leaf absolute

This is obtained from the true tea shrub, *Thea sinensis*, by treating the dry leaves with petroleum ether. Different firms sell it under different names. Owing to its dark colouration, subsequent distillation is often carried out, and a very pale product is then obtained having an herbaceous, rather sweet odour.

Green notes

Oakmoss (Mousse de Chene)

To these extracts one adds natural green or balsamic, synthetics of a green character and other synthetics which blend particularly well with the moss note – e.g. the isopropyl and isobutyl quinolines. These latter synthetics, associated with mosses, have formed the basis of some very great perfumes.

Oakmoss has a strong and persistent odour of humus and forest undergrowth, but it varies widely according to source and treatment. A product of great interest, it forms the basis of many perfumes. Formerly, oakmoss was used as a tincture. By distilling an essential oil such as bergamot in the presence of a certain quantity of the moss, a product may be obtained which is not exactly a bergamot or an oakmoss and has a new character of originality and fineness.

Absolute oil of fern

Although the odour of fern (fougere) is generally attributed to complex synthetic and natural substances. The fern absolute exists and is obtained from the rhizomes of male Aspidium. Its odour is sweet, woody and redolent of earthy humus.

Absolute oil of ivy

This essential oil, Ivy or Lierre, resulting from the treatment of ivy leaves with hydrocarbons, comes from raw material that is very widespread in France and Europe; it is a climbing plant which creeps over the ground or climbs along a wall or up a tree (*Hedera helix*).

Absolute oil of tree moss (Mousse d'Arbre)

These lichens are collected from all kinds of trees other than the oak, whose moss has a very fresh and pleasant top note. Tree moss absolute is used for blending with oak moss absolute, chiefly in order to lower the price.

Vetiver Absolute

Oil of vetiver is extracted from the rootlets of *Vetiveria zizanoides*, of the Graminaceous family, which is found in India, Malaya, Reunion, Brazil, New Caledonia etc. The best quality is that from India.

Balsamic Series

Resinous notes

Absolute oil of cypress

This absolute oil comes from the needles and sprigs of *Cupressus sempervirens*, and no doubt from other species such as *Cupressus lusitanica*; from which the concrete is extracted by benzene, a concrete that furnishes the absolute by further treatment with 95° alcohol. The absolute so obtained possesses an odour recalling that of conifers like the pine and to some extent has a slightly ambered background.

Absolute oil of fir needles

The products that come from France as Absolu d'Aiguilles are as a rule those from *Abies alba*. The concrete is prepared by treatment of the needles and twigs with benzene, followed by a second extraction with 95° alcohol, the odour is balsamic, with a suggestion of fruitiness and a coumarin note.

Everlasting (Immortelle) absolute

This comes from *Helichrysum angustifolium*. The concrete is obtained by volatile solvents and then the absolute with alcohol. The odour is powerful, warm, dry and very tenacious, recalling coumarin, but a little fruity and slightly honied.

Vanilla notes

Absolute oil of Vanilla

The plant that yields absolute oil of vanilla is *Vanilla planifolia*, a native of Mexico. To obtain the black pod with which we are all familiar, the exact treatment varies according to the country of origin. Many French perfumers still use tincture of vanilla, which they make themselves. The best qualities of vanilla pods are those from Reunion and Mexico.

Absolute oil of Heliotrope

Absolute oil of heliotrope, although the possibilities of producing it are great and its price is very reasonable.

The absolute is extracted by the process of hot oil enfleurage of the ends of the flowers. It is made in the Grasse region, and in odour is faithful to the perfume of the flower. Here it suggests both vanilla and anisaldehyde, and of course heliotropin, together with herbaceous and woody tonalities which recall the odour of hazel nuts.

Fruity Series

Absolute oil of *Fig Leaves* is made on a relatively small scale in the south of France. The plant is the *Ficus caria*, which produces also the fig fruit. The leaves are treated with petroleum ether for obtaining the concrete, which is then treated with 95° alcohol. The odour is sweet, fruity, honeyied, with a green, grassy tendency, slightly woody.

Animal Series

Amber notes

Ambrette seed absolute

This absolute oil is obtained from the seeds of *Hibiscus abelmoschus*, of the Malvaceae family, a herb that can reach a height of 2 m. Only the seeds are used. The plant is a native of India, but grows throughout the tropics.

The true absolute of ambrette from the concrete is not the ordinary commercial product, which is in fact obtained by redistilling and selecting the essential oil obtained by steam distillation. Its powerful, warm odour has a tenacious musk and amber character.

Angelica absolute

The treatment of this plant is carried out in Grasse; also in Holland and elsewhere. Its odour is powerful, pungent, herbaceous, and rather musk-like, after the manner of a macro-musk. It bears a slight similarity to that of celery, but is more fruity, less dry and with suggestions of opoponax.

Cumin absolute

The plant producing this absolute is *Cuminum cyminum*, also one of the Umbelliferaceae. It is the seeds that are processed.

Labdanum absolute

Labdanum has a typically balsamic odour, rather flowery, herbaceous, amber-like and very tenacious. The absolute derived from the species growing in Crete is the finest, the most amber-like and the least coloured.

The absolute is obtained from the concrete by benzene and extraction of the concrete by 95° alcohol.

Clary sage absolute

Clary sage or *Salvia sclarae* is harvested in Provence, in north Italy and Morocco. The concrete is obtained by petroleum ether from the leaves and the absolute by subsequent treatment with alcohol. The odour is an original, light note, slightly winy, with a muscat character. It has an amber and even slightly fruity background note.

Maritime notes

Seaweed Absolute

The experimental development of algae and seaweed odourants. One such product, prepared from several seaweeds, has a typically 'seaside' odour, recalling ozone, which is also the characteristic note of a synthetic product often termed 'sea aldehyde'.

The association of these two products can give interesting and original results, to obtain striking effects of atmosphere, space and sea: and perhaps of bathers and the odour of salt water on suntanned skins.

Musk notes

Costus absolute

From *Aplotaxis lappa* (Compositae), a native of India and particularly of the Himalayas, at altitudes of from 2000 to 4000 metres, comes costus absolute. The dried and crushed rhizomes are treated by benzene or petroleum ether to obtain the concrete. The absolute of the concrete is obtained by the conventional method.

Empyreumatic Series

Tobacco notes

Mate (Vert) absolute

The semi-solid dark green absolute extracted from the dry leaves of mate or *Ilex paraguayensis* and related species deserves to be more widely used in fine perfumery, especially as the leaves of the South American 'yerba mate' are widely available as the basis of a tea-like beverage. The odour is warm herbal-fruity, with a peppery phenolic shading and a patchouli-amber end note. This oil can give most interesting effects in fougere, foin coupe, chypre; tabac, leather; and mossy, green, patchouli, woody and aldehydic blends.

Melilot Absolute

This is extracted from *Melilotus arvensis* and *Melilotus officinalis*. It contains coumarin and methyl coumarin, and has a slightly fruity odour similar to that of mirabelle, with some resemblance to the tobacco note.

Tobacco Leaf Absolute

This comes from the *Nicotiana*. Its absolute recalls the smell of a cigar. It also bears some analogy to the body note of musk. Decolourised absolutes are available.

ESSENTIAL OILS, TERPENELESS OILS

These are the essential oils from which the tarpene components have been removed by extraction and fractionation, either alone or in combination. The optical activity of the oil is thus reduced. The most widely used of all natural perfumery materials, namely the essential oils—distilled from flowers, leaves, roots, fruits, seeds and, in some cases, whole plants. The old type of direct-fired still, heated with a naked flame, has been almost entirely superseded by the indirect type of steam distillation, which in turn can be modified by the use of a partial vacuum. Dry and destructive methods of distillation are also employed in a few specialised cases. Some oils are further modified by 'rectification', involving redistillation and possibly fractionation. A very restricted number of essential oils are expressed, not distilled; the most obvious examples being the cold-pressed (i.e. from the peel) citrus oils.

The terpeneless essential oils obtained from varieties of plants are summarised in Table 2.1 below:

Table 2.1. Summary of ternepenless essential oils.

Plants	Species	Constituents	Odour	Uses/Remarks
Agar	Agallocha	–	–	Distilled from fungus-infected wood
Ajowain	Ptychotis ajowan	Thymol, Cymene	Like thyme, but sharper, sweeter	little perfumery
Almond	Prunus amylgdalus varamara	Benzaldehyde (Circa 85%)	Warm and bitter sweet	–
Ambrette	Hibiscus abelmoschus	Lactone Ambrettolide farnesol, deoyl alcohol	Pungent, warm floral-musky vinous	–
Amyris	Amyris balsamifera	–	Woody-balsamic	used as a diluent of Haitian vetiver oil soaps
Angelica	Angelica officinalis	Lactone, cyclopentadecanolide	Earthy, musky woody, balsamic	–
Anise	Pimpinella anisum	Anethole methyl chavicol	Sweet, agreeable	Perfume and Flavour
Araucaria	Callitropsis araucarioides	Eudesmol, eudesmene, geraniol esters phenol	Sweet balsamic woody, rose-like	Soap perfume –
Arnica	Arnica montana	Thymohydroquinone dimethyl ether	Pungeut	Perfumery
Artemisia		–	Sweet, sharp-fresh camphoraceous herbaceous	Soaps, toilet water masculine perfumes
Atractylis	Atractylis lancea	Sesquiterpene alcohol probably eudesmol	Fresh warm, woody-spicy, buttery almost malty	Oriental, woody spicy, leathery and ambered composition
Backhousia	Backhousia citriodora	citral	Fresh and green shaded	
Basil	Ocimum basilicum	Linalool estragole d-camphor	etheral, herbaceous, green note, powerful,	de cologne as top note

(Cont'd)

Plants	Species	Constituents	Odour	Uses/Remarks
Bay	Pimenta recemosa	Chavicol, eugenol, methyl eugenol	powerful slightly, bitter, sugary	–
Bellary leaf oil	Neolitsea zeylanica	–	Sharp sweet like unripe mango	–
Bois De Rose	Aniba rosaeodora	Terpineol, nerol, geraniol, cineole, acetophenone	–	perfumery
Buchu leaf oil	Barosma betulina	Diosphenol	Minty, phenolic woody-root like	Gooseberry blackcurrant peppermint flavours
Cabreuva	Myrocarpus spp	–	Sweet-woody	–
Cajuput	Melaleuca cajuputi		Camphoraceous sweeter	
Calamus		Acorus calamus	Strange slightly	fantasy perfumes
Camomile	–	azulene hydro-carbons	spicy honey like	
Camphor	Cinnamomum comphora	Camphor	Pungent camphoraceous	soaps, household products
Cananga	Cananga odorata		Sweetless, cruder	perfumery
Caraway	Carum carvi	Carvone	warm pungent	Fine perfumery soap
Carda-mom	Elletaria Carda-momum	Cineole, borneol terpineol, esters	Slightly green-camphoraceous and later power-ful fresh, warm and woody quality	Flavours, incense
Carrot	Daucus carota	Methyl ionone	warm, spicy	Fantasy perfumes
Casca-rilla	Croton eluteria		spicy-woody	high class flavours, colognes
Cassia	Cinnamomum cassia	Cinnamic aldehyde, benzoic, salicyclic acid	Spicy, sweet warmer, woodier	perfumery
Cedar	Juniperus virginiana	Cedrol	Woody balsamic	perfumery
Celery	Apium graveolens	–	Fresh, sweet, herbaceous	Top notes
Cinna-mon	Cinnamonum zeylanicum	Cinnamic aldehyde, eugenol	Finer, more aromatic	Perfumery

(Cont'd)

Plants	Species	Constituents	Odour	Uses/Remarks
Citro-nella	Cymbopogon nardus	Geraniol, citronellal	Sweet, lemon-eugenol, esters, borneol, dipentene	Soaps rosy
Clove	Eugenia caryophyllata	eugenol, caryophyllene	warm, spicy	diluent, blending agent
Coria-nder	Coriandrum sativum	dextro-linalool linalyl acetate	fresh, woody-spicy, sweety	Flavours, perfumery
Costus	Saussaurea lappa	lactones, terpenes sesquiterpenes	complex fruity	Fine perfumery
Dill	Anethum graveolens	carvone		Flavouring of pickles sauces
Estra-gon	Artemisia dracunculus	estragole methyl chavicol	pungent, herbaceous	Perfumery, flavour, ingredient, soap, perfumes
Eucaly-ptus	Eucalyptus globulus	–	pungent	soap, cosmetic perfumes
Fennel	Foeniculum vulgare	limonene	camphoraceous phellandrene	Perfumes
Gera-nium	Geranium bourbon	rhodinol	rose-geranium fruity	–
Ginger	Zingiber officinale	Sesquiterpene C_9-C_{10} aldehyde	Sweet, warm spicy	perfumery
Ginger-grass	Cymbopogon martini	geraniol, esters	Fatty-sweet herbaceous	Soap, perfume
HO	Cinnamomum camphora	Camphor linalool		Soap, detergent
Lavender	Lavandela officinalis	linalyl acetate	Fresh	Perfumes
Lemon-grass	Cymbopogon flexuosus	Citral	verbenaceous, lemony, fresh	Cheap perfumery deodorant
Limes	Citrus aurantifolia	–	Sharp, fresh terpenic, citrus	Flavours perfumes
Marjoram	Thymus mastichima	–	Spicy, herbaceous	Food flavouring
Myrtle	Myrturs Communis	–	fresh, fruity camphoraceous	–
Neroli	Citrus aurantium L.	linalool, linalyl acetate terpineol, nerol, decyl aldehyde	Sweet, refreshing	Home made cosmetics, bakery, cologne
Nigella	Nigella damascena	–	wild strawberry	modern perfumery

(Cont'd)

Plants	Species	Constituents	Odour	Uses/Remarks
Opo-panax	Commiphora erythrea	–	Sweet, warm balsamic	powdery perfume
Orange			thin odour	household products
Orris	Iris florentina	myristic acid irone, furfural	wood-earthy	–
Palma-rosa	Cymbopogon martini	geraniol	sugary geranium-rose	–
Patcho-ouli	Pogostemon patchouli	Carvone	Sweet, powerful persistent	perfume
Pepper	Piper nigrum	aldehyde	fresh, woody	perfumery
Pepper-mint	Mentha piprita	menthol, methyl ester	menthol, minty	Flavours, after-shave lotion, cosmetics
Petit-grain	Citrus aurantium	–	Sweet orange	–
Bigarade	–	–	–	
Pimento	Pimenta officinalis	eugenol, cineole methyl eugenol,	Spicy	Spicy, savoury, flavours, sauces,
Pine	Pinus nigra, strobus, palustris	bornyl acetate	–	liquid bath oil
Rose	Rosa centifolia	–	Rosy fresh rosy	Rose water perfumery
Rose-mary	Rosmarinus officinalis	–	Fresh, sharp comphorated balsamic	Hungary water
Sandal-wood	Santalum album	–	aromatic, sweet rosaceous	source of santalol
Sassa-frass	Sassafrass albidum	Safrole	pungent, woody	beverage, confectionery
Silver fir	Abies alba	–	Fruity, balsamic	–
Snake-root	Asarum cana-dense	–	Warm, aromatic spice, mint	flavours perfume
Spear-mint	Mentha spicata	–	Citrus aldehyde type	Chewing gum dentrifices, soap
Lavender spike	Lavandula spica	–	Camphoraceous	Cologne
Tansy	Tanacetum vulgare	alcohols, ketones	herbaceous, camphoraceous	–
Thyme	Thymus vulgaris	Carvacrol	Warm, fresh	–
Terpen-tine	–	–	–	perfumery chemicals
Verbena	Lippia citriodora	citral	lemony	tisanes and herbal decoctions

(Cont'd)

Plants	Species	Constituents	Odour	Uses/Remarks
Vetiver	Vetiveria zizanoides	–	grassy woody	Chypre, fougere, perfume
Winter-green	Gaultheria procumbens	Methyl Salicylate	Fruity, sickeningly sweet	alleviation of rheumatic symptoms
Worm-wood	Artemisia absinthium	thujone, thujol esters, terpenes	herbaceous fresh	perfume
Ylang-Ylang	Cananga odorata	linalool paracresol, eugenol, benzyl acetate methyl benzoate	–	–
Zdravetz	Geranium macrorrhizum	–	sweet, rosy, woody	perfumery

Expressed Essential Oils

The citrus fruit oils may be conveniently considered as a separate group, because: (i) they are derived from trees of closely related species; (ii) they come from the fruits of these trees; and (iii) they are obtained by expression rather than distillation. If some grades of a few of them are in fact distilled, this is only because a cheaper everyday product is in demand, for purposes that do not call for the use of the finer quality perfumery oils. A case in point is their application in certain foods and beverages.

Citrus oils high in terpene content are much more prone to oxidation than most other essential oils. One of the best means of preservation is to pack and store hermetically in glass, using a stream of liquid nitrogen for filling the space between the surface of the oil and the closure with inert nitrogen gas. The containers should be kept in a dark, dry and well-ventilated place.

Bergamot

Citrus aurantium, subsp, bergamia (Citrus: Rutaceae), Grown almost exclusively in Calabria. The essential oil is obtained by cold expression from the peel or rind of the nearly ripe fruit. Originally the so-called sponge process was used, in which the worker exerted sufficient hand pressure on the pulp-free half-fruits to enable the sponge to soak up the mixture of oil and juice. The sponge was squeezed into an earthenware pot, and the oil is finally decanted. This process, described by Guenther as 'incredible drudgery', is said to have yielded some very fine oils. Now-a-days most of the oil is obtained by mechanical means. It includes the use of 'pelatrici' which rasp or grate the peel under standard conditions, hence releases the oil which is then carried by a stream of water, together with cellular debris etc., to a centrifugal separator.

Bergamot oil contains linalool, nerol, terpineol and dihydrocuminol, in a total of 40 to 50 per cent; linalyl acetate up to 40 per cent; 5 to 6 per cent of the odourless bergaptene; and traces of methyl anthranilate, aldehydes, hydrocarbons etc. The odour is fresh, sweet and ethereal, though evanescent, and has an inimitable quality which distinguishes it from the other citrus oils. The light orangeflower note found in good bergamot oils, and which gives them their cachet, is probably due to the traces of methyl anthranilate and the nerol content, as well as the predominance of linalool and linalyl acetate. (To one of us – F.V.W. – the peel note of bergamot suggests Guinea lemon oil, while the bouquet has an almost nutmeg-like quality due possibly to its terpenes.)

Bergamot oil itself; bergamot freed from its furocoumarin content (as a safeguard against the skin bronzing effect known as Berlock's Dermatitis); terpeneless and 'superabsolute' bergamot oils: all are of importance to perfumers everywhere.

Grapefruit

Citrus paradisi MacFadyen (Citrus). The grapefruit was originally called the shaddock, possibly after the mysterious Captain Shaddock, who may or may not have introduced it into the West Indies from China or Japan. The tree was subsequently cultivated in California and Florida, where its characteristics have been modified and the fruits improved. During this period the essential oil cold-expressed from the peel has shown a tendency to become milder and sweeter of odour. Even so, it remains very characteristic, and is an important addition to the highly useful range of citrus oils. Its odour is very refreshing, rather dry, typically citrus-like, and has a certain delicacy.

Distilled Grapefruit oil is available for use in low-cost perfumes, industrial deodorants etc., but bears no relationship to the vastly superior expressed oil.

Lemon

Citrus limonum (Citrus: Rutaceae). The lemon tree, probably a native of India, is cultivated in the Mediterranean countries, Guinea, Morocco and other parts of Africa, California, Florida, Brazil etc. California is the largest producer of the machine-pressed oil. Hand-pressing is carried out in Guinea and to some extent in Italy. Concentrated or deterpenated oil of lemon is used extensively in flavour work but is of relatively little interest to the perfumer. It is of course a source of naturally occurring citral, octyl and nonyl aldehydes and certain esters, but owing to the removal of limonene (which is predominantly present in the untreated oil) it is quite distinct in odour from natural lemon oil. The odour of the latter is fresh, clean and typical of the peel of cut lemons. A much inferior distilled oil finds some application in beverages and for the perfuming of household products.

Mandarin

Citrus recticulata Blanco var. mandarin. Very similar is the tangerine variety of the same tree. (Citrus: Rutaceae). Mandarin oil is expressed either by hand or machine in Italy, Spain etc. It contains about 1 per cent of methyl methyl anthranilate, which contributes considerably to its light, blurred note of neroli, coupled with decyl aldehyde, citral, citronellal.

The different essential oil producers have devoted much attention to the removal of terpenes and sesquiterpenes, and the fractionation of the oil, in order to increase the concentration of certain interesting components. So-called super-absolutes have been prepared by isolating the aldehydic portions and thus creating natural aldehydes which can be of great value to the perfumer. As for the perfumery factories, some have themselves produced these special fractions, concentrates and even new synthetics; for all that is required is a very small-scale production.

Orange

Bitter. (Bigarade). *Citrus aurantium*, subsp. amara. There are many varieties of the bitter orange tree and many growing regions. Odour and flavour vary accordingly. The main producers at the time of writing are Sicily, Spain, Guinea and the West Indies. Other production centres include Brazil, France and China. The oil is expressed mechanically or by hand. Inferior oils may contain additions of a distilled oil produced from the exhausted peel. Its underlying floral quality also helps to make it virtually indispensable in many Colognes, citrus-aldehydic, fresh-floral and fruity-floral types of perfume.

Orange, Sweet (Portugal)

Citrus sinensis. According to Swingle, it is probable that the sweet orange as we know it today was first brought from Southern China to Portugal, about 450 years ago. Hence the species, botanical name. The main producers of sweet orange oil (which like lemon oil attains a high annual tonnage) are Florida, California, Texas, Sicily, Calabria, Valencia, North Africa, Cyprus, Israel and Brazil. The oil is expressed by hand or by machine. An essential oil is also obtained by distillation, but its uses are mainly confined to low-grade flavours and certain household and industrial perfumes. It also serves as a basis for isolation and synthesis of other odourants, and to some extent as a diluent or adulterant of expressed oils.

The main constituent, quantitatively, of expressed sweet orange oil is limonene, which occurs to the extent of about 90 per cent. The decyl aldehyde content ranges from 1 to 2.5 per cent. Other trace constitutents include octyl alcohol, citral and linalool. The oil's odour is characteristic of sweet oranges and the flavour is pleasant.

As in the case of other citrus fruit, fractions and absolutes have been prepared which complete the perfumer's 'palette'. An orange with a rather special note is that from Jamaica. This is rather larger in size than the Sicilian variety. However, what distinguishes it is its special note. The oil in fact has the odour of sweet orange with a very slight and very pleasant suggestion of aniseed, which comes as a surprise and is able to create some interesting effects in perfumery.

These expressed citrus oils are not perhaps classifiable as true volatile oils, for they do in fact contain varying amounts of non-volatile matter. Even so, it would be splitting hairs to define them as anything other than true 'essential' oils. Much work has been undertaken on the initial and subsequent treatments of the citrus oils. Leaving aside the use in certain cases of antioxidants, there is also a whole gamut of concentrations, deterpenations, fractionations and blendings to be considered. Terpenes and sesquiterpenes have been variously removed. So-called absolutes and superabsolutes have been prepared, and special products made by combining (say) parts 2 and 5 of a fractional distillation in order to arrive at a certain result. This means, in effect, that a countless number of combinations can be made, in order to obtain high-quality products with novel odours.

Isolates, Esters etc.

Essential oils may also be considered as natural sources of alcohols, phenols and aldehydes etc. and, by means of relatively simple chemical treatment, of esters and other organic compounds. Many years before synthetic chemistry had advanced to its present stage of development, these simple processes of separation, extraction, esterification etc. had enabled some important additions to be made to the perfumer's array of odourants. There is of course an odour difference between odourants obtained entirely synthetically and those derived from natural sources (e.g. geraniol synthetic and geraniol ex palmarosa, ex citronella or – for that matter – ex geranium oil). Even if the spectrographs are all virtually identical, the perfumer's trained nose should perceive the distinguishing differences. All who have created perfumes know how easy it is for even a minute trace of certain products – contaminants, additives or naturally occurring substances – to change completely the character and appeal of a perfume blend.

One cannot afford to be indifferent to the selection of this or that geraniol: or of any other perfumery material. One must try out, smell and judge every product of its kind before using it. This ought in fact to be a general rule in all branches of perfumery. Perfumes are appreciated by the nose, and it is for the nose to decide. The following alphabetical list comprises those essential oils which are particularly valuable as a source of isolates and products derived therefrom.

Anise Oil

The principal constituent, anethole, may be obtained by chilling, crystallisation and filtration. It contains only trans-anethole, which is toxicologically acceptable, and it has the finest odour and flavour. Anethole from star anise oil is also acceptable in the first sense but has a less attractive odour. Cis-anethole is many times more toxic than the trans-isomer, but is sometimes marketed as technical anethole or as a constituent of artificial anise oils. Its use as a food flavouring should be avoided. Trans-anethole has a GRAS (Generally Regarded as Safe) rating and is of course widely used as a flavouring for liqueurs, confectionery, toothpaste etc.

Bois de rose oil (Brazil)

Linalool is extracted by fractional distillation and other processes. That obtained from the Brazilian oil contains both laevo and dextro isomers, whereas the Cayenne oil yields principally 1-linalool. Some users, however, prefer a linalool derived from a Cayenne oil, as being rosier and generally more agreeable in odour than those from other sources.

Camomile oil (German)

In addition to its odourous constituents, this oil also contains the blue colouring material, azulene, which is extracted from it and has proved of some interest in dermatology and cosmetics.

Cedarwood oil (Virginian)

This oil is the best natural source of cedrol, cedrene and cedrenol. The most interesting of these constituents, olfactorily if not economically, is cedrol; the odour of which varies according to its source and degree of purity. It has only a faint odour when pure but some examples have a warm note recalling methyl ionone though rather more woody.

Citronella oil (Java)

It has a high content of geraniol and citronellal. As a source of geraniol, Java citronella is inferior to palmarosa oil. Opinions on purely synthetic geraniol are divided, but some perfumers rate it higher than the palmarosa derivative. The citronellal extracted from Java citronella oil yields citronellol on reduction, and may also be used as a starting-point for hydroxycitronellal.

Clove oil

To obtain its characteristic constituent, eugenol, the oil is treated with potash solution, the supernatant portion being decanted and the potash derivative decomposed by dilute hydrochloric acid accompanied by refrigeration. The brown oil so obtained, when decanted, washed, dried and rectified under vacuum, yields a fine grade of eugenol.

Geranium oil (Bourbon)

It contains l-citronellol or rhodinol, in a quality superior to that of similar products obtained from Algerian or Moroccan geranium oils. Rhodinol ex Geranium Bourbon contains, moreover, the ketone menthone, which imparts to it quite a special note and one that, as we have seen elsewhere, can offer advantages in certain formulations.

Lemongrass oil

This oil is a notable source of citral, which is also obtained from Litsea cubeba oil etc. and by synthesis. The citral produced from lemongrass oil is of interest in itself and also as a starting-point for the ionones and methyl ionones.

Palmarosa oil

The geraniol derived from palmarosa oil, of which it is the chief constituent, is normally accompanied by nerol and other alcohols. Although it remains superior in quality to synthetic geraniol, the latter has made it much less attractive economically.

Sassafrass oil

It yields safrole, which in turn provides the perfumer with heliotropin, as a result of fractional distillation, isomerisation and oxidation.

One should also consider the widespread application of deterpenation. As is well known, essential oils contain monoterpenes and sesquiterpenes as well as aldehydes, alcohols, esters, ketones, lactones, phenols etc., and in some cases it is desirable to resort to deterpenation. Here we are generally concerned not with the complete removal of terpenes and sesquiterpenes but with their partial elimination, the object being to remove insoluble hydrocarbons and so increase the oil's solubility in alcohol. In this way, low degree alcohol used as a vehicle for less expensive toilet waters and lotions.

Another advantage is offered in the case of those oils whose terpenes rancidify and take on a less agreeable odour when oxidised. This does not apply of course to oils like vetiver and patchouli, which improve with age and the chemical changes that it brings, and for this reason should not be deterpenated. In regard to oil of vetiver, however, it is interesting to extract the mixture of sesquiterpenic alcohols known as vetiverol, which may then be esterified to give, the acetate.

Some essential oils are themselves subjected to esterification in toto. Vetiver oil may again be cited, and this (usually acetylated) oils adds a further note to the range.

Tinctures, Resins, Balsams

Tinctures

These are alcoholic extracts of natural perfumery or flavouring materials etc., in which the finely divided material is allowed to macerate or steep in alcohol, over a period of time, the solvent remaining in the finished tincture as a diluent. The strength of the alcohol is usually 95° and the period of time variable, ranging perhaps from some weeks to several months. In addition to the advice given elsewhere on the constitution of the containers in which tinctures are prepared, it may be stressed that brass and bronze taps should be suitably chromed or else avoided. Glass tubing should be of sufficiently heavy calibre, with as few joints as possible and these always in specially essential oil-resistant rubber or other material.

Infusions

Infusions are prepared in much the same way as tinctures, but with the application of more or less heat, the treatment usually involving reflux and a period of time varying according to the material and temperature. Infusions tend to be less mellow and true to type than tinctures, owing to the heat-induced reactions involved, especially with the alcohol. For this and other reasons the use of infusions is no longer common in perfumery practice.

Absolutes

Absolutes rather similar to those obtained from vegetable sources may also be prepared from animal substances (e.g. the famed 20 × times concentrate musk grain absolute). Strict equivalence must be observed when these are used to replace tinctures.

The more important products contained in this group of materials may now be referred to individually.

Ambergris

Ambergris occurs in lumps of varying size and weight, which are of a waxy, spongy texture, lighter than water, and of various colour shades from chalky white and silvery grey to dark brown and black. It is a pathological calculus, rather analogous to gall stones, in the stomach or, more rarely, in the bowel of the sperm or cachalot whale. Much more common in the male than the female, it is considered to result from the irritation caused by the ingestion of cuttlefish beaks. The whale sometimes expels these concretions by vomiting. They are thought to be usually black at this stage and to become progressively dark grey, light grey and white, on increasing exposure to sea water and light.

At one time ambergris was recovered only as flotsam washed up on the shores of the Indian Ocean, Persian Gulf or the Australasian Pacific. In more recent times the major finds have been made by whaling stations, when cutting up sperm whale carcases.

The so-called black amber or ambre noir is the least highly regarded type of amber. Its odour is often fishy and slightly fecal. The light grades are usually the best and it is these which go to make the tincture or the concentrated 'absolute'.

Ambergris has a discreet, smooth and pervasive odour which recalls not only the sea but also old books, old furniture and old cathedrals. Ambergris behaves rather like other products (nerolidol, for example) which though having an odour of relatively low intensity, nevertheless develop their effectiveness in the presence of other odourants, whose own odours they reinforce and with which they attain a characteristically mutual 'lifting' or exaltation. It is in this way, coupled with its exceptional tenacity and rounding-off properties, that ambergris induces a mutual reinforcement when accompanied by musk or civet or both. It also sustains floral and aldehydic notes and remains a traditional constituent of luxury perfumes.

Owing to its high cost and variable quality, ambergris should be evaluated and purchased only by a really competent expert. Much important research has been carried out on its constitution (notably by the Firmenich organisation and their collaborators) and some excellent synthetic and artificial ambergris-type products have become available.

The usual tincture is of 2 to 3 per cent concentration in 95° alcohol, the ambergris being pulverised for the purpose and allowed to remain in the solvent for a minimum period of three months, with constant stirring. Some compounders neutralise with ammonia. The tincture is filtered before use. Certain perfumers only use such a tincture after it has been macerated and aged for a very long time.

Castoreum

Castoreum is a secretion obtained from the beaver, which is nearly always the *Canadian beaver, Castor canadensis*. The name is also applied to the pear-shaped pouches or sacs in which the ointment-like secretion is found. This secretion becomes hard and brittle after the pouches have been sundried or smoked. Castoreum is used in the form of a tincture of 5 or 6 per cent strength in 95° alcohol, as a resinoid, or as a kind of concentrated absolute. Leading perfumery supply houses produce specialities of the latter type, one of which is Givaudan's Anhydrol Castoreum. To prepare the tincture, the whole pouch is often chopped and ground, covered with hot alcohol at 65° to 70°C, and allowed to macerate and infuse for one or two months, with intermittent stirring. Some processors prefer, where possible, to remove the adherent castoreum from the pouch when preparing this first and best tincture. A slightly inferior product is sometimes made by macerating the empty or exhausted pouches.

The odour of castoreum is very characteristic, suggesting gloves and Russian leather, and – to a lesser extent – galbanum and ripe figs. Its chief disadvantage in perfumery work is its dark colour. Small amounts have been said to impart 'lift' in raspberry, rum and other flavours.

Civet

Civet is a soft paste-like substance, obtained from Ethiopian sources in bullock, buffalo or zebu horns. It has a skatolic odour due to its content of methyl indole or skatole. Civet also contains the musk-like lactone, civettone. One of us (F.V.W.) possesses a sample of civet obtained – by very gentle treatment, incidently – from a healthy male civet cat kept in captivity in one of the world's most famous zoos. This civet has only a faint suggestion of skatole dend has much more in common with civettone.

Tinctures, Anhydrols etc. of civet are used, the former being usually an infusion of 3 to 6 per cent in 95° alcohol. The alcohol is warmed at first in order to allow the mass to be finely divided up during stirring. Tinctures should be matured for several months at least. Some perfumers have made 10 per cent tinctures of civet, in fine perfumery nothing is definate.

Musk

Musk is a granular substance, colour and texture similar to dried blood, which in fact is often used to adulterate it. Musk is secreted in the preputial glands of the Himalayan musk deer, of Assam, Nepal, Tibet and China. This gland or 'pod' is about 4 or 5 cm in length, and is covered in white and brownish hair in the centre of which is a small orifice. Musk is undoubtedly a sex-attractant to the female immediately prior and during the rutting season. The grains found inside the dried pods can be extracted by means of a bone curette or a metallic sampling device known as a sonde. Musk grains separated from the pods are also to be found on the market. They may have a fine, strong but slightly ammoniacal odour. Most expert purchasers seem to prefer to buy the whole pods, even though the latter may contain lead shot, paper and, worst of all, dried blood.

The Tonquin musk of which the famed 'blue skin musk' is typical, is generally recognised as being superior to that obtained from Assam and Nepal. It comes principally from Tibet and China. The blue skin is a membrane that lies between the grains and the leathery pouch or pod. If macerated with the grains it adds to the animal-leather odour. Pods should be round and slightly flattened, covered with rough, short hair. Inferior pods are often pear-shaped and covered in longer, redder hair.

The musk deer population has been severely depleted in the recent past by indiscriminate slaughter, e.g. by means of rifles equipped with telescopic sights. One sign of the animal conservation movement is apparent, however, in northern India, where attempts are now being made to farm musk deer and to obtain their musk without actually killing them. It is a historical fact that, in the remote past, the Chinese were in the habit of capturing the male deer, removing its musk gland and then allowing it to go free. Unless some form of conservation is practised, natural musk from the musk deer may well disappear entirely from the perfumer's shelves. It has already been largely replaced, except for the more expensive perfumes, by the enormous and exceedingly useful range of synthetic musks.

Musk tinctures are prepared from 2 to 6 per cent of musk grains in 90° alcohol. No heat is required and the equipment should be fitted with paddle stirrers. The addition of a little ammonia often facilitates matters, and small quantities of other substances are used to improve subdivision of the particles and so forth. We refer here to lactic acid (as cited in many old formularies), animal black, sugar and lactose. The inventor of the lactose procedure always found it best for the purpose, one learns. Animal black might well prove too active, by actually exerting a deodorising effect, which is precisely what is not wanted. The period of maceration is at least six months. A more concentrated musk at 10 per cent strength may be obtained. A lower-grade tincture may also be made from the empty pods and exhausted blue skins.

The odour of musk is so characteristic that it has given rise to the adjective 'musky'. Sweet, smooth, mildly animal and persistent, it not only acts as a fixative but greatly enhances the diffusive properties of perfumes and imparts added 'lift'.

All these odourants of animal origin have in common a mellowing, sustaining and diffusing action on perfume blends. They are in varying degree 'fixatives' and have a natural, warm animal character. The following tinctures may also be noted :

Ambrette seed

Ambrette seed, from which the essential oil is distilled, is also used to prepare a tincture: 100 grams of seed to 1 litre of 90° alcohol. The seeds are finely crushed and allowed to macerate in the usual way. It may be added that this tincture, while quite different from that of ambergris, nevertheless has something in common with it.

Chaulmoogra

Chaulmoogra (a fixed oil) is accompanied by a volatile fraction which yields a tincture of smooth, warm 'animal' type. Though little used, it could prove of interest in perfumery.

Orris

Orris, consisting of the powdered dried rhizomes of the iris, yields an interesting tincture when one adds to 25 per cent of it, placed in a container of stainless steel or tinned copper, alcohol of 95° strength, followed by gentle heating. The containers are kept well sealed and the powder stirred up in the alcohol everyday for at least a month. The tincture is finally decanted and filtered. Formerly these tinctures were made in hermetically sealed cylindrical apparatus provided with stirring blades moving countercurrently.

Vanilla tincture

Vanilla tincture is certainly superior for fine perfumery work to vanillin or ethyl vanillin. This applies equally to Bourbon vanilla and Mexican vanilla, even though these have in fact distinctive notes of their own. Some 12 to 15 per cent of vanilla pods, chopped or coarsely ground, is macerated in 95° alcohol and allowed to stand for at least two or three and preferably for six months. The alcoholic flavouring extract differs from the perfumery tincture in both concentration and solvent content. Vanilla resinoids and absolutes are also obtainable. Arctander refers to the 'very rich, sweet and true-to-nature odour' of a good vanilla absolute.

Balsams and Resins

These products are very important from the point of view of perfume fixation, because by their texture, which allows them to form on the surface of the tissues a sort of thin skin or film of varnish, the other perfumes which accompany them tend to be retained.

At one time these resins and balsams were conventionally used in the perfumery in the form of tinctures, but the latter have now been largely replaced by resinoids and absolutes. The solvents used in the preparation of resinoids include hydrocarbons, benzene, acetone, trichloroethylene etc., and in some cases non-volatile solvents and plasticising diluents. Clairs, Resinoines and other decolourised and specialised extracts are also available, including the Anhydrols. The fortunate perfumer still has a wide choice in this section of the raw material field. The principal balsams are as follows :

Benzoin

Four quite distinct types are used in perfumery. First and best is the so-called Siam Benzoin, a gum-resin obtained as a pathological secretion from the incised bark of a small tree, *Styrax tonkinense*, a native of Laos and Tonkin. Siam benzoin has a sweet, balsamic odour with a characteristic vanilla note. It comes to market as pebble-like 'tears' ranging in colour from ivory white to yellow-orange and yellow-brown. The lower grades may contain dust and debris, and are usually of the darker shades. The best grades should give reasonably pale tinctures, have a fine odour, and be almost completely soluble in alcohol (circa 95 per cent).

Benzoin resinoids, 'clairs' and absolutes are available. Perfumer may prefer to use a tincture prepared by macerating 20 per cent of the gum-resin in alcohol.

The Sumatra benzoin, derived mainly from Styrax Benzoides Craib, differs from the Siam type in containing a much higher percentage of cinnamic acid derivatives and a correspondingly lower percentage of benzoic acid derivatives. The Sumatran type also contains coniferyl alcohol. They both contain vanillin. Sumatra benzoin has a rather coarser odour and cannot be regarded merely as a less expensive substitute for the Siam type. It has its own applications and can be a very useful constituent of, and fixative in, floral-balsamic, fougere, new mown hay and other blends.

Copaiba

Various species of the Copaifera tree, a member of the Leguminosae family, yield *Copaiba oleoresin*, commonly known as Copaiba Balsam. The chief production area is Brazil, but smaller quantities come from Venezuela, Colombia and other South American countries. The trees, in the words of Guenther, 'may almost be compared with containers, being filled in the hollow centre with oleoresin', which is simply tapped and withdrawn. The odour of the balsam, which finds some use as a fixative and blending agent, is faintly woody.

Galbanum

Galbanum is the dried resinous exudation of *Ferula galbaniflua* and other closely related species. These are tall umbelliferous plants which grow in Iran, Lebanon and Turkey. The Persian or 'hard' galbanum has no interest as a perfumery material. It is from the 'soft' or Levant galbanum that the oil, resinoid and specially fractionated products are prepared. The odour of galbanum is very characteristic, but even so it varies considerably according to source. It has a woody, green and somewhat balsamic or resinous odour. Some samples have been said to suggest green peppers, while others have a sharp top note as of green applies.

The fashion for 'the green note' in perfumery has extended the use of galbanum. In very synthetic perfumes a touch of galbanum can sometimes bring about a remarkably 'natural' effect.

Labdanum

Cistus ladaniferus is found in Spain (Jaras) and France (Provence); also Portugal, Yugoslavia, Greece and Morocco. It is a perennial shrub of the Cistaceae family. The exuded resin is removed from the twigs and leaves by boiling, skimming, and allowing to harden in moulds. The odour is sweet, deep, balsamic and suggestive of ambergris. It has a slightly 'animal' note and is very tenacious. It finds wide use in perfumery as a resinoid, Clair, Anhydrol, oil and absolute. Its known constituents include acetophenone; a ketone $C_9H_{16}O$ with an odour recalling peppermint; eugenol and other phenols, traces of aldehydes, acetic esters, and the sesquiterpenic alcohol, ledol.

Myrrh

Myrrh and Frankincense have long been, to the perfumer, sources of that third gift of the wise men-gold. Myrrh is an exudate which forms naturally in various small trees and shrubs of the Commiphora species (Burseraceae).

The yield is increased by incisions made in the bark. The gum-resin is chiefly obtained in Somalia, Ethiopia, the Sudan and southern Arabia. It is one of the most ancient, if not the most ancient, of odourous resins known to antiquity.

It is used sometimes as a straightforward tincture, as a resinoid, oil and absolute. Among its constituents are cuminic and cinnamic aldehydes; esters of formic, acetic and other acids; eugenol and metacresol; limonene, dipentene, pinene; and probably two or more sesquiterpenes. For the so-called Bisabol-Myrrh, see under Opopanax.

Myrrh has a slightly bitter-astringent odour, mainly balsamic-resinous in character. Its bitter-sweetness recalls that of orangeflower and petitgrain.

Olibanum

Olibanum (Incense or Frankincense) is obtained by incision into the bark of various Boswellia species, of the Burseraceae family, which are found in Somalia and other parts of north-east Africa as well as southern Arabia. The famed historic uses of this oleo-gum-resin in religious ceremony have never restricted their widespread application in perfumes.

Olibanum resinoids, absolutes and oils are available. The odour of incense is well known and characteristic. The lemony green top note and smooth but pungently fresh and crisp incense note have led to its many uses in oriental, woody, powdery, spicy, aldehydic, citrus and floral perfumes.

Opopanax

The opopanax products marketed now-a-days vary considerably in odour notes, according to the source of supply; and the best advice one can give the creative perfumer is that he should decide upon one specific type or brand of opopanax and stick to it, buying always from the same house, as he would in the case of an admitted speciality or compound.

The usual natural source of to-day's opopanax is *Commiphora erythraea var*, glabrescens Engler, or bisabol-myrrh, which differs from ordinary myrrh in that it contains the sesquiterpene bisabolene. This opopanax has a deeper, more balsamic and less distinctive top note than myrrh. There are some opopanaxes, of course, which are merely blends of the commoner resins and balsams with small additions of crystalline and other synthetics.

Peru Balsam

The tree Myroxylon Pereira grows in Central America. The balsam is a pathological exudation resulting from the stripping, laceration and occasional burning of the bark. The crude means of production and collection employed in the chief source of origin, El Salvador, leads to some of the gummy exudation having a smoky odour. In earlier days stocks were sent off from Peru: hence the name of this balsam.

Purified balsams, alcoholic extracts, co-distillates and resinoids are prepared, as well as various types of Peru Balsam Oil, ranging from the complete oil, containing semi-solid, non-volatile fractions, and partial oils of light colour and less characteristic odour.

Peru balsam, which contains benzyl benzoate and cinnamate, farnesol, nerolidol, vanillin, resinotannol etc., has a smooth, sweet, typically balsamic odour, which blends admirably in compositions as diversified as rose, lily-of-the-valley, powdery and woody perfumes, oriental and spicy notes. It is an excellent fixative and forms long-lasting bases in conjunction with other balsams, resinoids, crystalline odourants, animal tinctures etc.

Styrax

Styrax or Storax or Liquidambar is a pathological exudation from the excoriated bark of *Liquidambar orientalis*, a tree of the witchhazel or Hamamelidaceae family. In addition to the crude Asian styrax there is also an American variety, obtained from Liquidambar styraciflua and produced in Honduras. Both types are purified and pre-heated before use. They reach the perfumer as the resinoid, absolute and oil.

Among the constituents of styrax are styrene, benzyl, cinnamyl and phenylpropyl alcohols; a variety of cinnamic esters including 5 to 10 per cent of styracine or cinnamyl cinnamate; traces of vanillin; and styrocamphene.

Styrax has a pleasantly balsamic odour. Its basic note is slightly 'animal' in type. The top note can be removed, but in any case tends to disappear with age and on blending. Styrax is used in lilac, hyacinth and many other perfumes.

Tolu Balsam

This comes from *Myroxylon balsamum* (Leguminosae), a tree closely related to that which yields Peru balsam. There has in fact been considerable controversy over the taxonomy of these two species or physiological forms. Tolu balsam is obtained by making deep, v-shaped incisions in the bark and using a calabash or cup to collect it. Columbia and Venezuela are the chief sources. The balsam is used as such or as Tolu balsam resinoid, oil or absolute. Both Peru and Tolu balsams are subject to some degree of adulteration with copaiba, colophony and similar substances.

The main constituents of Tolu balsam are identical with those of Peru balsam, but there is a higher content of benzoates and cinnamates. In addition, Tolu contains guaiacol, creosol and phellandrene. The odour of a fine quality Tolu resinoine, solid at room temperature, is very sweet and fresh, hyacinthine, and faintly creosolic and fig-like.

All these resins and balsams and the various products derived from them form the base. They are associated with synthetic odourants, crystalline substances, animal tinctures etc. In perfumes their tenacity remains and reinforces the general theme on which the perfume is built.

The natural infusions and tinctures has contributed to the supremacy of French perfumery. Even today inspite of modern techniques in the synthesis of perfumery chemicals, no perfumer will draw a picture of fine perfume without mention of natural essential oils, resin and balsams. It should be remembered that perfumery is not merely a branch of chemical industry.

Chapter 3

Raw Materials of Perfumes
(Synthetic Origin)

INTRODUCTION

The significant development in the last few years has introduced the commercial new musks and ambergris-like bodies. The use of myrtenal, myrtenol, esters, rose oxides, important jasmine odourants was appeared. Maillard derivatives, green and woody and empyreumatic odourants availability has increased recently. Most of the odourants are included in appropriate heading. This chapter deals with selected chemicals having specific odour groups and in particular some important pertaining to perfumes. It has been noted that the classification based on chemical headings (e.g., alcohols, aldehydes, esters, ethers, ketones) is difficult, hence they are grouped according to odour (e.g., scent of ambergris) or application (e.g., as a soap or textile perfumes).

In each group, unless otherwise stated, the first section (a) features those synthetic odourants and isolates which normally provide the characteristic odour of the perfume or at least contribute significantly towards it. In the second section (b) and as a supplement to what has already been said in Chapter 2, appear the corresponding natural odourants. The third section (c) consists of modifying and accessory odourants, including so-called fixatives. Other materials (d) capable of giving a special note to different varieties of a perfume (e.g. red rose, tea rose, white rose etc., in the case of rose perfumes) are tabulated in Table 3.1.

Rose, jasmine, orangeflower and lily-of-the-valley groups are included because of their basic importance not only for specific flower perfumes but as constituents of many other floral and fantasy perfumes.

Table 3.1. Different varieties of perfumes.

Synthetic Odourant	Natural Odourant	Fixative	Blenders
Rose Odourant	*Bulgarian Rose Oil*		
Rhodinol	Rose de Mai absolute	Wardia	Benzophenone
1-Citronellol	Geranium oil	Roselium	Bromstyrole
Geraniol	Zdravetz oil	and	Benzylacetate
Nerol	Guiacwood oil	many	and esters
Phenylethyl alcohol	Patchouli Oil	others	Cinnamic alcohols
Rose alcohol esters			Cinnamyl acetate
Dimethyl octanol			Citral
C_8 to C_{12} alcohols			Citronellal
and acetates			Citronellyl
C_8 to C_{11} aldehydes			Oxyacetaldehyde
Phenylacetaldehyde			Omega-Decenol
Phenylacetic acid			Dimethyl benzyl
phenylacetates			Carbinol and esters
Rose oxides			Diphenyl ether
β-Damascenone			Dimethyl
			phenylethyl
			carbinol
			Methyl diphenyl
			ether
			Ethyl cinnamate
			Menthyl acetate,
			p-Methyl phenyl
			acetaldehyde, musk
			ketone,
			phenylethyl acetal
			iso-Pulegol
			Eugenol, iso-
			Eugenol, Benzyl
			iso-eugenol
			Farnesol,
			Ionones,
			Linalool,
			Tetrahydolinalool,
Jasmine Odourants	*Jasmine absolutes*		
Benzyl acetate	Mimosa	Jasmones	Benzyl alcohol,
		Dihydrojasmone	Benzyl salicylate
Benzyl esters	Ylang-ylang oil	Jasmine	p-Cresol, cinnamic
		odoured	Amyl aldehyde
α-amyl cinnamic	Bois-de-rose oil	lactones	dimethyl acetal
aldehyde	Terpeneless		
	petitgrain oil		
Jasmone	Bergamot oil	Cis-Jasmone	α-Amyl hydrocin-
			namic alcohol and
			aldehyde

(Cont'd)

Synthetic Odourant	Natural Odourant	Fixative	Blenders
Rose Odourant	*Bulgarian Rose Oil*		
Indole	Celery seed oil	Noane diol-1; 3-acetate	alcohol and aldehyde
Dimethyl benzyl carbinol	Geranium oil	Methyl dihydro-Jasmonate	
Nonane diol-1 3-acetate	Peru balsam		
Hydroxycitronellal	Styrax resinoid		Methyl heptenone
Phenylethyl alcohol	Civet tincture		Eugenol and
and esters	Benzoira absolute		iso-eugenol
p-Cresyl phenyl acetate			phenylpropyl
Linalool, linalyl acetate			alcohol and aldehyde
Cinnamic alcohol			p-Methyl phenyl-1
Anisic alcohol			acetic aldehyde
ethyl anisate			Tetrahydro-p-methyl quinoline
Nerol, Methyl antranilate			Farnesol
and its hydroxycitronellal			
Schiff base			α-Terpineol
Acetates C_8 and C_{10}			
Alcohols C_8 and C_9			
Aldehydes C_8, C_{10}			
and pseudo aldehydes			
C_{14} and C_{16}			
Orangeflower and Neroli Odourants	*Orange flower absolute*		
Methyl anthranilate	Neroli Oil	Many proprietary specialities,	Ethyl, methyl (dimethyl)
Indole	Petitgrain Oil,	including artificial	phenylethyl and
Jasmone, iso-jasmone	natural and terpeneless	and reconstituted compositions	linalyl anthranilates
Alcohols C_{10} and C_{12}	Bois de rose oil	are of use in	Geraniol Methyl
Aldehydes C_8, C_9 & C_{10}	Camphor oil	wholly or partly replacing the natural essential oil and absolute	nephthyl ketone, Neryl geranyl and benzyl acetates
Linalool, linalyl acetate	Jasmine absolute natural and artificial		Nonyl aldehyde and acetate
Phenylethyl alcohol			Nonyl aldehyde and acetate
Nerol			Terpinyl and iso
Nerolidol			Butyl phenyl acetates
β-methyl napthyl ketone			β-naphthyl methyl
Aurantiol			acetaldehyde

(Cont'd)

Synthetic Odourant	Natural Odourant	Fixative	Blenders
Rose Odourant	Bulgarian Rose Oil		
Farnesol			Phenyl acetaldehyde
Benzaldehyde			Phenyl acetic acid iso-Eugenol isoButyl heptenone γ-undecalactone p-Tolyl aldehyde Amyl cinnamic aldehyde Methyl benzoate
Muguet Odourants	Rose de Grasse	phenylacetal-dehyde, dimethyl, and glyceryl	
Hydroxycitronellal	Sandalwood oil	acetals Cyclamen	
Citronellol	East Indian	aldehyde	
phenylethyl alcohol	Citronella oil Ceylon (traces)	Aldehydes C_{10}-C_{12} lauric	
Rhodinol		Amyl cinnamic	
Indole		aldehyde	
Dimethyl benzyl carbinol and acetate		Cinnamic alcohol	
Linalool		Ethyl linalool	
Lillial (Givaudan)			

Green Notes

Fresh green odours, complementary to floral odours as in a flower bouquet, have become an essential component of many fashionable perfumes. They are refreshing, evocative of nature, and serve admirably as a foil to sweeter and more exotic odours. The impetus for using them has also been stimulated by the discovery and production of naturally occurring 'leaf alcohol' and 'leaf aldehyde'. Nevertheless the more conventional odourants in this 'green' category are still widely used and, in their own way, are difficult to replace. In Group A(1) green category includes following odourants.

Group A(1)

Methyl Heptin Carbonate (M.H.C.)

The deceptively potent alkyne carbonates should be used with restraint and discrimination. M.H.C., with its fresh, penetrating and at first slightly fruity odour, is almost indispensable as an odourant of green, violet-leaf tonality. Used in a 5 per cent dilution, it not only gives excellent effects in violet, cassie, sweet pea, mignonette, stephanotis, gardenia, freesia and other perfumes - even including traces in certain rose and lavender bouquets - it also adds strength and character to the first emerging notes of chypre and fern. Its use in the shading of ionones and isoeugenol is also deserving of mention here. Its odour is long-lasting.

Methyl Octin Carbonate (M.O.C.)

Methyl Octin Carbonate (M.O.C.), the octin carbonate, has a rounder and fuller tone than M.H.C., which it can sometimes replace (or partly replace) with advantage. The isoamyl heptin carbonate has like M.H.C., been termed *Vert de Violette*.

Phenylacetaldehyde

Phenylacetaldehyde has a pungent green odour suggesting watercress, on a sweet background of hyacinth/almond/rose. It is indispensable as a constituent of spring flower perfumes. Relatively unstable, it should be used with discretion. Its dimethyl acetal, familiarly termed PADMA, has the characteristically green odour of fresh wet foliage, and is probably the most frequently used acetal in perfumery.

Hydratropic aldehyde

Hydratropic aldehyde has a powerful and penetrating green foliage odour of hyacinthine type. Much stabler than phenylacetaldehyde, which it may partly or wholly replace in given formulae, it is widely used for spring flower notes and as a constituent of soap perfumes. Its dimethyl acetal has a sweetish green fungal odour, reminding one of green walnuts. It is useful in woody, mossy and certain floral types.

p-isoPropyl phenylacetic aldehyde

para-isoPropyl phenylacetic aldehyde has a green, sappy, cortexal odour. *Group A(2)* contains some equally interesting items, most of which are either of natural origin or are synthetic equivalents. They result from a great deal of analytical work carried out on leaves, flowers, fruits and even roots (e.g. those of the radish).

Group A(2)

cis-3-Hexenal

It has a rather heavy green leaf odour and called as 'leaf aldehyde'.

cis-3-Hexenol or 'leaf alcohol'

It occurs naturally in Japanese mint, green tea, jasmine, geranium and other leaf and flower oils; also in raspberry, grape and other fruits. The odour is heady, powerful, of green leaves and grass.

cis-3-Hexenyl acetate

It is sharp, heady, green-fruity. For green folial top notes and to enhance fresh floral effects.

cis-3-Hexenyl formate

It is fresh green, with suggestion of cucumber and violet.

cis-3-Hexenyl salicylate

It is green-balsamic. Said to smell of carnations.

trans-2-Hexenal

It is also known as 'leaf aldehyde', particularly in the (U.S.A.). Widely distributed in nature. Powerful green-fruity odour.

Acetaldehyde-di-cis-3-hexenyl acetal

It is leaf alcohol acetal. Intensely green and powerful, with a slight woodiness. Stable in milled soaps.

Nonadienal

It is also known as violet leaf and cucumber aldehydes. Powerful, diffusive green odour. Used sparingly in a wide variety of compositions.

Nonadienal diethyl acetal

It is fresh cucumber-green odour. An interesting variant on the aldehyde.

Nonadiene-2-6-ol-l (nonadienol). Violet leaf or cucumber alcohol.

Dihydro-nor-dicyclopentadienyl acetate

It is verdyl acetate: Givaudan. It is fresh herbaceous green odour, persistent but very diffusive. Good stability. Useful modifier in chypre, fougere, aldehydic-woody, gardenia, muguet and lilac based compositions.

Natural flower absolutes which exhibit a green odour-shading include hyacinth, jonquil, narcissus and reseda. Rose leaf and violet leaf absolutes are to be noted. Green herbaceous notes are found in lavender, mint, rue and tansy oils; and greenish cortexal notes in the various grass oils.

Many proprietary compounds and specialities with green notes are available. One of the first to be introduced was Parmantheme (Firmenich). They vary in type from cucumber to lierre ('ivy') and from violet leaf to nasturtium and *vert de lilas*.

Modifying and accessory odourants cover a wide range and include some of the aliphatic aldehydes and acetals; and many esters such as geranyl and myrtenyl formates, heptyl isobutyrate, octyl acetate etc. – and at least one of the newer alkoxy-alkylpyrazines. The followings are modifying odourants :

Decyl aldehyde dimethyl acetal, Dimethyl benzyl carbinyl acetate, Methyl phenyl carbinol and its acetate, (Gardeniol II, styrallyl acetate), alpha-Methyl p. isoPropyl hydrocinnamic aldehyde, (cyclamen aldehyde), Nonadienyl acetate, Phenoxyacetaldehyde, Phenylacetaldehyde glyceryl acetal, p. isoPropyl hydratropic aldehyde.

Fruity Notes

Pear, pineapple, banana, apple and grape may be added to the list, although most of them are not perhaps as interesting as the foregoing. Fruity notes should in general be used in perfumery with restraint. Perfumes that remind one strongly of liqueurs or create an 'edible' atmosphere are not likely to attract and may prove nauseating. Nevertheless, most perfumers would agree with their eminent Parisian confrere, that: 'A finely liqueur-like note, slightly fruity, in the initial odour of a perfume is often an index to its success.' The following are the fruity notes available to the perfumer.

Peach

The peach odour can be used in all natural or synthetic moss notes, in compositions of the Chypre type - of a more or less powdery character - and in the gardenia note, where the effect is sensational, as in the very special case of 'nardo', a kind of Spanish tuberose. It was once popular in violet perfumes and blends well with many synthetic musks and spicy notes. Peach odourants include, gamma-undecalactone, delta-undecalactone, amyl, benzyl and ethyl esters, citrus, cinnamon, ginger oils, vanillin.

Apricot

Resembling peach but less forceful, the apricot odour suggests also almond, wine lees (ethyl oenanthate) and jasmine. It blends well with orange-blossom, neroli, syringa, honeysuckle, gardenia, broom and even lily-of-the-valley. Apricot odourants include, gamma-undecalactone, phenylethyl acetate, benzyl acetate, ethyl oenanthate, benzaldehyde, trans-anethole, amyl propionate.

Plum

The plum odour goes very well with the different natural or synthetic mosses and in all cases where moss plays a large part, as in the Chypres. Wherever ionones or methyl ionones are used, there is a place for the plum note. There is no typically plum-like aromatic but most artificial plum flavours contain almond and clove (benzaldehyde and eugenol) as modifiers of varying mixtures of fruity esters, sweet orange oil etc.

These first three fruity notes may give rise, in association, to a number of varied combinations providing very delightful effects in mosses, chypres etc. (Firmenich's Prunella has been suggested for use in chypres at the 1–3 per cent level. Many other fruity specialities of this and other types are commercially available).

Mirabelle

The odour of the mirabelle plum, with its special cachet, can be put to remarkable use in violet notes.

When proportioned with discernment and judgement it contributes, with moss and humus notes, towards giving violet an absolutely unique character of veracity.

Of course, in all compositions where ionone and the methyl ionones play a major part, the mirabelle plum can be tried with some chance of success, as well as in many flowery notes, aldehyde and chypre notes, etc.

Strawberry

The strawberry odour goes very well in rose compositions generally. It blends well into 'green' chypre and fougere. Wild strawberry notes have similar applications. The chief chemical component of most artificial strawberry compounds is of course ethyl methyl phenyl glycidate, which is sometimes accompanied by the so-called 'raspberry aldehyde' together with various esters and other modifying agents.

Raspberry

Raspberry aldehyde (a glycidate) and raspberry ketone (p.hydroxy phenylbutanone), as well as specialities designed to convey this particular fruity odour, find certain applications, according to their exact type, in violet, woody, fern, chypre, rose and jasmine compositions. Ernest Beaux regarded a raspberry shading in the top note of Bulgarian oil of rose as being almost a proof of fine quality.

Blackcurrant

Natural blackcurrant (*cassis*) and the corresponding artificial bases find limited application in certain jasmine, amber, boronia, violet etc. compositions.

Fig

The odour of figs is warm, honey-like and somewhat reminiscent of castoreum. It goes well into many rose, jasmine, fougere and leather-peau d'Espagne types.

Hazel

They find use in chypre, woody and powdery notes, and in lipstick perfumes. Chocolate notes, sometimes unintentional and due to the presence of vanillin and balsams, are also to be found occasionally in lipstick and other cosmetic perfumes.

Bitter Almond

The bitter almond note is chiefly used in spring flower perfumes, heliotrope, fern and new mown hay. In association with orange blossom one obtains an effect pleasantly recalling frangipanni. The natural oil of bitter almonds gives the finest note of this type, followed by best quality synthetic benzaldehyde.

Aniseed

Aniseed is fruity-spicy. Part of the aniseed note is given by estragole, which is found in oil of tarragon and in oil of basil - an aniseed note side by side with a green note. One is also reminded of caraway, carvone, limonene and isosafrole. Anise and related notes can prove very interesting, if used with discretion, in lilac, heliotrope, chypre and Parfum a la Marechale compositions.

Citrus oils

The citrus oils (Orange, Mandarin, Bergamot, Lemon, Limes, Grapefruit) are all well known for their usefulness in eau de Cologne and other citrus types, and in aldehydic perfumes and fresh 'top notes' generally. A certain type of Jamaican orange oil has a distinct aniseed shading. This and other types can blend extremely well with bergamot and lemon. It has been pointed out that essential oil of mandarin (and perhaps still more so, super-mandarin) like orange oil, supports aldehydes very well. These essential oils, 'infuse life and warmth into the dryness of the synthetic note of the aldehydes, while retaining their power and their tenacity, if their respective dosing is well carried out'.

Distilled oil of limes can give an entirely different character to a conventional type of Cologne etc. in which it is carefully used to replace other citrus oils. Interesting effects may be obtained by building round an accord consisting of oils of juniper berry and distilled lime in appropriate proportions.

Melon notes

Melon notes (e.g. 2-6-Dimethyl-2-Heptenal-7 and Methoxycitronellal) go well with cyclamen aldehyde, lauryl aldehyde etc. into cucumber lotion perfumes, 'green', sports and outdoor perfumes.

Angelica oils

Angelica oils are musky, woody, peppery and earthy, as well as fruity. They can be used to advantage in incense-based perfumes as well as in certain fern, chypre, woody and spicy blends.

Woody Notes

Pear, pineapple, banana, apple and grape may be added but they are not as interesting as the above. Fruity notes should in general be used in perfumery in control manner. The strong perfume may not be likely to attract but may prove nauseating.

Woody notes are often blended not only with floral notes but also with odourants of balsamic, 'animal' and, one might add, mossy and amber, fern and chypre types. Table 3.2 contains various types of woody notes.

Table 3.2. Various types of woody notes.

Synthetic	Natural	Fixative
Santalol and santalyl acetate	Sandal,	Sandela
Cedrol and cedryl acetate	Cedar	Vertafix
Vetiverol and vetiveryl acetate	Guaiac, agar	Vertenex
Amyris acetate	Cedrela	Acetyrisia
p.-tert. Butyl cyclohexyl acetate	Siam wood oils	Madrysia
Cuaiyl acetate	Vetiver	Cedris
Nopyl acetate	Patchouli oils	Truffe
Isobutyl and isopropyl quinolines		
Irones		
Methyl ionones		
Borneol		

Empyreumatic Notes

Empyreumatic, meaning pertaining to or having the taste or odour of slightly burnt animal or vegetable substances, covers a wide spectrum of odourants, ranging from such relatively crude products of destructive distillation as juniper tar, birch tar and rectified cade oils to substances having tobacco, leather, dried hay, caramel, malt, coffee, and 'cooked' odours (like those of fried potatoes or grilled steak). The new entrants to this group are the Mallard derivatives. They are obtained from the complex reaction between reduced sugars and amino acids involving the release of carbon dioxide and the development of brown colouration.

Some of the items in the following list could also be included under honey, coumarinaceous, mushroom and 'animal' notes, with which indeed they have something in common. Types of empyreumatic notes are shown in Table 3.3.

Table 3.3. Types of empyreumatic notes.

Synthetic odourant	Natural odourant	Fixative
p.methyl quinoline; tetrahydro p.methyl quinoline		Maillard derivative with nutty, smoky, chicory burnt
p.-tert. butyl metacresol	Amber	woody odours
p.-tert. butyl cyclohexanol and its acetate	birth tar	
phenylacetic acid; phenylacetates	cade oil	
2-methyl pyromeconic acid (Maltol)	tobacco leaf	
2-ethyl pyromeconic acid	immortelle	leather notes
4-tert. butyl pyridine	absolutes	tobacco, melilot,
Methyl furoate	broom absolute	hay, fern,
Hydroquinone dimethyl ether	olibanum	mushroom
Furaneol (Dimethyl hydroxy dihydrofuranone)	labdanum resins patchouli oil	

It has been shown that when loss of smell has been temporarily induced by anaesthesia, the first group of odours to become perceptible, with the recovery of the olfactive sense, is the empyreumatic or burnt group. In this respect such odourants take precedence over the merely fragrant, musk-like fruity and spicy groups.

A steadily increasing flow of additions to the currently available empyreumatic odourants can be expected, in view of the intensive research that is being carried out on the constitution of a whole range of cooked food flavours, from porridge to meat extracts.

Animal Notes

Animal notes primarily of those odourants which are of animal origin: musk from the musk deer, castoreum from the beaver, civet from the civet cat, and ambergris from the sperm whale. Next come to mind the many artificial chemical substitutes, including a great number of synthetic musks. Then there are the odourants of vegetable origin, such as the musk-like ambrette seed oil and the labdanum and other resins used in artificial ambergris compositions. Finally, there are odourants that suggest the animal smell of the clean human skin. Among these being costus and cumin seed oils, tincture of chaulmoogra (fixed) oil, lauric and other fatty aldehydes, irone and the ionones, nerol, various honey odours etc.

Principal odourants of animal origin

The principal odourants of animal origin and their less expensive, less rare substitutes are listed below :

Musks, Synthetic

(1) The nitro-musks comprise musk ambrette, musk ketone, musk xylol, musk tibetine.

The macrocyclic musks, to which group belongs muscone, the active odour principle of natural musk, are represented by :

Cyclopentadecanolide (Exaltolide), cyclohexadecenolide (Ambrettolide), cyclopentadecanone, oxahexadecanolides, ethylene brassylate etc.

The indane group, also of considerable commercial importance, contains such musks odourants as :

4-acetyl-6 tert. butyl-1, 1-dimethyl indan

5-acetyl-1, 1, 2, 3, 3, 6-hexamethyl indan.

Among the tetralin type musks, note should be made of Givaudan's Versalide and Polak's Frutal Works' Tonalid.

Volume 2 of Arctander's *Perfume and Flavour Chemicals* contains in its Table 3 a list of over 80 synthetic Musk Odourants. '*The Musk Odour*', by J. Pickthall, F.R.I.C., is also worth consulting.

(2) The classic natural musk is that obtained from the scent pouch or 'pod' of the small musk deer, *Moschus moschiferus*, which inhabits the wooded mountain slopes of Tibet and the Tibetan border. The inaccessibility of the terrain has so far saved this creature from extermination. Musk rat oil, formerly obtained from the scent glands of the Louisiana musk rat (*Ondatra zibethicus rivalicius*), seems to have disappeared from the market. It contains cyclopentadecanone and dihydrocivettone - ketones closely related to muscone - together with dihydrocivettol and normuscol.

Other oils with musky odours include ambrette seed, angelica seed and root, costus, valerian and sumbul root oils.

(3) Many proprietary and speciality musks are available. Apart from the justly celebrated Exaltolide (Firmenich), there are Moskene and Versalide (Givaudan), Phantolide (P.F.W.), Ambrettozone (Haarmann & Reimer), Celestolide and Galaxolide (I.F.F.), Musk R-1 (Naarden) and Oxalide (Takasago). See also Arctander's list.

Civet, Synthetic

(1) Civettone and related ketones. Also mixtures of phenylacetic acid and esters, tetrahydro-p.methyl and other quinolines, skatole, synthetic musks.

(2) Civet is a soft fatty substance of characteristically foecal-musky odour, obtained as a glandular secretion from the civet cat, *Viverra civetta*, chiefly in Ethiopia. It is sold in Zebu horns, the contents of which represent the total production from one animal over a period of four years. During this time the cat 'will have undergone 400 to 800 painful scrapings of its glands'. Teasing the caged cat is regarded as a means of increasing its output of civet.

(3) In addition to civet concretes and absolutes a number of artificial civet compounds are widely offered as less expensive substitutes or part-replacements for the natural product. The following formula represents one type of approach to this problem.

Civet Art. (V.W.)	
Phenylacetic acid	4.0
isoButyl phenylacetate	7.0
Benzyl phenylacetate	13.5
Amyl phenylacetate	16.0
beta-Naphthyl ethyl ether	8.0
Musk ketone	3.0
Tolu resinoid	4.0
Amyl caproate	1.0
Geranyl butyrate	2.0
Geranyl valerate	3.0
Styrallyl valerate (methyl phenyl carbinyl valerate)	1.0
Benzyl valerate	5.0
Skatole	0.5
Tetrahydro paramethyl quinoline	2.0
Total	100.0%

Other potential constituents include indole, musk xylol, civettone, 10-oxahexadecanolide; methyl and metacresyl phenylacetates; p.tert. butyl phenol and p.methyl quinoline; amyl and benzyl salicylates. The characteristic warmth, smoothness and diffusive properties of natural civet tincture are not easy to imitate.

Ambergris, Synthetic

The furan known commercially as Fixateur 404, and discovered by the Firmenich laboratories in the course of their research into the odouriferous constituents of ambergris, is still an outstandingly important and useful product. Its dry, woody, earthy ambergris odour is more powerful and tenacious than may at first appear. It needs to be used with quantity. The dosage at 0.1 to 0.2 per cent of a perfume concentrate, though satisfactory no doubt in some cases, might well prove three or four times too strong in others.

The older type synthetic reproductions of the ambergris odour relied largely on labdanum and other resins such as, Labdanum resin, Olibanum R., Liquidambar and styrax R., Benzoin R. absolute, Tolu and Peru balsams, Oakmoss absolute, Clary sage oil, Sandalwood oil, Patchouli oil, Vetiver oil, Cypress and Costus oils, Civet.

Chaulmoogra oil

It is a fixed oil cold-expressed from the seeds of *Taraktogenos kurzii* King yields on treatment with alcohol over a long period a tincture with a smooth warm, slightly sebaceous odour, suggesting ambergris and the human skin. It has been considered capable of importing warmth and secondary sexual attraction to perfumes. Chaulmoogra oil has long been used, of course, in the treatment of leprosy.

Ambrophore and Grisambrol (Firmenich) and Ambrarome (Synarome) are typical of the fine quality ambergris substitutes available to the perfumer.

Castoreum Synthetic

Chemical research has facilitated the development of artificial replacements or part-substitutes for natural musk, civet and ambergris, but no comparable work has yet been carried out into the elucidation of the constitution of castoreum. Obtained from the preputial follicles of the beaver, *Castor fibre*, this has a considerably more restricted field of application than musk, civet and amber.

Among the usual constituents of artificial castoreum type compounds one finds rectified birch tar oil, creosol, p.ethyl phenol, p.cresol and methyl p. cresol, borneol and bornyl acetate, anisyl acetate, ethyl anisate, methyl benzoate and cinnamate, dihydrocarveol. Fig concrete, prepared from ripe figs, suggests the underlying fruity-balsamic notes of castoreum.

Recently introduced classic perfumery chemicals are amyl salicylate (trefle), methyl nonyl acetaldehyde (used with other fatty aldehydes in Pompeia etc.), the ionones, leathery notes (as in *Cuir de Russie*), hydroxycitronellal (still one of the most widely used of floral type odourants), nerolidol and farnesol; cyclamen aldehyde, synthetic musks; nonadienal and nonadienol (cucumber, violet leaf notes); methyl anthranilate and its hydroxycitronellal Schiff base (orangeflower); various green and woody types, rose oxide, dihydro methyl jasmonate, cis-jasmone, and nootkatone.

The firm of Naarden International issued to the perfumery industry a 'fragrance palette' for 1973, together with a Fragrance Profile Chart of 250 selected odourants, odour panel-evaluated and computer-calculated. Among the synthetic odourants of outstanding interest to perfumers are the following :

Hexyl benzoate (woody-green-balsamic)

Ethyl pelargonate (nonanoate: for natural floral-fruity top note effects in rose, freesia, tuberose etc.). Citral and its dimethyl acetal (the latter a link between 'green' odours and the lemon note of the citral itself).

Citronellyl ethoxylate (sweet, musky, ambrette odour)

Delta-Decalactone (creamy, coumarinaceous musky) and delta-Dodecalactone.

The Salicylates

Amyl salicylate has been one of the most important and distinctive of synthetic odourants since centuries. The widespread applicability of the isoamyl and other salicylic esters has in fact inspired some perfumers to classify a whole group of perfumes as being of the 'salicylate' type.

IsoAmyl salicylate

Most important of the esters, despite the fact that it has not so far been found in nature, is the isoamyl, with its powerful and persistent herbaceous, woody odour which to some people suggests clover (trefle) and orchids, and is also extremely useful in fern (fougere), mossy types and new mown hay (foin coupe); florals such as genista, hyacinth, cyclamen, camellia; carnation and wallflower; chrysanthemum and mimosa; chypre, spicy, woody and certain oriental blends. Many carnation compounds, especially the less expensive ones, contain high levels of amyl salicylate. This ester can easily be used to excess, owing to its property of developing or intensifying in odour after it has been incorporated in a blend. (See also Nerolidol.)

Benzyl salicylate

Next in importance, quantitatively at least, is the benzyl ester. This has a mild sweet floral-balsamic or floral-woody odour. Benzyl salicylate finds considerable application in perfumery as a solvent and blender, particularly for synthetic musks, and as a 'fixative' and diluent in many floral and non-floral bouquets.

A peculiarity of benzyl and hexyl salicylates is that their odours, though distinctly apparent to some people, are not so apparent—and may even seem totally absent—to others.

Isobutyl salicylate

It has a characteristically flowery, clover-herbaceous odour, and is useful for modifying the amyl salicylate note. It blends well with clary sage, lavender and many herb and wood oils, musk ambrette and the other nitro-musks, coumarin, resins and balsams etc.

Methyl salicylate

It has typical odour of wintergreen oil, is also present in sweet birch oil and, as a minor constituent, in oils of tuberose, ylang-ylang, clove and cassie. Used sparingly, it can prove valuable in artificial imitations of cassie, tuberose, ylang, mimosa, gardenia, reseda, castoreum, and in spicy bouquets, Russian leather etc. It finds less restrained use in flavours for toothpaste, mouth washes and chewing gum.

Ethyl salicylate

It has an odour very similar to that of the methyl ester but milder, suggesting meadow sweet (*Spiraea* or *Filipendula ulmaria*). It blends well with methyl benzoate, oakmoss, cuminic and anisic aldehydes, methyl naphthyl ketone etc.

Phenylethyl salicylate

It is a crystalline odourant with a mild, sweet, balsamic floral note, suitable for use in spicy rose, carnation and other floral bouquets. It can assist in rounding-off otherwise crude odours and blends well into many formulae of hyacinth, ylang, lily, lilac and even honeysuckle types. A good grade is free from phenolic notes.

Other salicylates, of less importance, include the Hexyl, Linalyl, Geranyl and Methyl Phenyl Carbinyl esters.

Cerbelaud perceived a relationship between the salicylate group and mimosa, cassie, violet, orris, honey and mushroom; while Arctander associates the salicylates, at a distance of course, with the odours of anise, fennel, parsley, basil, tarragon and sassafrass. The salicylates certainly blend admirably with many floral and herbaceous odourants–as well as with patchouli, vetivert, sandalwood etc., as in fougere.

Acetates

The acetates are widely used by perfumers in essential oils. The important acetates are listed below :

IsoAmyl acetate

It has a volatile fruity odour of pear-banana type. Traces impart piquancy and lift to fruity-floral, oriental and the heavier flower types. The normal ester has also an earthy shading in its top note and is therefore interesting in violet compositions.

Anisyl acetate

It recalls that part of lilac which suggests hawthorn. Used in odours of these types, e.g. lilac, acacia, honeysuckle. Can also give 'powdery' effects.

Benzyl acetate

One of the most widely employed esters, used to 'push' the note in both floral and fantasy types. A constituent of jasmine oil, it naturally enters into the composition of jasmine perfumes, as well as in perfumes of a rosy and lilied (muguet) character. Used also in flavours: GRAS rating with FEMA.

Bornyl acetate

A synthetic much employed for its pine-fir odour. It has also a slightly 'animal' and camphoraceous note. The isobornyl ester has a less animal, rather more delicate odour.

IsoButyl acetate

It is used in strawberry, raspberry and other fruit flavours. (GRAS – FEMA.) Employed in traces to contribute to the lift and top note effects in tea rose, spring flowers, etc.

Cedryl acetate

It is chiefly used in woody notes; also oriental, tobacco, leather (cuir) types.

Cinnamyl acetate

It is found in cassia oil, has a warm balsamic-floral odour and notable fixative properties. Useful in hyacinth, rose, woody, spicy, and many floral and oriental blends.

Citronellyl acetate

It imparts freshness when used in small quantities in rose, muguet, lavender; flowery and aldehydic chypre blends etc.

p-Cresyl acetate

Its pungent narcissus odour and somewhat urinaceous undertone, is used with restraint in narcissus, hyacinth, lily and other floral compositions.

Dimethyl benzyl carbinyl acetate

It gives a flowery lift. Its odour is rosaceous, greenish, suggestive of muguet.

Dimethyl octanyl acetate (tetrahydrogeranyl acetate)

It is valuable in compositions with a dominant rose or lavender note.

Dimethyl phenyl carbinyl acetate

It has a strongly rosaceous-green note that recalls the odour of bois de rose Brazil.

Ethyl aceto-acetate

It has fresh, sweet, natural top note effects. Also used in flavours (GRAS rating).

Ethyl hexyl carbinyl acetate (3-nonanyl acetate)

It has an interesting fruity-floral, rather waxy odour. Very versatile, it can be used to good effect in rose, carnation, lavender, fougere, chypre – and even Cologne and modern fantasy compositions of many kinds.

Ethyl linalyl acetate

A more floral odour than that of linalyl acetate, but also with a faint fruitiness. It is used to support bergamot oil and has already found numerous applications in perfumery.

Eugenyl acetate

Freshens and 'pushes' carnation and related spicy notes. The isoeugenyl ester is a good fixative for carnation perfumes and can exert a mellowing effect on green and herbaceous compositions.

Geranyl acetate

Odour rosy, woody, fruity. A constituent of lemon and petitgrain oils etc., it is widely used in perfume compounding.

Linalyl acetate

It varies very considerably according to its origin (*ex bois de rose* or shiu synthetic etc.). Blends well with lavender, bergamot, petitgrain etc. in numerous compositions.

Epoxy linalyl acetate (linalool oxide acetate)

It has a fresh floral odour, slightly woodier than that of linalyl acetate. Very useful in lavender and related types.

Menthyl Acetate

It has a rather herbaceous, sweetish odour of a floral-rosy oriental type. It blends well with citrus notes, lavender water etc., and gives 'lift' in toilet soap perfumes. Used also in flavours. (GRAS rating with FEMA). Menthyl propionate is somewhat similar and also of interest.

p-Methoxy phenylbutyl acetate

Like the parent carbinol is useful in fruity-floral types: jasmine, tuberose, gardenia, freesia etc.

Methyl hexyl carbinyl acetate

Like many members of the carbinol series, this ester has a somewhat geranium-like odour and finds application in warm herbaceous complexes: lavender, chypre etc.

Neryl acetate

It has the qualities of nerol but its odour is rather more suggestive of citrus-orangeflower and not quite so rosaceous. It is very useful in flower notes, notably rose, peony, magnolia–even neroli and jasmine.

Octyl acetate

Its fatty-floral, somewhat green odour, is used in a number of flower perfumes, including jasmine, neroli and orangeflower, rose, chypre, fougere.

Phenylethyl acetate

Its has a very sweet, rosy, honeyed-floral odour – recalling red rose and peach. Useful also in flavours (GRAS rating by FEMA).

Phenylpropyl acetate (hydrocinnamyl acetate)

It has a fresh, balsamic, rather styrax-like odour. Useful in lilac, lily and floral bouquets. (GRAS rating by FEMA).

Rhodinyl acetate

It is rosaceous, rather fruity and slightly green odour. Valuable in rose, muguet, carnation and other perfumes and used in flavours (GRAS rating).

Santalyl acetate

It is useful for 'pushing' the top note of sandalwood.

Styrallyl acetate (methyl phenyl carbinyl acetate)

It has a green, hazel-nut odour, which also suggests apricot, and is characteristic of gardenia. Used also in tuberose, muguet, hyacinth and some types of jasmine.

Terpinyl acetate

It recalls the odours of linalyl acetate and bergamot oil, but is more 'ordinary'. It is useful, nevertheless, to impart zest to certain compositions, including those with a linalyl acetate note.

Trichloromethyl phenyl carbinyl acetate

Valuable for creating a rosy, powdery base. Known also as 'rose crystals' etc.

Vetiveryl acetate

It has the woody-rooty-powdery odour of vetiver oil but also the sharper edge, aptly described by Arctander as sweet-and-dry. Blends well with ionones, fatty aldehydes, resins and balsams etc. A classic perfumery material.

The Fatty Alcohols, Aldehydes and Acetals

The term 'aldehydic perfumery' has become synonymous with 'modern perfumery,' as higher aliphatic aldehydes were incorporated in perfumes in significant quantity. Alcohols both individually as well as members of corresponding group and acetals are briefly considered here. The higher fatty aldehydes are chiefly valued as forceful and distinctive top note constituents, although their effect may sometimes persist throughout the entire evaporation period of the perfume in which they are used. The corresponding alcohols have in general less distinctive, less violent odours of similar types. The acetals are less prone to oxidation and polymerisation than the aldehydes from which they are derived, but in the process of conversion the characteristic aldehyde note is always modified and sometimes virtually lost. The important alcohols used are described below in Table 3.4.

Table 3.4. Fatty alcohols, aldehydes and acetals.

Chemical Alcohols	Source	Odour	Use
C_8 (octyl, caprylic)	Sweet orange grapefruit	Rosy-waxy	Rose perfume
C_9 (nonyl, pelaorgonic)	Sweet orange oil oakmoss	rosy	orange, neroli, roseda, reseda gardenia
C_{10} (decyl, capric)	Sweet orange oil ambrette seed	rose with orange note	rose, orangeflower orris, jonquil, lilac jasmine compounds
C_{11} (undecylenic)	Leaves of Litsea, odourifera	herbaceous, flowery, fruity,	tuberose, gardenia, green notes
C_{12} (lauric dodecyl)	Oil of lime	sweet, sugary flowery than aldehyde	tuberose, reseda, muguet exotic, wood, fantasy compositions
Fatty aldehydes			
C_7 (heptanals, oenanthic aldehyde)	Oils of Ylang-ylang hyacinth	Sweet top note acidulated and fruity	Orangeflower
C_8	Citrus oil, lemongrass, rose	rosaceous fruity with jasmine note and waxy background	Rose, jasmine
C_9	Citrus oil, orris ginger	Sharp, fatty, rose orris, orange	tuberose, rose jasmine, orris, heliotrope
C_{10}	Citrus oil, pine needles, coriander cassie, orris	Orange-rose-lemon on a waxy background	rose, jasmine, neroli cyclamen, cassie, lilac
C_{11} (undecylenic)		rose, tuberose orris, fruity	tuberose, woody, and fantasy
C_{12} (lauric)	citrus oil, rue, pine	green herbaceous, fatty, tuberose violet leaves	tuberose, reseda cassie, violet, fern, new mown hay, chypre, pine

(Cont'd)

Chemical	Source	Odour	Use
Alcohols			
C_{12} (methylnonyl acetic)		Orange-amber-moss	tuberose, orchid honeysuckle, sweet pea
Pseudoaldehydes			
C_{14} (peach aldehyde)		peach	fruity, chypres
C_{16} (ethyl methyl phenyl glycidate)		strawberry	rose centifolia
C_{18} (Coconut aldehyde)		coconut	gardenia, tuberose exotic, oriental, fantasy compositions
C_{19} (benzyl isobutyrate)		fruity-floral suggesting pineapple	woody notes based on vetiver, spicy-floral perfume
C_{20} (ethyl p-methyl β-phenyl glycidate)		raspberry, in association with ionones	modern chypres
Acetals			
C_7 to C_{12} (dimethyl and diethyl acetals)		lemony green	gardenia, tuberose perfume

Though not derived from the aliphatic aldehydes certain other acetals may for convenience be mentioned here. They include the acetals of citral, the chief disadvantage of which is odour variability due to impurities. Uses: Cologne types, 'green' perfumes, soap compounds. The dimethyl and diethyl acetals of phenylacetaldehyde are widely used for their fresh green, hyacinthine notes. The glyceryl acetal is also of interest. Hydroxycitronellal acetals may be used in lily-of-the-valley compositions but are relatively weak odourants.

Dimethyl acetal of hydratropic aldehyde (alpha-methyl phenylacetaldehyde): Has interesting walnut-mushroom shadings and is useful in tuberose, mossy and woody notes. The diphenylethyl acetal can give interesting effects in rose, gardenia, magnolia. champak and oriental blends. Anisaldehyde dimethyl acetal recalls lilac and elderflower. Amyl cinnamic aldehyde dimethyl acetal is useful as a component of jasmine compounds. Heliotropin acetals have been produced but have aroused little interest.

The approximate amount of aldehydes and higher fatty alcohols for perfume are given in Tables 3.5. and 3.6.

Table 3.5. Fatty alcohol and aldehyde in flower perfumes.

Perfume	Chemical % aldehyde	Amount in grams
Cassie	C_{10} (10)	2.5
Cyclamen	C_{12} lauric (10)	10
	C_{12} MNA (10)	5
Daphne	C_{12} (lauric) (10)	20
	C_{12} MNA (10)	10
Gardenia	C_{10} (10)	12
	C_{11} (enic) 10	12
	C_{12} (lauric) 10	16
Heliotrope	C_8 (10)	5
	C_9(10)	5
Honeysuckle	C_{12} (MNA) 10	30
Jasmine	G_{14} (peach) 10	30
	C_{12} (lauric) 10	20
	C_{12} (MNA) 10	10
Lilac	C_{12} (lauric) 10	5
Magnolia	C_9 (100)	5
	C_{10} (100)	5
Orangeflower	C_7 (10)	15
	C_{10} (10)	6
Peachblossom	C_{14} (peach) 10	15
	C_{16} (strawberry) 10	5
Rose	C_{10} (10)	5
	C_8 (10)	5
	Alcohol	
Cassie	C_{10} (10)	3.5
Orange flower	C_{10} (100)	10

Table. 3.6. Amount of aldehyde and alcohol in fantasy perfume.

Perfume	Chemical % Aldehyde	Amount in gms.
Modern type	C_{12} MNA (10)	20
	C_{11} (enic) (10)	15
	C_{10} (10)	10
French type-1	C_9 (10)	15
	C_{10} (10)	12.5
	C_{11} (10)	10
	C_{12} lauric (10)	12.5
type-2	C_{12} lauric (10)	30
	C_{10} (10)	15
type-3	C_{12} lauric (10)	10
	C_{12} MNA (10)	40
type-4	C_{12} (MNA) (10)	10
Woody type-1	C_{11} (10)	35
	C_{12} (lauric) 10	15

(Cont'd)

Perfume	Chemical % Aldehyde	Amount in gms.
type-2	C_{12} (MNA) (10)	15
	C_8 (10)	30
	C_9 (10)	25
Eaux de	C_{10} (10)	4
Cologne	C_{12} (MNA) (10)	5
	Alcohol	
French type-4	C_9 (100)	15

Acetophenone

Though rather crude and violent, this ketone of sweet mimosa-almond odour, can perform useful service when used with restraint in mimosa, lilac, new mown hay etc.

Amyl cinnamic aldehyde

Fresh and full, though somewhat fatty odour, which intensifies and becomes sharper during the evaporation period. Invaluable in jasmine and many other floral compositions. (GRAS rating for flavours.)

Anisic alcohol

Powdery, rosy, lilac-vanilla. Blends well with other components of jasmine, lilac, muguet, sweet pea and gardenia. (GRAS rating for flavours.)

Anisic aldehyde

Sweet floral, hawthorn type of odour. Useful in aubepine, new mown hay and lilac types etc., but principally in aldehydic compositions. Important chemical used in perfumes are listed in Table 3.7.

Table 3.7. Important Chemicals used in Perfume.

Chemical	Odour	Uses
Anisates		
Ethyl anisate	Sweet fruity,	lilac, carnation, ylang, jasmine, fern, chypre
Methylanisate	Sweet, herbaceous anisic	mimosa, hawthorn, sweet pea magnolia, new mown hay, lime
Anthranilates		
Ethyl anthranilate	Sweet, flowery, grape-like	neroli, orangeflower, acacia,
methyl anthranilate	pungent, fruity, orange-flower	flowery composition with orangeblossom
Phenylethyl anthranilate	rosaceous	honeysuckle, gardenia
iso-propyl anthranilate	neroli, hawthorn, honey	mandarin, jasmine
Benzoates		
Amyl benzoate	pungent treffle-orchid type balsamic	clover, orchis, tobacco, oriental composition, good solvent for nitro-musks

(Cont'd)

Chemical	Odour	Uses
iso-butyl benzoate	powdery-rosy, mild green note	amber, incense, floral type
Cinnamyl benzoate	spicy-balsamic note	oriental and floral woody types
Benzophenone	balsamic-rosy, faint metallic note	Soap, low cost perfume
Benzylidene Acetone	pungent, sweet	Sweet pea, neroli, hyacinth, lavender blend
Butyrates		
Citronellyl Butyrate	fruity rosy odour	rose, muguet
Citronelly isoButyrate	sweet rosaceous	
Geranyl Butyrate	fruity	rose, peony, cassie, lavender lipstick, apple, peach, apricot flavours
Nonyl isoButyrate	floral-fruity	Perfumery, flavours
Phenylethyl Butyrate	fruity-rosy-plum like	perfumery, flavours
Phenylpropyl Butyrate	fruity-balsamic	jasmine, narcissus, hyacinth, apricot, peach, mango flavour
Styrallyl Butyrate	fruity-floral	jasmine, gardenia
Caproates		
Ethyl caproate	fruity-herbaceous	modern flower
Geranyl caproate	rose-geranium fruity	Oriental composition
Cinnamates		
Benzyl Cinnamate	balsamic note	floral and fantasy perfumes face for body and talcum powder
isoButyl Cinnamate	ambered	modifier, blender, in chocolate cherry, plum
Cinnamyl Cinnamate	sweet, mild, tenacious odour of cinnamon	fixative and blending
Ethyl Cinnamate	Chypre note, woody incense	flavours
Linalyl Cinnamate	lily	perfumes
Methyl Cinnamate	pungent, balsamic-fruity	Parfum ideal, incense-woody composition
Phenylethyl Cinnamate	rosy, honeyed	floral notes such as rose, muguet, lilac
Cinnamic Aldehyde	spicy, balsamic	Balsamic, oriental hyacinth floral type
Citral		
Citronellal	lemon-rosy-citronella	soap, perfume
Citronellol	Sweet, rosy-floral	rose, muguet
Formates		
Anisyl formate	sweet, heady, hawthorn-syringa	lilac, tuberose, gardenia, Ylang, vanilla flavour

(Cont'd)

Chemical	Odour	Uses
Benzyl formate	sharp and fruity jasmine	perfumes, flavours.
Cinnamyl formate	woody	blends with sandal, patchouli vetiver, banana
Citronellyl formate	rosy-leafy	perfumery
Ethyl formate	green and rosy note	flavours
Geranyl formate	tea rose	rose, orangeflower, lavender, Cologne
Linalyl formate	citrus-bergamot	Chypre, fougere
Phenylethyl formate	rosaceous green,	flavours
Phenylpropyl formate	Sweet honeyed green	hyacinth, narcissus, chypre
Rhodinyl formate	Leafy green, rosy	top note rose
Terpinyl formate	bergamot and terpineol	woody, green, fern, chypre, lavender and cologne
Phenylacetales		
Amyl phenylacetate	animal odour	Cosmetic perfume
isobutyl phenylacetate	eglantine	basis for orchidee
p-cresyl phenylacetate	musky	lily, narcissus, hyacinth
Ethyl phenylacetate	honey	perfumes, flavours
Methyl phenylacetate	smooth honey	versatile odourant
Phenylethyl phenyl-acetate	rosy	hyacinth, spring flower type
Propionates		
Amyl propionate	fruity	top note, flavours
Benzyl propionate	jasmine	fruit flavours
Citronellyl propionate	sweet, rosy	honeysuckle, syringa
Geranyl propionate	Oriental rose	floral composition
Linalyl propionate	lily-of-the-valley coriander	clary sage oil, in freesia
Phenylethyl propionate	dried rose petal muguet	ester

Coumarin

This is one of the important crystalline fixatives which form the basis of persistent perfume notes. The others include the synthetic musks and heliotropin. It blends well with the balsams and forms with bergamot oil the basis of numerous fougere notes. It also blends well with lavender, amyl salicylate, oakmoss etc.

p-Cresyl amyl ether

It has a flowery, green odour, and gives excellent service in hyacinth, narcissus, gardenia.

Dihydrocarveol

Rosy-animal note suggesting muguet but with a less pleasant acetylenic shading. Can give interesting effects in some fantasy types.

Dihydromyrcenol

It has a citrus-floral odour suggesting lemon and lime. If stabilised against polymerisation can be very useful, particularly in soap and detergent perfumes.

Dimethyl benzyl carbinol

It can give interesting effects in lilac, muguet and some rose notes.

Dimethyl hydroquinone (hydroquinone dimethyl ether)

Sweet, penetrating cut grass odour. Useful for tobacco or herbaceous effects in perfumery, usually in small quantities.

Dimethyl octanol (tetrahydrogeraniol)

Powerful rosy odour, blending well in most rose and allied types. (GRAS rating for flavours.)

Dimethyl phenyl carbinol

Odour: a very rosaceous linalool. Useful in rose, muguet, woody, powdery chypre types etc.

Diphenyl methane

Pronounced odour of geranium leaves. To be used with due restraint.

Diphenyl oxide

Rosy-geranium leaves is a valuable product for imparting to rose compositions the green note of the calyx so essential for a true reproduction of the odour of the flower. It must, however, be used with due caution and in traces only.

Estragole (methyl chavicol, iso-anethole)

It occurs in tarragon oil. Used in small amounts in chypres and 'green' notes; also in eaux de Cologne to impart a piquancy to the top note. (GRAS rating for flavours.)

Ethyl-Hexyl carbinol (3-nonanol)

Herbaceous and earthy in odour. One sometimes needs an earthy, humus-like note of undergrowth in perfumes of the fougere, exotic woody and certain oriental types. This product is well adapted for use in such cases.

Ethyl linalool

It has a softer and rather more flowery odour than that of linalool.

Ethyl vanillin (Vanillal, Bourbonal)

The odour of commercial 'ethyl vanillin' is sweeter than that of vanillin and rather nearer to that of natural vanilla. It also has less tendency to cause discolouration problems. Widely used in perfumery, and in flavours. (GRAS rating.)

Eugenol

The principal constituent of clove oil. Enters into the composition of many spicy, woody and eastern notes, carnation etc. (GRAS rating.)

isoEugenol

Carnation odour but softer than that of eugenol and less clove-like. Extensively used in perfumes for its smooth, long-lasting flowery effect. (GRAS rating.)

Fixateur 404 (Firmenich)

It deserves special mention as one of the most interestingly persistent and adaptable of modern innovations. (See also Animal Notes.)

Geraniol

The odour and performance of geraniol vary greatly according to its source (ex citronella, palmarosa or through synthesis). Geraniol of one grade or another is widely used in all kinds of composition. (GRAS rating by FEMA.)

Heliotropin

Odour strongly suggestive of heliotrope. Heliotropin is extensively used in all kinds of perfume, from 1 per cent or even less to 10 per cent or more of the perfume concentrate. In the lower concentrations it is chiefly detectable in the *fond*, on dry out, but at the higher levels it can make itself felt in the top note. It is strongly indicated for lilac, for new mown hay when associated with coumarin, and for tobacco notes. Widely used in flavours. (GRAS rating.)

Heptanal glyceryl acetal

Sweet earthy-fungal can give interesting effects in floral compositions requiring a hint of undergrowth odour. Has also been used in flavours requiring a mushroom note (GRAS rating.)

isoHexenyl tetrahydro benzaldehyde

It is an interesting product of pungent, clean, citrus-floral odour. Useful in many compositions, including muguet, lilac, cologne, violet, woody and other types.

Hexylcinnamic aldehyde

The odour rather similar to but greener and initially sharper than that of Amyl Cinnamic Aldehyde. Increasingly used in the creation of jasmine, tuberose, gardenia and other floral compositions. (GRAS rating for flavours.)

Hydroxycitronellal

It has a sweet flowery odour suggestive of lily-of-the-valley and lime blossom. Finds extensive use in spring flower and many other perfumes. Employed in moderation in flavours (GRAS rating). Should be free from citronellal and potentially irritant traces of sulphite.

Indole

Of natural occurrence in jasmine, neroli, jonquil, pseudo-acacia, wallflower etc. Used sparingly for its floral effects. Care must be taken to avoid its proneness to cause colour reactions with aldehydes. (See Arctander and others.) GRAS rating for flavours.

Ionone alpha

It has a marked odour of violets. Used widely in perfumes of many types.

Ionone beta

It has a softer violet odour than the alpha isomer and is more reminiscent of Parma violets.

The ionones vary considerably according to the firm that manufactures them, because of the traces of auxiliary bodies and isomers present in the course of fabrication. But also because of the occurrence of alpha, beta and gamma isomers in various admixtures. The same complexity occurs with the methyl ionones. It is therefore necessary for the perfumer to make a very careful and comprehensive study of these important odourants. This must of course be done gradually and cautiously, in view of the well-known property of the ionones of causing olfactory fatigue and temporary anosmia. (GRAS rating.)

isoJasmone

Jasmine odour which blends especially well with those of jonquil and neroli.

Linalool

It is a constituent of many essential oils: 1-linalool occurring in Brazil and *Cayenne bois de rose* (the latter being more rosy in odour), while d-linalool occurs in coriander oil as coriandrol, which is still more rosaceous than the others. (GRAS rating.)

Menthone

It exists in oil of *Geranium bourbon* and in rhodinols that are incompletely rectified. It has a rather minty-rosy-woody odour and is excellent for 'pushing the rose note'. (GRAS rating.)

Methyl acetophenone

It has a pungent odour of hawthorn, mimosa type. Widely used in spring flowers, fern, chypre, woody, mossy types; also in cassie etc., for special effects. It has also a limited use in flavours. (GRAS rating.)

methyl p-Cresol (paracresol methyl ether)

Pungent, rather crude hyacinthine odour. If used with restraint can give excellent service in narcissus, ylang and lilac notes etc. (GRAS rating.)

Methyl eugenol (eugenol methyl ether)

It has a sweet, warm, rather spicy carnation odour, blending well with rose odours, which it can effectively accentuate. (GRAS rating for flavours.)

Methyl heptyl ketone

Odour flowery, green-herbaceous. Used in lavender, Cologne etc.

Methyl hexyl ketone

The normal (not the isomeric) ketone is useful for reinforcing lavender and new mown hay notes.

Methyl ionones

These occupy a very important place in perfume composition. According to the place of the methyl group in the ionone nucleus, products of very different odours are obtained: methyl ionone alpha, beta, gamma etc. Commercial products having yet other notes consist of mixtures of isomers. The odour gamut ranges from flowery-violet and slightly ambered to orris and woody.

Methyl methylanthranilate

It occurs naturally in mandarin (tangerine). It has an odour of orangeflower type but greener. Finds numerous applications in perfumery, e.g. in cosmetic perfumes. Used also in tangerine, orange, grape and other flavours. (GRAS rating.).

Methyl naphthyl ketone

Sweet and tenacious orange blossom odour.

p-Methyl quinoline

It has a powerful and penetrating odour that commends it in bases of greenish 'animal' type, e.g. in association with such modern marine-like odourants as Aldehyde Mer or Absolu Marin; also in oakmoss and tabac compositions.

Muguet aldehyde

It is a name given to citronellyl oxyacetaldehyde and citrylidene acetaldehyde; also to isohexenyl tetrahydre benzaldehyde.

Nerol

It has a remarkably fine not ettending towards neroli and rose. It occurs fairly widely in nature and finds widespread current use in floral odours. (GRAS rating.)

Nerolidol

It occurs in oils of neroli, ylang, petitgrain and in Peru and Tolu balsams etc. Considered alone, its odours shows up but little. It is one of those bodies, not very odourous in themselves, which develop unsuspected properties when actually incorporated in a composition. A curious phenomenon and one as yet unexplained.

Phenylacetic acid

It has honey-animal, notably persistent odour. Very useful for honeyed notes, artificial civet and rose perfumes. Also in traces in many other perfumes. (GRAS rating for flavours—usually in honeyed, chocolate and some fruity types.)

Phenylacetic aldehyde

A pungent, even brutal hyacinthine odour. Finds useful application in hyacinth, lilac, rose, 'green bouquet' and numerous other compositions.

Phenylethyl alcohol

The principal odourous constituent of rose water. One of the most widely used of perfumery raw materials. The blender supreme. (GRAS rating for flavours.)

Phenylpropyl alcohol

The odour of this alcohol is sweet and balsamic, suggesting styrax and hyacinth. Used in floral, oriental and fantasy types (e.g. cuir).

Phenylpropyl aldehyde (hydrocinnamic aldehyde)

Useful in hyacinth and tobacco notes; also lilac, cyclamen, reseda etc.

Piperonyl acetone (heliotropyl acetone)

It has a very sweet-floral-woody odour. Finds application in muguet, lilac, mimosa. (Also GRAS rating for flavours.)

Rhodinol

It is obtained from oil of geranium. Its typical rose odour varies considerably according to source and treatment. Sometimes traces of menthone are left in and these exalt the rose note. A fine commercial rhodinol has a warm red rose odour of great distinction.

Safrole

The main constituent of sassafrass oil. It has been said to impart a special cachet to chypres, when incorporated in traces. Its odour is slightly anisic, with warm spicy-woody undertones.

Santalol

The main and characteristic alcohol of oil of sandalwood. Because of its fineness it is essential for many luxury perfumes; chypres, woody and oriental notes etc., to replace the oil – but with corresponding restraint in view of the difference in strength. Traces are used successfully in certain flavours. (GRAS rating.)

Skatole (methyl indole)

A constituent of natural civet and of civet artificial.

Terpineol

The terpineols, of which the alpha, beta and gamma isomers are known to perfumers, exist in a number of natural oils. The terpineol currently marketed is the alpha isomer containing a variable proportion of beta and gamma. It is widely used in certain compositions as a freshening agent, e.g. lilac and muguet, and is a major constituent of many soap perfumes and disinfectants.

Terpinolene

It has a vague odour of pine, less harsh than that of pinene. It is widely used in perfumes for household products. Good grades are also used in lime flavours. (GRAS rating.)

p-Tolyl acetaldehyde ('Syringa aldehyde'.)

Milder, sweeter and easier to use than the related phenylacetaldehyde, but less green.

Trimethylcyclohexanone

A petrochemical product. It has a powerful odour suggesting labdanum and honey. Blends well with phenylacetates, ionones, patchouli etc. Cinnamyl Tiglate has a woody odour and finds use, for example, in cuir and similar perfumes. Citronellyl Tiglate has an agreeably rosy odour. Used in rose, tabac etc.

Geranyl Tiglate has a pleasant rose-geranium odour, slightly fruity. Used in 'reconstructed' geranium; also in lavender and oriental notes.

Benzyl valerianate (or valerate)

A fruity, musky, almost rosaceous odour. Used in oriental notes.

Citronellyl valerianate

It has a more rosy odour than the benzyl ester. Used in the perfuming of tobaccos.

Geranyl valerianate

Rosy, fruity odour. Used in the perfuming of tobaccos.

Linalyl valerianate

Recalls the fruity part of the lavender fragrance. Some also see in it a resemblance to apricot. Goes well with lavender and pine bouquets.

Vanillin

The main constituent of natural vanilla. Its odour is drier than that of the vanilla pod. One of the most valuable and widely used perfumery and flavour materials, it is produced from lignin, guaiacol, eugenol, safrole etc. (GRAS rating.) Some of the best known French perfumes contain liberal proportions of vanillin.

It seems that there is continuous supply of synthetic perfumes, a more precise knowledge of aroma chemistry, which will lead to replacement of scarce natural oil. The more chemical with natural fragrance will create greater opportunity for perfumer to introduce novelty in his fragrance creation.

Chapter 4

Classification of Odours
and Odourants

INTRODUCTION

Thousands of materials with various characteristic odour, odour intensities and different properties are used producing flavours and perfumes. It is, therefore, necessary to classify these raw materials into groups, and subgroups. McCartney expert in olfaction and odours, has pointed out that, perfumers often invent systems of their own for private use. Fourcroy in 1798 pointed out classification made by private user is arbitrary, uncertain, and fragile because our sensory impression and olfactory origin are not fixed, permanent or equal in all men at one time or in one individual at all times. A perfumer's classification is based on a perfumer's expertise and experience. For perfumer this classification is of greater significance and practical, than the non-perfumery classification based on botanical, chemical or psychological considerations.

The odour classification has been carried out by Rimmel, Piesse, Zwaardemaker, Heyninx, Henning, von Skramlik, Henderson, Matteotti and Crocker put forward their 'semi-quantitative evaluation of odours,' which made possible the accurate description of any odour by the simple device of four digit number. These digits represents fragrant (sweet), acid (sour), burnt (empyreumatic) and caprylic, remarkable simplification. They consider that these four elementary odours are the 'principal'. The maximum intensity of each of these elementary odours is arbitarily allocated the number 8, so that Tonquin musk, whose code number is given as 8476, is top rated as 8 fragrance, with a moderate 4 for acidity, 7 for burntness, and 6 for its supposed 'caprylic' tonality. Similarly 'rose' is coded as 6423.

Of much greater interest to the latter are some of the classifications, both of complex odours and individual odourants, made by perfumers.

98

These are necessarily subjective but cannot be lightly dismissed, because of this, as unsatisfactory. They not only fill an immediate need but are often extremely reliable.

A clear distinction must made to classify odours, e.g. as floral, woody, balsamic and so on, and attempts to classify the actual odourants or constituents of complex odours, as for example into top note, middle note and base note constituents. Both systems are useful. It is probably more convenient to look first at some current classifications of odourants. Poucher's classification is based on a subjective assessment of the relative duration of evaporation of some 330 perfumery chemicals, essential oils and other odourous materials. Obviously this published list could be very considerably extended, as Poucher has himself suggested. The main criterion of this type of classification is relative volatility, which may be regarded as the vapour pressure at ordinary temperatures, but in practice, owing to the complex character of essential oils and flower absolutes etc., it is vastly more satisfactory to use a subjective method of comparing odourants. Poucher accomplished this by examining each material, or an appropriately standardised dilution, by means of smelling slips. When carrying out tests he 'had to decide on what should be the end point of each odour. The characteristic note of some natural products may be fleeting while the residual smell lingers on. But since each aromatic substance is employed primarily for its typical odour note. He decided to check and re-check the point at which this distinguishing feature disappeared. Moreover, he had to place a time limit on these substances of longest duration, such as patchouli and oakmoss, and he gave them the figure or coefficient of 100. To those that evaporated in less than one day he gave the coefficient 1, and to the others 2 to 100'. Thus, eventually, the Poucher classification as published comprised over 300 items, each of which was distinguished by a coefficient ranging from 1 (e.g. amyl acetate) to 100 (e.g. ambergris extract, vanillin and vetivert).

Poucher further subdivided his materials into those with coefficients ranging from 1 to 14, which he called Top Notes; those with coefficients ranging from 15 to 60 (Middle Notes); and those from 61 to 100, which he termed Basic Notes or Fixers. The following are the examples of top note constituents :

amyl acetate	amyl salicylate	lavender oil
methyl benzoate	petitgrain oil	methyl octin
lime oil, distilled	bergamot oil	carbonate

The examples of middle note are :

Bulgarian rose	heliotropin	indole
	phenylethyl isobutyrate	dimethyl hydroquinone
rose demai	clary sage oil	anisic alcohol
Jasmine	angelica seed oil	citral

The examples of the basic notes are amyl phenylacetate, natural cinnamic alcohol, methyl naphthyl ketone, civet absolute, hydroxycitronellal and cyclamen aldehyde to a long list of resins, balsams and crystalline materials (coumarin, vanillin, artificial musks), aldehydes (amyl cinnamic, methyl nonyl acetic, phenyl acetic), all rated at 100, together with ambergris, castoreum, patchouli, pepper, sandalwood and vetivert.

These subdivisions are useful and enhance the value of the list as a source of general reference. They do not, however, supply a ready answer to the questions: 'What is a top note? What is a fixative?' As Poucher himself has hastened to observe, there are occasions when longer-lasting odourants are used in such a quantity as to raise them temporarily into a higher category. He in fact illustrates this point by giving two simplified but characteristic formulae for a Lilac and a Hyacinth perfume respectively. Among other features, the former contains 1 per cent of a 10 per cent dilution of phenylacetic aldehyde while the latter contains 10 per cent of the pure, undiluted aldehyde. There is no doubt that this aldehyde acts as a top note, despite its persistent character, when it is utilised in a dominant proportion.

The comparison of results of different authors finds many discrepancies which indicate difference in opinion. However, there is broad area of agreement which can be noted from Tables 4.1 and 4.2.

Ellmer's classification appears to have been the first of its kind to be published. His results match fairly closely to those subsequently obtained by other perfumers, although few will agree with his remarkably high persistence values for rosemary oil and cuminic aldehyde: nor does he appear to have given sufficient attention to the basic notes as a class. Jean Carles, on the other hand, seems to have over-emphasised the importance of the basic notes, when he writes that they 'will serve to determine the chief characteristic of the perfume, (for) their scent will last hours on end and will be essentially responsible for the success of the perfume, if any'. In the same way he deliberately over-simplified the structure of a perfume by considering it almost as a definite, neatly defined architectural entity instead of a dynamic, changing, imbricated composition, not merely existing in space but simultaneously evolving and fluctuating in time.

Table 4.1. Odour persistence of perfumery materials.

Top notes	Middle notes	Base notes
Methyl salicylate	Petitgrain oil, Paraguay	Styrax oil
Amyl acetate	Geranium oil, Bourbon	Lovage oil
Benzyl cinnamate	Marjoram oil	Elecampane oil
p-Methyl acetophenone	Cinnamon oil	Phenylacetaldehyde
Dill oil	Citronellol	Sandalwood oil
Anise oil	Nerol	Santalol
Borneole	Methyl anthranilate	Vanillin
Terpineol	Aldehyde C_9	Rosemary oil
Linalyl acetate	Methyl heptin carbonate	Cuminic aldehyde
Bornyl acetate	Angelica root oil	
Anisaldehyde	Eugenol	
Lemon oil	Citronella oil, Java	
Bergamot oil	Geraniol	
Phenylethyl alcohol	Amyl salicylate	
Sage oil	Geranyl formate	
	Clary sage oil	
	Cinnamic aldehyde	
	Patchouli oil	
	Cinnamic alcohol	

The importance of the basic note or accord or group of accords cannot, be denied. Carles gives some extremely useful information on this point. There are few good perfumes, and certainly no characteristically modern ones, that depend chiefly for their appeal and individuality upon those constituents which have a low volatility and high tenacity. The three fundamental parts of a perfume are: the head, the body and the base. The relative size, strength and general assembly and behaviour of these essential parts depends upon a number of interrelated factors. Possibly the aptest name ever to have been given a perfume is Arpege, because an arpeggio effect, in which the notes of a chord are played successively instead of simultaneously, is so admirably descriptive of perfume behaviour.

According to some perfumers odours are homogenous and oscillating. Thus a world-famous floral bouquet perfume oscillates between a homogeneous accord consisting of a fresh jasmin note sustained by an aldehyde and bergamot oil etc., on the one side and, on the other, a warmer lower chord composed of ylang-ylang blending into a residual note of vanilla and incense.

Table 4.2. Classification according to volatility (Carles).

Top notes *Very volatile products* *lacking tenacity*	*Modifiers* *of base notes (i.e.* *products of inter-* *mediate volatility* *and tenacity*	*Base notes* *(products of low* *volatility and* *high tenacity)*
Amyl acetate	Basil	Methyl Ionone-Ionones
Bois de Rose	Terpineol	Absolute Orange Flower
Linalool	Petitgrain, Paraguay	Clary sage
Phenylethyl acetate	Galbanum	Amyl salicylate
Lemon	Verbena	Absolute Jasmine
Lavender	Thyme	Benzyl salicylate
Bergamot	Geranyl acetate	Cedarwood
Orange	Juniper	Aldehyde C_{16}
Coriander	Tansy	Aldehyde C_{18}
Tarragon	Phenylethyl alcohol	Sandalwood
Laurel nobilis	Geraniol	Artificial Musks
Petitgrain from the	Absolute Lavender	Absolute Oakmoss
lemon tree	Citronellal	Vetiver and derivatives
	Neroli	Patchouli
	Rose Bulgarian	Calery
	Ylang	
	Geranium	
	Aldehydes C_8, C_9, C_{11}, C_{12}	
	Cloves	

The type of classification of odourants shown in Tables 4.1 and 4.2 is a most useful guide. However, one must not overlook when resorting to such classifications, that the behaviour of some of the materials involved is more complex than its mere position in a relative volatility table might indicate. This brings us back to reconsider what exactly we mean when we talk about top notes. Firstly, there are the very volatile 'true' top notes, for example, ethyl and amyl acetates, ethyl aceto-acetate, methyl amyl ketone, phenyl-ethyl acetate, linalool, and the citrus terpenes. There are also many relatively non-volatile odourants which, in addition to their long-lasting character, have also pronounced top note effects. In this group may be cited, as common examples: musk ambrette (as distinct from the ketone or xylol), the macro-cyclic musks, ethyl vanillin, methyl nonyl acetaldehyde, gamma-undecalactone, and Fixateur 404. It will be seen that all these have penetrating odours and it is perhaps this penetrating or piercing quality rather than mere pungency which also gives the Middle Note odourant, indole, the emergent force and character of a top note.

Though complex, essential oils are also capable of being assessed on the basis of relative volatility and must therefore be included in such comparative tables.

The practical value of a chart showing odourants grouped according to their relative volatilities is that it can serve as a guide to formulation. It has been observed that a perfumer beginning work on a Lilac perfume can quite simply extract from the complete chart a number of odourants (in this case as few as nine) which will give him the foundation of his perfume. These he will subdivide into the three main categories. Thus he quickly arrives at a top note section comprising benzyl acetate, terpineol and phenylethyl alcohol; a middle note section consisting of heliotropin and anisic aldehyde; and a basic section containing cinnamic alcohol, hydroxycitronellal, isoeugenol and phenylacetic aldehyde. He will then proceed in the usual way, comparing his perfume, from time to time, with the natural flowers that he is trying to imitate or with some other lilac perfume that he aims at copying. At this stage he will doubtless think of other odourants, in order to improve the natural character of his experimental blend, to shade the odour into Pink Lilac or some other specific type, or to convert the floral base into a more sophisticated blend rather than a simple floral composition. A reference to the chart is useful stimuli in many instances and also offers information enabling new accords to be elaborated.

'The persistence of perfumes depends on the coefficient of useful work, i.e., the nature of the evaporation and the value of the threshold concentration. This stability increases with the coefficient of useful work and with reduction in the threshold concentration. These questions, i.e., the change in the rate of evaporation and the proper use of threshold concentration values should be carefully studied by perfumers and research institutes'.

Unfortunately the margin of subjective error is even wider here than in the assessment of relative volatility and persistence. At times it is so enormously wide as to represent an obstacle which is simply not negotiable. The threshold value of perception of an odourant is that at which its odour, in given conditions, is only just perceptible, so that further dilution causes it to disappear. The threshold of recognition is obviously somewhat higher in each case.

Many lists have been published of odourants arranged according to their relative odour intensities which are not reliable for general acceptance.

This matter of odour strength, like that of most other aspects of odourant behaviour (e.g., diffusiveness), is one that the practising perfumer is obliged to take into account and deal with in the light of his own experience. In many cases he will have to proceed on trial-and-error lines.

CLASSIFICATION BASED ON PHYSICAL CHARACTERISTICS

From classifications based on physical characteristics we run to those based on similarity of odour. The first 18–group classification is given in Table 4.3.

Table 4.3. Classification based on odour.

Group	Odour	Plant	Group	Odour	Plant
1	Rose	rose	10	Lemony	lemon
2	Jasmine	jasmine	11	Herbaceous	lavender
3	Orange	orangeflower	12	Minty	menthol
4	Tuberose	tuberose	13	Anisic	aniseed
5	Violet	violet	14	Almondy	bitter almond
6	Balsamic	vanilla	15	Musky	musk
7	Spicy	cinnamon	16	Ambered	ambergris
8	Caryophyllaceous	clove	17	Fruity	fruit
9	Camphored	camphor	18	Santalaceous	sandalwood

Many perfumers, using this as a model, (Table 4.3.) have devised similar classifications. Billot, in 1948 classified odours into eight main groups. Since then the classification in close association with the Society Technique des Perfumers de France, has remained unchanged except addition of one i.e., nine groups. The classification of odours is given as under : (i) Floral; (ii) Woody; (iii) Rustic (*agreste*); (iv) Balsamic; (v) Fruity; (vi) Animal; (vii) Empyreumatic; (viii) Repulsive; and (ix) Edible.

The last two groups, repulsive and edible, have in principle little or no application in perfumery, but they have been included in an attempt to provide a classification comprising all kinds of odours. The sub-division in each series into different notes given as under :

Group 1 : Floral Series

(i) Rosaceous (rose); (ii) Jasmine-like (jasmine); (iii) Hyacinthine (hyacinth); (iv) Lilac-like (lilac); (v) Orangeflowery (orange blossom); (vi)Tuberose-like (tuberose); (vii) Violet-like (violet); and (viii) Resedaceous (mignonette).

Group 2 : Woody Series

(i) Spruce-fir (pepper); (ii) Santalaceous (sandalwood); (iii) Caryophyllaceous (clove).

Group 3 : Rustic Series

(i) Menthol (peppermint); (ii) Camphoraceous (rosemary); (iii) Herbaceous (lavender); (iv) Green (violet leaves); (v) Lichenous (earthy odours, oakmoss); and (vi) Leguminous (methyl heptenone).

Group 4 : Balsamic Series

(i) Vanilla-like (vanilla); (ii) Olibanaceous (incense, balsams); (iii) Galbanaceous (galbanum); (iv) Resinous (pine resin).

Group 5 : Fruity Series

(i) Hesperidean (bergamot); (ii) Aldehydic (fatty aldehydes); (iii) Almond-like (bitter almond); (iv) Anisic (aniseed); (v) Fruity types (peach, plum, strawberry); and (vi) Chocolate, cocoa.

Group 6 : Animal Series

(i) Musky (musk); (ii) Castoreum (castoreum, leather); (iii) Skatolish (civet); (iv) Maritime (seaweed); and (v) Ambered (ambergris).

Group 7 : Empyreumatic Series

(i) Smoky (birch tar) and (ii) Tobacco-like (tobacco).

Group 8 : Repulsive Series

(i) Caseinous (cheese); (ii) Alliaceous (garlic); (iii) Fishy (fish); (iv) Stagnant; (v) Butyric (rancidity); (vi) Mircinian (mud); (vii) Musty; (viii) Pyridinic; (ix) Mercaptan; and (x) Carbylamine.

Group 9 : Edible Series

(i) Buttery note (fresh butter; diacetyl) and (ii) Broth.

(1) This classification has been arrived at by taking into account the fact that formerly perfumers with five or six floral and some other notes, all natural, managed to reproduce all the other flower or other notes, rather like using the three primary colours where, by mixing them, almost all the others can be reproduced.

(2) The categories of notes have been kept to a minimum and the analogy of odour has been used in its widest sense.

(3) The aldehydes are included in the fruity series because they have indeed an odour similar to that of fruit; others have a very distant resemblance to fruit, but the special part they play in compositions generally relates them to the other notes of the fruity series.

(4) Although the repulsive and edible series may appear useless in perfumery, it is possible that for imparting a certain nuance some items (like dimethyl and dibutyl sulphides) might be found quite interesting in a very small dosage.

It is not possible to define limits of any class of odour. Even if we set up a class of flowery odours, there would not be considerable agreement on exactly where flowery stops. We agree – even though some odours are obviously classifiable in more than one group. Thus an essential oil with a predominantly woody odour can also be balsamic, spicy or even fatty. And a dominantly fruity odour may also be floral, spicy or even animal in quality.

Nevertheless, when dealing with odourants and mixtures of odourants one is bound to think of them in relation to one another; hence the necessity to proceed by analogy and thus, for the sake of convenience and clarity, to classify them according to their degree of similarity or some other significant characteristic.

Chapter 5

Creating a Perfume

INTRODUCTION

To create new thing is satisfactory work for perfumer. Therefore in many panel discussions proposal for producing novel perfume have come forward. Few desired to make a mystery of it, a sort of arcane divination, almost as though they considered the simple truth to be unworthy of such a specialised subject. To produce novel perfume, quality perfume requires perservance, in depth of knowledge of perfumery chemicals, imagination and trained mind.

Working within a Framework

Price

Perfumery is a very different art from painting and sculpture and commercial craft such as creation of hand of jewellery. It operates within relatively strict limits of price. The topmost perfumes having high price ceiling and very low a limit might not succeed the object of pleasing whatever section of public the perfume is designed to reach. Even if perfumer, is working with a reputed organisation or has renowned his work, has to consider the necessity of working out price level acceptable to buyer. One must not exaggerate the importance of cost factor because it will put restriction on the performer's general freedom of action.

The most important difference of perfumery as an art is complete absence of amateurism. In case of sculpture and painting a large number of amateurs or dilettantes are found. As an amateur it is very difficult to learn the art, craft and technique of perfumery. Even if the few amateurs available due to heavy demand they tend to become involved sooner or later as professionals.

Packaging

Packaging is the important activity, that contribute to the acceptance of a fine perfume. It involve the designing of bottles, the choice of casket or carton, the label, the colour scheme, the general *ambiance* — and, of course, the name. But the utmost importance has to be given to the perfume itself as, even with the prettiest and most luxurious presentation, the perfume must fulfil all the demands made of it in agreeable odour, originality, tenacity, strength etc., or else it will soon disappear from the market.

One has known perfumes that had a strikingly successful impact because of their pack and publicity but which, owing to their inherent defects as perfumes, fell off alarmingly in sales as soon as the initial publicity died down. On the other hand (and this is perhaps more important) there have been many perfumes of considerable merit introduced, over the years, only to rank as failures because of poor packaging, ineffective marketing and insufficient distribution.

Inspiration

The perfumer, in fact, is inevitably one of a team. Like all star performers, he is apt to have his idosyncrasies. His inspiration can come from many sources, not the least of which may be conversations with his colleagues - or even with rival perfumers. On some occasions, it is true, he may be absorbed entirely in his own thoughts and impressions, work out certain basic accords and proceed to devise a perfume that is truly 'himself' — but even then he owes much, like any other artist or craftsman, to the thoughts and teachings of those who have preceded him. On other occasions, however, the inspiration will definitely come from outside.

THE PRACTICAL FOUNDATION

The first important point to remember for obtaining new perfumes is that, one should have complete study of natural and synthetic materials, and their characteristic odours should be fixed in olfactory memory. It is necessary to have knowledge of the differences that exist between high-grade and inferior products and the new products appearing on the market. But how is all this knowledge to be gained?

All that need be added at this juncture is that perfume creation is essentially bilateral: one side is intuitive, aesthetic, imaginative, subjective, while the other is technological, methodical, and to an increasing extent objective and scientific. Neither side can operate successfully alone.

The perfumer's inspiration comes from many quarters, from the laboratory, the perfumery boutique, the garden, the boudoir; from the beach, theatre or fashionable restaurant; even from books. The talented perfumer, not only qualified by technical expertise but also endowed with an imagination, may well have some specially attractive woman in mind when ideas come to him, or he may merely be excited by life in general. But the perfumer at this stage will not be thinking in poetic or musical or literary images: his ideas will come to him as top notes and *accords*, as aldehydes, esters, lactones—a touch of this, a background of that, a hint of something else, something subtle and unusual, and beyond that initial phase the need to improvise, to discover, to round off and perhaps, at last, achieve a new climax of originality and seduction. Yet these are words and he, at that time, will be thinking in odours.

Inspiration: Response to Stimuli

Different perfumers react to different stimuli. Henri Robert of Chanel suggested that odour-diary should be maintained by perfumer during their travels. On returning back to the laboratory perfumer with the notes in odour diary, attempt to create in memory the various olfactory impressions that has been received. To inspire is to breathe, in and this the perfumer does, receiving his inspiration from all quarters. 'A perfumer should be most concerned with expressing his personality in his compositions. His perfumes should incorporate an ideal – his ideal – and should therefore be characteristic, so that they cannot be confused with any other compositions. Later on, when he has gained more experience, the perfumer must see to it that each perfume combines strength of odour with diffusive power, without dispensing with the fineness and purity of the odour. He will have to devote his special attention to the volume of odour, which is usually more important in use and is often harder to achieve than, say, stability of the odour. This is practically guaranteed if one limits oneself to simple compositions and takes careful heed of the reactions of the various ingredients.'

Here one notes the essential association of artistic and technological factors. The former may perhaps range, as in painting, from the representational art typified by the simple flower perfume through impressionistic and expressionistic concepts down to those which are almost purely abstract; while the technological side of the perfumer's craft will control the actual form and outcome of his efforts.

Trial and error

Irrespective of source, the perfumer's feelings and ideas must eventually result in action, which can be accomplished by traditional trial-and-error methods. One such procedure consists of forming a nucleus of a few essential oils and synthetics, and to proceed by tentative additions of further consituents. As one can add in this context but not subtract, many false starts and fresh beginnings may thus be involved. There are two ways to carry out trial and error method. In first case one may work up a base suitable for incorporation in an already established composition, so that when it is properly adjusted and added all that is necessary to round off the harmonious finished composition is the addition of a few final touches here and there. Another method is to work on a base, self-made or purchased from outside, by adding different products to it, little by little, in accordance with the type of perfume that one is setting out to create.

This method of making perfumes can only be practised satisfactorily, of course, after the perfumer has acquired a long and intimate experience of the effects of mixing odouriferous substances together. When the mixtures or blends are produced perfumer must observe very attentively and acutely all the little olfactory phenomena which takes place. These were the only methods of producing perfumes and it is still used today.

Colours

Creating a perfume is analogous to painting. A painter knows that if he would like to get certain shade e.g., flame, he just gives touch of yellow to saffron to obtain effect. Similarly the touching effect can be obtained in mixture of odours wherein the part played by induction is enormous. Nevertheless, in the mixture of colours we have the conventional decomposition of light passing through a prism into the primary colours: violet, indigo, blue, green, yellow, orange and red, that are reduced to three: blue, yellow, red, which by being mixed enable all the others to be obtained. This process is not so complete with odours. Then there is the Chevreuil disc which aids the process of mixing colours. We have nothing comparable for mixing odours.

Until we have a reasonable theory of olfaction, enabling us to conduct similar work, we shall remain more or less subject to the trial-and-error methods of empiricism. That is the reason why so many research workers have suggested practical methods relating to the technical side of perfume composition and paid little attention to the indispensable artistic inspiration. As their views collectively present many points of interest, we attempt to look at them briefly but critically in the paragraphs that follow.

JEANCARD ON COMPOSITION

As per the idea of composition proposed by the Jeancard, the perfume will have a prelude and a finale, which will frame the motif and bring it out. We have here, he added, a distillation at ambient temperature and pressure, where according to their vapour pressure, components will become dissociated and volatilise, undergoing a true fractionation. This evolution, governed by the laws of physics, is a function of the molecular weights and vapour pressures, which express the volatility of the mixture.

Highly volatile odourants with an evanescent effect can have their action modified and made more tenacious by being admixed with a less volatile component. This has been clearly demonstrated in the cases of methyl amyl ketone and benzyl acetate in alcoholic solution, with and without benzyl benzoate. We must not overlook the formation of azeotropes, i.e., a mixture of substances which distil as though they in fact formed only one single substance. In theory it is possible, when there are only two or three substances, to calculate the formation of azeotropes, but the picture becomes too complicated when there are a considerable number of substances present.

Many years ago, it was noticed that the great perfumes which had a lasting success behaved on evaporation almost always in an identical way. It indicate that there must be, in all of them, a similar azeotropic evaporation that would show up in an evaporation curve. An experiment was conducted. First collect considerable quantities of different well known perfumes having the qualities sought and recognised, originality, atmosphere (character), power tenacity, diffusiveness etc. Extract the perfume concentrate using pentane and separate leaving behind a residue which resembled closely that used at the time of making the perfume. Evaporate the solvent separated at the ambient temperature, 15–20°C and relevant curves obtained for temperature against time. It was found that all these curves were almost identical or at least their outlines were very similar.

While in nature, i.e. in the essential oils, azeotropes must undoubtedly exist, the complexity of these phenomena becomes quite formidable when one has to deal with mixtures of many different substances, as opposed to mere binary systems. Here the empirically derived experience of the perfumer has to take over in judging a priori the substances to be associated and the relative proportions that will be required to obtain a desired note.

LIMITATIONS OF FORMULAE

For those who have come into possession of a formula and think they are going to have success with it, they should remember that a formula is only of maximum value, if the person using it knows exactly the origin and name of the supplier of each of the raw materials and it is, moreover, olfactorily necessary for every raw material in the formula to be identical with that used in the creation of the perfume.

As per Felix Cola: 'A perfume is composed like a poem, a symphony. It is a happy inspiration which decides the substances that have to be used and the respective proportions of the constituents which have to be employed.' He also put forward that the proportions are gradually modified to obtain a better balance. Then, an attempt is made to simplify the formula, while still retaining its special character; for if the number of constituents is unduly increased, the perfume will be deprived of personality. It becomes grey and undistinguished – what a painter calls *grisaille*.

In other words, the finest of the natural essential oils seem to consist of several basic organic substances which form the typical note, but this characteristic basic odour is rounded off by the presence of infinitesimal quantities of a large number of organic substances, related or not to the first ones, which combine to give fineness, distinction and sweetness, which characterise the quality of these essential oils.

Otto Gerhardt, expert seems indispensable that a perfumer should have above all else an artistic talent, experience, and complete knowledge of the development of the composition. He should of course have a full knowledge of the raw materials (natural and synthetic) and way to use them. In this case, an olfactory memory plays a leading part.

THE PERFUMER'S KNOWLEDGE OF CHEMISTRY

A knowledge of chemistry helps perfumer to understand perfume mainly from synthetic products in better manner. There is of course no incompatibility between the fact of having studied chemistry and that of having artistic tastes. Perfumers who have not had scientific training have usually endeavoured later on to acquire at least a minimum of chemical knowledge, after realising that they lacked it and were thus at a disadvantage. This does not mean that any chemist can become a perfumer: far from it. And it is quite certain that very many people who have never studied chemistry have been excellent perfumers.

For perfumer it is necessary to remember the composition of the principal types of perfume, so as to have a comprehensive knowledge of possible combinations. At the same time he should know the majority of the natural and synthetic odourants. He must have an excellent and extensive olfactory memory, which has recorded all the knowledge he mentions and which will enable him to identify and copy each special note, floral or otherwise, just as occasion demands.

Frank Atkins wrote in 1935-36 a series of articles on the Use and Abuse of Fixatives, in which he dealt with fixation as an integral part of perfume composition. Marcel Billot, in a long and heavily documented review of Fixation in Perfumery, also looked at fixation in the general content of composition.

The Russian perfumer, contributed on fixation and other aspects of perfumery. According to him: (i) The persistence of perfumes depends on the coefficient of useful work, i.e., the nature of the evaporation and the value of the threshold concentration; (ii) The stability increases with the coefficient of useful work and with reduction in the threshold concentration; (iii) The change in the rate of evaporation and the proper use of threshold concentration values should be carefully studied by perfumers and research institutes; and (iv) Qualities demanded of the perfumer are the ability to conceive his objective clearly, to persist in his purpose, to have immense patience, and to have the requisite wide knowledge of his medium.

'The perfumer must have an intimate knowledge of a thousand aromatic substances. He must be familiar with the odour value of each, its source, and the characters that determine its quality, as well as its floral type and blending range. Moreover, he must also select and use each one of them skilfully to compose, reinforce and shade his basic note until it becomes a symphony of fragrance. But this is not all; for his creation is to be employed as a perfume for cosmetics and soaps he must also have a profound knowledge of their composition and the effect each of their raw materials will have upon every constituent of his original fragrance, which involves cunning substitution and fresh blending to achieve ultimately the same finished note.'

Creativity in the Art of Perfumery

Creativity in the Art of Perfumery is based on the following principles :

(1) Chose and work out a harmony or a pleasing combination of harmonies: original, agreeable and carefully completed.

(2) Perfect the top note or *depart* with light components which harmonise with the base.

(3) Complete the basic note so that it harmonises with the general, characteristic note of the perfume as a whole.

Carles' Method

Jean Carles, a perfumer of high repute opposed to the idea of 'exceptional nose'. There are no doubt people whose sense of smell is naturally keen, but the highly developed olfactory sense can only come from education and training, and particularly from education of the olfactory memory (just as for other memories). No perfumer can exist without an olfactive memory.

Carles conducted a study of both contrasting and similar odour of each family which suggested that each perfumer should prepare for his own use an empirical classification of this kind, listing all raw materials, natural and synthetic, according to their relative volatility, i.e. as (i) top notes— or very volatile products lacking tenacity; (ii) base modifying notes – or products of intermediate volatility and tenacity; and (iii) base notes – or products of low volatility and high tenacity.

Carles make a very important point, that the skilled perfumer, like the skilled musician, should be able to 'smell' his perfume by setting it down on paper and thinking about it. The preliminary 'blending' thus goes on in the mind, although it obviously has to be put to material proof, tested and modified, before it can be regarded as satisfactory.

The stepwise method of Carles' to produce perfume is (i) First find combination of two or three odourants capable of providing the basic note of a perfume; (ii) Examine them rapidly and make a mixture of two substances which is possibly an attractive blend or accord; (iii) Perfect them quantitatively. Assume these two odourants as 'A' and 'B'. Prepare series of a trial using different ratio of 'A' and 'B' as under :

Serial No.	'A' Parts	'B' Parts	Total
1	9	1	10
2	8	2	10
3	7	3	10
4	6	4	10
5	5	5	10

From these series perfumer can select what seems best to him. In case this series fails to provide best perfumer may prepare series of solutions having reverse proportions. i.e. :

Serial No.	'A' Parts	'B' Parts	Total
1	1	9	10
2	2	8	10
3	3	7	10
4	4	6	10

In many cases he would presumably make less than the suggested five trials. The basic group obtained by this method may consist of two, three, four or even five substances, although the fact should not be forgotten that an excessive increase in the number of constituents of such a basic accord will involve the risk of 'extinguishing' or 'blanketing' a perfume, leading to the production of *grisaille* or a merely nondescript odour.

After preparing a mixture of odourants, add modifier or modifiers. The modifier is a substance of intermediate volatility and tenacity, chosen to enhance the character of the basic accord and to sweeten, subdue or round off any harshness exhibited by the latter. As Carles points out, most basic accords – and he instances one composed of 6 parts of oakmoss absolute and 4 parts of an ambergris/cistus compound – may have a rather unpleasant top note, which soon gives way to a pleasant, long-lasting note. The chosen modifier – in this case, a rose compound – is added for the dual purpose indicated and then is itself surmounted, as it were by a top note that further develops the character of the blend. At each step the perfumer may proceed either by experience or by trial-and-error tests of the type used in his selection of the basic accord. The best example shows that an oakmoss absolute and a certain Ambergris 162 B form a more balanced accord in the ratio of 6:4 than in any other ratio. This simple blend is further improved by the addition of 1 part of musk ketone. The modifier, Rose d'Orient 2644, is shown to function best in the proportion of 3 parts, while the top note of this highly simplified Chypre perfume consists of 4 parts of sweet orange oil and 1 of bergamot. As a very general guide Carles suggests the following :

Top notes (*notes de tete*)	20–25 per cent
Body notes (modifiers)	20–30 per cent
Base notes	40–55 per cent

These indications are not of course binding, but will be modified according to the character of the materials used and the type of perfume that it is intended to make. Musk ambrette, for example, is one of those basic constituents which also emerge as part of the top note. Nevertheless, as Carles points out, a perfume containing 20 per cent of bases, 30 per cent of modifiers and 50 per cent of top notes would lack tenacity, since the percentage of bases would be relatively too low as compared with that of the more volatile modifiers and top notes. Therefore, the proportions are selected so as to obtain a balanced evolution during evaporation.

Merit of Carles' system

It enables a balanced nucleus to be formed in a simple, straightforward way, and for that nucleus to be modified or extended at will. Thus he starts by illustrating his theory with a Chypre theme based on a carefully ajdusted accord of oakmoss absolute and an ambergris speciality. He subsequently suggests that this might be replaced by accords of oakmoss/patchouli/vetiver, oakmoss/methyl ionone/vetiver, or oakmoss/ciste-labdanum/musk ketone. Any of these basic accords might likewise be modified by the addition of products other than rose, e.g. orangeflower absolute, jasmine, lily-of-the-valley or carnation. The top note will naturally be modified accordingly. He suggested modifications as below :

---

1. *Chypre modification A*
Sweet organge
Bergamot
Orange flowers 1103
or absolute, colourless
Absolute oakmoss
Ambergris 162 B
Absolute jasmine
Musk ketone

2. *Modification B*
Bergamot
Laurel nobilis
Angelica seeds
Juniper berries
Muguet 113
Absolute oakmoss
Vetiver
Patchouli
Ambergris 162 B
Aldehyde C_{14}
Absolute Jasmine
Musk ketone

3. *Modification C*
Bergamot
Lemon
Linalyl acetate
Jasmine 1103

(Cont'd)

Geranium African

Orange flowers 1103

Aldehydes C_9, C_{10}, C_{11}

Absolute oakmoss

Gardenia Invar

Styrallyl acetate

Vetiver

Ambergris 162 B

Musk ketone

The important feature of Carles' method is that certain odourants are regarded as 'accessories', which in general are products to be used in small quantities and to impart special effects. These products can not be used in large amounts in basic accords or as conventional modifiers. Addition of traces of these products in a formulation results in complete change in the character of the latter and imparts to it unique cachet. e.g., aldehyde C_{12} (MNA) and C_{14}, styrallyl acetate, isobutyl quinoline, galbanum, cascarilla and the like. There are sometimes, of course, exceptions to this rule of moderation. We quote: 'Aldehyde C_{12} (MNA), for examples, proves to be an exception and it should be known that some products such as Geranium give most successful blends with as much as 50 per cent of it. The advantages which may be derived from the use of accessory products are therefore readily apparent...'

Carles advised his young pupils against a quest for maximum equilibrium. 'Dominantly effective notes in perfumes should be neither feared nor deliberately avoided. Such "faults" have quite often been responsible for tremendous success.' Sharp contrasts have their place in fine perfumes, as in the other arts. An excess of polish or rounding-off can, at worst, result in a subdued, characterless composition.

The point has been made that Carles' method may overstress the importance of the basic notes, but this need not be so if sufficient interest is taken in the top notes and the accessories.

Raymond Kling, perfumer had some interesting notions about aids to creativity. His ideas were coloured by his early training as an engineer and he never ceased seeking for a methodical mode of procedure. Like Carles he started with the notion of an accord of two, three or four odourants which would together provide a homogeneous base. This basic accord was obtained by creative artistic induction or by trial-and-error tests extended almost 'ad' infinitum with an extraordinary degree of patience.

Kling classified odourants into – top notes, basic notes, exalting or lifting agents and proposed new concept called as "place creators", which is defined as a substance which, in an already established composition, insinuates itself among the constituents and forms a place to enable odourant substances of a different nature from those existing in the composition as originally conceived, to "hang on to" this place creator like touches of bright paint against the grisaille of a background cloth. The substances of relatively mild or moderate odour and having great viscosity, with rather low volatility, such as hydroxycitronellal, nerolidol, benzyl benzoate etc., are included in the class of place creators.

Although our understanding of odours and odour phenomena remains limited, certain factors such as tone, volatility, intensity, volume and tenacity can to some extent be studied.

Tone

Tone is the general character of an odour. If the note is acidulated, acute or 'pointed' (e.g. lemon oil or octyl aldehyde) it is said to be sharp. If mild and even, like the odour of rhodinol or geraniol, it is medium. If warm, a little muffled or heavy, like sandalwood oil or Peru balsam, it is described as low.

Volatility

Volatility means evaporation or the more or less rapid sublimation of an odouriferous substance. It is proportional to the vapour pressure of the substance, and inversely proportional to its molecular weight. This factor may be expressed by the formula :

$$m = P \sqrt{\frac{M}{2\pi RT}}$$

where m is the rate of evaporation in g/cm^2/sec., P the saturating vapour tension in atmospheres of the substance at T° absolute, M the molecular mass in grams, and R the constant of the law of perfect gases which, in the system of atmospheric units and cm^3 is equal to 82. The formula thus becomes :

$$m = P \sqrt{\frac{M}{515\,T}}$$

It has been confirmed that the theoretical figures are matching with practical figures obtained by evaporation in air. It has been found that the classification of the products is the same in both cases. The result is that, to keep to simple practical data, it is sufficient to determine figures that are proportional to the volatility of each substance and characterise each of them.

The coefficients of volatility is carried out in the following manner. Carles' method is different from this one. In this process evaporation is conducted in similar condition i.e., same room, place, time, day, temperature (about 18°C) and free from draughts. The capsule is weighed with a standard uniform weight for each of the substances whose volatility is to be measured. First of all we observe how evaporation takes place, eliminating for the purposes of measurement the terpene notes and the base or residual notes. Only the characteristic note of the product is taken for effecting the measurement.

Now take dry extract in a small flat bottomed capsule having dimensions about 5–6 cm diameter and 1–2 cm depth. Note the weight and entering time. As soon as no further characteristic odour is perceived, the capsule is re-weighed and the time noted.

With all these figures noted a calculation is made of the weight of the evaporated product and the time necessary for this evaporation. Taking the quotient obtained by dividing this time period by the evaporated weight, we have the 'coefficient of volatility', which must be differentiated from persistence or tenacity. In the composition, the substance of appropriate volatility can be selected as required.

The study of volatility makes it possible to specify and state the three fundamental properties, which are of great importance in perfumery are intensity, volume, and atmosphere.

Intensity

It is the unprovable psycho-physical proposition Weber-Fechner law states that the intensity of the sensation is proportional to the logarithm of the stimulus. It is unprovable precisely because the sensation is not at present capable of being measured. Billot and Appell have both produced variants on this law and thus helped to elucidate the rather complex matter of odour intensity. Appell, indeed, has set out to measure the 'absolute intensity' of odours and as a result has listed over 250 odourants, including natural and synthetic materials, in order of intensities rated from the minimum of 1 (rosacetol or 'rose crystals') to 10 (citronellyl oxyaldehyde, skatole and violet leaves absolute), 12 (2-hexenal) and the maximum 15 (crude nonadienal). We may note that all the aliphatic aldehydes from C_7 to C_{12} are shown in the Intensity 8 or 9 groups. Angelica, bay, cardamom, costus and camomile oils are all to be found under Intensity 9, together with methyl heptin carbonate, rose oxide and some of the paramethyl quinolines. Hydroxycitronellal and ionone are grouped under Intensity 4, in the company of lavender and lemon oils, bergamot, cyclamen aldehyde and musk ambrette. Ylang-ylang and orangeflower absolute come under Intensity 6, jasmine and rose absolutes and styrallyl acetate under Intensity 7; and benzyl acetate under Intensity 2— in the company of benzyl salicylate, guaiacwood acetate and phenylethyl alcohol. Guaiacwood oil is assigned the same intensity rating as lemon oil, and sandalwood oil the same as citral.

It is evident, as Appell states, that a distinction must be made between what he terms the absolute intensity, i.e. 'the intensity factor of the substance itself', which 'must be distinguished from the psychological response'. Elsewhere, in the same series of papers, he observes that 'we must make a distinction between apparent intensity, i.e. the psychological response to odour as perceived by the sense of smell and true or intrinsic intensity of the odourant. The first is the intensity of the sensation; the second is the intensity of the stimulus.' He adds that, whatever we measure or deduce, 'we do not of course know the nature of this (intensity) factor or stimulus'.

Referring to the Purkinje phenomenon, which in the study of colours relates to the great difficulty of comparing, for example, the intensities of red shades with those of blue shades, Billot emphasises the comparable difficulty of comparing equal or disparate intensities of two odours of completely different character.

At the present time the perfumer has no practical alternative, in the matter of judging or estimating comparative intensities, to prolonged experience based partly on trial-and-error. The theoretical approach to the subject is nevertheless well worth pursuing.

Volume

The concepts of 'volume', 'atmosphere' and even 'lift' or exaltation tend to merge into one another and are frequently confused. Volume refers in general to the fullness or amplitude of a perfume. It is analogous to the volume of sound that fills a room. It somewhat resembles the 'body' of a wine. A good perfume has volume; a poor one is *tenu* or 'thin'. Roudnitska says, volume in this sense is an aspect of form and is not to be confused with strength or effective range (*portee*). The concept of volume is more applicable to blended perfume compositions than to simple odourants as such, although some of the latter, as for example essential or flower oils, will have more volume than, say, a relatively 'flat' synthetic like linalyl acetate. Billot has suggested that evaporation curves of individual odourants should be plotted as a function of time. The angular coefficient of each curve will, in his view, represent the volume. He adds that the curves of a blended perfume can enable the perfumer to see at what moment its major characteristic portion evaporates, and thus to form a positive idea of the perfume's structure, with a view to making it more homogeneous by suitably filling in any gaps. This would merely serve as an aid to what a skilled perfumer would do almost automatically.

Atmosphere

The 'atmosphere' of a perfume has been described as its more or less pronounced faculty of dispersion. In practice it is the perfumed wake which a woman using perfume leaves behind her. It is to some extent possible to measure its value comparatively as follows: We measure the rate of propagation of the molecules carrying the odour through a long tube of fairly thin diameter. These different times will provide coefficients proportional to the dispersion, thus characterising the atmosphere, the shortest time representing the greatest atmosphere.

Closely associated with the idea of atmosphere is that of effective range (*portee*), that of 'lift' or exaltation; and Francis Bacon's division of flowers and plants into those which 'do best perfume the air' and those whose scents are detectable only at close quarters. Bacon mentions the wide ranging fragrance of double violets, wallflower, clove pinks, lime blossom and honeysuckle. To this list E.S. Maurer has added hawthorn, lilac, heliotrope, evening primrose, mignonette and nicotiana. Maurer considers that the aerosol-like, zephyr effect of these natural perfumes must be due not only to the obvious and admittedly potent osmophores which they contain, but also to some kind of 'carrier-wave' mechanism which operates simultaneously from the radiating flower centre and may involve the emission of odourous or inodourous volatile materials concurrently with the more potent and characteristic odourants. As carriers or boosters he tentatively suggests certain terpenes, formates, acetates, and simple aliphatic ketones, a preference being given to products of high mineral oil solubility. He adds that a perfumed spray made up on these lines operates best in an atmosphere of high humidity.

An Indian writer has similarly observed that a balanced blend of all the known constituents of a natural jasmine perfume would not in fact exhibit the airborne fragrance of growing jasmine flowers. He thinks that the flowers in the living plant 'generate ...some gaseous lifters or organic compounds of low boiling point and high vapour pressure, which blend with the odour of the flowers and carry it away to a distance'. His thoughts on artificial substitutes for these natural exaltants tend in the direction of ethyl alcohol, methyl and ethyl formates, methyl acetate, etc. He also mentions the chloro-fluorinated propellants used in aerosol perfumes.

At the other end of the scale are the tenacious, long-lasting synthetic musks of the cyclopentadecanolide type. A Firmenich booklet, referring to the nomenclature of Exaltone and Exaltolide, states that 'this designation comes from the fact, which was immediately evident, that large-ringed ketones and lactones possess an exceptional power of exaltation on perfumes'.

To predict the 'lift' of a perfume by reference to its individual constituents is very difficult. One may consider, however, the exalting properties of such esters as isoamyl acetate, ethyl aceto-acetate, ethyl oenanthate and pelargonate (nonanoate) and p.tert. butyl cyclohexyl acetate; and such terpenes as myrcene, d-limonene and pure alpha-phellandrene. One might note that methyl heptin carbonate, anisyl acetate, methyl nonyl acetaldehyde and other perfumery aldehydes, and of course the synthetic musks are odourants of lower volatility.

The elucidation of the problems involved in ensuring adequate lift or exaltation await further developments in olfactory research. Meanwhile the empiric approach must make what use it can of the simple test that we have described or smelling the sprayed perfume in a spacious smelling chamber, as suggested by Carles. (It must be remembered, however, that a perfume allowed to evaporate from a surface is in a different physical form from that which it assumes when sprayed into space and that different odour effects can be expected from space spraying).

Tenacity

Tenacity, is the ability of a perfume to last a longer or a shorter time on its support, is somewhat complex. The nature of the support plays a significant part. Thus a fur retains perfume well, whereas a silken fabric does not. Wells has investigated the tenacity of perfumes on the human skin. Two aspects of tenacity or persistence have to be considered :

(1) Strength of persistence, which is the relative force of a perfume after a limited period of evaporation (2 hours, for example) expressed as a percentage of its original odour; and

(2) The duration of persistence, which is expressed by the number of hours or days which elapse until no further odour is perceived.

In the first case the note remains in general unchanged over the period, whereas in the second case it may have become completely different.

All determinations of this kind should be performed under identical conditions (time, temperature, same paper, same quantity of perfume etc.). A disc of filter paper is taken. On each disc is placed the same standard quantity of perfume. It is allowed to evaporate until nothing more is smelled, and the time taken is noted. This time represents for each substance the total tenacity. It might be possible to note the specific tenacity, which would be the time of evaporation from the start and throughout the time of evaporation, while the characteristic note remains, until it stops and changes to another note.

Fixation

'A fixative is simply a substance whose odour lasts a sufficient length of time. The idea of fixing a perfume in order to retard or prevent the evaporation of its volatile constituents is chimerical. There are no means of fixing the odour of top notes, of linalool for example, whatever the nature of the fixative or mixture of fixatives.' Further it was observed : 'If the evaporation of these volatile bodies (such as amyl and propyl acetates) is slowed down a little by tricks of blending, the fact is certain that this slowing down will be of short duration and will not respond to the perfumer's idea of fixation.

The odd fact is that Bassiri outlines some simple experiments with ethyl alcohol, diethyl phthalate and benzyl benzoate, very similar to those with which Pickthall appeared to arrive at opposing conclusions. Poucher had almost simultaneously carried out yet a third series of tests of this kind.

While the traditional high boiling, low vapour pressure 'fixatives' with little or no intrinsic odour (e.g. diethyl phthalate, benzyl benzoate) are undoubtedly useful materials their utility, is strictly limited. Although they exert a retarding action on the evaporation of the more volatile constituents of a perfume, their mere addition to a blend will not necessarily make the desired odour more lasting. Such an addition may in fact unbalance or exert a deadening effect on the entire complex. There is a place for these fixatives in the formulation of perfumes, and especially of those of a volatile, evanescent character, but for a properly constructed formula the perfumer must pay maximum attention to basic accords and graduated persistence. 'If he is a master of his craft,' writes Poucher, 'he will ... select his aromatics in accordance with their lasting properties, rather than rely upon dubious additions which might, in effect, impair his work of art.'

Billot's conducted study on the Problem of Fixation which should be noted, and a comparison made of the various fixatives listed by Bassiri, Maurer and Cerbelaud. The last-named divides fixatives into eight separate classes, and these are still worth studying even if the theory uniting them has become rather out of date.

Systematic Recording of Data

The first draft or sketch of a perfume, no matter whence or how the inspiration or creative impulse comes, must subsequently be reinforced by patient and often painstaking technique. It is here that properly recorded information can facilitate the flow of ideas and save both time and effort. Billot's method of card-indexing essential information has been expounded at some length. In summary, each odourant used by the perfumer has a large card allocated to it. The name of the product provides the heading. Then are recorded, in order: its note and the series to which it belongs, its nature (i.e. essential oil, absolute, resinoid, synthetic, composite base, etc.); its concentration, tone, volatility, tenacity, colour and dermatological behaviour, etc. The card has also space for 'uses' to be entered and solubilities, as well as 'analogies'.

All these cards are filed in a card index in alphabetical order, separating the natural from the synthetic products by making two different card indexes. A third index could be made with the bases and specialities. To complete this information, however, there are registers which gather together for each fundamental property, as enumerated above, the list of products classified according to their places in these properties. There are five of these registers :

(1) Register of the tone of the note – three classes: sharp – medium – basic, and in each of these classes the list of products connected therewith.

(2) Register of volatilities – from 1 to 100, 100 being the figure for the least volatile product. Against each number is inscribed the list of substances having such volatility.

(3) Register of Intensities – in sections from 30 to 120, the greatest intensity being represented by 120 :

$$30 - 60 - 90 - 120$$

In each of the sections a list of substances with intensity corresponding to that section.

(4) Register of volumes – in sections of 15 of the angular coefficients of the evaporation curves up to 90 :

$$15 - 30 - 45 - 60 - 75 - 90$$

In each section a list of substances having a volume within that section.

(5) Register of Atmospheric Diffusive Values, in sections of proportional coefficients from 10 – 30 – 50 – 70 etc., the lowest figure representing the largest 'atmosphere' or the most marked diffusive properties. In each of the sections, a list of substances belonging to that section.

Finally, for tenacity (persistence), the coefficient is indicated on the card. For this property, it is not deemed necessary to have a register.

By way of illustration the original paper shows models of the index cards, although each firm or individual can easily modify them according to his own ideas. Admittedly the compilation of such indexes and registers involves a good deal of time and effort, but it systematises masses of what would otherwise be heterogeneous and perhaps inaccessible information and, in the long run, can be both helpful and time-saving. It is a task, moreover, that may well be entrusted to an assistant or apprentice, provided that the work is conscientiously carried out and well controlled.

Chapter 6

Flower Perfumes and their Formulation

INTRODUCTION

For the perfumer flowers and perfume has immence importance. There are three different reasons cited: (i) They are stimulus, a point of reference, pleasure source and provides immense information; (ii They are irreplaceable even with drawback and higher cost; and (iii) Many flower notes or floral accords have become essential structure. Flowers and their perfumes are still of very considerable and even basic importance to the perfumer. In the first place they are a stimulus, a point of rerefence, and a source of pleasure and invaluable information. In the second, some of them are still irreplaceable raw materials, despite any disadvantage that may attach to them in respect of scarcity or cost. And, in the third place, many flowers notes or floral accords form part of the essential structure of even the most advanced, and up-to-date fantasy perfumes.

The prime object of this chapter is to obtain the true note of each flower using the various constituents for each flower and provide illustrations of complete formulae. Here it is necessary to point out that such formulae can only be considered as the sum of the actual materials used in preparing them in the first place; and that subsequent reproduction must involve olfactory adjustments in order to give the desired result. Let us begin with the more familiar perfumes based on flower notes.

ROSE

Rose notes differ considerably one from another. Besides the basic constituents for red rose, *rose damascena*, and *rose centrifolia* naturally occurring esters, few aldehydes and acetals are required as auxillary notes.

The traces of many other odourants creates significant effect (Table 6.1). Into the composition of specialised rose notes, e.g., those of the tea rose, Banksiana, Zephyrine Drouhin and Maréchal Niel types.

Table 6.1. Effects of some odourants.

S. No.	Type of Rose Notes	Main Constituents	Special effects
1.	Red rose	Rhodinol, phenylethyl alcohol, α-ionone, nerol	Rose absolute, Bulgarian otto Bulgarian geranium oil
2.	Rose damascena	Rhodinol, geraniol, ex palmarosa oil, phenylethyl alcohol cinnamic alcohol	Natural oils and esters
3.	Rose centifolia	Citronellol, geraniol, phenylethyl alcohol, rhodinol	C_9 aldehyde citral
4.	Tea rose	Citronellol, phenylethyl alcohol, geraniol	guaiyil acetate menthone, tuberose absolute
5.	Rose Marèchal Niel	Geraniol ex palmarosa oil, citronellol	isoeugenol, benzoin, sandalwood oil
6.	White rose	Rhodinol, phenylethyl alcohol, benzyl alcohol, linalool.	bergamot, phenylethyl acetate

Bulgarian rose has played very creative role in almost all rose perfumes. It is often required but due to higher cost pure Bulgarian rose oil's use is restricted. This necessitates research into compositions 'de coupage', i.e. diluents or extenders, and these must be the best possible for the purpose. By using certain raw materials judiciously one can in fact arrive at some very interesting extenders which will blend well with the natural oil, giving an excellent quality of end product at a reasonable price. The raw materials in question are relatively few in number. One thinks of: phenylethyl alcohol, oil of geranium – such as the Geranium incolore de Grasse, geraniol, rhodinol, l-citronellol; and very small quantities of nerol and farnesol.

An example is given below in forumla 1. This is a convenient point to emphasise that no formula can be better than the type and quality of its individual constituents. It is therefore essential for perfumers to make their own trials and adjust their final formulae accordingly.

Formula 1
Rose Extender 1   *Parts by weight*

Geraniol ex palmarose, 1st quality	1.50
Oil of Grasse Geranium, *incolore*	0.50
Phenylethyl alcohol	2.50
Linalool ex bois de rose, 1st quality Brazilian	1.20
Rhodinol ex African geranium oil	1.20
1-Citronellol	3.00
Nerol, pure	0.07
Farnesol, pure	0.03
Total	10.00

With the aid of this 'extender', following blends can be produced.

Formula 2
Odourants

	Rose Blend		
	No. 1%	*No. 2%*	*No. 3%*
Rose Bulgarian	75	50	25
Bulgarian absolute oil of rose	–	4	3
Rose extender 1	25	46	72
Total	100	100	100

The above table shows that three different rose formulae having Bulgarian rose tonality and different prices can be easily produced. Formulation over a reasonably wide price range is thus facilitated. Some formulae of actual rose perfumes are given in Formulae 3–6.

Formula 3 (Rose F)

Constituents	*Parts*
Rhodinol, 1st quality	40.0
Ionone alpha	23.0
Rose-2	6.0
Phenylethyl alcohol	12.0
Bois de rose Brazil	12.0
Rose de Grasse absolute	6.5
Jasmine absolute	0.5
Menthone	trace
Total	circa 100

Formula 4 (Rose H)

Constituents	*Parts*
Rhodinol, 1st quality	10.0
Phenylethyl alcohol	20.0

Cinnamic alcohol	6.0
Geraniol	20.0
Citronellol	10.0
Linalool	6.0
Hydroxycitronellal	5.0
Linalyl acetate	4.0
Eugenol	4.0
Rose de Grasse absolute	4.0
Geranium (African) oil	10.0
Phenylethyl acetate	1.0
Total	100.0

Formula 5 (Rose rouge)

Constituents	Parts
Violettone A (Firmenich)	18.0
Rose-1	4.5
Rose de Grasse absolute	13.0
Phenylacetic acid	0.4
Lorena (Firmenich)	4.5
Phenylethyl alcohol	9.6
Rhodinol	50.0
Total	100.0

Formula 6 (Rose blanche)

Constituents	Parts
Rose-1	13.50
Sandalwood oil	6.30
Patchouli oil	6.30
Tuberose	5.50
Tubereuse (Synarome)	5.00
Orris resinoid	5.00
Jasmonis (Givaudan)	3.50
Benzyl acetate	2.00
Geraniol synth, pure	12.00
Lorena (Firmenich)	9.00
Phenylethyl alcohol	11.00
Iraldeine 100% (de Laire)	9.00
Rhodinol P (Rhone-Poulenc)	11.50
Aldehyde C9	0.10
Aldehyde C10	0.15

Aldehyde C12 (lauric)	0.15
Total	100.00

Where specialities are cited under the names of specific firms, this simply means that no firm makes and offers exactly the same end product as the others. In each case one must take into account the fact that chemicals included in a process as trace impurities are not likely to be present in exactly the same pattern or proportions in a competitive product.

JASMINE

Rose and jasmine are the most important flower notes used in perfumery. Singly or together, they provide a conventional floral background for a great diversity of perfumes. A wide range of formulae for 'jasmine artificial' have been resulted after conducting many experiments based on analytical research of jasmine oil. The basic constituents of these formulae include, benzyl acetate, amyl and hexyl cinnamic aldehydes; benzyl alcohol, formate, salicylate and other esters; indole and derivatives; phenylethyl alcohol, dimethyl benzyl carbinyl acetate, hydroxycitronellal, linalool, linalyl acetate; esters of propionic and butyric acids; Peru balsam etc. The use of jasmine absolute *châssis* may be made to sweeten formulae for artificial jasmines which may be crude and synthetic. Jasmine absolute *châssis* is the absolute obtained by petroleum ether or benzene extraction from jasmine flowers that have previously been treated by the enfleurage process but which have nevertheless retained some of their perfume. This must not be confused with the actual absolute of jasmine d'enfleurage.

It is a principle in fine perfumery that natural materials should be used to sweeten and soften the odours of synthetic blends. When costing considerations prevent the more liberal use of 'naturals', the result of judiciously incorporating even small quantities will usually prove conclusive. (Formulae 7–10).

Formula 7 (Jasmine 1)

Constituents	Parts
Benzyl acetate, pure	28.0
α-Amyl cinnamic aldehyde prima (Descollonges)	20.0
Jasmine chassis absolute ex benzene (Schmoller & Bompard), 10% in benzyl benzoate	10.00
Benzyl propionate	4.00
Paracresyl isobutyrate	3.02
Dimethyl benzyl carbinyl acetate	1.06
Benzyl alcohol	9.02

Linalyl acetate	2.00
Methyl anthranilate	0.04
Isojasmone (Descollonges)	0.05
Bois de rose oil, Brazil	6.00
Geraniol (synth., chem. pure, Firmenich)	1.06
Iraldeine 100% (de Laire)	2.00
Allofloral (Allóndon-Firmenich)	6.75
Ylang-ylang oil	2.04
Indole, 10%	2.08
Total	100.00

Formula 8 (Jasmine 2)

Constituents	*Parts*
Benzyl alcohol	11
Linalyl acetate ex bois de rose	11
Benzyl acetate	19
Benzyl benzoate	6
Ethyl phthalate	3
Jasmine châssis absolute, ex benzene (Schmoller & Bompard)	6
Laurine (Givaudan)	10
Phenylethyl alcohol	8
Linalool	9
Bergamot oil, sesquiterpeneless	4
Peru balsam	5
Ylang-ylang oil	3
Benzyl salicylate	5
Total	100

Formula 9 (Jasmine 22.0, Fantasy 3)

Constituents	*Parts*
Phenylethyl alcohol	22.0
Benzyl acetate	45.5
Cinnamic alcohol	3.5
Methyl anthranilate	2.0
Diethyl phthalate	5.5
Citronellol	3.5
Geraniol (synth, chem. pure, firmenich)	5.8
Ylang-ylang oil	2.0

Cananga oil	1.0
Isoeugenol	0.5
Benzyl salicylate	2.5
Ethyl salicylate	0.2
Cedrenol (ex cedarwood oil)	3.0
Virginia cedarwood oil	3.0
Total	100.0

Formula 10 (Jasmine Base 5)

Constituents	*Parts*
Cinnamyl acetate	0.8
p-cresyl isobutyrate	1.1
Methyl anthranilate	0.3
Benzyl alcohol	18.0
Benzyl acetate	18.0
Linalool	4.6
Benzyl propionate	3.1
Terpineol pure (BBA)	3.1
Amyl cinnamic aldehyde	15.3
Methyl eugenol	1.5
Linalyl acetate	7.5
Raldeine D(Givaudan)	3.8
Hydroxycitronellal	3.1
p-Methyl methylsalicylate	1.5
p-Cresyl phenylacetate	0.7
Phenylpropyl aldehyde	2.0
Aldehyde C.10	0.15
Indole	0.45
Tolu tincture	15.0
Total	100.00

Formula 11 (Jasmine base 5a)

Constituents	*Parts*
Jasmine Base No. 5	70.0
Jasmine d'Espagne (Givaudan)	13.0
Jasmine Frutal (Polak's Frutal works)	13.0
Jasmine châssis absolute ex benzene	
(Schmoller & Bompard)	4.0
Total	100

ORANGEFLOWER AND NEROLI

The difference between Orangeflower and Neroli are well known to the perfumers. Both are frequently used, and it could even be said that there are relatively few compositions into which the orangeflower note does not enter. Neroli is particularly valuable in eau de Cologne and related notes. Here are some (Formulae 12–13).

Formula 12 (Orange flower A)

Constituents	Parts
Linalyl acetate	7.1
Petitgrain oil, terpeneless	33.4
Flonol 160 (de Laire)	14.3
Indole 10%	4.7
Eugenol	2.2
Geraniol synthetic	5.6
Geranyl acetate	3.7
Brouts absolute (Robertet)	19.0
Ylang-ylang oil	1.2
Phenylethyl alcohol	4.7
Lentisque (Mastic) oil	0.8
Methyl anthranilate	2.4
Citral	0.9
Total	100.0

From the above example Brouts Absolute is extracted from petitgrain bigarade water or a mixture of this with orangeflower water.

Formula 13 (Neroli A)

Constituents	Parts
Petitgrain oil pays (Grasse)	31.5
Methyl anthranilate	25.0
Eugenol	18.5
Geraniol	6.0
Phenylethyl alcohol	4.25
Linalyl acetate	7.3
Indole 10%	4.7
Aldehyde C.8 10%	2.75
Total	100.00

This formulation is useful for blending with natural neroli, in certain applications, as a diluent. A suggested proportion is 1 part of natural neroli oil to 2 parts of Neroli A.

Flower perfumes such as honeysuckle, syringa, gardenia, and narcissus not only has orange blossom note but also has fruity note of apricot.

VIOLET

The classic note of violet is favourite of ladies. There are two examples of violet perfume. The simple violet (Nice or Toulouse) may be based on a blend of methyl ionone, alpha ionone, orris absolute or resinoid, violet leaf absolute and phenylethyl alcohol. A floral note, projected as it is by a trace of MNA aldehyde, is imparted by jasmine or cassie. The double or Parma violet is based on varying proportions of beta ionone, alpha ionone and methyl ionone, together with orris resinoid or concrete, a little phenylethyl alcohol, methyl octin carbonate, a trace of vetiver, and a small quantity of hydroxycitronellal and even, sometimes, bergamot. Very small amounts of violet leaf absolute may be included, but the green leafy note of Parma violets is much less accentuated. For imparting the necessary sweetness to the base of Parma violet, use may be made of a plum or mirabelle note.

Wood violets, with their rather earthy background, may be simulated by the inclusion of very small quantities of a pure grade of amyl acetate, which helps to provide the odour of damp earth and humus. Other humus notes may be similarly employed. (Formulae 14–15).

Formula 14 (Violet 1)

Constituents	*Parts*
Ionone alpha10	
Iralia (Firmenich)	10
Irisone extra (Givaudan)	10
Jasmine No. 2	20
Ysminia (Firmenich)	5.0
Jasmonis (Givaudan)	2.0
Violet leaf absolute	2.0
Orris absolute or Irophore (Firmenich)	1.1
Ylang ylang oil	1.0
Ylang ylang absolute	0.6
Cassie absolute	0.55
Rose No. 1	0.1
Neroli oil	0.06
Phenylethyl alcohol	1.0
Bitter orange oil	0.6
Oakmoss absolute (Robertet)	0.16
Folione (Givaudan)	0.03
Linalool	0.5

Bois de rose Brazil	3.0
Cinnamylal (Givaudan)	4.5
Heliotropin	2.0
Musk ambrette	0.1
Musk ketone	0.5
Vanillin	0.2
Civet tincture	16.5
Peru balsam tincture	8.5
Total	100.00

Formula 15 (Violet 2)

Constituents	Parts
Violet No. 1	60.0
Irisone Beta (Givaudan)	6.0
Iralia (Firmenich)	10.0
Cetone V (allyl ionone, Givaudan)	9.0
Ylang ylang absolute	4.0
Irophore (Firmenich)	1.0
Violet leaf absolute	2.6
Cassie absolute	0.8
Mimosa absolute	1.3
Jasmine absolute	1.2
Mirabelle (Firmenich) 1%	3.5
Total	100.00

The absolute oil of Parma violets, which had a delightful odour, is no longer commercial. The following formula given in formula 16 yields a product of fine odour which may be utilised as a 'reconstituted' absolute oil of Parma violets.

Formula 16 (Parma Violets Absolute Art)

Constituents	Parts
Violet leaf absolute (Schmoller & Bompard)	1.8
Irophore (Firmenich)	4.0
Violettone B extra fine 100% (Firmenich)	4.0
Parmanthene (Firmenich)	0.2
Benzyl benzoate	90.0
Total	100.0

The basic formula that follows gives a product of fine quality, which can be used in a multitude of different combinations in order to give effects approaching those of a little violet bouquet. The delicate odour of earth and humus that is so often found in a bunch of violets is mixture of amyl salicylate and pure amyl acetate, etc. (Formula 17).

Formula 17 (Violet base)

Constituents	*Parts*
Paracetone (Givaudan)	24
Beta-iso methyl ionone	32
Irisone extra incolore (Givaudan)	24
Beta methyl ionone	3
Boronia absolute	1
Ionone alpha	16
Violet leaf absolute	10
Orris absolute (Robertet)	10
Total	120

The concrete and absolute oils of Boronia, which contain ionone beta in its natural state, are much sought after in violet compositions, where they soften any harshness and add a tendency towards the Parma violet note.

ACACIA

This is not in fact a true acacia. The cassie is true species represented in perfumery. This perfume is widely accepted in warmer climates. It has fragrant and orangeflower like odour of the tree *Robinia pseudoacacia*. The composition of acacia has basic odourants such as, phenylethyl alcohol, cinnamic alcohol, benzyl acetate, hydroxycitronellal, methyl naphthyl ketone, methyl anthranilate, linalool, terpineol, anisic aldehyde ex anethole, and just a touch of apricot.

BROOM

Broom, genista or *genet* is a shrub very popular in the south of France and in Spain; particularly in Catalonia, where the hills of Barcelona and the Tibidabo are covered with *retama* (the Catalan name for broom) at the approach of summer. From the perfumery point of view, the base is one of orange blossom: methyl anthranilate, fleur d'oranger, etc., associated with methyl paracresol, methyl acetophenone, anisaldehyde and Grasse absolute of genet. Other constituents include alpha ionone, terpineol, linalool, phenylethyl alcohol, heliotropin, synthetic musk and olibanum.

CARNATION

There are three main types of carnation: (i) The carnation of Nice has a rather rosy, honeyed odour; (ii) The garden pink has a spicy odour recalling cloves and pepper; and (iii) The so-called sea pink (not to be confused with *Armoria maritima*) has the basic odour of carnation plus a note of phenylacetic aldehyde. In all carnations oleoresins are used as fixatives therefore they have slight resinous note.

Types of Carnation

Oeillet de nice

Here the basis is a blend of eugenol, isoeugenol and methyl eugenol, in varying proportions, accompanied by phenylethyl alcohol, benzyl salicylate, and sometimes amyl salicylate, vanilla, vanillin or ethyl vanillin. This base is sweetened with rose notes from rhodinol or rose de Mai absolute, carnation absolute or concrete and, above all, tuberose absolute. Unfortunately, as the latter has become very expensive, it must usually be replaced—as by a blend of synthetic tuberose with the natural material, or by a good ylang-ylang oil or ylang-ylang absolute. As aids to fixation there are benzyl isoeugenol, heliotropin and so-called heliotropin amorphous (a mixture of heliotropin, vanillin, anisic and benzoic acids).

Garden pink

Garden pink, known as *Oeillet mignardise*, has a slightly more clove-like odour. This indicate that the use of more eugenol and less isoeugenol in the base, together with a trace of clove oil. The vanilla base will preferably contain methyl vanillin (veratraldehyde). Oils of pimento, bay or black pepper will add characteristic note.

Sea pink

This small pink, which has a simple flower whose colour is actually mauve, is found growing wild on dunes by the sea. Its strong odour may be imitated by adding to a garden pink formula amyl salicylate, phenylacetic aldehyde and, if desired, a trace of hyacinth absolute.

Carnation notes have a vital role in all kinds of perfumery. Even they can be fully exploited in the area of masculine perfumes. It is not necessary to go for too familiar spiced–cologne complexes. A straightforward carnation note is much better for use. Some such formulae are shown in formulae 18–19.

Formula 18 (Base Claveline 1)

Constituents	Parts
Eugeno	51.35
Cananga oil	25.7
Ylang-ylang oil	2.55
Rose	5.6
Benzyl acetate	2.55
Orris absolute	4.25
Ceylon cinnamon oil	0.5
Nutmeg concrete	1.5
Nutmeg oil	1.0
Benzyl benzoate	3.45
Heliotropin	1.55
Total	100.00

Formula 19 (Base Claveline 2)

Constituents	Parts
Eugenol	61.5
Ylang-ylang oil	4.6
Cananga oil	4.0
Rose	4.6
Cinnamic alcohol	16.7
Benzyl acetate	2.5
Vetiver oil, Java type	1.25
Nutmeg oil	0.65
Labdanum absolute (Lautier)	0.20
Heliotropin	4.0
Total	100.00

CYCLAMEN

C. persicum and *C. europaeum* important species of cyclamen are scented.
The basic elements responsible for scents are hydroxycitronellal, linalool,
phenylethyl alcohol, terpineol, cinnamic alcohol, amyl cinnamic aldehyde,
styrallyl acetate, rhodinol, phenylacetic aldehyde, ionone, Lilial, Lyral and,
finally, the essential cyclamen aldehyde and some of the classic 'freshness'
components.

The interesting variations of the composition of this perfume could result
into novel and appreciating formulae. Think, for example, of its association
with a group containing cyclopentadecanolide, aldehyde MNA, incense and

vanilla. This type of association might well provide a very attractive ensemble. The cyclamen odour and its possible derivatives have not yet, in our view, been fully exploited – especially if the cumin note is one of the elements used in its formation.

FOUGERE (FERN)

The fern or *fougere* may be one of several thousand species, but in any case it will rarely have an odour other than a mild and vague effluvium of humus, greenery and earth. This perfume has been created by imaginative perfumers. Most of them have a more or less pronounced note of verdure and moss. Fern perfumes are divisible into two main groups : (i) based on the triad bergamot, Tonka bean tincture or coumarin, and a product of musky odour; and (ii) in which linalyl, terpinyl, geranyl and above all bornyl and isobornyl acetates, give a compound of more or less rustic or woodland character.

Example 20 (Fougere 1)

Constituents	*Parts*
Tree moss absolute	2.0
Bergamot oil	10.1
Lavender oil	10.1
Rhodinol	10.1
Rose M.S.	6.1
Patchouli oil	3.1
Geranium absolute incolore (Grasse)	4.1
Methyl ionone	12.15
Cyclopentadecanolide	1.0
Petitgrain oil, Paraguay	2.0
Coumarin	14.3
Musk ketone	6.1
Heliotropin	6.1
Phenylethyl salicylate	3.1
Methyl salicylate	1.0
Eugenol	1.0
Jasmine No. 1	3.1
Clary sage oil	1.35
Amyl salicylate	2.0
Musk tincture	1.0

| Tobacco (I.F.F.) | 0.2 |
| Total | 100.00 |

Formula 21 (Fougere 2)

Constituents	*Parts*
Musk ketone	0.4
Patchouli oil	1.0
Vetiver oil, Java	1.0
Lavender oil, Mitcham type	6.5
Lavender absolute	1.6
Bergamot oil	35.0
Linalyl acetate	2.2
Terpinyl acetate	2.0
Coumarin	31.4
Cyclopentadecanolide	0.2
Tobacco (I.F.F.)	0.2
Oakmoss absolute, Yugoslav	0.8
Mousse de Chene 1026 (Roure-Dupont)	0.2
Alliantone (Givaudan)	1.3
Allofloral (Firmenich) 10%	0.2
Rose No. 3	4.3
Palmarose oil	1.1
Geranium oil	4.7
Neroli A	4.7
Myrtle oil	0.2
Cuir 1073 (Firmenich)	0.8
Bornyl acetate	0.2
Total	100.00

When it is desired to impart a fruity note other than the classic peach, apricot, prune or mirabelle, one such note that can be extremely successful is strawberry, and this has in fact been incorporated in the above formula.

GARDENIA

The gardenia odour is reproduced using several components of tuberose. However, emphasis is given to the rose like tonalities. Thus in the first place consideration is given to ylang, jasmine, benzyl salicylate, phenylethyl alcohol, an orange blossom note, etc. Secondly, one turns to dimethyl benzyl carbinol, acetyl isoeugenol, hydroxycitronellal, rose de Mai absolute, benzyl acetate, terpineol, etc. And, above all, there is styrallyl acetate, which provides the peculiarly green and fruity note of gardenia, and gamma nonyl lactone.

No doubt that gardenia has been considered as a flower but it has prominent role in perfumery. It contributes fragrance to such perfumes as Millot's Crepe de Chine. Absolute oil of gardenia has been produced and is, when available, a remarkable and attractive product.

Formula 22 (Gardenia 1)

Constituents	Parts
Dimethyl benzyl carbinol	4.0
Styrallyl (methyl phenyl carbinyl) acetate	9.5
Neroli	2.5
Isoeugenol	4.0
Rose de Mai absolute	1.0
Ylang-ylang oil	7.0
Benzyl acetate	4.0
Jasmine absolute	1.0
Lilol (Givaudan)	6.0
Ionone alpha	10.0
Hydroxycitronellal	20.0
Phenylethyl alcohol	8.0
Phenylacetic aldehyde 50%	1.5
Citronellol	4.5
Cinnamic alcohol ex styrax	7.0
Coumarin	2.5
Heliotropin	7.5
Total	100.00

Formula 23 (Gardenia 2)

Constituents	Parts
Musk ketone	2.5
Musk ambrette	1.0
Jasmine de Provence (Descollonges)	1.0
Jasmine absolute	1.0
Rose de Mai absolute	3.5
Jonquil absolute	1.0
Tuberose absolute	3.0
Opopanax L.G. (Givaudan)	16.9
Muguet	10.2
Styrallyl acetate	12.0
Hydroxycitronellal	15.0

Ylang-ylang oil	5.0
Cinnamic alcohol ex styrax	2.0
Phenylethyl alcohol	2.0
Gardenia (Givaudan)	23.9
Total	100.0

At current raw material prices it would obviously be essential to replace most of the natural absolutes with the best 'artificials' available within costing limits.

HAWTHORN

The hawthorn perfume has a base of anisic aldehyde along with anisic alcohol and methyl acetophenone. One can use hydroxycitronellal, cinnamic alcohol, citronellol, linalool, terpineol for middle note and heliotropin and coumarin for lower note.

The noted English horticulturist, Graham Thomas, listed '*Crataegus*, all species' (including of course the common may or hawthorn, *C. oxyacantha*) among thirty or so flower scents that he considered 'unsuitable for sensitive noses'. The perfumer may here and in similar cases improve on nature by ignoring the methylamine or other off-notes present in the actual flowers and create, as it were, an idealised version of the natural floral odour.

HELIOTROPE

The heady, delightful and characteristic perfume of the heliotrope (*Heliotropium peruvianum* and other spp.) is duplicated chiefly with heliotropin and heliotropin amorphous; the latter being a mixture of heliotropin with vanillin etc. With these are associated benzoic and anisic acids, phenylethyl alcohol, ylang-ylang oil, rhodinol, cinnamic alcohol, anisic aldehyde, benzaldehyde (traces) and benzoin resinoid. A synthetic of hazel nut odour, used discreetly, will help to give an impression of the real flower. The slight note of greenery is obtainable with an oakmoss or, better still, a mignonette absolute or base. Another useful constituent is paramethyl hydrocinnamic aldehyde.

HONEYSUCKLE

The delightful fragrant of flowers are honeysuckles, *Loincera fragrantissima*, *L. brachypoda*, *L. caprifolium* and *chevrefeuille* species. Their general odour, sweet and diffusive but only slightly honeyed, can be simulated with basic constituents such as: muguet, orange blossom, rose and jasmine odourants; cinnamic alcohol, phenylethyl alcohol, terpineol, citronellol, linalool, hydroxycitronellal, paracresyl phenylacetate and anisic aldehyde. Small quantities of methyl cinnamate, methyl naphthyl ketone, phenylethyl acetate, iseugenol, and ylang-ylang, enter into consideration. To ensure the correct emphasis on the fruitiness of the natural flower odour, use may be made of an apricot note.

Formula 24 (Chevrefeveuille 1)

Constituents	*Parts*
Petitgrain oil, Paraguay	5.0
Neroli	6.55
Brouts absolute (Robertet)	1.0
Ionone alpha	4.85
Anisic aldehyde	2.5
Linalool	3.0
Terpineol	10.0
Geraniol	5.1
Hydroxycitronellal	3.8
Lilac	5.1
Phenylacetic acid	1.0
Bergamot oil	5.0
Benzyl acetate	5.1
Jasmonone	5.0
Chevrefeuille	35.0
Musk ketone	1.25
Musk ambrette	0.75
Total	100.00

HYACINTH

The odour of the hyacinth flower is often harsh and strident. Variants biased towards the rose or orange blossom are more acceptable. The basic odourants are phenylacetic aldehyde, hydroxycitronellal, cinnamic and phenylethyl alcohols, terpineol and phenylethyl acetate. The piquant note of hyacinth is due to dimethyl benzyl carbinol and its acetate, bromstyrole (traces), methyl octin carbonate, cinnamyl acetate; and, for rounding off the composition, when costing limits permit, hyacinth absolute. There was at one time, incidentally, an absolute of wild hyacinths (bluebells) available, but despite its remarkable odour properties it proved too expensive, owing to the high cost of gathering the flowers even before they were treated.

Hyacinth bases have found quite a number of applications in currently popular perfumes.

Formula 25 (Hyacinth 1)

Constituents	Parts
Lilac	10.0
Jacinthone (Givaudan)	20.0
Cinnamic alcohol	20.0
Neroli	5.0
Jasmonone	10.0
Jasmine châssis ex benzene	4.0
Hyacinth absolute	5.0
Phenylacetic aldehyde 50%	6.0
Hydratropic aldehyde	3.0
Dimethyl benzyl carbinol	2.0
Hydroxycitronellal	10.0
Rose de Mai absolute	5.0
Total	100

If it is found that the peculiarly piquant note of the natural flower, compared with the odour of horseradish, is not sufficiently emphasised, an added trace of bromstyrole will impart the desired effect.

IRIS

The scent of the iris flower is quite different from that of the dried iris (orris) root, and it varies according to species. Sometimes one observes a note of orange blossom, associated with the odour of orris root; sometimes almost the odour of hyacinth. The bases that are usually employed to duplicate this elusive floral scent are the various ionones and methyl ionones, terpineol, orris absolute and resinoid, anisic aldehyde, myristic acid, etc.

Now-a-days iris flower bases are rarely used. However, it is worth to use for producing such compositions that would no longer resemble iris, instead add to the number and novelty of the solid blending notes available.

LILAC

There is a good deal of variation in odour among the different species and forms of lilac. Persian lilac, for example (*Syringa persica*), resembles hyacinth, while purple lilac comes closer to heliotrope and vanilla; while yet another species recalls hawthorn. The basic components of a lilac perfume are heliotropin, cinnamic alcohol, terpineol, phenylethyl alcohol and hydroxycitronellal. Special notes are given by paracresyl acetate, phenylacetic aldehyde, anisic aldehyde and cinnamyl acetate. To emphasise the floral quality use is made of rose absolute, rhodinol, jasmine, etc. And, to impart 'thrust', benzyl acetate. The green stalk note is always to be observed in a spray of freshly cut lilac; and this may be contributed to the composition by *Vert de Lilas* notes: phenylacetic aldehyde dimethyl acetal, etc. The special note of *Persian lilac* is conveyed by means of a natural or synthetic hyacinth.

Since many years Lilac has become an essential element in some of the classic semi-floral, floral bouquet and sophisticated perfumes. Lilac really came into its own after discovery of hydroxycitronellal.

Formula 26 (Lilac base 1)

Constituents	Parts
Hydroxycitronellal	18.0
Cinnamic alcohol	1.7
Terpineol	7.4
Benzyl alcohol	17.0
Linalool	1.0
Phenylethyl alcohol	18.0
Styrax resinoid	2.55
Jasmine	9.1
Phenylacetaldehyde	0.5
Anisic aldehyde	7.0
Indole 10%	2.55
Heliotropin	15.2
Total	100.00

Formula 27 (Lilac 2)

Constituents	Parts
Lilol (Givaudan terpineol)	10.0
Phenylethyl alcohol	18.0
Hydroxycitronellal	33.5
Cinnamic alcohol (ex Styrax)	9.0
Menthol	0.7
Phenylacetic aldehyde 50%	0.1
Anisic aldehyde extra	0.5
Peru balsam	1.0
Jasmine	8.0
Heliotropin	11.0

Ethyl phthalate	8.2
Total	100.00

LILY-OF-THE-VALLEY

One may readily distinguish between the odours of *Muguet des bois* and green-house-grown lily-of-the-valley. A somewhat rough-and-ready way of compounding a lily-of-the-valley consists of mixing a lilac with a fresh rose compound. By gradually and systematically altering the proportions of both of these constituents, it is easy to arrive at a note similar to that of lily-of-the valley. Two features of the delicious *Muguet des bois*, the variety found in the woods, are difficult to reproduce: namely the sharp and very fresh lemony top note and the slightly green note at the end. The basic muguet constituents are :

Hydroxycitronellal, rhodinol, citronellol, terpineol, linalool, indole. To these must now be added the relatively recent para-tertiary Butyl alphamethyl hydrocinnamic aldehyde (e.g. Givaudan's Lilial) and 4 (4 Hydroxy-4-methyl pentyl)-3-cyclohexene-10-carboxyaldehyde (Lyral of I.F.F.). Both have become important additions to the perfumer's shelves. Lilial is particularly notable for its extreme persistence and unusual diffusive properties. Also deserving of mention is citronellyl oxyacetaldehyde (I.F.F.'s Muguet Aldehyde).

The sharp lemony note *de depart* can be obtained with fatty aldehydes, certain lactones, lemony esters, etc. Violet leaf absolutes, some of the 'greener' oakmosses, phenylacetaldehyde and its acetals, and other odourants may be used to give the necessary verdant background effect. Tarragon (estragon) has been suggested for a green note. Some of the carbinols and their esters are also of interest. To impart the impression of freshness of lily-of-the-valley, use can be made of traces of Ceylon citronella oil as 'citronellyl lactone'.

Formula 28 (Muguet No. 1)

Constituents	Parts
Citronellol	33.7
Rhodinol (Rhone-Poulenc)	11.3
Phenylethyl alcohol	24.7
Laurine (hydroxycitronellal, Givaudan)	14.6
Benzyl benzoate	6.7
Ethyl phthalate	4.5
α-Amyl cinnamic aldehyde	2.25

Indole 10%	2.25
Total	100.00

Cuminic aldehyde is included in next formula due to typical muguet note touch. This hint of cuminal has contributed to the success of a muguet compound. For a different effect it may be replaced by a good grade of cyclamen aldehyde. Lilial, Lyral, etc. may also be introduced on an experimental basis.

Formula 29 (Muguet 2)

Constituents	*Parts*
Hydroxycitronellal	34.5
Rhodinol	20.7
Phenylethyl alcohol	13.8
Alcohol C_{11} (enic)	2.0
Jasmine d'Espagne (Givaudan)	6.9
Linalyl acetate	3.5
Heliotropin	6.9
Terpineol	3.5
Cuminic aldehyde	1.35
Jasmonone No. 4a	6.85
Total	100.00

LINDEN (LIME BLOSSOM)

The lime (linden, tilleul) blossom has a captivating and powerful odour. A synthetic perfume sold under the name of lime is nothing more than hydroxycitronellal with something of a 'bouquet'. It is, however, quite possible to reproduce the odour of lime blossom, on a basis of hydroxycitronellal, linalool, linalyl acetate, ethyl linalool and its acetate, phenylpropyl alcohol, terpineol and farnesol. Small quantities of ethyl anisate, anisic aldehyde, cyclamen aldehyde, p.-methyl acetophenone, phenylethyl phenylacetate, geraniol and orange blossom etc., may be added. Persistence is accentuated with heliotropin, Peru balsam, etc. Aldehyde (lauryl) gives good results in the top note.

Formula 30 (Tilleul 1)

Constituents	*Parts*
Hydroxycitronellal	15.0
Linalool ex bois de rose	8.0
Farnesol	6.0
Bergamot oil	6.0
Tilleul absolute (Grasse)	18.0
Nerol	8.0

Geranium absolute incolore (Charabot)	3.0
Rose de Mai absolute	4.2
Lavender absolute	4.5
Tonka bean hyperessence (Charabot)	1.5
Petitgrain oil, French	8.5
Orangeflower	2.5
Jasmine absolute	6.15
Jasmine	6.0
Peru balsam resinoid	1.5
Phenylethyl phenylacetate	1.0
Aldehyde (lauric)	0.15
	100.00

MAGNOLIA

The thick foliage is intensely perfumed with enormous white, fleshy flowers, splashed with a few mauve or pink patches. What is most striking about the perfume of this flower is the contrast between the heavy, warm base note and the very lemony and fresh top note. The base note may be simulated with ylang and cinnamic alcohol, together with vanilla, vanillin or ethyl vanillin, with a dash of labdanum or incense. The body consists of hydroxycitronellal, nerol and various rosy notes. Finally, the top note is given by lemon, citral and alpha-amyl cinnamic aldehyde. Magnolia provides a good starting point for the development of new and original compositions.

MIGNONETTE (RESEDA)

The characteristic note of reseda or mignonette is obtained with methyl octin or heptin carbonate, methyl nonyl ketone, methyl nonyl acetaldehyde, dimethyl hydroquinone, phenylpropyl acetate, phenylethyl acetate and styrallyl acetate. The components of the body and base are: geraniol, alpha ionone, hydroxycitronellal, rhodinol, cyclamen aldehyde, heliotropin, coumarin and guaiyl acetate. Also of interest are diphenylethyl acetal, phenylethyl isothiocyanate, sweet basil oil and boronia absolute.

The following formula (Formula 31) incorporates some alternative suggestions.

Formula 31 (Reseda 1)	
Constituents	*Parts*
Reseda (Firmenich)	10.0
Reseda absolute hyperessence (Charabot)	27.1
Cyclamen aldehyde	10.9
Muguet	27.4

Orangeflower	0.5
Jonquil absolute (Robertet)	1.0
Rose	1.0
Ylang-ylang oil	3.15
Supercitron 50% (Lautier)	0.8
Superportugal (Lautier)	1.0
Jasmine	10.9
Jasmine absolute	2.1
Methyl ionone delta	1.5
Violet leaf absolute	0.15
Coriander oil	1.0
Musk ketone	1.5
Total	100.00

The mignonette note, blends well with rose and the already popular 'green' notes, and deserves to be considered for the elaboration of new-style perfumes.

MIMOSA

The base of mimosa is paramethyl acetophenone, isobutyl salicylate, phenylacetaldehyde and anisyl alcohol. These form the characteristic note, together with absolute oil of mimosa. There are three types of absolute oil: absolute from the flowers, absolute from the flowers and leaves, and absolute from the leaves. The body of the note is obtained with hydroxycitronellal and cinnamic alcohol, and the base consists of coumarin. Lemon, alpha amylcinnamic aldehyde and fatty aldehydes contribute to the top note.

NARCISSUS

The white flower of the valley narcissus, is extracted and supplied under the name of *Narcisse des Montagnes*. Benzyl acetate, cinnamic alcohol, hydroxycitronellal, terpineol, methyl anthranilate, ionone, eugenol, aurantiol, ylang-ylang and tuberose is used for producing adour of narcissus. The characteristic paracresylic shading of the narcissus note is then provided by paracresyl acetate and phenyl acetate, metacresyl acetate and phenylacetate, octyl acetate and phenylacetic aldehyde.

Formula 32 (Base Narceine 1)

Constituents	Parts
Benzyl acetate	12.0
Phenylethyl alcohol	20.0

p.Cresyl phenylacetate	7.5
p.Cresyl acetate	3.0
p.Cresol	0.5
Neroli	6.0
Phenylacetic aldehyde	2.5
Hydroxycitronellal	20.0
Ylang-ylang oil	19.0
Narcisse des Montagnes absolute	2.5
Musk ketone	5.0
Musk ambrette	2.0
	100.0

Formula 33 (Narcissus 2)

Constituents	Parts
Isoeugenol	6.0
Eugenol	6.1
Bois de rose oil, Brazil	2.4
Anisic aldehyde	1.2
Base Narceine	6.0
Phenylethyl alcohol	14.5
Terpineol	7.3
Hydroxycitronellal	14.5
Cinnamic alcohol	3.6
Methyl ionone	3.9
Bergamot oil	7.3
Neroli	7.3
Benzyl acetate	7.3
Jasmine	7.3
Mastic (lentisque) oil: Robertet	0.5
Musk ketone	1.2
Indole 10%	2.4
Heliotropin	1.2
Total	100.0

The narcissus note has been stylised in some well-known and successful perfumes, and lends itself admirably to this type of work.

NARDO

Nardo is a Spanish flower quite similar to the Tuberose. Its perfume is heavier than that of the tuberose from the south of France, and above all its background note is much more fruity than that of the tuberose and the French gardenia. Nardo perfume is very fashionable in all the Spanish-speaking countries. Its bases are the same as those found in the tuberose and the gardenia, with a few lighter constituents than in the first two: benzyl acetate, cinnamic alcohol, linalool, hydroxycitronellal, terpineol, cyclamen aldehyde and, above all, there is more accent on the peach note.

NEW-MOWN HAY

The natural odour of new mown hay, though delightful and exhilarating, is not really a flower perfume, although the small blossoms of such flowers as woodruff and alyssum may contribute to it. In some respects it has points in common with fern or fougere and with clover or trefle, but essentially its odour is that of coumarin and tonka beans. There are several interesting foin coupe absolutes and a foin coupe distilled oil available; and, also of Grasse origin, flouve absolute, extracted from the flowering tops of a special grass, known as *flouvre odorante*. Other constituents of new-mown-hay perfumes may include lavender, lavandin, clary sage and rosemary oils; acetanisole, dimethyl acetophenone, ethyl anisate, anisic aldehyde and alcohol, dimethyl hydroquinone, heliotropin, linalool; and small amounts of sandalwood, patchouli, olibanum.

NICOTIANA

Nicotiana tabacum, known as nicotiana or tobacco flower, perfume is most pronounced in the evening. Its odour may be simulated by introducing variations on the jasmine theme, i.e. by adding to a typical jasmine flowers base such lily components as Lilial or Lyral, cyclamen aldehyde, etc., together with small quantities of p.cresol methyl ether or p.cresyl acetate.

OPOPANAX

This is not a flower perfume but is included here because of its general applicability, like many flower perfumes, in forming part of the background structure of a complex composition. In this respect is much resembles the 'Amber Art.', so frequently found as a perfume formula constituent. Opopanax resins and resinoids come in a variety of types and are of diversified and sometimes unknown origins. Many opopanaxes sold under the name are nothing but mixtures, even though some are exceedingly well contrived.

The following formula for an Opopanax Art (i.e. 'artificial') is for a blend that has given consistently good results in many different applications.

Formula 34 (Opopanax Art. L)

Constituents	Parts
Coumarin	9.3
Vanillin	17.6
Heliotropin	5.7
Musk ambrette	3.4
Musk ketone	2.3
Patchouli oil	1.7
Labdanum brun absolute (Lautier)	4.6
Mandarin oil	2.3
Bergamot oil	9.3
Petitgrain oil, French	1.1
Castoreum tincture	3.4
Olibanol (Robertet)	1.1
Hydroxycitronellal	4.6
Geranium oil, African	1.1
Vetiver Bourbon oil	4.6
Sweet orange oil, Guinea	4.6
Tolu balsam tincture	5.7
Diethyl phthalate	17.6
Total	100.0

ORCHID (ORCHIDEE)

Orchid, poppy and clover (trefle) are examples of perfumes that owe more to the perfumer's imagination than to nature. The base of orchid in perfumery is a combination of the salicylates, formed by the amyl, isobutyl and benzyl esters – the relative proportions varying. Very small amounts of the methyl ester may also be incorporated. Besides these products, there are several others worthy of special mention, i.e. alpha ionone, ylang-ylang, bergamot, neryl butyrate, rose components, cyclamen aldehyde, terpineol, paracresyl phenylacetate, aldehydes etc.

PANSY

Pansy (pensée) represents for thought. Its scent, is light and fresh, reminding one more name heart's-ease. The perfumer chiefly recognises in this lowly member of the violet family, Viola tricolour, a violet odour with a rather special green note.

The foundation of a simulated pansy perfume was essentially similar to that of the violet. It included alpha and beta ionones, methyl ionones, neryl acetate and, optionally, orris absolute. These are associated with phenylethyl alcohol, rose, jasmine; violet leaf absolute, the alkyne carbonate and other green notes; and a sweet persistence is assured by including a bouquet of 'crystallines' harmonising with the general tone.

PEONY

Peony (Fr. *pivoine,* Germ. *pfingstrose*) is not such a well known fragrant flower, perhaps, as it deserves to be. Its varieties have in the main two distinct types of odour. One of the red peonies is not very pleasant-smelling and recalls in fact the harsh green scent of the dahlia. On the other hand the white peony, with a few pink patches, and the completely pink peony, have a deliciously fresh odour. When smelling this perfume one has often had the impression of being in the presence of an excellent citronellol. The basic products for simulating the peony odour are primarily the rose-like odourants: cintronellol, rhodinol, phenylethyl acetate, dimethyl benzyl carbinol, the butyrate and isobutyrate esters of the rose alcohols, and linalool and ethyl linalool. For the base, there are the derivatives of guaiac wood oil, alpha ionone, styrax, ylang-ylang oil, heliotropin and vanillin. The green note may be given by styrallyl acetate associated with paracresyl phenylacetate and cinnamyl acetate. The peony is yet another of those flower types which could contribute rather more than they currently do to novel odour combinations.

PHLOX

Phlox, a herbaceous plant which has become well acclimatised in Europe. It includes many odourous species and varieties. The typical phlox is the one having scent of violet and heliotrope. This can be simulated by using a violet base rich in methyl ionone, fixing it with heliotropin and adding a touch of a hazel nut or *noisette* compound. Although the phlox deserves further investigation as a potential flower component in perfumery, the fact must not be overlooked that many people find its perfume disagreeable.

SWEET PEA

The sweet pea odour is highly regarded in Anglo-Saxon domestic circles. Basic odourants for its simulation include: phenylethyl phenylacetate, benzyl salicylate, benzyl isoeugenol, methyl acetophenone, and the various orangeflower notes. Hydroxycitronellal, cinnamic alcohol, terpineol, linalool and traces of rose de Mai absolute – are added.

Formula 35 (Pois de Senteur 1)

Constituents	Parts
Coumarin	2.85
Neroli	2.0
Rose	4.3
Sandalwood oil, East Indian	5.7
Bergamot oil	5.7
Verbena oil, art	0.85
Lavender oil	1.4
Ionone alpha	1.4
Jasmine d'Espagne (Givaudan)	8.6
Sweet orange oil	8.6
Cinnamic alcohol	14.3
Hydroxycitronellal	11.45
Phenylethyl alcohol	11.45
Rhodinol	4.3
Phenylacetic acid	3.4
Methyl anthranilate	2.3
Heliotropol (Firmenich)	2.85
Jasmine enfleurage absolute (Lautier)	5.7
Phenylethyl phenylacetate	2.85
Total	100.00

If this type of compound seems a trifle harsh, a small addition of Rose de Grasse absolute will round off the edges and enhance the desired honeyed note.

Formula 36 (Sweet pea 2)

Constituents	Parts
Rose	13.2
Sandalwood oil, Mysore	2.3
Bergamot oil	10.6
Hydroxycitronellal	13.3
Phenylethyl alcohol	13.3
Benzyl acetate	1.3
Wardia (Firmenich)	5.3
Methyl ionone	4.3
Jasmonone	16.0
Pois de Senteur (Firmenich)	15.6
Vetiver oil	0.8
Benzyl propionate	4.0
Total	100.0

SYRINGA (PHILADELPHUS)

The base of the so-called syringa perfume (the true syringa being lilac) is methyl acetophenone. This syringa or mock-orange, which is in fact *Philadelphus coronarius*, has an odour that resembles orange blossom and has, in its finer varieties, a fresh lemony top note.

'The components, are methyl acetophenone, hydroxycitronellal, cinnamic alcohol, ylang-ylang, alpha-ionone, phenylethyl alcohol, products with an orangeflower note, terpineol, anisic aldehyde, isoeugenol. The freshness of the *depart* is imparted by lemon oil, fine quality citral and/or octyl or decyl aldehyde.

The perfume of the mock-orange, and particularly the lemon-scented variety, is heady, fresh, very agreeable. This is a floral odour that could be effectively used as the flower bouquet, or part of the flower bouquet, of new and attractive perfumes.

Formula 37 (Syringa 1)

Constituents	Parts
Phenylethyl alcohol	15.0
Folia (Bertrand freres)	2.0
Methyl anthranilate	0.5
Isoeugenol	5.0
Syringa aldehyde (p.-tolyl acetaldehyde)	4.0
Anisic aldehyde ex anethole	0.5
Bois de rose oil, Brazil	2.5
Muguet	2.5
Cassie absolute (Robertet)	1.5
Hydroxycitronellal	15.0
Jasmonis (Givaudan)	13.5
Jasmonone	15.0
Jasmine absolute	2.5
Amyl salicylate	4.0
Heliotropin	9.0
Indole 10%	5.0
Aldehyde	2.5
Total	100.0

TREFLE (CLOVER)

Trefle or clover, like orchidee (orchid), is one of those important half-floral, half-herbaceous odours founded on the salicylates. One of the famous classic perfumes of the past was the Trefle Incarnat of Piver. Trefle compounds are conventionally based on the salicylates in association with coumarin, lavender, bergamot; with suitable flower notes and aldehydes.

Formula 38 (Trefle 1)

Constituents	Parts
Amyl salicylate	20.0
Isobutyl salicylate	22.0
Benzyl salicylate	8.0
Bergamot oil	12.0
Coumarin	17.0
Isoeugenol	2.5
Vetiver oil	1.5
Ionone alpha	2.5
Ylang-ylang oil	2.5
Oakmoss absolute	0.5
Methyl nonyl acetaldehyde (1%)	0.6
Musk ambrette	2.2
Lavender oil	2.0
Rose oil art (Bulgarian type)	1.5
Jasmine absolute art	2.5
Hydroxycitronellal	2.5
Birch bud oil	0.2
Total	100.0

TUBEROSE

The tuberose, has nevertheless found widespread use in perfumery. This applies both to the natural absolute and to 'tuberose artificial'. As a dominant theme it is especially appreciated in eastern countries and in territories under Spanish influence. More generally it finds application as a floral support capable of conferring body and persistence. Its basic components are: Ylang-ylang oil, phenylethyl or phenylpropyl alcohol, benzyl salicylate and acetate, benzoin resinoid, Peru balsam, tuberose absolute, tuberose pomade absolute and – this is indispensable – methyl salicylate or oil of wintergreen in very small quantities. To give a flowery touch one uses lauryl aldehyde or other aliphatic aldehydes, in association with orange blossom, jasmine, and a mere suggestion of styrallyl acetate.

Formula 39 (Base Tuberose 1)

Constituents	Parts
Jasmine absolute ex chassis (benzene)	1.6
Tuberose absolute	1.2

Wintergreen oil, natural	4.0
Methyl salicylate	8.0
Benzyl acetate	24.5
Benzyl alcohol	22.5
Methyl anthranilate	22.2
gamma Nonyl Lactone	8.0
Peach synthetic (Givaudan)	8.0
Total	100.0

Formula 40 (Base Tuberose 2)

Constituents	Parts
Tuberose absolute	1.0
Hydroxycitronellal	29.5
Benzyl benzoate	3.0
Isobutyl salicylate	10.0
Methyl salicylate	3.0
Ionone alpha	20.0
Aldehyde (lauric)	10.0
alpha-Amyl cinnamic aldehyde	7.0
Heliotropin	16.5
Total	100.0

VERBENA

Men pleasing odour is Verbena. It is simple, refreshing and comes midway between what one may term 'hygienic' perfumes. It is used in toilet preparations for personal hygiene, and the more frankly perfumed toilet waters of the Cologne type. Verbena oil, is distilled from the true verbena.

Formula 41 (Verveine 1)

Constituents	Parts
Supercitron (Lautier)	3.5
Grapefruit oil	15.55
Cassie synthetic (Givaudan)	7.2
Geraniol extra	9.1
Geranium Bourbon oil	6.0
Verbena oil, Grasse, rectified	31.1
Citral extra	15.55
Verbenol (Firmenich)	3.0
Bergamot oil	3.0
Sandalwood oil, East Indian	1.2
Opopanax	4.8
Total	100.00

WALLFLOWER

In nineteenth-century perfumery, especially at the beginning, the wallflower (giroflee) enjoyed great favour. The odour of this flower is fresh and agreeable. The base consists of eugenol, isoeugenol and acetyl isoeugenol, associated with benzyl salicylate, amyl salicylate and anisic aldehyde, to which there may be added small quantities of oils of cloves, pimento, bay, or even pepper, and a trace of methyl paracresol. The 'filler' consists of terpineol, hydroxycitronellal, cinnamic alcohol, phenylethyl alcohol, linalool, geraniol, rhodinol, nerol, dimethyl benzyl carbinyl acetate. There may be added an ionone or its derivatives and for the flowery touch, Lilial (Givaudan) and neroli too. The lower notes comprise heliotropin, vanillin, ethyl vanillin, methyl naphthyl ketone, and a little mace or mastic oil or guaiyl acetate.

The wallflower note, orientated towards the basic carnation-salicylate-auhepine base, could well contribute to the development of an interesting perfume.

WISTARIA

The wistaria, wisteria or glycine, grown for its climbing habit and its attractive drooping racemes of blossom. It has a sweet and diffusive odour of rather honeyed type, suggesting orangeflower, lily and mimosa. The odourants chiefly entering into the composition of artificial wistarias are hydroxycitronellal, benzyl salicylate; geraniol, phenylethyl alcohol and other rose alcohols; ylang-ylang, mimosa absolute, isoeugenol, terpineol, methyl naphthyl ketone, methyl anthranilate, isobutyl phenylacetate, paramethyl acetophenone, p.-methyl totyl ketone, p.cresyl acetate.

YLANG-YLANG

The essential oil of ylang-ylang is one of the most widely used of natural materials in perfumery. In old days, there were three types of natural products: ylang from the Bourbon islands of Madagascar, ylang from Manilla and cananga from the Java islands. To-day ylang from Manilla has almost entirely vanished. It had a very special note, more like orange-blossom than ylang from Reunion and lighter, too, with the base more like tuberose. Ylang is so useful that there have been numerous attempts at synthesising it. Some 'artificials' are in fact quite close in odour to natural ylang. For Bourbon ylang, the basic products are; methyl paracresol and paracresol. These very powerful products are only present in small quantities. Associated with methyl benzoate, benzyl benzoate, eugenol and isoeugenol, they form the characteristic part of this note. The body consists of terpineol, geraniol, linalool and alpha ionone.

This original note has certainly not been fully utilised; it is powerful and heady, especially in the case of Manilla ylang, which is less violent but otherwise similar. Here must be added an orange blossom note (natural or synthetic), methyl anthranilate, and a tinge of methyl salicylate or oil of wintergreen. As a rule, when it is desired to give a 'bouquet' to a composition, it is considerably more advantageous to use the natural essential oil – and, moreover, the finest quality that costing will permit.

There are very few fine quality perfume compositions in which ylang-ylang does not or could not enter as a valuable constituent. Here is a formula for an artificial version of Ylang-Ylang Bourbon:

Formula 42 (Ylang 1)

Constituents	Parts
Bois de rose oil, Brazil	10.5
Linalool ex bois de rose	15.7
Linalyl acetate	8.65
Benzyl acetate	11.35
Vanilla tincture	10.55
Methyl paracresol	6.9
Palmarosa oil	8.6
Nerol (Lorena, Firmenich)	2.6
Benzyl alcohol	4.3
Orris concrete (10% in benzyl benzoate)	5.2
Isoeugenol	4.35
Peru Clair (Lautier)	3.5
Coriander oil, Russian	3.5
Methyl benzoate	1.75
Estragon oil (P. Chauvet)	0.85
Methyl anthranilate	0.8
Methyl salicylate	0.8
Cananga oil, terpeneless	0.1
Total	100.00

Chapter 7

Sophisticated/Fantasy Perfumes and their Formulation

INTRODUCTION

Perfumes of Fantasy (as we may call them for want of a better name) are roughly divisible into six categories :

(1) Floral perfumes;

(2) Aldehydic perfumes;

(3) Chypre types;

(4) Oriental types;

(5) Green types; and

(6) Perfumes with a distinguishing dominant note.

The independent notes that characterise perfumes in each separate category, and tonality also plays an important part in all these types. This may be, in each group, one of three, namely light, medium and heavy.

The formulae for general and specific nature can be given. Here is a representative of the very interesting bergamot-vanilla bases :

Bergamot oil	100
Vanillin	5
Aldehyde C_{12} (MNA) 10%	1
Aldehyde C_8, 10%	0.3

In the formula that follows the cedrat oil, once a natural product, may be replaced by the now conventional artificial blend of lemon, bergamot and orange oils.

Formula 1

Constituents	Parts
Aldehyde C_8	0.50
Aldehyde C_9	1.25
Aldehyde C_{10}	2.50
Aldehyde C_{11} (undecylenic)	1.25
Aldehyde C_{12} (lauric)	2.50
Aldehyde C_{12} (MNA)	1.50
Bergamot oil	12.50
Bergamot oil, sesquiterpeneless	6.20
Superorange (Lautier) 50%	10.00
Supercitron (Lautier) 50%	1.90
Lemon oil	1.25
Cedrat oil	3.75
Neroli	3.10
Folia (Bertrand freres)	0.60
Ylang-ylang oil	37.50
Opopanax L.G. (Givaudan)	1.25
Costus absolute	0.60
Musk tincture	2.50
Rose	6.20
Sandalwood oil	3.15
	100.00

As orange curacao oil can not be obtained which has characteristic note hence the following formula is useful.

Formula 2 (Curacao Base)

Constituents	Parts
Bitter orange oil (Guinea)	37.50
Bitter orange oil, deterpenated	7.45
Anethole	4.65
Citral	3.70
Coriandrol (from Russian coriander oil)	2.70
Aldehyde C_8	2.00
Aldehyde C_9	5.00
Aldehyde C_{10}	7.00
Aldehyde C_{11} (undecylenic)	18.30
Aldehyde C_{12} (MNA)	3.00
Aldehyde C_{12} (lauric)	8.70
	100.00

Leather or 'cuir' bases have continued to be popular and are widely utilised in perfumery at the time of writing. They are indispensable in certain fashionable perfumes and are particularly of interest in perfumes for men.

Formula 3 (Perfume Cuir base 1)

Constituents	*Parts*
Bergamot oil	1.5
Framboisis (de Laire)	7.5
Coromia (Givaudan)	40.0
Corinal (Firmenich)	14.0
beta-Methyl ionone	4.5
Jasmin	2.4
Mandarin oil	1.5
Cedrol	12.0
Lilac	9.0
Musk ketone	1.8
Musk ambrette	1.8
Neroli	0.9
Coriandrol	0.3
Cedrat oil	0.8
Vanillin	1.5
Dimethyl phthalate	0.5
	100.0

The woody note is often employed in association with other notes, not only to confer its own characteristics but to soften and round off the finished composition.

Formula 4 (Perfume woody base 1)

Constituents	*Parts*
Linalyl acetate	27.00
Bergamot oil	12.35
Rose de Grasse absolute	12.20
Jasmin absolute	6.90
Bois de rose oil, Brazil	6.00
Lilac	5.70
Muguet	5.25
Vanillin	4.95
Coumarin	4.95
Sandalwood oil	3.90
beta-Methyl ionone	2.70
Neroli oil	1.95

Bitter orange oil, Guinea	1.95
Labdanum absolute	1.20
Patchouli oil	1.05
Vetiver oil, Bourbon	1.05
Hyperessence oranger (Charabot)	0.90
	100.00

The moss note plays a very important part in perfumery. It sustains the head notes of 'green' compositions, which are currently fashionable; but long before this it found other applications.

In addition to the newer 'green' odourants marketed as pure chemicals, others have been introduced as proprietary bases. In most instances these bases have been well contrived, so as to permit their introduction into standard formulae without causing any undue imbalance or interference. They nevertheless give rise to effects that are quite distinctive and different from those that preceded their introduction. The auxiliary notes so introduced may of course be pleasing or displeasing, but the possibility of producing novel and interesting effects cannot be overlooked.

As per taste moss note may be similar to fruity. Moss Base M.2 can very easily be reorientated in this fashion.

A certain number of odourants are capable of suggesting the sea and the open air, faintly suggesting the odour of ozone. This type of perfume could well be the subject of tomorrow's success story.

Formula 5 (Parfume QH 1)

Constituents	*Parts*
Lilac	50.00
Hyacinth absolute	0.10
Aldehyde C_{12} (MNA) 10%	0.25
Aldehyde C_{12} (lauric) 10%	0.10
Citronellyl lactone (I.F.F.) 10%	0.05
Amyl salicylate	5.00
Sandalwood oil	1.55
beta-Methyl ionone	5.05
Benzyl acetate	3.90
Rose	4.00
jasmin absolute	2.00
Cedrat oil	7.00
Bergamot oil	5.00
Ylang-ylang oil	10.00
Musk ambrette	1.00

Musk ketone	1.00
Vanillin	2.00
Ambrophore (Firmenich)	2.00
	100.00

The next flower perfume (AO) is characteristic of a kind of bouquet which had a great success at the end of the nineteenth century. Bouvardia, which has a wide acceptance to perfumer, is essential part of flower perfume. Therefore, the example for a base given below in formulae.

Formula 6 (Bouvardia BM)

Constituents	*Parts*
Ylang-ylang oil	20.00
Geraniol extra	5.00
Jasmin absolute	3.75
Base Tuberone	2.50
Neroli	12.50
beta-Methyl ionone	25.00
Bois de rose oil, Brazil	25.00
Hyperessence orangeflower (UOP)	6.25
	100.00

Formula 7 (Parfum AO)

Constituents	*Parts*
Mimosa flower absolute	1.7
Rose de mai absolute	3.4
Hyperessence orangeflower	5.0
Narcisse des Montagnes absolute (Camilli)	2.0
Base tuberose	4.0
Jasmin absolute	3.4
Jasmonone	1.0
Ylang-ylang oil, Manilla	3.0
Heliotropin	2.0
Acacia (Firmenich)	8.5
Bouvardia BM	4.0
Bouvardia 100% (de Laire)	3.0
Bergamot oil	10.0
beta-Methyl ionone	8.0
Anisic aldehyde	6.0
Methyl anthranilate	1.0
Linalyl acetate	2.0

Ionone alpha	2.0
Ambrophore (Firmenich) 3%	20.0
Civet tincture	10.0
	100.0

Formula 8 (Parfum F)

Constituents	*Parts*
Ylang-ylang oil	12.70
Jasmin absolute	3.40
Narcisse des Montagnes absolute (Camilli)	2.10
Lilac	21.50
Paracetone (Givaudan)	22.70
Gardenia 9058 (Givaudan)	10.25
Heliotropol (Firmenich)	4.20
Jasmin	4.20
Grapefruit oil	5.10
Bergamot oil, sesquieterpeneless	3.40
Neroli	3.40
Base cuir	2.10
Coumarin	1.70
Boronia absolute oil	1.00
Rose	1.00
Pollenol (Firmenich)	0.50
Hyperessence reseda (Charabot)	0.15
Hyacinth absolute (Polak's frutal)	0.15
Aldehyde C_{12} (MNA) 50%	0.05
Phenylacetaldehyde dimethylacetal 50%	0.05
Cyclopentadecanolide	0.35
	100.00

Formula 9 (Parfum HB)

Constituents	*Parts*
Jasmin	7.50
Bergamot oil	10.00
Lavender oil	7.45
Opaslaunax (Ets. Hasslauer)	7.50
Methyl anthranilate	6.70
Orange oil	4.10
Lilac	3.60
Rose	3.00
Methyl ionone	2.20
Jasmin absolute	2.20

Muguet	2.00
Neroli	1.40
Claveline base	1.30
isoEugenol	1.30
Sandalwood oil	1.30
Heliotropin	1.30
Ambreine (Firmenich)	0.80
Vanillin	0.80
Coumarin	0.75
Orangeflower water absolute	0.80
Ylang-ylang oil	0.65
Anisic aldehyde	0.45
Cedarwood oil, Virginia	0.40
Merisia (Descollonges)	0.40
Orris absolute P.V.	0.20
Benzyl acetate	0.20
Zdravetz oil, 10%	0.90
Cinnamon oil, Ceylon 10%	0.40
Framboisis (de Laire) 10%	0.20
Peru balsam tincture	13.00
Civet tincture	13.00
Musk (tincture)	4.20
	100.00

ALDEHYDIC PERFUMES

Formula 10 (Parfume VN)

Constituents	*Parts*
Bergamot oil	5.30
Bitter orange oil	4.25
Aldehydic base	3.00
Ambreine (Firmenich)	3.25
Methyl ionone	4.25
Cinnamic aldehyde	3.40
Jasmin de Provence (Descollonges)	3.00
Petitgrain oil, French	2.20
Jasmin absolute	2.20
Sophora (Givaudan)	2.80
Lavender	1.40
Neroli	1.15
Jonquil absolute (Charabot)	0.50
Hydroxycitronellal	2.70
Sandalwood oil	0.60
Benzoin gomodor (UOP)	0.60

Opopanax vrai resine (Hasslauer)	3.00
Elemi resinoid	2.00
Rose	2.20
Civet tincture	30.60
Musk tincture	12.45
Peru balsam tincture	5.15
Castoreum tincture	4.00
	100.00

Formula 11 (Parfum C)

Constituents	Parts
Rose	0.40
Lilac	4.20
Jasmin	6.70
Neroli	0.15
Hydroxycitronellal	6.30
Terpineol	3.20
Cinnamic alcohol ex styrax	0.50
Methyl ionone	3.20
Amarante (Givaudan)	5.00
Oakmoss absolute	1.25
Opopanax	3.75
Base	2.60
Liquidambar	0.10
Mousse de Saxe (de Laire)	5.00
Thyme oil	0.30
Vetiver oil, Java	1.90
Patchouli oil	3.00
Sandalwood oil	3.00
Coriander oil, Russian	4.00
Cedrat oil	1.00
Bitter orange oil	2.00
Lavender oil	3.00
Bergamot oil	10.00
Linalyl acetate	5.00
Ambreine S (Firmenich)	4.50
Dimethyl phthalate	0.60
Ethyl vanillin	0.80
Musk ketone	0.50
Musk ambrette	1.00
Cyclopentadecanolide	3.00
Coumarin	3.00

Aldehyde C_{11} (undecylenic)	0.42
Aldehyde C_{12} (lauric)	0.36
Aldehyde C_{12} (MNA)	0.27
Tincture of civet	1.50
Tincture of castoreum	8.50
	100.00

Fairly frequent reference is made in these formulae to the natural resin, liquidambar. This balsamic product of very dark bituminous aspect is not used as such but is appropriately treated to yield three different fractions. The first of these has the strongly 'gassy' styrene odour; the second has a labdanum odour and an 'animal' note; while the third has the animal note in an even more marked degree. In fine perfumery only the second and third fractions are normally used and formulae are called as liquidamber II and liquidamber III. The raw material can be suitably treated and used in perfumery.

Formula 12 (Parfum SP)

Constituents	*Parts*
Opopanax	2.20
Opopanax (de Laire)	0.90
Cyclamen aldehyde	0.10
Amyl salicylate	0.25
Costus resinoid (Camilli)	0.15
Bergamot oil	3.20
Patchouli oil	2.40
Grisambrol (Firmenich)	0.80
Labdanum oil (Schmoller)	0.25
Moss Base	0.40
Oakmoss absolute superessence (Yugoslav)	0.90
Liquidambar	0.05
Mousse (Roure Dupont)	2.15
Musc (Rhone-Poulenc)	0.10
Musk ketone	0.60
Cyclopentadecanolide	0.15
Ethyl vanillin	0.10
Thibetine (Givaudan)	0.60
Animalis (Synarome)	0.30
Civette (Rhone-Poulenc)	0.10
Elecampane oil	4.50
Sandalwood oil	0.20
Methyl ionone	3.20
Jasmonone	5.70

Gardenia (Givaudan)	19.70
Gardenia	4.50
Cardamom oil, Naardenised (Naarden)	0.25
Hydroxycitronellal	2.40
Corona (Roure Dupont)	1.70
Hyacinth absolute (Polak's Frutal)	1.60
Phenylacetaldehyde	0.05
Camomile absolute (Schmoller)	0.15
Lavender oil, Barreme (42% esters)	1.30
Linalool ex bois de rose	4.50
Ylang-ylang oil	0.25
Styrallyl acetate	2.10
Aldehyde C_8	0.10
Aldehyde C_9	0.10
Aldehyde C_{10}	0.60
Aldehyde C_{11} (undecylenic)	0.60
Aldehyde C_{12} (lauric) 50%	0.70
Opopanax resin tincture	6.30
Peru balsam tincture	20.00
Vanilla tincture	1.10
Castoreum tincture	2.70
	100.00

CHYPRE TYPES

Chypre perfumes have always found great favour among women and men. The characteristic note of the Chypre is a safrole or isosafrole. Chypre has also conventional notes of bergamot, oakmoss, civet, linalyl acetate, amyl, salicylate, rose, neroli, vetiver, sandalwood, labdanum and the little rustic or herbal notes of red thyme, tarragon and basil.

Many Chypre variants have been prepared, each with its slightly different note or emphasis which distinguishes it from the other members of its class. This can be referred as 'special Chypre', which is safe to produce. During the past few years aldehydic chypres have made headway. The following examples are given as typical of this class.

Formula 13 (Parfum)

Constituents	*Parts*
Rose	1.80
Ysminia (Firmenich)	15.00
Jasmine absolute	15.00
Oakmoss absolute superessence,	

Yugoslav (Schmoller)	0.60
Bergamot oil	4.20
Oakmoss absolute (Camilli)	4.35
Jasmine	4.00
Geranium sur rose oil	1.80
Methyl ionone	11.30
Vetiver oil	3.60
Sandalwood oil	1.30
Linalool ex bois de rose	1.80
Dianthine (Firmenich)	7.60
Eugenol	0.75
Hydroxycitronellal	0.75
Gardenia 9058 (Givaudan)	4.00
Costus absolute 10%	2.10
Mace oil	0.15
Florizia (Firmenich)	3.15
Tincture of Musk, 3%	7.60
Tincture of Civet, 3%	1.00
Musk ambrette	0.70
Musk ketone	1.40
Coumarin	0.30
Vanillin	0.15
Aldehyde C_{10}, 1%	0.80
Aldehyde C_{11} (undecyclenic) 1%	1.20
Aldehyde C_{12} (MNA) 10%	3.60
	100.00

The following Chypre base is typical aldehyde.

Formula 14 (Base Chypre H)

Constituents	*Parts*
Coumarin	8.70
Vanillin	4.60
Ethyl vanillin	2.90
Heliotropin	1.80
Methyl ionone	1.20
Musk ketone	0.75
Rose	0.60
Orange oil, bitter, Guinea	4.50
Geraniol extra	2.90
Bois de Rhodes oil (Chiris-UOP)	1.00
Noisette (de Laire)	0.20
Sandalwood oil	2.80

Benzoin Supergomodor (Chiris-UOP)	0.45
Liquidambar	2.90
Labdanum Clair (Lautier)	5.80
Linalool ex bois de rose Cayenne	0.75
Linalyl acetate ex bois de rose	2.90
Terpinyl acetate	8.80
Benzyl acetate	2.90
Vetiver acetate	0.09
Estragon (tarragon) oil 5%	0.54
isoButylquinoline 5%	0.75
Ysminia (Firmenich)	1.20
Bergamot oil, sesquiterpeneless	34.97
Bergamot oil	6.00
	100.00

Modified chypre perfume formula with a peach type of top note is given below.

Formula 15 (Parfum V.W.)

Constituents	*Parts*
Ysminia (Firmenich)	3.00
Wardia (Firmenich)	2.00
Benzyl acetate	9.00
Orange oil, sweet	3.00
Jasmine absolute	1.25
Vetiveryl acetate	4.50
Cedryl acetate (Givaudan)	4.00
Sandalwood oil (Mysore)	2.50
Lavender oil, Barreme 42% esters	6.00
isoEugenol	1.75
Amyl salicylate	1.50
Bergamot oil	12.00
Lemon oil, Guinea	3.50
Methyl ionone	5.00
Ylang-ylang oil	3.50
Oakmoss decoloree (Robertet)	3.50
Patchouli oil	2.50
Petitgrain oil, Paraguay	4.00
Indole	0.15
Citral	0.50
Aurantiol	4.00

Dimethyl benzyl carbinol	3.00
Hydroxycitronellal	3.00
l-citronellol	2.00
Geranium extra	2.50
Fennel oil	0.75
Black pepper oil	0.75
Coumarin	2.60
Musk ketone	1.50
Civettone	0.20
Ambrettozone (Haarmann & Reimer)	0.40
Ambrarome Absolute (Synarome)	0.25
Clove bud oil	0.30
Aldehyde (pseudo) C_{18}, 10%	0.40
Aldehyde (pseudo) C_{16}, 10%	0.20
Aldehyde C_{14} ('peach'), 10%	5.00
	100.00

ORIENTAL PERFUMES

The creation of oriental perfumes is carried basically by using balsams and all kinds of resins, e.g., Peru balsam, benzoin, opopanax, olibanum, labdanum, etc., accompanied by musk and civet, etc., natural and artificial. The conventional crystalline 'fixatives' are also used according to taste. The desired objective is to obtain a perfume with warm and even seductive tonalities. The typically oriental base is 'floralised' by means of heady flower odours such as jasmin, tuberose, magnolia, gardenia, honeysuckle, syringa, etc.

Due note must be taken of the fact that all the basic raw materials are heavy in character and slow to be released into the atmosphere. It is therefore necessary to use, with them, a sufficiency of citrus type oils, etc., acetates and other synthetics of a light nature, in order to entrain and disperse them.

By way of formula we give a base that may serve as a starting point for these types of perfume.

Formula 16 (Base S.H.)

Constituents	*Parts*
Tonka beans hyperessence (Charabot)	1.75
Peru balsam rectified (Hasslauer)	3.30
Castoreum anhydrol (Givaudan)	2.55
Ethyl vanillin	2.20

Bergamot oil	51.25
Orange oil, bitter (Guinea)	8.70
Orange oil, sweet	8.70
Lavender oil, Barreme	4.40
Ondatrol (Descollonges)	0.75
Linalyl acetate ex shiu oil	16.40
	100.00

Here is another oriental perfume :

Formula 17 (Parfum J)

Constituents	Parts
Coriandrol (Robertet)	10.85
Vetyrisine (Firmenich)	7.30
Opaslaunax (Hasslauer)	7.30
Humuscol (Descollonges)	5.82
Vanillin	3.90
Heliotropol (Firmenich)	0.35
Coumarin	3.65
Sandalwood oil	3.65
Lavender oil, Barreme, 42% esters	3.65
Bergamot oil	3.65
Coriander oil	2.91
Gardenia (Givaudan)	3.65
Neroli	2.54
Supercuracao (Lautier)	2.59
Clary sage (Chiris-UOP)	1.48
Jasmin absolute	1.48
Patchouli oil	0.74
Musk ambrette	0.74
Ylang-ylang oil	0.37
Rose	0.37
Wintergreen oil	0.14
Peru balsam rect. (Hasslauer)	2.50
Cyclopentadecanolide	0.92
Anhydrol Castoreum (Givaudan)	0.37
Civettone 100%	0.37
Musk tincture	28.71
	100.00

'GREEN' PERFUMES

For some years there has been a growing demand for perfumes with a more or less pronounced note of greenery or verdure. It could of course be said that the original perfumes with a type of green note were the ferns or fougeres. Other fougeres were quick to follow, and all helped to popularise the fern note, which is really one of forest trees and undergrowth rather than a truly green leaf note.

Here is a hyacinth-green base which can be of considerable service in the creation of green perfumes.

Formula 18 (Hyacinth Green Base)

Constituents	*Parts*
Hydroxycitronellal (Cyclosia, Firmenich)	3.60
Phenylethyl acetate	1.80
Phenylethyl alcohol	43.00
isoEugenol	3.50
Cinnamic alcohol	16.66
Phenylacetaldehyde 50%	0.85
Phenylacetaldehyde dimethyl acetal	0.50
Folial (Firmenich)	0.05
Hyacinth absolute	1.25
Benzyl acetate	1.80
Phenylethyl formate	0.50
Terpineol	3.10
Rhodinol	1.20
alpha-Amylcinnamic aldehyde	0.30
Paracetone (Givaudan)	0.20
Rose	0.06
Jasmin absolute	0.20
Hosaldeine (de Laire)	0.79
Framboise (Firmenich) 10%	0.24
Dimethyl phthalate	20.40
	100.00

DOMINANT NOTE TYPES

Dominant notes are perfumes which have a well-defined characteristic note yielded by a specific synthetic odourant or a group of two or three odourants, natural or synthetic. It is easy enough to obtain in perfumery what, in painting, one terms a *grisaille*: in other words a soulless overall tint, which shades off into a heavy and nondescript mist. Here is a formula in which the dominant note is formed by an accord of four principal odourants: ambergris, Arhenol, boronia and cassis :

Formula 19 (Parfum)

Constituents	Parts
Boronia absolute oil (Plaimar)	3.75
Cassis bud absolute incolore	3.00
Oakmoss absolute incolore	2.50
Jasmine absolute Butaflor (Robertet)	1.25
Vetiver oil, naardenised (Naarden)	7.38
Carrot oil, naardenised	0.07
Paracetone (Givaudan)	9.80
Benzoin Supergomodor (Chiris-UOP)	2.00
Rose	1.87
Cinnamon oil, naardenised (Naarden)	1.25
Clary sage oil, naardenised	3.50
Clary sage oil, natural	3.50
Sandalwood oil	4.00
Hydroxycitronellal	4.50
Arhenol (Firmenich)	4.50
Pollenol (Firmenich)	3.00
Ambrarome (Firmenich)	7.38
Ondatrol (Descollonges)	3.50
Alcohol C_{11} (undecylenic)	0.62
Aldehyde C_{10}, 10%	0.03
Linalool ex bois de rose Brazil	3.75
Bergamot oil, sesquiterpeneless	15.50
Mandarin oil	0.75
Moskene (Givaudan)	0.25
Ethyl vanillin	2.50
Benzyl acetate	2.25
Opopanax tincture	4.00
Castoreum tincture	3.60
	100.00

The Ambergris Note

When it comes to using this composition for preparing a perfume of chosen concentration, the quantity of tincture of ambergris that should be added is calculated in the proportion of 100 parts of the concentrate to 73.5 parts of ambergris tincture—with the balance made up of perfumery alcohol. As an example, suppose that what is wanted is a perfume containing 15 per cent of concentrate. Then to 15 parts of the concentrate we add 11 parts of tincture of ambergris, and complete the formula by adding a sufficiency of alcohol at 95° to arrive at a total volume of 100 parts.

Tincture of ambergris at 3 per cent concentration may be replaced by much less expensive synthetic odourants. In this particular case excellent results, possibly even superior to those yielded by the natural product, may be obtained with a 3 per cent tincture of Ambergris Synthetique (Givaudan).

Another formula in this category is given below:

Formula 20 (Parfum BM)

Constituents	*Parts*
Arhenol (Firmenich)	5.40
Sophora (Givaudan)	20.00
Cattleya (Givaudan)	3.60
Jonquille alpha (de Laire)	6.00
Hydroxycitronellal	23.00
Orangeflower	0.30
Folia (Bertrand freres)	2.00
Methyl ionone	1.00
Paracetone (Givaudan)	3.00
Cinnamic alcohol ex styrax	9.78
Benzyl acetate	2.00
Linalyl acetate	4.00
Phenylethyl alcohol	0.80
Rose	3.20
isoEugenol	0.80
Citral	0.40
Orange oil, bitter	4.00
Orange oil, sweet	1.60
Cedrat oil	2.00
Aldehyde C_{12} (MNA)	0.03
Aldehyde C_{11} (undecylenic)	0.02
Patchouli oil	0.27
Coumarin	3.00
Ethyl vanillin	0.80
Jasmonone	3.00
	100.00

The formula for the finished perfume is made as follows, chilled and filtered.

Formula 21 (Perfume concentrate as above)

Constituents	Parts
(Parfum BM)	15 kilos
Tincture of musk	4 litres
Alcohol at 95°	81 litres

Many simple, small-scale chemical processes may be profitably used by the perfumer or his associates. One thinks of the fractional distillation of Virginian cedarwood oil, for example, or the interesting end-product of distilling bergamot oil over oakmoss, just as the larger supply houses have distilled geranium over rose flowers and petitgrain over orangeflowers. The separation of a determined portion or portions of a known essential oil can also yield an attractive result. Nothing should be overlooked, in fact, that may contribute a novel and perhaps an inimitable note to a new perfume.

Manufacturing Processes

It is necessary to select best quality of formulation as the concentrate prepared should be better. Take utmost care in choosing, buying and storage of raw materials. e.g., Citrus oil. It deteriorates rapidly. Hence its stock should be used accordingly and note deterioration. The place of storage should be dry, shaded and ventilated and have fire protection.

Storage should preferably be in glass, although internally lined aluminium cans provide one of the permissible alternatives. Some oxidation-prone essential oils are often treated on the production site with traces (e.g. 1 part in 20,000) of antioxidants, but the practice of adding such materials should be kept to a minimum and not knowingly encouraged in fine perfumery. A first-grade fresh oil is always superior to a comparable treated oil.

A few odourants, e.g. myristic aldehyde, should not be stored in the cold. This particular aldehyde tends to polymerise if kept below room temperature.

First make typed note on Cellophane-wrapped card and clipp it on holder having movable wide flat strip specially designed for the actual line to be weighed. The design and material of containers and utensils should be appropriate. It means that they are fabricated from good quality tin plate, with spouts for drip-free pouring and suitable for steam bath or hot plate.

The weighing has to be accurate specifically in case of small weights, or 10, 5, 1 per cent solution of ethyl alcohol, benzyl benzoate, diethyl phthalate and other diluent. A microprocessor based electronic balance having facility for zero setting, tare weight facility which facilitate direct weighing.

Almost all the mixtures are prepared in cold. In certain cases heat is required to enhance solubility e.g., flower absolutes and fatty aldehydes. In this case heat is applied for dissolving and then blend is cooled.

In olden days mixing is carried out by hand methods. Modern methods of liquid mixers are more efficient. At the beginning the concentrate is matured on storage, then reduced or diluted with alcohol; chilled, filtered and filled into bottles.

The finished perfume is prepared by mixing the concentrate with equal quantity by volume of the alcohol and dilute with proper amount of distilled water. Then cool it completely and add concentrate and alcohol with stirring. This process presents colloidal precipitation of concentrate and easily solubilise all the constituents. After 24 hours, the alcoholic batch is chilled at –0.4 to –0.6°C and filtered, using asbestos wool and diatomaceous earth as filter aid.

Alcoholic Strengths

The perfume content of Colognes, eaux de toilette, eaux de parfum and other fancifully named 'dilute perfumes' may range from 3 to 8 per cent, although in some few cases this figure has been found to exceed even 10 per cent. There seems little point in marketing as a toilet water what is, in fact, a perfume – unless its water solubility has been artificially enhanced for the purpose.

Perfumes themselves range from the perhaps old-fashioned 'low' of 10 per cent to 15, 20, 25 and even, quite irrationally, 35 per cent. The alcoholic strength may lie between 85° and 95°.

Inexperienced perfumers, and some captains of industry in the perfumery world who are not perfumers, seem occasionally to believe that higher perfume concentration means enhanced strength and tenacity. This is not so. Sometimes the tenacity may be slightly increased but, beyond a certain optimum level, the strength is not increased. Indeed, too high a concentration lowers the tonality of a perfume and flattens it out, curtailing its ability to expand. With a very high concentration dull, non-diffusive perfumes result, and one of the most valued properties of a good grade of ethyl alcohol is wasted.

Control

Adequate checking and control methods are of course essential at all stages of perfume production: from the intake and assay of raw materials, through storage and processing to filling and packaging. The variability of certain materials because of batch or climatic deviations, etc., is an additional obstacle in the way of producing a uniform series over a period of time. The perfumer should be closely concerned with the provision and use of standard samples for odour checking and comparison. These standards must be kept in optimum conditions, i.e. where possible in sealed ampoules or under nitrogen. Watch should also be kept on the exact matching of colour shades.

It is frequently necessary to dilute 95 per cent alcohol to 90, 85, 80, 70 per cent or even lower strengths. To determine the amount of proof alcohol required is calculated using following formula :

$$\text{Amount of proof alcohol} = \frac{\%\ \text{alcohol desired} \times \text{Total volume}}{\%\ \text{proof alcohol}}$$

$$= \frac{\%\ \text{alcohol desired} \times 1000}{95}$$

e.g., prepare 90% alcohol using 95% proof alcohol.

$$\text{Amount of proof alcohol required} = \frac{90 \times 1000}{95}$$

$$= 947.3\ \text{ml.}$$

It means 947.3 ml of 95% alcohol is diluted to 1000 ml using 52.7 ml of water.

80 per cent alcohol results from diluting 842 ml of 95 per cent alcohol to 1000 ml. 70 per cent alcohol is obtained by diluting 737 ml of 95 per cent alcohol; 60 per cent by diluting 632 ml, 50 per cent by diluting 526 ml, and 45 per cent by diluting 474 ml to 1000 ml with purified water.

Contraction of volume and rise of temperature occur when ethyl alcohol and water are mixed. The final adjustment of volume when preparing these dilutions, is therefore made at the same temperature as that at which the 95 per cent alcohol is measured, i.e. about 20°C.

Chapter 8

Colognes : Perfumes for Men

INTRODUCTION

Since two and half centuries *eau de Cologne* has been described as the most popular perfume. As far as predecessors of this perfume, the only name appear is *eau de Cordova* and *eau de la Reine de Hongrie.*

In June 1960 an international celebration of the birth of *eau de Cologne* was held at Santa Maria Maggiore. The *Encyclopaedia Roret* for 1834 contained details of a number of patented formulae for *eau de Cologne.* The following formula of the J.M. Farina type for *eau de Cologne* given below is published in latter edition.

Formula 1

Constituents	*Parts*
Bergamot oil	6 kg 200
Lemon oil	3 kg 100
Neroli oil	0 kg 800
Clove oil	1 kg 600
Lavender oil	1 kg 200
Rosemary oil	0 kg 800
Alcohol (90°)	100 litres

This mixture was allowed to stand and mature for at least 30 days before filtration.

Fresh melissa herb 10 kg, rosemary 5 kg, finely reduced orris root 1 kg are macerated in 25 litres of 95° alcohol and 4 litres of water for 12 hours. Distillation is then effected and the extract so obtained blended with 310 grams of bergamot oil, 250 grams each of lemon and sweet orange oils, 60 grams each of neroli and petitgrain oils, 120 grams of lavender oil and 25 additional litres of 95° alcohol. The batch is left for one month and then filtered.

Francesco La Face, the world's leading authority on Italian citrus oils, has discussed most of the raw materials utilised in classic Colognes. First and foremost are the citrus oils or Hesperidaceae: bergamot, lemon and sweet orange, followed by orangeflower, neroli and petitgrain derived from the bitter orange tree. Next come the herbal oils: lavender, rosemary, melissa (*balm*) and clary sage. Many other ingredients, wormwood, calamus, nutmeg, hyssop, caraway, aniseed, cinnamon and clove are used in traces. Here we would point out that the clove or carnation note should be regarded as important and even indispensable in a Cologne base.

Even if one was current access to the types of material actually used in the original classic Colognes, one could not use them in the same way, i.e. by maceration, infusion and distillation with alcohol, followed by the addition of essential oils and floral waters and a period of maturation. Many efforts have been made to 'translate' what is known of the older, classic formulae into modern practice. From these the following have been selected:

Formula 2

Constituents	per cent		
	A	*B*	*C*
Bergamot oil	33	20	35
Lemon oil	18	20	25
Sweet orange oil	25	20	20
Lavender oil	6	7	7
Rosemary oil	5	5	5
Petitgrain oil, bigarade	8	13	0
Neroli oil	3	10	6
Clary sage oil	0.5	—	—
Clove oil	—	—	2
Melissa oil	—	5	—
Benzoin resinoid (Siam)	1.5	—	—

Natural unadulterated oils of cedrat, melissa, verbena and limette are now-a-days difficult if not impossible to obtain. Mere traces of other oils were often included, in order to impart a distinctive cachet. Among them were rose oils, peppermint oil and, in addition to those spicy and herbaceous oils already mentioned by Fenaroli: angelica, thyme, cardamom, fennel, cumin and juniper.

In current perfumery usage one finds personal preferences among perfumers for minor additions of clove, nutmeg or caraway. Thyme, hyssop, estragon and myrtle have their adherents.

Variations among the major citrus constituents may include the introduction of mandarin, grapefruit and lime oils. Citral-verbena notes emphasise the essential freshness of the compound. Petitgrain is of course invaluable. Linalool, linalyl acetate and ethyl linalyl acetate tend, like oil of lavender, to add depth and richness, but they should not be used to excess. Accentuation of the rose or rose-geranium motif does not affect the initial odour or effect so much as it does the residual odour on the skin. Even a first-class product of this type (and there is at least one very attractive blend) leaves behind, on the skin, an odour that is frankly and persistently rosaceous.

Classic *eau de Cologne* is a fresh and harmonious blend of predominantly citrus oils. Its fragrance is exciting, refreshing, altogether delightful and of short duration. That is its nature : it cannot be changed without losing this unique combination of qualities. If we try to prolong its brief existence by modifying the formula, which is a comparatively simple one, or adding so-called fixatives, we merely succeed in altering its character: it is no longer a classic eau de Cologne. It can be made more sophisticated, but only at the expense of its exhilarating freshness. Flowery notes can be blended with it and the results are sometimes extremely attractive, but whatever their merits, they are not, of course, true eaux de Cologne.

Varying the Cologne note can nevertheless prove to be an educational and rewarding task. There are two main approaches. One is to preserve the Cologne character while introducing a certain amount of novelty and variety. The second is to use the Cologne note as part of a distinct and different blend, so that while the effect of the note is still felt, its individuality is merged and subordinated in the main design.

Green, hyacinthine top notes will sometimes blend effectively into Cologne compositions. Ethyl acetate (about 0.2–0.4 per cent) tends to 'lift' the top note. Methyl nonyl acetaldehyde may be present at about 1 per cent of a 10 per cent solution, or at rather higher levels in modern ambered or sophisticated Colognes. Decyl aldehyde is also useful. At the other end of the scale one considers duration of odour and 'fixation'. It is sure that, a really well-fixed, long-lasting Cologne would not be a Cologne at all. Where ambergris tincture or artificial ambers, etc., are present in small quantities, they serve in this instance more as blending and homogenising agents than as fixatives. Traces of decolourised oakmoss can give some interesting effects.

The analysis of *eau de Cologne* was carried out using live human skin as testing ground instead of conventional smelling slips. The test on the smelling slip showed that hesperidean oil is short lived. Added citral imparts improved persistence and Grasse verbena oil enhances freshness. Sweet and bitter orange oils each behave differently, as one would expect, but both are good and they can usefully be employed in admixture. Lime oil is forceful and too characteristic at first but soon fades. Clary sage lasts well; so also does rosemary. Coriander is an extremely good skin perfume but high proportions spoil the effect of a Cologne smelt on a slip or a handkerchief. Of the shading notes, we liked the odour on the skin of estragon, thyme, hyssop, nutmeg and caraway. There is only one of the nitro-musks that remains sweet and stable on the skin: that is musk ambrette. Of the lower notes, labdanum, Peru balsam and benzoin are of interest.

It will be appreciated that a toilet water must first of all be attractive in the bottle and in its first evaporative phase outside the bottle. It is not judged primarily by its odour on the skin, but this latter is nevertheless an important factor, especially in a type of perfume that is often liberally applied to the skin surface.

Many of the early Cologne formulae or traditional 'botanical mixes' of citrus, spice and herb oils are very similar to the Benedictine and related types of liqueur; especially if one substituted sugar for tincture of benzoin. The dual purpose of the benzoin was to act as a fixative and ensure an opalescent effect when the Cologne was poured into water.

TOILET WATERS

Toilet waters consist of perfume oil, alcohol, water and — occasionally— glycerine. In .case of glycerine there are two contradictory opinions.

(1) It retards volatilisation of the perfume from skin.

(2) It accelerate perfume evaporation from skin.

However, the presence of water generally increases the persistence of odours on the skin.

Many Colognes and toilet waters with 50° alcoholic strength which is minimum permissible are available on the market. Such low alcoholic Colognes have solubility problems. It dealt with by careful selection of deterpenated oils, relatively soluble synthetics, etc., appropriate testing, chilling, filtration and even, in extreme cases, by using solubilising agents and co-solvents to give stable, clear pseudo-solutions in very dilute alcohol or even in water.

Isopropyl alcohol has been used at various periods as a partial substitute for ethyl alcohol as a perfume vehicle but it lacks the vinous quality and other attractive properties of the latter.

Alcohol remains the ideal perfume solvent, with its extremely mild, smooth odour which blends so well with perfumery materials. It has boiling point of 78°C hence not too volatile, it permits a satisfactory evaporation of the perfume and at the same time conserves the fixative elements which ensure the tenacity that is generally expected of a good perfume.

The odour of perfumery grades of ethyl alcohol must be as neutral as possible. The principal sources are as follows :

(1) Grape juice. Even after careful rectification, alcohols from fermented grape juice always retain a mildly fruity character, which makes them generally unsuitable for use in fine perfumes although they find useful application in eau de Cologne.

(2) Grain, such as rice and wheat. When rice alcohol is well rectified it is one of the best qualities for use in perfumes. It is very neutral. Poorly rectified rice alcohol is not so suitable and, from the flavour point of view, it has a slight greasiness on the palate, suggesting coconut oil.

(3) Molasses. This alcohol is widely used in spirituous flavours but, again, it tends to entrain the odour of its original source, i.e. cane sugar. Sugar beet alcohol, if obtained from the fermentation of a thoroughly purified beet juice and subsequently rectified, can be an excellent product.

(4) Other vegetable sources : e.g. potato, tapioca, maize.

(5) Synthesis : This is much used in the U.K., chiefly on the grounds of comparative cost.

In France, manufacturers of fine perfumes use rice alcohol or a good grade of beet sugar alcohol, with wine alcohol reserved for Colognes, etc. The quality of the alcohol and of the denaturants incorporated in it, has obviously a decisive bearing on the quality of the finished perfume. Denaturants are intended either to 'mark' the alcohol (i.e. make it analytically identifiable by the excise authority) or to make it unpleasant to drink. Extreme toxicity is usually avoided nowadays, because some heavy drinkers are not in fact deterred by any additive, however dangerous or nauseating. The use of denaturants in alcohol is disadvantages to perfumer. Instead he will prefer the use of suitable blend of essential oil or odourants. The denaturants used are propyl and methyl alcohols and diethyl phthalate.

The boiling points and vapour pressures of these denaturants tend to modify the resulting perfume, even when the purest grades are used. The last word, of course, is with the individual governments and their excise authorities, and they usually demand mixtures of two or more substances including propanol, methanol, diethyl phthalate, quassin, brucine, sucrose octa-acetate, etc.

MODIFIED COLOGNES

So far we have glanced only at relatively minor variations of the Cologne theme, in which the latter still occupied its distinct and characteristic primacy. The next and more dramatic development, however, consisted of the introduction of perfumes and toilet waters in which the Cologne note, though present, was nevertheless subordinate.

This sustained fashion led also to a variety of other diluted perfumes, and only time will tell whether its effect on the perfumery industry has really been wholly beneficial.

One should not expect that satisfactory blending will result by merely adding a proportion of some floral or fantasy concentrate to an ordinary Cologne. The entire formula must be properly balanced to achieve the desired effect, which is that of a usually subdued but still refreshing citrus floral water, blossoming into something more definitely suggestive of, say, gardenia, mimosa or jasmin.

Felix Cola's suggested formulae for fancy or fantasy Colognes were published about forty years ago, but some are still of interest. For a *Gardenia Cologne* he merely suggests the addition to each kilo of a classic Cologne concentrate 30 grams of styrallyl acetate and 5 g of phenylacetaldehyde. The suitable gardenia compound and blender, if mixed in different proportion with the Cologne base, one could get best results. From that beginning one could proceed to finalise the formula. The same observation applies to his *Jasmin Cologne*, which is excessively simplified with its addition of 50 grams of benzyl acetate and 40 g of amyl salicylate and musk ambrette; for *Ambered Cologne*, ambreine, Bulgarian rose oil, vanillin, heliotropin, benzoin and musk ambrette; for *Carnation*, 50 g isoeugenol, 40 g eugenol, 20 g vanillin, 40 g phenylethyl alcohol; and for *Chypre Cologne*, oakmoss absolute 25 g, vetiver 50, patchouli 25, sandalwood 60 and coumarin 75. His *Rose Cologne* suggestion consists of geranium oil 30 g, rhodinol 50, phenylethyl alcohol 100 and geranyl acetate 25. The remaining suggestions relate to Russian, Fantasy, Sweet pea, lilac, mimosa and narcissus Colognes.

PERFUMES FOR MEN

Men rarely used perfume of any kind in the nineteenth century. Today the picture has changed completely. There seem to be two main tendencies in perfumes for men. One is for men who really appreciate perfumes, the *aficionados* of odour or *olfactifs*. These select a perfume that pleases them, from the whole available range of perfumes (i.e. those intended for women as well as for men) and use it discreetly for their own satisfaction. The other tendency, the conventional one, is based on the assumption that certain more stylised and vigorous perfumes help to affirm an air of clean-cut masculinity. The perfumer who is asked to devise a new perfume or after-shave lotion for men is naturally more interested in this second tendency. We offer the following suggestions.

First of all, there is the eau de Cologne with a more or less concentrated 'green' note. Then there are the ambered Colognes and a whole potential range of Colognes with herbal notes, i.e. rosemary, lavender, etc. There are also blends shaded with a tobacco note and those that are still more tobacco-like, strong and dry.

Carnation has always been a masculine favourite and can, from the perfumery point of view, give results of elegance and distinction. Somewhat related to this is the odour of leather and these (*cuir*) perfumes exist in considerable variety.

Lavender : Here also there are many varieties, from pure lavender with a touch of rose to ambered lavender and English lavender, the latter shaded with lemon, very fresh and agreeable. An English formula follows :

Formula 3 (Lavender)

Constituents	per cent
Lavender oil, Mitcham	60.0
Lavender oil, French (40–42%)	25.0
Bergamot oil	3.0
Labdanum tincture	5.0
Oakmoss tincture	4.0
Geraniol	2.0
Lemon oil	1.5
Benzyl acetate	1.0
Musk ketone	1.2
Neroli	0.5
Vanillin	0.3
Coumarin	0.25
Indole (5% dilution)	0.25

Alcohol, 95°	2296.0
Distilled water (approx.)	800.0
	3200.00

Verbena : Fresh and restful. Also attractive, though not so refreshing, is ambered verbena.

Violet : It is well worth considering if used with discretion. It should normally 'impregnate' the final composition and not reveal its presence as a bold exhalation.

A wide appeal to men are perfumes and lotions based on ylang-ylang and cananga. Eau de Cananaga carries with it an air of oriental mystery and yet, technically, can be readily developed into a dryer note than one would ordinarily associate with the oils of cananga odorata.

Rural, countrified types of *eau de Toilette* may be compounded on such lines as the following :

Formula 4 (Town and Country H.M.)

Constituents	*per cent*
Bergamot oil	28.5
Lavender oil	60.0
Clary sage oil	2.0
Rosemary oil	2.0
Thyme oil, red	1.5
Serpolet (wild thyme) oil	6.0
	100.00

Formula 5

Constituents	*per cent*
Thyme oil	1.0
Savory oil	1.0
Serpolet oil	7.0
Rosemary oil	5.0
Bergamot oil	42.0
Lavender oil	13.0
Pine oil	4.4
Silver fir oil	4.3
Coumarin	0.2
Aldehyde Mer (Laserson & Sabetay)	0.1
Tolu balsam tincture	22.0
	100.0

A sufficient quantity of either of these two concentrates is used in order to give the desired effect in the finished *eau de Toilette*. Chypres of a sufficiently vigorous and not too sweet type may also be considered. Fougere (fern) and foin coupe (new mown hay) have always been considered as classically masculine types. Vetiver can give some very interesting results indeed, in association with some of the modern green notes. A pure musk note can be of interest, but always provided that it is used with discretion.

Geranium can give attractive effects when the green note of the leaf is emphasised. Patchouli lends itself to a perfume for men, in which the overall odour is peppery, powdery—and not too strong. Harking back to the leather notes, a reconsideration of Peau d'Espagne should be made.

Formula 6 (Peau d'Espagne)

Constituents	per cent
Bergamot oil	32.0
Bois de rose oil	20.0
Vetiver oil	2.0
Hydroxycitronellal	15.0
Ionone alpha	1.0
Oakmoss resinoid	1.0
Musk ambrette	4.0
Musk ketone	2.0
Labdanum resinoid	4.0
Olibanum resinoid	5.0
Sandalwood oil	2.0
Sweet birch oil, terpeneless	1.0
Indole (10% sonl. in linalyl acetate)	1.0
Phenylacetic acid	0.5
Cinnamic alcohol	9.5
	100.0

Among the green notes, the once very popular reseda (mignonette) note should not be forgotten. Trefle or orchidee, based on amyl salicylate, could serve as a model for a masculine perfume suggesting the famous Trefle Incarnat but, in this case, dryer and perhaps less delicate, less sweet. Sandalwood blends well with Virginia cedarwood and other woody notes and offers many possibilities in perfumes for men, but always provided that excessively 'perfume-y' effects are avoided.

Here are a few other suggestions: the camellia note, based on Manila ylang-ylang, methyl ionone and orris, with suitable 'masculine' characteristics introduced into the blend.

Rondeletia, that ancient accord of carnation and lavender, can very readily be adapted for use as a masculine *eau de Toilette*.

The analysis of British market was carried out by Wells. He offered a number of approaches to formulation in a survey of perfumes and perfumed toiletries for men. According to Max Stoll, men have sweet and women sharper odour.

He also suggested that it is probably more profitable for the perfumer to leave the difficult field of physicological and psychological speculation and confine himself to mere associations of ideas. Many other perfumes can be adapted for masculine use, by revising the formulae, toning down notes, that are too sweet and perfume–y and introducing other notes to give a sharper edge.

Chapter 9

Olfaction and Gustation:
The Sense of Smell and Taste

INTRODUCTION

Perfumery is much more an art than a science. Some knowledge of olfaction would enable the good perfumer to become a still better practitioner of his art. Be this as it may, there is no doubt that a profounder knowledge of olfaction must eventually lead to advances on all fronts, including perfumery chemistry and even the practice of perfumery itself.

Dr. R.H. Wright, an expert in olfaction, imagines a dividing line. At one end you and other end is the outside world. The line represents the interface between you and all the things that are outside you. The two or three square inches of that interface situated high up in each nostril are the space where air-borne molecules of various kinds interact with specialised sensory receptors so as to generate sensation of smell. As Kare says, taste and smell are specialised receptors for communications with the chemicals in our environment.

A living organism communicates with the chemistry of its environment in the essential activities of respiration, ingestion of food and fluid and, in many species, reproduction. Environment, age, disease and nutritional state are some factors that modify olfactory and gustatory perceptions.

OLFACTION

John H. Kenneth, pioneers into the modern concept of the odour-stimulus and the receptor mechanism of the nose and brain, pointed out that the reactions of a subject to different smell stimuli depend on the quality and quantity of the substance producing the stimulus, and on the physiological and psychological condition of the subject. The complexity of the chemical and physical factors involved in the process of smelling is as bewildering as the complexity of the reactions invoked, a complexity which dates back to the era when life itself evolved and began to become aware of its chemical environment. Leaving aside what may be termed purely physiological reactions such as olfactory reflexes, vasomotor effects, etc., and problems of fatigue, both physiological and psychological, mental reactions to smell stimuli may be grouped as direct and indirect.

The direct reactions are mainly the perception of the quality and quantity of a smell, whereas indirect reactions are the associations subsequently evoked.

The nose is a physico-chemical laboratory in miniature and can distinguish tens of thousands of different odours. There is no scientific instrument which has this selective ability. Perfumers and professional odour evaluators, as well as tea and coffee tasters, cheese and wine tasters, have developed faculties that are beyond the scope of any instrument, however sensitive, although it is often an advantage that their findings can be checked against and supported or amplified by appropriate instrumentation.

Mechanism of Perception

M.G.J. Beets has rightly pointed out that, two types of informations are resulted from the physical interaction of molecules of individual food constituents with the receptor sites on the tongue and in the nose may be called the primary information patterns of olfaction and gustation.

The olfactory system can be described in very simple terms as consisting of the receptor membrane where the information originates, the olfactory nerve in which it is transported, and the olfactory bulb in which it is processed for delivery to the higher centres. The gustatory system has its peripheral receptors on the receptor cell membrane in the taste buds. Their innervation is less simple than it is in the olfactory system. In the brain several regions are involved in taste but, adds Beets, there is no simple gustatory analogue of the olfactory bulb. It is nevertheless assumed that the initiation, transport and processing of information occur in both systems according to closely related principles.

It may seem surprising that an elementary sense such as that of smell, depending on stimuli initiated by chemical substances which often have an exceedingly simple molecular structure, and mediated by nervous pathways which in certain respects are a good deal less complicated and tortuous than those of other senses, is still so little understood in anatomical and physiological terms. But this only serves to emphasise how incredibly complex is any problem, whose basis is neurological. The history of neurological research makes it abundantly clear that investigations into the functions of the nervous system have always been, and will certainly continue to be, a slow and arduous process.

It is possible that initially odourants must be adsorbed by the cell membrane which according to Davies theory must penetrate and puncture. G. Ohloff regards this theory favourably since it accords with the principles of cell physiology and may permit the olfactory process to be interpreted quantitatively, although he thinks that it fails to account at all adequately for the effects of the molecular properties of odourants. Here, however, he warns that 'the correlation between the molecular profile or morphology of odourous substances and the quality of the odours constitutes a crude and, in many cases, an inadmissible simplification of the complicated olfactory process'.

Odourous and Odourless

Why is it that some substances have an odour while others do not? Olfactory interaction takes place when a population of odourant molecules is carried by a current of air to the thin layer of mucus covering the receptor sites. The concentration of molecules available for interaction depends on odourant volatility. This suggests that there is only an upper limit for the molecular weight of odourants. This is indeed what we observe. Even chemicals with very low molecular weights, such as formaldehyde, elicit olfactory responses, frequently accompanied by trigeminal effects. We do not know any odourants with molecular weight much beyond 300.

In a provocative essay entitled 'Odourous and Odourless', Bassiri takes the matter of odour and non-odour a stage further; almost into the metaphysical world of Bishop Berkeley. While conceding that a significant vapour pressure, a certain volatility, is a condition necessary to but not alone sufficient for the manifestation of odour, he nevertheless quotes examples of materials which, though normally considered odourless, may be shown to possess odourous properties in certain abnormal conditions. He concludes, that 'the odour activity of a chemically defined body cannot be detected to the same degree, or in the same way, by animal species which possess different organic tissues, which are differently detailed and which confer upon their hosts olfactory capacities which are not equivalent when physiologically compared.' He quotes Le Magnen's dictum that 'odourless molecules are odourless only because of the absence of a receptive organ which can differentiate their odour' and observes The division of substances into two classes, odourous and odourless, is purely arbitrary and has no sound basis in reality. Evaluation by individuals or even by groups cannot be relied upon to determine the absence of all odour-activity in a substance considered odourless. The odour of a product is a biological consequence which cannot pre-exist in a molecule, though obviously the conditions which make the biological consequence possible do pre-exist.

The apparent emission of odour by non-volatile substances such as sugar, salt, stones, rocks, clays and metals. The most satisfactory explanation seems, in the majority of these cases, to be the presence in the non-volatile material of minute traces of volatile contaminants, which may even result from human handling or be adsorbed from natural surroundings.

What Makes a Substance Smell?

When studying olfaction and theories about olfaction, we are always and inevitably brought back to an admission that problems of odour and its perception will only be solved when, 'the nature of the energy carried by the material and the conditions of its intervention in the olfactory process are more fully understood. In essence, this theory claims that the size and shape of the molecule govern the type of odour, whereas Wright's theory contends that the inherent vibrational frequencies of the molecule are the causative factor.

It has been investigated that 'there exist correlations between odour and frequencies, since frequencies describe molecular structure and, indirectly, all other properties of odourants.'

In regard to his vibrational theory, Wright observes: 'The examination of the infrared absorption spectra of compounds which evoke specific olfactory responses has confirmed an association between the two. A particular stimulus pattern may depend on the absence of certain frequencies as much as on the presence of others.'

Olfaction is the chemical sensing process that enables a trained dog to discriminate between the body scents of two persons; that enables a homing salmon to make its way back to the very gravel bed where its life began; that enables a honey bee to recognise the other members of is hive; and that enables closely related flies to breed true even when crowded together.

For a substance to generate a unique odour sensation, it must present and impress upon the organism a unique set, or pattern, or combination of attributes, and this pattern must be transmitted to the central nervous system as a neural discharge of equivalent specificity. There is no way for the neural response to be more specific than the stimulus.'

A Coding System for Olfactory Specificity

Subtle variations in the sensation imply equally subtle variations in the stimulus. This means the olfactory specificity cannot stem from a simple molecular attribute such as the length or weight of the molecule, or the size of its dipole moment, or its tendency to form hydrogen bonds. It is hard to conceive of any such simple physical or chemical dimension that could differ reliably in enough ways for us to discriminate a really large number of odour sensations. Therefore, the odour must correlate with a definable set of qualities so that the set for one substance can be different from the set for another, even though some elements of the two sets may be the same. This is how we use an alphabet of 26 letters to spell an effectively unlimited number of words. The alphabet is one of the most important human inventions of all time, and nature seems to have invented something of the same sort when evolving an olfactory coding system. We are still a long way from breaking the code; but we know there must be one and we are beginning to have some idea of how it may work.

How Does a Perfumer 'See' Odours?

But where and how, you may ask, does the perfumer fit into this fascinating but rather esoteric world of facts that are often difficult to interpret and theories that change almost annually. The answer is that he rarely fits into it at all. His preoccupations lie elsewhere. It is probable, however, that he 'sees' odours as non-simple, non-pure, non-primary; rather as one might visualise a three-dimensional form and be able to look at it, mentally, from different vantage points. Perfumers would, we feel, be inclined to agree with the description put forward by Beets.

The interaction of a population of odourant molecules with the olfactory receptors produces a pattern of types of olfactory information. Each type makes its own quantitative contribution to the total pattern. Sometimes one of these types predominates (e.g. in musks) but this is not always the case. The most odourants produce a complex pattern of types of olfactory information. This is processed in the olfactory system and projected to the higher centres of the brain where it is translated into an equally complex odour sensation. The principle is that 'a pure odourant invariably produces a complex information pattern and a complex odour sensation.'

It may be compared, for example, with Gaze and Keating's tentative hypothesis that light stimuli passing from the retinal ganglion to the brain depend on a 'systems matching' mechanism rather than on specific cell-to-cell matching. The 'incorrigibly plural' nature of odours, however distinctive they may be and however chemically pure their source. The pure and relatively simple ester, neryl acetate, is described under no less than eleven different headings (ranging from fresh, citrus, rose and lavender to fruity, earthy, powdery and sweet). Rose oxide and ionone alpha, both of which are distinctive and readily recognised by perfumers, are each given twelve different odour-descriptions. Methyl salicylate, with its familiar wintergreen odour, is included under nine odour-descriptions. The nitro-musks are described under five headings and indole (the 'purest' odour in this sense) under a mere four.

It will be noted that even the most monolithic odour has different facets. It is more readily conceived of as a pattern than as a single entity. This idea of a 'systems matching' pattern, taken in conjunction with what we know about olfactory thresholds, offers at least a possible explanation of what brings about an apparent change of odour when certain substances are diluted. The validity of this contention may be tested against such information as that which is available for that matter, based on the perfumer's familiarity with odours. The subjective and idiosyncratic nature of odour appreciation is also in keeping with the idea of complex odour patterns; and the same observation similarly applies to the phenomenon known as parosmia.

Abnormalities of Smell

The subject of abnormalities of the sense of smell is too specialised to be dealt with here, but we would do well to recall that anosmia is the term used for inability to detect odours; partial or specific anosmia to inability to smell certain odours; and parosmia to a state in which certain odours are perceived but not, apparently, in the same way as by a 'normal' person. In other words the odour perceived by the parosmic subject is qualitatively different from that perceived by the normality. Parosmia of one degree or another may well be an extremely common state attributable to sensory imperfection. Few if any of the cases cited would seem to indicate a truly pathological dysfunction, when identifying a given odour-pattern, but merely a slight shift of emphasis and interpretation. The line between what is normal and what is abnormal in the field of smell perception is not at all plain. A certain amount of abnormality is perfectly normal.

On the subject of odourants and receptor sites, it has been investigated that odourant molecules as having 'numerous mutually independent or overlapping odour-active molecular profiles' and corresponding receptor sites, with which they interact 'physically and reversibly'. This would seem to have much in common with the expressed views of Beets and also if one were to alter 'profiles' to 'vibration frequencies'.

GUSTATION

Gustation is closely associated with olfaction specifically in the field of flavours and the sense of taste. The study of the sensation of taste is simple compare to that of smell. In case of gustation there are four different characteristics of taste, viz-sweet, salt, sour and bitter. By definition in terms of specific chemical substances, sweet is considered to be the characteristic quality of sucrose, salt, the quality of sodium chloride, sour, the quality of citric or tartaric acid, and bitter, the salient quality of quinine sulphate.

The sense of taste is nearly always affected by such modifying factors like odour, texture, temperature and even colour. Few people other than professional tasters (i.e. of wine, cheese, coffee etc.) cultivate the analytical use of this sense with, the explicit intention of abstracting mentally what is due to the sensitivity of the tongue alone from the total complex.

Taste receptors are to be found in the taste buds located mainly on the upper surface of the tongue, on the edges and in the rear and on the upper palate. Regions predominantly sensitive to sweet and salty tastes are found in the front of the mouth and those similarly sensitive to sour and bitter tastes at the back. The tip of the tongue in children is said to be very sensitive to sweetness. Electrophysiological and biochemical investigations have also been carried out into the mode of reception of tastes, into the sour taste, about which it is concluded that 'the sour taste is induced by binding of hydrogen ion to a phosphate group of phospholipids in the gustatory receptor membrane.' As in olfaction, the stimulant molecules reach the receptor sites by way of a supernatant aqueous phase. Here, also, an absorption-desorption process obviously comes into operation. Volatility not being an essential feature of taste stimulants, the size and weight of the molecule may be quite large, and this is certainly the case with some of the bitter tastes.

The Complexity of Flavour Perception

If one looks at flavours schematically, attributing a sweet taste to the presence of sucrose, a salt taste to sodium chloride, an acid taste to citric acid, and a bitter taste to certain toxic substances, one must attribute the flavour of a foodstuff to a whole assembly of complex mixtures. And those mixtures are not limited to the four categories of taste. They constitute a subtle balance between the non-volatile bodies of the particular taste (small in number, but large in mass) and the volatile bodies responsible for the odour of the food (large in number but small in mass). A third class of compounds act as potentiators between the first two categories. By acting together, these substances enable a specific taste to be perceived or modified. Proteins and allied substances, the main constituents of foodstuffs and certain amino acids, should be placed in this category. Cheese is an example, too. Its flavour is thought to be closely dependent on certain free amino acids.

Taste-Sensitivity, Intuition and Introversion

Professor Roland Fisher's raised question, whether sensitivity to taste is not merely one of the manifestations of the general sensitivity of the entire organism. He found that those who had a high sensitivity of taste in relation to quinine, the substance used to test sensitivity, were the people who liked good food and who generally might be called gourmets. On the other hand, those who showed a poor reaction to weak concentrations of quinine were the sort of people who would eat anything and did not appreciate culinary refinements. The same applied to tobacco habits. Those people who were hypersensitive to taste included the lowest percentage of smokers. Whatever might be the physiological connection, male smokers seemed more liable to lose their faculties of taste with advancing age, more so than non-smokers, and more so than women smokers. It is also considered, that the sense of taste was closely linked to the individual's sense of intuition and degree of introversion. We need not wonder, then, at the durability of the old tag: 'De gustibus non est disputandum.'

Chapter 10

Applications of Perfumes

INTRODUCTION

Odour is part and parcel of the life today. A good progress has been made in odour and its character of urban life throughout the world has been astonishingly extended and intensified. The metropolitan cities are closely associated with perfumery industry. Perfumes have diverse applications in many areas, new applications are introduced everyday. It is difficult to compile list of applications, however, the following list may serve the purpose.

Perfumes used in Industrialised Society

1. *Household Products*

   Soaps and synthetic detergents

   Miscellaneous cleansers

   Disinfectants

   Polishes

   Paints

   Adhesives

   Air (space) deodorants etc.

2. *Personal Products*

   Cosmetics; make-up preparations

   Toilet and beauty preparations

   Perfumes, toilet waters

   Pharmaceutical preparations etc.

3.   *Industrial Products*
     Dry cleaning
     Leather and rubber articles
     Artificial leather
     Linoleum
     Plastics
     Printing inks, perfumed board and paper
     Textiles

4.   *Agricultural Products*
     Insecticides
     Insect and animal repellants
     Animal baits and attractants
     Veterinary products
     Cattle feeds
     Farmyard 'masking' odourants etc.

5.   *Flavours*
     Beverages
     Foods
     Medicines

PERFUMES FOR SOAPS

Perfumes of outstanding performance in soap are the exception rather than the rule, adding as a corollary that many perfumes which give excellent results in a variety of other preparations will be failures in soap. It will probably be conceded by most perfumers and soap makers that their most successful perfumes have been the result of considerable trial and error. Years of experience will have shown which individual ingredients give strong and lasting effects in soap, but in certain combinations or blends even these otherwise successful items will fail to produce the anticipated effects.

Factors Influencing Soap Perfuming

There are four factors which are responsible for soap perfuming to reduce time and effort in production. They are : (i) psychological and aesthetic; (ii) economical; (iii) technical and (iv) chemical. The fourth term is loosely used in this context to cover physico-chemical and biochemical as well as straightforward chemical reactions.

The governing factors are, or should be, those of a psychological, aesthetic and artistic character. While it is most desirable that the soap perfumer should have a sound scientific knowledge of the probable behaviour of aldehydes, ketones, esters and so on, when they are incorporated in a soap base.

It is even more essential that he shall be an artist in perfumery, able to understand, anticipate and satisfy changing fashions in the public taste for perfumes. In addition to this, he ought to be capable of selecting and using his raw materials to the best economic advantage. The techniques that he employs should be adopted and modified in such a way as to save him the maximum amount of time and trouble, both during the creation and compounding of the perfume and its subsequent shelf-testing in soap. The techniques has been described below.

An experiment for compounding was carried out in the following manner.

Take 10 ml graduated (1/10th) pipette. Start the work with 5 pipettes steeping in alcohol and another 5 standing in a draining rack. 1 ml pipette graduated 1–100th and 10 ml graduated 1–100^{ths} and 50 ml measuring cylinder are also useful. All the liquids are measured in volume and solids are in weight. The volumetric measurement for liquid is more economical, rapid, easy and convenient. The only error incorporated is due to values of specific gravities translating into volume e.g., A product having specific gravity 0.85, 1 ml of A is equivalent to 0.85 gm. An error may also be introduced when pouring or pipetting viscous liquids. These errors can be eliminated by using 50:50 or weaker dilutions of such materials or eutectic mixtures (equal parts of cinnamic alcohol and methyl cinnamate). The dilutions, to maintain the odour intensity, may be made in other 'active' constituents of the formula rather than with alcohol, diethyl phthalate or similar solvents. The problem of contamination is best dealt with by using a fresh pipette for each raw material and by keeping current stocks of the latter to a minimum, i.e. in bottles ranging from about 15 ml to 100 ml capacity. It is most desirable to avoid the use of small bottles that have a neck orifice too narrow for the easy introduction of the pipette. Steeping alcohol should be changed frequently.

The soap perfumer is never allowed to forget that the perfume: soap system presents certain unique and sometimes unpredictable behavioural phenomena. Therefore perfume should be tested in a standard soap base, and not merely on smelling-strips, at all stages of its development. The test is often taken at the level of 1 per cent in soap; and may similarly test partial blends in soap before finally completing his perfume and testing that. The small perfumed soap tablets may be examined immediately, stored wrapped or unwrapped and then re-examined, or subjected to acceleration tests with U.V. light etc. In this way it is possible to assess odour and colour effects and record them over a prolonged period, filing the information so obtained for future reference.

Sfiras pointed out that properties of perfume in soap are retained due to various physico-chemical phenomena. He specifically mentions the adsorption of perfume by the soap; evaporation of perfume; the autoxidation of both soap and perfume; and the reactivity of soap due to the equilibrium: In the following reaction :

$$RCOONa + H_2O \rightleftharpoons RCOOH + NaOH$$

In regard to the first two factors, Pickthall's work shows that some perfumes will tend to remain in the aqueous phase; some will probably be adsorbed on the outer surface of the soap micelle; some may be absorbed between the methyl tails; and some apparently become orientated in the micelle, forming a more or less stable complex.

Esters

Pickthall pointed out that the effect of acetylation is to increase strength of odour. With the esters there is no association, hence no solubilisation, and as a direct consequence the esters give in soaps a more intense odour than do the alcohols. Decyl alcohol gives a feeble odour, decyl acetate a strong odour; geraniol a feeble odour; tetrahydrogeraniol, which is a saturated compound and thus more soluble, gives a very weak odour; while geranyl acetate gives a very strong odour. So if we mix geraniol and geranyl acetate in a soap, it is the acetate which will dominate, whereas in an alcoholic solution the reverse will hold good. These results are valid for so-called anhydrous soaps which actually contain a certain proportion of water (5–10 per cent).

Esters vary considerably, one from another, in their stability and odour yield in soap; but as a group they are valuable constituents of soap perfumes. Among the carbinol esters styrallyl (methyl phenyl carbinyl) acetate is outstanding.

Alcohols

Alcohols tend to give uniformly lower odour values in soap than do the corresponding esters, but for the same reason they usually remain well 'fixed' in the soap and their odours though mild are persistent. Fatty alcohols have weak odours, but unsaturation in the chain improves the odour performance. Of the terpene alcohols the most important are linalool and terpineol. Anisic alcohol is useful in lilac and muguet.

Ketones

Ketone group performance varies considerably. The substances with a good odour yield in soap are the ionones, benzophenone, p-methoxy and p.-methyl acetophenones and ethyl amyl ketone.

Aldehydes and acetals

Many aldehydes — aromatic, terpene and aliphatic – are widely used in soap perfumery, despite their reactivity and relative instability as a class. Cyclamen, amyl cinnamic and lauric aldehydes are among those which can give very satisfactory results. For improving the stability of aliphatic aldehydes they are mixed with corresponding alcohol in order to form hemi-acetals.

Dimethyl and diethyl acetals cannot be used as simple substitutes for the corresponding aldehydes. Their odours are different and their performance not always encouraging.

Other constituents

Several ethers are distinguished by their satisfactory odour effects in soap. Examples are amyl benzyl ether, p.cresol methyl ether and diphenyl oxide. Terpenes can give quite interesting results. As one would expect, the odour yield of phenols ranges from moderate to poor. Eugenyl acetate gives a stronger odour than eugenol or isoeugenol, but it discolours and is not entirely stable. The effect of methylation, as in eugenol methyl ether, is to enhance the strength of odour. Of the lactones one may note such useful items as gamma-nonyl lactone, gamma-undecalactone and coumarin.

A typical formula contained large percentages of such natural products as neroli, cassia, cloves, geranium, lavender, patchouli, rosemary, sandalwood and vetiver, with only small proportions of coumarin and a few other synthetic aromatic chemicals and isolates. The wide range of synthetic odourants with standard quality and stabilising cost are available to the perfumer. This has simplified the selection of products of good odour and colour stability for creating new perfume.

Soap perfumery fashions

It is essential for the creative soap perfumer to keep in touch with changing fashions throughout the world.

A selection of new odourants available few years ago was discussed at a meeting of the *Society Technique des Parfumers de* held at France with a theme entitled 'Changing Fashions in Soap Perfumes'. In forties the white toilet soaps with restrained 'clean smelling' odours were fashionable. This has resulted into creation of more colourful and more pronounced, distinctive and even exotic perfume soaps. The colour in soap helps to conceal discolouration caused by odourants, which is otherwise unacceptable in white soaps. The higher concentration of perfumes serves the skin after bath provided odourants used are not irritating to skins.

At the end one can recommend that soap perfumes should frequently consist of the smallest possible number of ingredients and that, above all, each odourant should be selected for its strength of odour and stability in soap base. What is not wanted is a formula containing 'passengers' which do not contribute effectively to the final result and which reduce the perfume's overall intensity.

PERFUMING SYNTHETIC DETERGENTS

The perfumes used in blended synthetic detergents have a dual role :

(1) They mask the inherent odours of the raw materials.

(2) They impart a type of fragrance which meets the special requirements of the finished product.

There are, in general, five different types of off-odour associated with detergent materials, and these must be masked when they become perceptible :

(1) Odourous substances such as petroleum derivatives occurring with alkyl-aryl sulphonates; also slight petroleum odours present in non-ionics.

(2) Fatty odours originating from aliphatic alcohols found in alkyl sulphates.

(3) Sulphur odours sometimes originating from such detergents as nonyl phenol sulphonate made from propylene trimers and tetramers, which may contain sulphur derivatives.

(4) Odours formed as the result of slight oxidation of certain alkyl-aryl detergents. To avoid this condition, various antioxidants have been recommended.

(5) Odours resulting from pyrolytic action when detergents are overheated during the drying processes.

Formulator will aim at a combined job of odour-masking and definite perfuming. Here he is faced with the type and end-use of the detergent. So-called light duty detergents for washing dishes should have clean, fresh odours that do not conflict with food odours and do not persist after the detergent has been used. Nobody wants a perfume to adhere to china and glassware after these utensils have been washed. For this purpose lemon odours are popular. A more sophisticated perfume is used for light duty textile detergent, used for washing delicate fabrics and feminine garments. This perfume leaves behind faint but pleasant residual odour. It is necessary to concentrate on the more alkali-resistant odourants for heavy duty detergents and products of high pH. Then there are detergent-disinfectants to be considered, in which convention may restrict the choice to pine, lavender, rosemary.

Stability of Perfumery Chemicals

The experiments to determine the stability over a period of one year, in an alkaline perborate-containing detergent, of a number of conventional perfumery chemicals is given below : Mix 100 mg of each odourant with 100 mg of detergent containing 5 per cent of sodium perborate. The pH of 1 per cent solution at 20°C was 9.5. After one year each odourant sample was examined for odour. Prepare 1 per cent solution freshly and compare with one year old solution. The sample under test is again re-examined after boiling for 30 minutes and for one hour. The following odourants showed good stability under all these conditions.

Cineole, linalool, menthol, linalyl acetate, phenylethyl acetate, methyl salicylate, acetophenone, methyl ionone, musk ketone, p.methoxy and p.methyl acetophenones, anethole, diphenyl oxide, cyclamen aldehyde. (Anethole undergoes limited oxidation but is otherwise satisfactory. Vanillin is transformed to its sodium salt, loses odour intensity, and acquires a rubber-like note.) The methyl and ethyl ethers of beta-naphthol could be added to this list, as well as most of the odourants once used in perfuming the so-called cold process soaps.

Examples of a lingering perfume for adding to wash-day fabric cleaners and a more evanescent perfume for use in a dishwashing detergent have been suggested. The former contains from 3 to 10 per cent of such long-lasting odourants as styrax, amyl cinnamic aldehyde and cedryl acetate, together with 10 per cent of heliotropin; while the second formula is based largely on relatively transient esters and unobtrusive alcohols. The 15 per cent each of benzophenone and lauric alcohol, together with smaller quantities of cyclamen aldehyde, musk ketone, heliotropin, coumarin and oil of patchouli rounding off the complex with phenylethyl acetate and other rose-type esters and alcohols are suitable as perfume for household laundry detergent. Many alternatives are available to the practising perfumer.

Among the household cleaning aids which are now-a-days sent to market with the added incentive of a pleasant perfume are scouring powders, carpet and other specialised cleansers, starches and bleaches. Lavatory bowl cleaners are also perfumed, whether they are based on alkalies or on sodium acid sulphate. Liquid cleaners based on ammonia and window cleaners, sometimes incorporating both ammonia and isopropanol, may likewise be perfumed—and, indeed, compounded perfume compositions of good stability and all-round performance are specially produced and marketed for all these purposes.

PERFUMED DISINFECTANTS

Perfumes can in fact exert a dual function in disinfectants and antiseptics; and in similar fashion can act as preservative agents in a wide variety of products, ranging from embalming fluids to cosmetics. Odourants as a class possess antifungal and antibacterial properties. The various groups of perfumery chemicals, of which tested, exhibit the following order of decreasing antibacterial activity :

aldehydes < alcohols < acids < lactones < ethers < ketones < esters < acetals.

Of the few phenols tested eugenol was the most active. It is curious, perhaps, to find acetals at one extreme of the list and the parent aldehydes at the other. Benzaldehyde and undecylenic aldehyde inhibited growth of *B. subtilis*, *E. coli*, *S. aureus* 10390 and *S. aureus* Ox-H at concentrations ranging from 1 in 500 to 1 in 2000. Methyl acetophenone also showed a broad spectrum of effectiveness at 1 in 500. Octyl and decyl alcohols, cinnamic aldehyde, heptyl formate, cedrol, methyl heptin carbonate and eugenol were highly effective against some but not all of the bacteria cultured in these tests.

The more widespread use of essential oils and perfumery chemicals as preservatives, antiseptics and disinfectants is restricted due to cost. The essential oils are by their nature volatile and their effectiveness tends to diminish with time. Even so, they are frequently used to support the main antibacterial, antifungal action, thus helping to make the product in which they are used both more attractive in odour and less liable to micro-organic attack.

Chlorophenolic disinfectants have commonly been improved in odour by the incorporation of odourants like terpineol, benzyl acetate, diphenyl methane, diphenyl ether, terpinolene, safrole, bornyl acetate; together with such essential oils as citronella, lemongrass, steam-distilled pine, camphor and cinnamon leaf. In the following formula of basic perfume compound, the terpineol acts not merely as an odourant but also as a solvent or coupling agent and as an auxiliary antibacterial agent :

Formula 1

Constituents	*per cent*
Terpineol	50
Lemongrass oil	10
Pine oil, steam-distilled	10
Benzyl acetate	8
Terpinolene	7
Citronella oil	10
Bornyl acetate	5

Disinfectants based on quaternary ammonium compounds, e.g. cetrimide, are particularly well adapted for perfuming. In developing perfumes of this type it is necessary to study the solubilisation of perfumery materials, the relative 'emergence' of odours from mainly aqueous solutions, and the mechanism of complex formation in mixed cationic/nonionic products.

Effective covering of the odour is said to have been achieved with the aid of two alternative mixtures consisting of 1 gram of either musk xylol or nerolin II (ß-naphthol methyl ether, or yara yara) dissolved in 10 mls of carbon tetrachloride, 15 mls of sulphonated castor oil and 1.5 grams cetyl alcohol, made up to a total of 100 mls with water. The dilution required is 1 ml of either mixture to every 100 mls of 1 per cent hypochlorite solution. Both blends were reported to ensure adequate odour masking, coupled with resistance to the chemical action of the hypochlorite, non-toxicity in the required concentration, and non-interference with bactericidal action.

Some simple basic perfume formulae suitable for addition in small proportions to a non-reacting disinfectant base reads as follows :

Formula 2

Constituents	*per cent*
Citronella oil	25
Thyme oil, red	25
Sassafrass oil, artificial	25
Terpineol	25

Most perfumers would probably desire to bring such a formula up to date, by part replacement of some of the constituents and the introduction of other, newer synthetics; or at least by adjusting the stated proportions in order to get the optimum odour-effect.

PERFUMING THE AIR

In olden days the products like cinema sprays, perfumed sickroom sprays, specialised air disinfectants, deodourisers, and wick type air fresheners were used. The latest development include room sprays of the aerosol air freshener type. Also in the same general category may be included incense cones, pastilles or papers; the joss sticks known as agarbatties; and even fumigant candles and lavatory deodourant blocks.

Perfumes for aerosol air fresheners and similar purposes may have a light, refreshing floral odour, or they may be 'green', minty, herbal, spicy or woody. They offer considerable scope, in fact, for the enterprising and ingenious perfumer because they should be fairly persistent (as covering or masking agents for kitchen smells) and yet light and unobtusive. Citrus and lime blossom notes can be attractive. Pine, pine needle, cedarwood and cedarleaf oils have the requisite pungency, and blend well with other useful odourants such as terpineol, terpinyl acetate, borneol, bornyl acetate, thyme, rosemary, spike and lavender.

The leading aerosol air freshener was at one time a blend of orange, cologne, verbena, lavender and clove. Others that achieved varying degrees of popularity were lemon and pine, lavender and pine, lemon and eucalyptus, lavender, honeysuckle, jasmin, floral and pine, oriental, jasmin-lilac, rose and spicy-floral. The leading wick type room deodourants were said to have lemon-pine and camphor-pine odours.

The theory behind the development and use of air fresheners is not as simple and uncontroversial as might be supposed. Three types of odourant come into consideration :

(1) Those that mask or cover less pleasant odours, those that temporarily inactivate or anaesthetise the olfactory nerves and thus lessen sensitivity to malodours;

(2) Those that act by pairing with specific malodourants to lessen sensitivity to malodours;

(3) Those that act by pairing with specific malodourants to lessen the combined odour-intensity.

Most of the odourants used in practice come into the first category. A few, such as formaldehyde, acetaldehyde are to be found in the second category. Those in the third group, being limited by the specificity of their action, are similarly limited in number. 'Some of the more interesting pairs of odours compensating each other are: skatole and coumarin, ethyl mercaptan and eucalyptol. The classical studies of Zwaardemaker, also claimed that musk could 'neutralise' the odour of bitter almonds, cedarwood the odour of rubber.

Room deodourants containing formalin (solution of formaldehyde) need also a covering agent capable of disguising the formaldehyde odour. Perfumery supply houses, aware of this need, have perfected perfume blends stable in the presence of the aldehyde and possessing such odours as fruity, heliotrope, sweet vanillin, jasmin, lavender, lilac, pine. These are normally used at the rate of 1/2 oz to the gallon of formalin. Similar blends are specially compounded as 'covers' for the odours of isopropyl alcohol, acetone and para-dichlorbenzene—as used in various household cleaners, polishes, lavatory deodourants etc.

INCENSE AND FUMIGANTS

Incense and fumigants represent the oldest method of perfume release. The burning of fragrant woods, spices, resins and balsams, a practice which in itself gave rise to the word perfume (*per fumum* = through, smoke). The burning of these fragrant materials has persisted through the ages and is extremely widespread geographically. Among the odourants commonly used in incense cones, joss sticks, agarbatties, Armenian papers and fumigants of all types—often in association with charcoal, saltpetre and a mucilaginous binding agent are Olibanum (incense, frankincense), benzoin, myrrh, labdanum, styrax, mastic, elemi, Tolu, Peru and other oleoresins and balsams; sandal, agar, deodar, cedar, cascarilla and other woods; cloves and cinnamon bark; vetiver, patchouli, sandal, cedar and various other essential oils; and, finally, a number of crystalline synthetics of sweet, persistent odour, including vanillin, coumarin, heliotropin, rose crystals and the nitro-musks.

PERFUMED CANDLES

The most popular type of scented fumigant is currently the so-called fragrance candle. Candles which give off a pleasant aroma, whether they are lit or unlit. The perfume, which must be soluble in the wax-stearic acid candle base, may be present in proportions ranging from 2 to 5 per cent or even, exceptionally, 7 or 8 per cent.

PAINTS AND POLISHES

The paint industry has introduced a host of new and useful products during the second half of the twentieth century. The problem of paint's inherently malodourous character still remains. The worst offenders are solvents and thinners, oils and resins, as well as certain of the more specialised constituents such as anti-skinning agents. Vanillin and coumarin are much used in paint and lacquer 'deodourants' (i.e. as masking agents and reodourants). Sweet woody bouquets, violet-vanilla, spicy-vanilla, oriental and sweet citronella compositions, have found favour in this respect.

The following composition for a paint deodouriser has been suggested: amyl butyrate 6, benzyl acetate 21.6, phenylethyl alcohol 31.2, citral 5.6, geraniol 20, phenylacetic acid 7.8, coumarin 7.8 parts respectively. While this is of some interest as a specific reference.

Too much perfume, or perfume imperfectly compatible with the formula, may spoil the consistency, levelling, drying or even the adhesion or durability of the paint. The perfume should therefore have a good odour yield, even on dilution. Its various components should be able to mask the more volatile matching of the solvents and the more persistent lower notes of synthetic resins, oils.

A typical perfumery supply house markets a series of solvent masking odourants and marketing series of resin-oil masking agent. It also provides special odourant perfumes for aerosol points and lacquers and masks for sealers, primers and paint removers.

Pine, cedar, lavender and citronella types of perfume are among those used for the improvement of furniture polish odours. Among the odourants commonly used are safrole, bornyl acetate, terpinolene, terpineol, steam-distilled pine oil, spike and lavandin oils, coumarin, musk xylol. Shoe polishes still tend to smell of bitter almonds but cuir de Russie, lavender, pine, woody and lilac compounds are all on offer to the more enterprising leather polish manufacturers.

OTHER HOUSEHOLD PRODUCTS

Other household products that call for specialised perfuming are adhesives of various types (e.g. with minty lilac or carnation-clove compounds), laundry starch, para-dichlorbenzene blocks, lavatory bowl cleaners, plastic sheeting, paper tissues and dry cleaning fluids. With the advent of micro-encapsulated perfumes the range of 'reodourisation' has been further extended to cover reactive products in which some normal, unprotected odourants would be unstable.

PERFUMING COSMETICS

Perfumes devised for use in cosmetics and toilet preparations have to be dermatologically acceptable as well as aesthetically appealing and chemically and physically compatible.

Much has been written on the subject of perfumes that can give rise to primary irritation and those that may provoke reactions of an allergic nature on certain hypersensitive skins. Perfumes and perfumed cosmetics are now-a-days increasingly subjected to stringent dermatological 'prophetic patch', 'repeated insult' and other specially devised tests before they are distributed for sale to the general public. A noted French allergist, describes these tests as, 'Allergic reactions due to cosmetics and especially to perfume oils are extremely rare, and notably so if one relates them to the amounts used daily.'

The first practical experiments in determining the relative mildness on the skin of various perfumery materials can be carried out. W.A. Poucher converted irritant lilac formula into two milder versions, by omitting some but not all the aldehydes, replacing a phenol with a phenol ether, and increasing the relative proportions of alcohols and esters.

Adapting the Perfume to the Vehicle

All perfumers know, the fact, that odourants and combinations of odourants exhibit variable odour characteristics according to the medium in which they are used. 'A given composition, when incorporated in a hair oil, reveals a quite different odour from that which it has when used in a powder or as a simple alcoholic solution. A similar type of odour shift occurs when the perfumed product is put to use, i.e. when the cream or lotion is applied to the skin or the hair oil to the hair.

The same phenomena can be described in rather different terms. The problem faced by the manufacturer who produce wide range of cosmetics with same odour is that, each product has different effect on the perfume. In order to achieve the same odour effect (e.g. with a water base, a water-alcohol solution, an oil-in-water emulsion, an anionic shampoo, a water-in-oil emulsion and a clear oil base) it is necessary to balance the perfume composition for each and every preparation.

In some cases the variation in odour, using a simple standard formula in several different vehicles, may not be excessive, but in most cases some special modifications will have to be made in the standard perfume formula in order to ensure a constant level, throughout the series, of characteristic odour effect and odour intensity. The odour yield or relative intensity of anisic alcohol or phenylethyl alcohol in a watery base is much lower than that of the more water-insoluble amyl salicylate or geranyl acetate. It has been shown, in two specially devised formulae, how two parts of coumarin in a mineral oil base will have to be replaced by six parts in a water base, if the same apparent odour effect and intensity are to be obtained. The respective figures for phenylethyl alcohol are 10 and 30, but in the case of amyl salicylate the effect is reversed and they become 10 and 1.5, while for geranyl acetate they are 10 and 0.6, and for alpha-ionone 2 and 0.3.

Perfuming Creams

Pickthall worked on the surface activity of odourants. The study revealed that according to type and structure of chemical, they interfere more or less and sometimes deleteriously with the structure of cosmetic emulsions e.g. Thus 'terpineol is an extremely active substance and can make or break an emulsion. It has an adverse effect upon certain water-in-oil emulsions and in many cases will break a cream in which wool wax is the emulsifying agent.' Moreover, 'benzyl alcohol tends to solubilise or peptise emulsifiers and greatly change the nature of an emulsified product'. The knowledge of reactivity and behaviour of cosmetic emulsifying agents like, anionic, cationic and nonionic makes elimination of pitfalls and formulate cream perfumes.

Apart from dermatological and physico-chemical considerations, the perfumer has also to take into account the type of perfume to be used in a cream and the level at which it should be incorporated. Simple floral perfumes are not so widely used as formerly, although rose-based bouquets, being satisfactory from so many points of view, are still justly popular. The proportion of any perfume used in a cream is best arrived at by joint decision of the perfumer and the cosmetic chemist, but in general it will lie between 0.15 and 0.6 per cent of the total formula.

Perfuming Powders

The perfuming of cosmetic powders is one of the most difficult tasks confronting the perfumer. Powders are usually mixtures of materials, each with different structures and adsorption/absorption characteristics. The separate constituents may and often do differ from source to source or even from delivery to delivery. Talc is a notable case in point. Changing the grade of talc can readily modify the perfume odour and increase or decrease its intensity. Such phenomena call for the closest co-operation between perfumer, cosmetic chemist and the quality control laboratory.

When a perfume is intimately distributed throughout a powder of fine particle size, the exposed surface may be very great and the likelihood of evaporation and oxidation correspondingly increased; particularly when a pack has been opened and the contents exposed. This calls not only for careful selection of perfumery materials and expert compounding but also for equal care in the choice of re-sealable containers such as talcum canisters. If plastic packs are used in place of metal, adequate tests have to be carried out to ensure that the plastic is one which does not allow too marked a loss to take place through it. The rate of vapour loss of individual odourants through the plastic walls of a container varies, with the result that a blended perfume subjected to loss of this kind soon becomes unbalanced and of altered character.

Talcs, kaolins and diatomaceous earths have their own, often earthy odours, and an inadequate powder perfume will sometimes allow this inherent off-odour to 'show through'. Much pre-testing of odourants in these bases is often desirable before compounding and testing perfumes for use in face, talcum, body and baby powders. Typical proportions of perfume range from 0.2 per cent for baby powders to 1.5 or even 3 per cent for face powders.

Perfuming Lipstick and Nail Lacquer

Lipstick perfumes must to some extent be flavours, but the general aim is to make them have a pleasant taste without actually suggesting flavour-appear. There are many odourants with a bitter or otherwise objectionable flavour, and these are to be avoided.

The banana or pear odour of nail lacquers based on nitrocellulose and volatile solvents is not invariably liked, but the masking of this odour is difficult in view of both the volatility and the strong odour of the vehicle. It is relatively easy to devise a perfume which will persist on the nails after the lacquer film has dried. The nail varnish manufacturer prefers to have a nail lacquer having pleasant odour in the bottle and while drying on the nails. On drying, however, the film formed on nail is odourless.

Hair Preparation Perfumes

The hair constitute depilatory formulae and thioglycollates. This creates disadvantages as they are reactive with some perfumery chemical, liberates sulphurous odour and alkaline condition unsuitable for odourants. Some helpful suggestions are to be found in the perfumery literature, relating to the behaviour of some 200 odourants in ammonium thioglycollate solution, and to the suggested perfuming of depilatories with combinations of camphoraceous woody and herbal notes, rosemary, patchouli, spicy, ambered and musky notes etc.

Shampoos must have a pleasant odour in the bottle and also leave behind an attractive aura, more or less persistent, on the dried hair. Some perfumes have a marked and, of course, undesirable softening effect upon hair lacquer films. Perfumes for use in hair sprays should not contain excessive amounts of odourants (e.g. terpineol) having a solvent action on the resin content; they should not, moreover, contain more than a limited proportion of high-boiling constituents, as these can unduly prolong the drying time.

Bath salts and bath oils are often highly perfumed. The former may have to be alkali-resistant.

Perfumed Aerosols

The traditional handkerchief perfume consists of a number of odourous constituents dissolved in alcohol. Its aerosol or pressure-packed equivalent exists as an alcohol-odourants solution in association with a liquefied propellant, usually of the halogenated hydrocarbon type. It is necessary to take into account the influence of the propellants, constituents of aerosol, effect of value material and metal container on the stability of perfume package. In a chromatographic study of the action of propellants on the perfume constituents of aerosols, it has been found that only certain aldehydes (hydroxycitronellal, decanal and undecanal) underwent chemical modification, viz. acetalisation, under the conditions of testing, and then conventionally stabilised with nitromethane. Hydrocarbons, alcohols, carbinols, esters, ethers, ketones, lactones and nitro-compounds are apparently unaffected. Even the modified aldehydes remained unchanged when the nitromethane stabiliser is replaced by 1 in 10,000 of morpholine, o.-tolyl biguamide or diphenylamine.

In one of the most comprehensive studies of aerosol perfumes, it is suggested that propionates, butyrates, valerianates, phenylacetates and lemon oil should in general be avoided, and ionones and orange oil chosen with care.

Perfumes in non-alcoholic vehicles include the perfumed candles already discussed; also other solid perfumes such as perfumed pommades, solid Colognes, and the briquettes used as moth repellants and to impart an elusive odour to linen cupboards. The perfumed pommade was the forerunner of the solid cream sachet and this, in turn, gave rise to the liquid cream sachet. Liquid sachets may contain up to 10 or 12 per cent of perfume oil, and this naturally gives rise to problems of stability. The perfumer will avoid readily hydrolysable odourants and other materials likely to favour emulsion breakdown and discolouration, and in this may be assited by keeping the emulsion on the acid side, at a pH between 5 and 6. So-called solid Colognes usually have a high content of menthol.

PHARMACY AND MEDICINE

Perfumes have in fact been used for centuries in medical and pharmaceutical (not to say alchemical) practice. Certain odourants are of special utility, quite apart from their odouriferous character, as antiseptics bacteriostats, fungistats, carminatives, anodynes, mild local anaesthetics and cooling agents, fumigants, inhalants, counter-irritants etc. It has been discovered that some essential oils promote expectoration by direct stimulation of the respiratory tract. The most effective oil for this purpose was anise, the runners-up being oils of turpentine, pine, eucalyptus and lemon.

The impact of perfumes and odours upon medicine dates back to the times when the burning of malodourous substances around a sick person was thought to be a means of driving out demons, and when odourants were used as disinfectants, carminatives, primitive tranquillisers and, of course, embalming agents. Odour can be helpful, too, in diagnosis. Some diseases have distinctive odours. Attention has been given to the psychological and physiological aspects of odours and odourants, but even so the subject is still largely unexplored and could well repay serious investigation.

INDUSTRIAL PERFUMES

Perfumed gloves go back through the ages. Perfumed textiles are as old, in origin, as the wax and resin treated cere cloths used in embalming. Perfumes were devised for covering the off-odours imparted by dyes and other additives to crush resistant valvet.

Perfumes for Textiles

Perfumes for textiles should be substantive to the material to be treated and yet sufficiently volatile to be readily perceived. That some odourants are remarkably persistent in these conditions may be appreciated when smelling a face cloth that has been used for soap-and-water washing for a deliberately prolonged period of several weeks. Notably persistent, for example, are many of the synthetic musks, patchouli, gamma-undecalactone and methyl nonyl acetaldehyde. Some compounds may have to withstand unusually high processing temperatures and resist several washes.

The main function of masking agents devised for addition to dry cleaning fluids is to cover solvent odour and leave behind a faint residual perfume on the treated fabric or garment. In all cases adverse reactions between odourants and fabrics, dyes, finishing and proofing agents etc. must be avoided.

Perfumed Ink and Paper

The perfuming of printing ink has been given a good deal of attention and special perfumes have been devised to blend well with all kinds of printing inks without detriment to the various printing processes. The proportion of perfume tends to range between 0.3 and 0.5 per cent of the quantity of ink used, in the case of absorbent newsprint, to 1 per cent for normal coverage on most papers, or even 5 per cent where a specially perfumed effect is required.

Paper in many forms may be subjected to the perfumer's techniques not only paper perfumed by means of printing ink but spray-perfumed cards, programmes, stationery; as well as waxed wrapping papers, towels, tissues, toilet paper etc. One of the most important uses of masking odours in the paper industry is in gummed papers.

Widely used in nearly all industries, gummed tapes have, normally, an almost universally unpleasant odour and taste. A number of companies have endeavoured to eliminate either or both of these undesirable factors, and with the help of the perfume and flavour chemist have been achieved satisfactory results. Masking odours of a peppermint or spearmint, sassafras, vanilla or spice type are commonly used and can be prepared in either oil or water-soluble form, depending upon the manufacturing process. In such instances it is usually an animal base glue that is the olfactory culprit, and this adhesive is normally applied hot... at around 130°F to 140°F, although some go as high as 350°F for a brief, finishing process. This requires an odour and/or flavour that will withstand such temperatures.

The masking compounds must possess unusual stability, as the product is sometimes stored for as long as two years before actual use. Clove oil, safrole and methyl salicylate are sometimes used but most satisfactory are specifically developed compounds that give the proper coverage and stability together with the distinctiveness of a tailor-made product. Cost is an important factor and must be kept to a minimum. Compounds developed can usually be applied in a proportion of 0.1 per cent. Tapes used in the food industry, particularly, are often masked, owing to the unpleasant odour associated with the animal glue involved and for this application usually a neutralised effect, rather than a perfumed one, is desired.

Unwanted odours and flavours must also be considered in connection with paper wraps and cartons and other forms of food packaging.

Masking Malodours

The perfuming of wallpaper as a means of masking or offsetting any inherent malodour of the paper or adhesive is probably a more attractive idea than that of seeking to impart a noticeable and persistent odour by this means.

Synthetic rubber emulsions, widely used in industry, are distinguished by their offensively ammoniacal odour. This is frequently masked by specially prepared compounds. Natural rubber products such as rubber glue, rubberised textiles, carpet and upholstery backings and shoe components, are also improved by the skills of the industrial perfumer. The same applies to plastics and other man-made materials and fabrics.

PERFUME IN AGRICULTURE

Agricultural and horticultural products, including fertilisers and insecticide sprays as well as cattle feeds and veterinary products, have been improved upon by the perfumer's art.

Most odour-masked fertilisers are more likely to be found in horticultural consumer type packs, put up specially for home gardening, rather than in the average farmyard. Big farms have associated problems.

Liquid fertilisers in bottles and powder fertilisers in bags or cartons may simply include a proportion of odour-masking compound or else may give a suggestion of new mown hay, apple loft, fougere or moss odours.

Perfumed Insecticides

The perfuming of insecticides is not likely to affect their insecticidal activity, but this factor must nevertheless be considered. In most cases it is necessary to mask the disagreeable odours of the active agents and/or solvents. Pheromone research has led to the experimental production of both insect attractants and insect repellents, based of course on species-specific odourous substances or compounds. Moreover, some essential oils have been proved to act synergistically, when added to fly sprays, by increasing the normal effectiveness of the active agent. *Backhousia myrtifolia, Zieria smithii, Melaleuca bracteata* and *Huon pine (Dacrydium franklinii)* are suitable essential oils for use as activators for increasing the effectiveness of pyrethrins in fly sprays. Oil of Backhousia is said to have proved more efficient than seasame oil, a widely accepted activator. Perfumery supplier, if requested, incorporate such oils in a pleasantly attractive perfume compound.

MISCELLANEOUS USES

It was a practice, half century ago, to use sprays into cinemas in order to add another dimension. Since that time special devices have been used to produce 'scented movies'. These include mechanical sprays for injecting perfumes into the ventilating system, aerosols and electronic precipitators.

The introduction of spray-dried and micro-encapsulated perfumes and flavours has further extended the scope of perfume application.

By the encouragement of good smells life will become a fuller and lovelier thing. To which may perhaps be added the afterthought 'Perfumes that persist on the skin; perfumes that persist on fabrics'. Our greatest good fortune, however, may be that perfumes, even the most persistent, are volatile by definition and sufficiently evanescent in the conditions of use.

FLAVOURS AS PERFUMES

Finally one comes to the important use of odourants, both natural and synthetic, in the preparation of flavourings. The flavour of a food or beverage is a perfume plus a taste. That is why in a book on perfumery one must glance, at the subject of flavours. Taste is perceived by receptor spots or 'buds' on the tongue and elsewhere in the mouth, whereas the odour part is perceived by the nose and, 'most of the discrimination attributed to the palate is really performed by the olfactory membranes'. The statistical occurrence of the true gourmet among perfumers and flavourists, food and wine tasters and chefs de cuisine, is likely to be at a higher level than one would find in the traditional ranks of bibulous clubmen and boardroom epicures.

Properly speaking the term 'flavouring' should be used to describe a substance employed to give a flavour. Yet another definition: 'a flavouring is a substance... which has predominantly odour-producing properties and which possibly affects the taste'. It is no mere coincidence that the world's most active and knowledgeable producers of synthetic odourants for perfumery also occupy approximately the same pre-eminent position in the field of artificial flavourings.

Biological and toxicological factors necessarily restrict the number of odourants that can be regarded as permissible for use in flavourings.

Among the flavourings dealt with in detail are fruit, beverage, liqueur, cordial, and confectionery flavours, herbs, spices, sauce and curry flavours, tobacco, toothpaste and pharmaceutical flavours and flavours for all types of 'convenience foods' and protein replacements.

Chapter 11

Packaging of Perfumes

INTRODUCTION

The elegant packing makes marketing easier. The essential factor in presenting perfume such as toilet water or cologne is attractive package. The technology and aesthetic attitude has reduced the problems of packing.

Another important feature of packing is that, the other component parts of pack should also be attractive and have public acceptance. What will the public accept for purchasing and what will they not accept? It is difficult, for example, to get the public to accept a pack in which the bottle is presented horizontally, i.e. is lying down. Why is this so? The reason is probably a fear that some of the perfume will spill and be lost, but there are other rejections and predilections for which there is no such ready explanation. One comes up against the imponderables of fashion and taste. Why is it that certain shapes of bottle simply will not sell? No obvious law regulates these preferences and only experience and flair can guide the designer's choice.

The major disadvantage of the packaged perfume is that the bottle can give no idea of the contents and affords no means of appreciating them—until, of course, it is opened. Hence the utility of tiny sample sizes and counter-displays featuring the perfume packed in an atomiser-dispenser. To attract the customer's attention, perfumers throughout the ages have tried to make the outward appearance of the pack as luxurious and conspicuous as the perfume deserves.

Packaging in the Past

In the fifteenth and sixteenth centuries perfumes were contained in cut crystal bottles, which were often encased in precious metal (gold, silver gilt, silver), finely enwrought. The whole attached to a wrought chain of the same metal and carried cross-wise or by a short chain on the wrist, with the purse which elegant women also carried.

In the centuries that followed, the bottles were of precious metal or decorated porcelain. Private and public collections contain many objects of this kind, some of them very fine and luxurious.

It was really only in the nineteenth century, however, when the large perfumeries were started, that bottles of original shape were produced for this purpose, each belonging specifically to a certain perfume.

Glass or crystal bottles assumed the most varied forms, following the inspirations of fashion and the general artistic tendency of the times. These shapes were more or less felicitous, although some of them may now appear to us a little tasteless and clumsy.

On these bottles were glued labels showing the name of the perfume and that of the perfumery house which had created the perfume. These labels, sometimes multicoloured, were apt to state the extraordinary properties of the product. Their dimensions varied. Thus for toilet waters, lotions and eaux de Cologne, they were often rather large and covered the whole surface of the bottle. For extracts, where the bottles were of smaller size, the labels were smaller, very luxurious and in most cases gilded.

The coffrets or boxes in which the bottles were encased were often quite luxurious, with embroidered materials or printed silk fabrics vying with one another in luxury and sumptuousness.

THE IMPACT OF AEROSOLS

The presentation was greately influenced due to development in bottle making. Aerosol packs raised the problems of valves, propellant reactions, and the use of metal cans or heavy and plastic-coated glass containers. An attempt was made to use neutral, odourless gases such as carbon dioxide and nitrogen. Labels then began to disappear and give way to printed decorations on the metal cans. With the coming of reduced internal pressures, glass containers with thinner walls became practicable. Concurrently the processes of printing on glass and productive of some fine decorative effects were much improved.

As regards plastic bottles, due account has to be taken of their permeability, which is in general unbalancing and sometimes actually denatures a perfume. Even the use in perfumery works of plastic pipes and tanks is often ruled out for this reason.

In case of lables whether they are still used or direct printing takes their place, the tendency has for some time been towards simplicity. This is partly due to fashion and the streamlining tendency in production and partly, also, to the continual increase in the cost of manpower and man-hours. Jobs that are done by hand simply have to be minimised, especially when one is trying to refute the allegation that perfumes are excessively expensive luxuries making fabulous profits for their makers. Here we should add in parentheses that the manufacturing perfumers themselves, by frequently referring to and exaggerating the costliness of their raw materials and so forth, are largely to blame for this absurd invention.

PERFUME IDENTITY

The packaged perfume can be identified by the name, the bottle, the label, the carton or coffret, and the outer wrapping.

The Name

The name establishes the identity. The names of perfumes have been divided into five groups.

(1) Those refer to flowers, e.g., *Rose jacqueminot*, *Vera violetta*.

(2) Those refer to love, e.g., *Amour-Amour*.

(3) Homage to person, a regime or fashion, e.g., Zibeline.

(4) Those which have foreign names or evocative of exotic places or romantic occasions e.g. Nuit de Noel, Quadrille.

(5) Which are more abstract, relating to letters and numbers, e.g., No. 5s N, S, Y.

The name must appeal to the class of customer for whom the perfume is intended. It should suggest the personality of the perfume, be relatively simple, short, original and easy to pronounce, even by foreigners (if export is to be considered).

Whoever chooses the name should appreciate the fact that the name is the perfume and is of extreme importance. Another fact that has to be faced is that in these days, in most so-called advanced countries, there exists a long list of names already registered as trade marks. The chosen name may not therefore be registrable. If not, another has to be chosen insted and then, perhaps, another and yet another. It is often wise to start with five or six alternatives, no matter how enthusiastic one may be about the name originally selected.

The Bottle

Bottles are mostly of glass, fitted with metal or plastic closures and, additionally, with either plastic plugs or liners made of cork-and-metal or plastic material. Ground glass stoppers are now-a-days but rarely encountered, this being due to their involving too much hard labour and also to their suffering from the defect of 'rattling', i.e. not fitting perfectly. Atomiser packs must also be noted.

In principle every large perfumery house has its own art department, whose function it is to be responsible to and with the management for all aspects of package design. This service involves the preparation of mock-ups or models of bottles, cartons, coffrets, labels etc. and it is necessarily accompanied by a technical service which pays close attention to all the practical details of presentation: screw threads of bottle necks and closures, dimensions of the boxes, quality control of all accessories, consideration of the quality of the glass. Rules and standards are laid down or devised, and these are then followed in collaboration with the purchasing department and the various suppliers.

Objectionably alkaline glass must in any case be avoided. In warm and humid climates it can result in bottles losing their polish and developing matt or frosted patches. The obsolete practice of packing in straw or contact with some qualities of board or even paper can similarly lead to matt or irridescent patches. Before use, bottles should be vigorously rinsed in demineralised water and thoroughly drained and dried.

Apart from the artistic merits of its line or shape, a perfumery bottle should always be steady on its base and not top heavy. The surprising fact is that this simple rule is sometimes broken.

Atomisers form a separate category. The internal gas pressures which they sustain make it essential for the glass containers to by cylindrical. Fancy shapes are out. Decoration is by printing on glass, this being the only process that can conveniently be used. The stainless metal closures, which are now available in various colour shades as well as in white or gilt, can be embellished with engraved or relief designs.

Labels

Despite the advances made in direct printing on to bottles, fine quality labels are still widely used. Their style and finish have never reached a higher standard than they reach today; and this particularly applies to very small labels meticulously designed and finished to the strictest specifications.

Labels, fabrics, board and paper that are likely to be exposed to daylight in window displays or on the user's dressing-table should be tested for their tendency to fade. This forcibly applies to many colour shades, as well as to gilded and silvered finishes. Such tests can be conducted in a solarium or by using arc lamps with selected electrodes always, of course, under standard conditions.

The adhesive used on metallised labels should be carefully chosen so as not to interact with the metal or varnish. Manufacturers should also beware of varnishes that can be dissolved or made tacky by spilt perfume.

Cartons and Coffrets

Each and every packaging material should be pre-tested and shelf-tested for faults. Some plastics, for example, show a marked disposition to warp when subjected to high humidity. Natural cork has the disadvantage of variability. Every aspect of a package should be studied as a possible source of off-odours. Nothing should be taken for granted. A rubber gasket in an aerosol container may contribute a disagreeable mercaptan note. Residual solvent odours from a label's ink or varnish, or from its adhesive—even the smallest items—must be scrutinised as potential trouble makers. A product in a clear glass bottle is subject to the influence of light, and it must be remembered that some perfumery materials, in these conditions, will oxidise, resinify or discolour.

External Wrapping

The perfume coffret containing the precious bottle is often protected by being enclosed in an outer carton fabricated from white or grey chipboard, which in turn is enveloped in a decorated paper outer-wrap. This paper may be suitably embellished with the name of the perfume, the name of the firm, and sometimes a coded reference number. It should be identifiable at a glance, and should in fact be an attractive and recognisable advertisement for both the perfume and its makers.

The instantly identifiable signs such as large letters, bands of different colours are all helpful to the heavily stocked retailer in the management and arrangement of wares. A good pack is one that inspires confidence in the product; that assures the user that the same perfume or one like it cannot be had under another label at half the price. Eccentric packs make a sensible woman think that the manufacturer takes her for a fool. There is one thing certain, and that is that a perfect pack is as rare as a really good perfume: which means that there are not very many on the market.

PERFUME AND THE WORLD OF FASHION

Whoever decides upon the perfume and the pack ought, in any case, to be familiar with the market, and this particularly applies in the case of an elegant perfume. Hence, indeed, came the many successes of the *couturier* and *couturiere*, the high-class dressmaking establishments, the trend-setters. The worlds of fashion and art have to be frequented and appreciated, and their reactions and suggestions carefully taken into account but the immediate interpreter should be the perfumer, not the marketing executive, the advertising accounts man or the industrialist promoter.

The Question of Colour

Most fine quality perfumes are green or yellow, whether artificially tinted or not. At least one well-known perfume is blue, but the former fashion for perfumes and toilet waters in mauve, violet, pink etc. is no longer acceptable. Perfumers with little experience sometimes decide to add no colouring matter, but the fact is that a colourless perfume looks insipid, while a perfume with a natural colour will tend to vary in shade from one batch to another unless it is corrected each time by small additions of a suitable dye. In this way one guards against complaints from customers who see a colour change as indicative of a decline in quality.

THE MEANS OF ADVERTISING

Press Publicity

One extraordinary feature of the selling of fine perfumes is that this has hitherto been achieved almost entirely without the aid of memorable commercial publicity. We say 'commercial' because writers and journalists have more than compensated for that deficiency. Books on advertising and motivation research seem to give very little space or attention to perfumery marketing.

Samples and Models

All forms of publicity should be considered on their merits for each particular case; the usual media, such as magazines and journals, newspapers, television and commercial radio; window displays, even perfumed cards and theatre programmes. The distribution of small samples of a perfume can provide excellent publicity but it is costly and can be wasteful unless it is handled with discernment by the retailers.

Finally there is a direct form of showing one's wares to the public, which consists of vapourising them in theatres, cinemas or other suitable places to the public. A variation of this method is the presentation of a perfume on a female model, just as a gown is presented on a mannequin. This method is particularly suitable for the introduction of a new *couturier* perfume during a showing of the latest season's fashions, so long as each day is confined merely to the showing of one perfume. This idea can be carried further by suggesting that special rooms or cubicles, or even travelling air-conditioned units, could be devised as a means of perfume advertising or collaborative perfumery propaganda, the perfume being sprayed into the space on to selected wall or other surfaces and subsequently removed by modern air conditioning in order to prepare the space for the next perfume. In this way, a whole series of perfumes might be presented in sequence. 'During the brief intervals required for deodourisation, music could be played and time allowed for conversation. This might be an amusing as well as a novel way of encouraging the appreciation of fine perfumes.'

Chapter 12

Testing of Perfumes

INTRODUCTION

The quality of the perfume is adjudged on the basis of analysis of its every constituent. The process involves two basic steps : (a) Sampling and (b) Analysis. The analysis is sub-divided into physical and chemical analysis. The physical analysis include, estimation of colour, specific gravity, viscosity, optical rotation, residue, solubility, melting and boiling ranges, olfactory assessment and freezing point. The chemical analysis include, determination of water content, alcohol and acid values, ester value, carbonyl value, and content, phenol, cineole, petroleum and mineral oil, chlorine and heavy metal content.

SAMPLING

The mode of sampling varies with the physical state of sample and from which it has to be drawn. In case of liquid product the sampling also depend upon number and capacity of containers. The following precautions should be taken while carrying out sampling.

(i)   The sampling device should be cleaned and inert e.g. aluminium or glass is most suitable.

(ii) If samples are drawn from different containers of same products, all of them should be properly mixed.

(iii) The sampling container should be completely filled, sealed airtight after sampling with suitable labels.

(iv) Store samples in cool and dark place and if required in inert atmosphere.

(v)   Samples should be protected from moisture, dust and spilage.

Sampling Equipment

Three types of sampling equipments are available :

(i) Closed type (divided or undivided) : The closed type of sampling tubes are used for heterogeneous, semi-liquids, free-flowing powders and pastes.

(ii) Open type : The open type sampling tubes are used for homogenous samples.

(iii) Sampling scoop : The samples in the form of powders, crystals, tablets and lumps if broken into pieces can be sampled.

Closed type (Undivided)

It consists of two concentric metallic tubes closely fitted into each other throughout their entire length, so that one tube can be rotated within the other. Longitudinal opening of about one-third the circumference are cut in both tubes (Fig. 12.1).

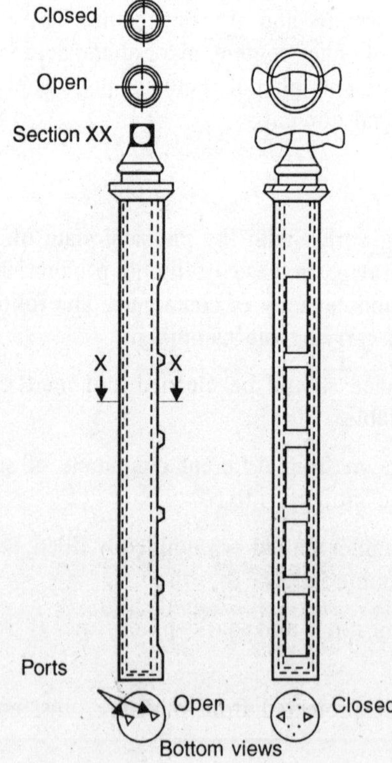

Fig. 12.1. Closed type sampling tube, undivided.

In one position, the openings in the two tubes coincide; the sample tube is open when in this position and admits the material. By turning the inner tube through an angle of 180°, it becomes a sealed container. The inner tube may have a diameter of 20 to 40 mm and is undivided along its length to serve as a single container.

The two concentric tubes shall be provided with V-shaped ports at their lower ends, so placed that the material contained in the equipment can be drained through them, when the longitudinal openings are in line.

The length of the equipment shall be such as to enable it to reach the bottom of the container being sampled.

The equipment is inserted closed, the material is admitted by opening it, and finally it is closed and withdrawn.

Closed type (Divided)

It is also of metal and has D-shaped cross-section. It is provided with compartments along its length and is opened and closed by means of a closely fitting shutter which moves up and down throughout the entire length. It may be from 25 to 50 mm wide (Fig. 12.2).

The equipment is inserted closed, the shutter is pulled out to admit the material, and the tube is then closed and withdrawn.

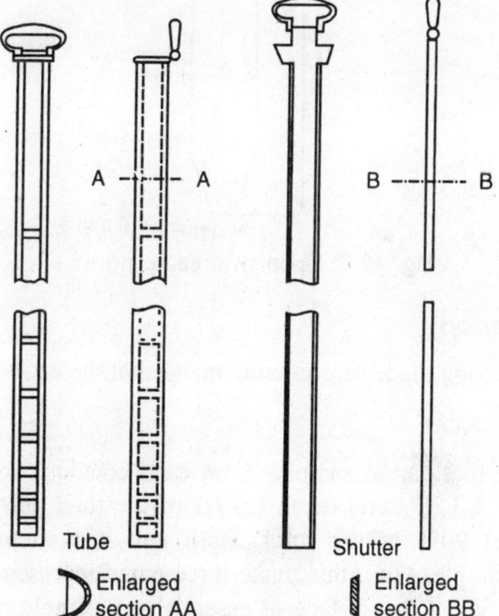

Fig. 12.2. Closed type sampling tube, divided.

Open type

It is made of metal or thick glass, and may be of 20 to 40 mm diameter and 400 to 800 mm length. The upper and lower ends are conical and narrow down to 5 to 10 mm diameter. Handling is facilitated by two rings at the upper end (Fig. 12.3). For taking a sample, the equipment is first closed at the top with the thumb or a stopper and lowered until the desired depth is reached. It is then opened for a short time to admit the material and finally closed and withdrawn.

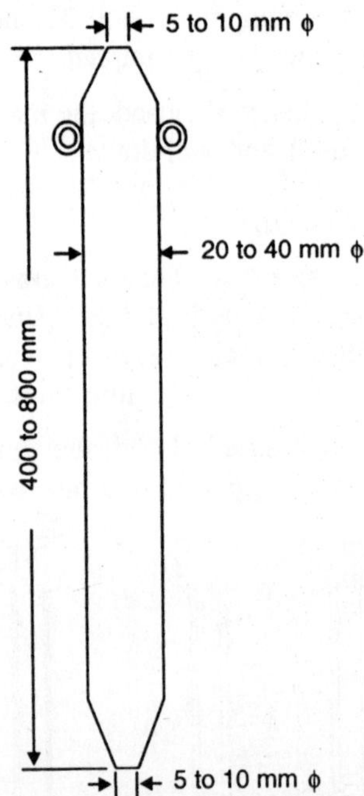

Fig. 12.3. Open type sampling tube.

Sampling scoop

The sampling scoop made of a suitable metal shall be as shown in Fig. 12.4.

Sampling from tanks

First withdraw five partial samples from each container at (a) 10% of the total depth (b) 1/3 of total depth (c) 1/2 of the total depth (d) 2/3 of the total depth (e) 90% of the total depth and mix them to obtain one homogenous sample. From this make three equal representative sample. In case of impurities at the surface of essential oil, sample it separately, mix it and obtain three representative sample.

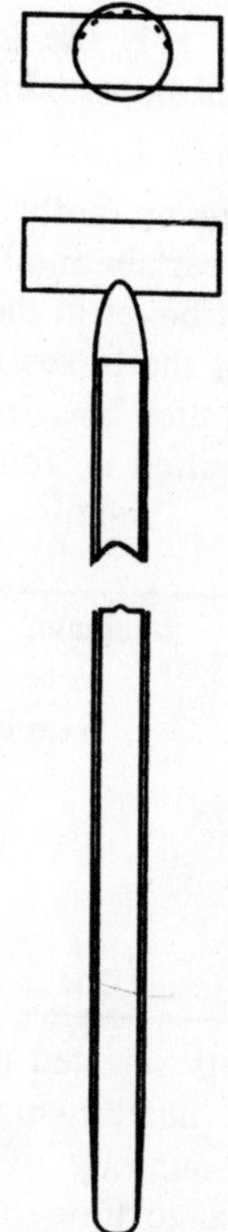

Fig. 12.4. Sampling scoop.

The composite sample can be obtained by taking equal quantity of material from the three samples obtained and mixed. The quantity of the sample should be sufficient to carry out triplicate analysis.

All the individual and composite samples shall be divided into three equal parts, one for the purchaser, another for the supplier and the third to be used as a referee sample. For packing of individual and composite sample and for referee test samples the procedure are given below in preparation of test samples.

The number of containers to be selected for this purpose (n) shall depend on the size of lot (N) and shall be in accordance with Table 12.1.

The containers shall be selected at random. In order to ensure the randomness of selection, following procedure may be adopted :

(1) If the lot consists of individually packed containers, then starting from any container in the lot, count them in any suitable order as 1, 2, 3......

up to r and so on where r is the integral part of N/n. Every rth container thus counted shall be withdrawn till the requisite number of containers is obtained.

(2) If the lot consists of cartons or cardboard boxes each containing more than one container, then a certain number of boxes (not less than 5 per cent of the total number of boxes in the lot wherever feasible) shall be chosen first. From each of the boxes so chosen, approximately equal number of containers shall then be selected so as to obtain the required number of containers specified in Table 12.1.

Table 12.1. Scale of sampling.

Size of the Lot (N) (No. of Containers)	Minimum No. of Containers to be Selected (n)
1 to 3	Each container
4 to 20	3
21 to 60	4
61 to 80	5
81 to 120	6
121 and above	One is every twenty

From each of the containers selected according a small representative portion of the material shall be drawn with the help of the sampling equipment. The approximate quantity of material to be drawn from a container shall be nearly equal to thrice the quantity required for testing indicated below.

Criteria for conformity

For individual samples

For each of those characteristics, which have been determined on the individual samples the mean ($\bar{x}$) and range (R) of test results shall be calculated as follows :

$$\text{Mean } (\bar{x}) = \frac{\text{The sum of test results}}{\text{Number of test results}}$$

Range (R) = The difference between the maximum and the minimum values of the test results.

If the specification limit for the characteristic is given as a minimum, then the value of the expression ($\bar{x} - KR$) shall be calculated from the relevant test results. If the value so obtained is greater than or equal to the minimum limit, the lot shall be declared as conforming to the requirement of that characteristic. If the specification limit for the characteristic is given as a maximum, then the value of the expression ($\bar{x} + KR$) shall be calculated

from the relevant test results. If the value so obtained is less than or equal
to the maximum limit, the lot shall be declared as conforming to the
requirement of that characteristic. If the characteristic has bilateral
specification limits, then the values of the expression $(\bar{x} \pm KR)$ shall be
calculated from the relevant test results. If the values so obtained lie
between the two specification limits, the lot shall be declared as conforming
to the requirements of that characteristic. The value of the factor K referred
shall be chosen in accordance with Table 12.2. given below, depending
upon the acceptable quality level, namely, the percentage of non-conforming
containers that could reasonably be tolerated.

Table 12.2. Values of K for achieving different acceptable quality levels.

Acceptable Quality Level	Value of K
Not more than 3 per cent defectives	0.4
Not more than 1.5 per cent defectives	0.5
Not more than 0.5 per cent defectives	0.6

For declaring the conformity of the lot to the requirements of all other
characteristics determined on the composite sample, the test results for
each of the characteristic shall satisfy the relevant requirements given in the
product specification.

Examples :

In case of oil of peppermint total alcohols as menthol (per cent by mass)
is to be tested on individual samples and the value of K has been chosen
as 0.4. In the case of pine oil 'Specification of pine oil', distillation yield (per
cent by volume) and total alcohols as terpineol (per cent by mass) are to
be tested on individual samples and the value of K is 0.5. The conformity
of the lot to the specification requirement in these two cases shall be judged
as follows :

Product	Characteristic	Type	Specification Requirements	Criteria for Conformity
Peppermint oil	Total alcohols as menthol, per cent by mass	1	75, Min.	$(\bar{x} - 0.4\,R) \geq 75$
	–do–	2	45, Min.	$(\bar{x} - 0.4\,R) \geq 45$
Pine oil	Distillation	Below 185°C	5.0, Max.	$(\bar{x} + 0.5\,R) \leq 5.0$
	yield, per cent	Below 200°C	25.0, Max.	$(\bar{x} + 0.5\,R) \leq 25.0$
	by volume	Below 230°C	95.0, Min.	$(\bar{x} - 0.5\,R) \geq 95.0$
	Total alcohols as terpineol, per cent by mass	—	70, Min	$(\bar{x} - 0.5\,R) \geq 70$

Packing and labelling

All containers should bear labels showing at least the following information to guarantee the authenticity, indentity of the sample and sealed :

(1) Sample number;

(2) Nature and quantity of the product;

(3) Name of the owner or his authorised representative;

(4) Date of sampling;

(5) Number, kind and marking of the containers;

(6) Signatures and names and, if necessary addresses of the interested parties or their authorised representatives; and

(7) Signature and name of the sampling supervisor.

Instructions regarding costly materials

Small containers are generally used for packing costly products. The bulk sampling depends on the number of containers used; the combined partial samplings, however, shall not exceed the quantities necessary for a normal analysis. The interested parties shall agree in advance as to the size of the bulk sample and the manner in which it shall be done.

ANALYSIS

The representative sample obtained using above procedure is brought to the laboratory for physical and chemical analysis.

Preparation of the Material for Physico-chemical Analysis

Remove as far as possible any visible water present in the sample by decanting, then add about 10 per cent by mass of neutral, freshly ignited and powdered magnesium sulphate and shake the mixture vigorously from time to time during a period of 2 hours. Filter through paper and store if necessary in clean, dry, airtight, opaque, non-absorbent containers, preferably of glass or of metal on which the sample has no action. Fill the sample containers leaving a little air space to allow for expansion and immediately seal after filling. Use only new good quality corks or glass stoppers or screw-lids in the case of glass containers and new good quality corks or sore stoppers in the case of metal containers. To prevent contact with the sample, aluminium foil or other suitable material may be wrapped round corks. Protect samples from heat.

In the case of perfumery materials which are solid or partly solid at room temperature (oils of rose and guaiac wood) warm the material just sufficiently to liquefy if and proceed as above, maintaining the temperature so that it remains liquid throughout. The colour of some dark coloured perfumery materials may be lightened by shaking them intermittently for 10 minutes with 1 per cent (*m/m*) of tartaric or citric acid. Filter and proceed as above.

Physical Analysis

The original containers selected for sampling and the test samples drawn, as the case may be should be opened and the condition of the content noted for colour, clarity and by note. Then the following physical properties of the sample were determined.

Estimation of relative density

Estimation of relative density by Pyknometer.

Principle

The relative density of perfume is very important as it plays vital role in the measurement. It is defined as the ratio of the density of the the liquid material to the distilled water at 27°C. It is dimensionless.

Apparatus

Pyknometer or specific gravity bottle, water bath and analytical balance, thermometer.

Procedure

First clean and rinse pyknometer or specific gravity bottle with alcohol and dry it. Take weight of empty specific gravity bottle with stopper. Fill the specific gravity bottle with distilled water and weight it. Empty the bottle, dry it and filled it with material under the test, and weight it. Take all weights at 27°C.

Calculation

The relative density (d) $= \dfrac{m_3 - m_1}{m_2 - m_1}$

m_1 – Weight of empty specific gravity bottle at 27°C

m_2 – Weight of specific gravity bottle with water at 27°C

m_3 – Weight of specific gravity bottle with test material at 27°C

Determination of freezing point

Principle

Either the constant temperature, or the maximum temperature, observed when the oil, in a supercooled liquid state, liberates its latent heat of fusion. Slow and progressive cooling of the essential oil. Observation of the variations in temperature as the oil passes from the liquid to the solid state.

Apparatus

Calibrated thermometers, test tube, stout-walled test tube rubber stopper.

Procedure

First fill the container with water, melting ice, or freezing mixture so as to obtain a temperature 5°C less than the freezing point of material. Fit the stout-walled test tube through rubber stopper. Place 10 ml of essential oil in the test tube, dip thermometer in it and place it in stout-walled test tube and allow the temperature to fall further up to 2°C. (Fig. 12.5).

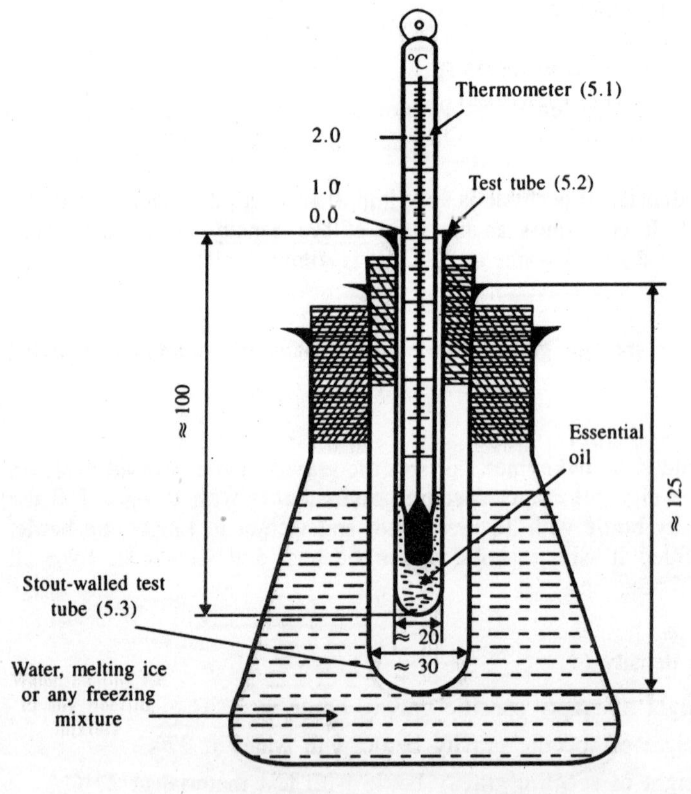

(All dimensions in milimetres)

Fig. 12.5. Suitable apparatus assembly for determination of freezing point.

Seed the oil with a trace of the solidified oil, obtained in the preliminary test, and stir vigorously with the thermometer, taking care to avoid the adhesion of particles to the walls of the tube. Observe the temperature variations as carefully as possible.

Note the observed temperature when the temperature curve in relation to time shows a maximum value, or a constant level for at least 1 min.

Remove the test tube from the apparatus and again liquefy the essential oil. Repeat the determination until two consecutive results do not differ by more than 0.2°C. Average the two last readings to obtain final figure.

Results

The freezing point is the maximum temperature observed at the end of the test. It is expressed in degree celsius to one decimal place.

Determination of refractive index by refractometer

Principle

Refractive index is the ratio of the sine of the angle of incidence to the sine of the angle of refraction, when a ray of light of defined wavelength passes from air into the material kept at constant temperature. The wavelength is 589.3 ± 0.3 nm corresponding with D_1 and D_2 lines of sodium spectrum. The reference temperature is 27°C except those which are not liquid at this temperature, in which case a higher temperature (say 30°C) depending on the melting point of the material shall be used.

Apparatus

Refractometer, light source (Sodium light).

Procedure

Pass a stream of water through the instrument so as to keep it at the temperature at which the readings must be made. This temperature shall not differ from the reference temperature by more than ± 0·2°C. Before placing the material in the instrument, bring the test sample to a temperature similar to that at which the measurement must be made. Make readings only when the temperature is stable.

Calculation

The refractive index n_D^t at the specified temperature is given by the following formula :

$$n_D^t = n_D^{t'} + 0.0004\,(t'-t)$$

where

n_D^t is the reading taken at the working temperature t'.

Determination of solubility in ethanol

Principle

An essential oil is said to be soluble in V volumes and more of ethanol of given strength, t when the mixture of 1 volume of the oil in question with V volumes of that ethanol is clear and stays so after further gradual addition of ethanol of the same strength up to a total of 20 volumes.

An essential oil is said to be soluble in V volumes of ethanol of given strength, t and to become cloudy when diluted in V' volumes, when the mixture of 1 volume of the oil in question with V volumes of that ethanol is clear, becomes cloudy after further gradual addition of $(V'-V)$ volumes of ethanol of the same strength, and remains cloudy after further addition of ethanol up to a total of 20 volumes.

An essential oil is said to to be soluble in V volumes of ethanol of given strength t, and to become cloudy when diluted in V'' volumes, when the mixture of 1 volume of the oil in question with V volumes of that ethanol is clear, becomes cloudy after further gradual addition of $(V'-V)$ volumes of ethanol of the same strength, and again becomes cloudy after further addition of $(V''-V')$ volumes of ethanol of the same strength.

An essential oil is said to be soluble with opalescence when alcoholic solution is being diluted shows a bluish tinge, this even being similar to that of the standard of opalescence, freshly prepared.

Apparatus

Burette, Measuring cylinder, constant temperature device, thermometer.

Reagents

Dilute solution of ethanol : Solutions of 50, 60, 70, 80, 90 and 95 per cent (v/v). *Standard solution for opalescence :* Add 0.5 ml of 0.1 N silver nitrate solution to 50 ml of 0.0002 N sodium chloride solution and one drop of 25 per cent (m/m) nitric acid. Stir the solution and allow it to stand for 5 minutes.

Procedure

Measure 1 ml of the material into the cylinder either with a pipette or by weighing to an accuracy of $\pm$ 5 mg. Place the cylinder and its contents in the constant temperature device maintained at a temperature of $27 \pm 0.2°C$. Using the burette, add the diluted solution of ethanol of known strength, which has previously been brought to a temperature of $27 \pm 0.2°C$ by increments of 0·1 ml until the material is completely dissolved, shaking frequently and vigorously during the addition of the solvent. When the solution is perfectly clear, record the volume of ethanol solution added. Continue to add the solvent by increments of 0·5 ml up to 20 ml, and keep shaking after each addition. If the solution becomes cloudy or opalescence before a total volume of 20 ml of ethanol has been added, record the volume of ethanol solution added at the point where cloudiness or opalescence appears and, if applicable, the volume at which it disappears. If a clear solution is not obtained when 20 ml of solvent has been added, repeat with the next higher concentration of ethanol solution given in Table 12.3.

Table 12.3. Dilutions of ethanol by volume and by mass.

Dilutions: ml of ethanol in 100 ml of the mixture (% V/V) to ± 0.1 %	Volume of distilled water at 20°C to be added to 100 ml of ethanol 95 % (V/V) at the same temperature within ± 0.1°C, for the preparation of the corresponding dilutions	Mass of ethanol 95 % (v/v)	Mass of water to be added	Limits of relative density and the apparent specific gravity $\dfrac{20}{d}\ \dfrac{27}{d}$ $\dfrac{}{20}\ \dfrac{}{4}$
(% v/v)	(ml)	(g)	(g)	
50	95.8	46	54	0.931 6 to 0.927 9 0.932 0
60	62.9	56.4	43.6	0.910 5 to 0.906 4 0.911 9
70	39.1	67.6	32.4	0.886 8 to 0.882 5 0.887 3
80	20.9	79.6	20.4	0.860 4 to 0.856 0 0.861 0
90	6.4	92.7	7.3	0.830 3 to 0.825 8 0.830 9
95	0.0	100	0	0.812 3 to 0.808 0 0.813 1

Calculations

If V is the volume in ml, of dilute solution of ethanol of dilution t to obtain a clear solution.

If V'' is the volume, in ml, of dilute solution of ethanol of dilution t to obtain cloudiness after the clearness, if it occurs; and

If V' is the volume in ml of dilute solution of ethanol of dilution t at which cloudiness disappears; the solubility of the material in the dilute solution of ethanol of strength t will be 1 volume in V volumes, with cloudiness between V' and V''' volumes.

If the solution is not entirely clear, but only opalescent record whether the opalescence is 'greater than, equal to' or less than that of the standard solution.

Determination of optical rotation and specific rotation by polarimeter

Principle

The optical rotation. The angle through which the plane of polarised light is rotated by a layer 10 cm thick of the material under test at a specified temperature. The wavelength of radiation used should be $589\cdot3 \pm 0\cdot3$ nm.

The specific rotation. The optical activity of the material in a solution determined at a particular temperature and spectral line of light source.

Apparatus

Polarimeter, monochromatic light source, polarimeter tubes, thermometer.

Procedure

Optical rotation. Switch on the monochromatic light source and wait till full luminosity is obtained. Determine the zero error of the instrument. Fill the polarimeter tube with the material, ensure that air bubble is absent. Place the tube in the polarimeter and read the dextrorotatory (+) or laevorotatory (−) optical rotation on the scale of the instrument. After carefully adjusting the analyser so that both the halves of the field when viewed through the telescope, should show illumination of equal intensities.

When the room temperature is outside the limits 17 to 27°C or when the material contains large amounts of highly optically active constituents, use a jacketed polarimeter tube provided with water circulation at 27°C. By means of a thermometer inserted in the central opening check that the temperature of the material in the tube is 27 ± 1°C. Record the results as the average of at least three readings; the readings should agree within 0.08°(5'). Make zero error correction if any.

Specific rotation. Dissolve the material in rectified spirit reserving a separate portion of the solvent for a blank determination. Make at least five readings of the rotation of the solution at $27 \pm 1°C$ Replace the solution with the reserved portion of the solvent, make the same number of readings and determine the average which shall be zero point value. Subtract the zero point value from the average observed rotation, if the two figures are of the same sign or add, if opposite in sign, to obtain the corrected observed rotation.

Calculation

$$\text{Specific rotation } (\alpha) = \frac{100\,a}{l\,p\,d} = \frac{100\,a}{l\,c}$$

where

a = the corrected observed rotation in degrees of the solution at temperature $t°$, using D line of sodium or yellow green line of mercury as the case may be;

l = the length in mm of polarimeter tube;

d = the relative density of the material at the temperature of observation $(t°)$;

p = the concentration of the material expressed as grams of active substance in 100 g of solution; and

c = the concentration of material expressed as the number of grams of active substance in 100 ml of solution.

Determination of melting point and melting range

Principle

The temperature at which the material melts and becomes liquid throughout as shown by the formation of a definite meniscus is known as the melting point of the material. The temperature between which the material collapses against the side of the tube and the temperature at which it is liquid throughout, is known as the melting range of the material.

Apparatus

Capillary tube, thermometer and glass heating vessel.

Procedure

Spread a small quantity of the finely powdered material in a thin layer and dry at a temperature below its melting point/melting range in an oven, or in a vacuum desiccator over sulphuric acid for 24 hours.

Transfer a quantity of the dried powder to a dry capillary tube and pack the powder by tapping the tube on a hard surface so as to form a tightly packed column of 2 to 4 mm in height. Attach the capillary tube and its contents to the thermometer so that the closed end is at the level of the middle of the bulb, and heat in the heating vessel, regulating the rise of temperature during the first period to 3 deg per minute. When the temperature reached is 10 deg below the expected melting point/melting range, of the material being tested, the heating of the vessel is adjusted so that the rate of rise in temperature is 1 to 2 degree per minute.

Results. For melting point, note the temperature reading at which the liquefaction of the material occurs. For melting range, note the temperature reading at which the material collapses against the side of the tube and material remains liquid throughout.

Determination of boiling (Distillation) range

Principle

This method specifies a method of test for determining the boiling (distillation) range of perfumery materials.

The range of temperature between which a liquid boils or the percentage of the material that distill between two specified temperatures is determined. The lower of the two temperatures is the corrected thermometer reading when the first five drops of distillate have been collected, and the upper temperature is the corrected reading when the per centage specified has been collected. This method covers the distillation of aromatic perfumery materials of relatively narrow boiling ranges between 30°C and 250°C.

Apparatus

Distillation flask, water condenser, receiver, asbestos sheet, heating mantle thermometer.

Procedure

Assemble the apparatus as shown in Fig 12.6. Carefully measure a 100-ml sample of the material to be tested in the 100-ml graduated cylinder at room temperature and transfer to the distillation flask, draining the cylinder at least 15 s. This is preferably done before mounting the flask in position in order to prevent liquid from entering the side arm. Connect the flask to the condenser and apparatus, assembled as shown in Fig. 12.7. Do not rinse out the graduated cylinder used to measure the sample for distillation, but place under the lower end of the condenser tube to receive the distillate. Heat the flask slowly, especially after ebullition has begun, so as to allow the mercury column of the thermometer to become fully expanded before the first drop distills over. Regulate the rate of heating so that the ring of condensing vapour on the wall of the flask reaches the lower edge of the side arm in not less than 90s, and preferably approximately 120s, from the start of the rise of the vapour ring. The total time from the start of heating until the first drop falls into the receiver should be not less than 5 nor more than 10 minutes. Avoid major changes in heating rate. Even operation is best gained through experience with the method. When distillation starts, adjust the receiver to allow condensate to flow down its inner wall to prevent loss by spattering; then adjust the heater to continue the distillation at the rate of 5 to 7 ml/min (about 2 drops/s). Maintain this rate, and continue the distillation to dryness.

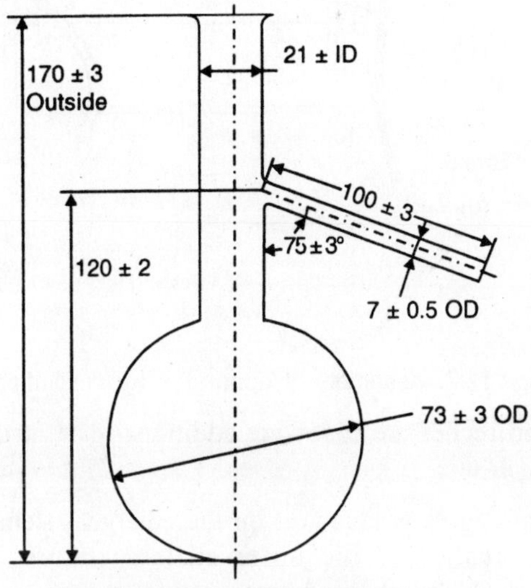

All dimensions in millimetres.

Fig. 12.6. Distillation flask.

Record the temperature reading when the first drop of distillate falls into the receiving cylinder. Take additional readings when 5 per cent, 10 per cent, each additional 10 per cent up to 90 per cent, and 95 per cent of the sample have distilled over. Take a final reading when the liquid just disappears from the bottom of the flask. When testing crude materials, a decomposition point, rather than a dry point, may be obtained. When a decomposition point is reached at the end of a distillation, the temperature will frequently cease to rise and begin to fall. In these case, take the temperature at the decomposition point as the maximum temperature observed. The decomposition point may also be indicated by the appearance of heavy fumes in the flask. Should that occur, record the temperature at the time the bulb of the flask becomes substantially full of fumes.

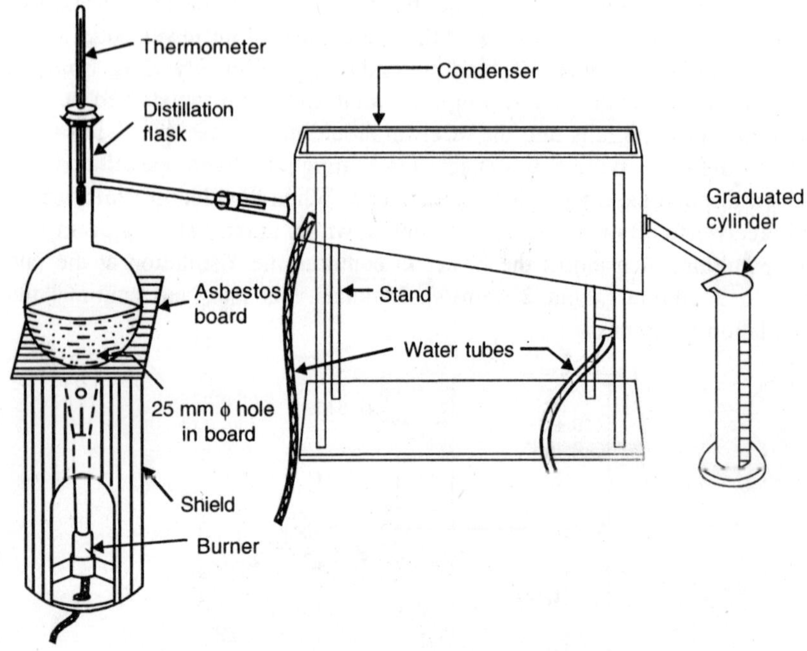

Fig. 12.7. Assembly of apparatus for distillation test.

Observe and record the following additional data at the time and place of the distillation test.

(1) Number of degree graduations on the emergent stem from top of cork to the graduation at the observed temperature, and the average temperature of the exposed stem.

(2) Correction for inaccuracy of the thermometer.

Emergent stem corrections: Any convenient mercury, etched glass stem thermometer with a range including 20 to 40°C may be used for determining the temperature of exposed stem of the distillation thermometer. Calculate the correction for emergent stem of the thermometer as follows :

$$C = KN\,(T - t)$$

where

$C =$ Number of degrees celsius to be added to the observed thermometer reading;

$K =$ Correction factor for the thermometer = 0.000154 = difference between coefficients of expansion for glass and for mercury;

N = Number of degree graduations on the stem exposed;

T = Observed reading of the thermometer in degrees celsius; and

t = Average temperature of the emergent stem in degrees celsius.

Method for olfactory assessment of natural and synthetic perfumery materials

Principle

Natural and synthetic perfumery materials, such as essential oils, aromatic chemicals, etc., are used primarily for their odour appeal. Although the analytical characteristics which are commonly determined may provide some assurance regarding the chemical purity of an odouriferous substance, they do not necessarily indicate the 'purity' of odour. Hence, olfactory evaluation has been practised for centuries and, in the perfumery trade, it has been formed the basis of acceptance or rejection of odouriferous materials. In spite of the importance attached to this subject, there is no uniform method for odour evaluation nor has a standard procedure yet been formulated in any country. This standard was formulated with a view to introduce a uniform method of test for olfactory assessment of natural and synthetic perfumery materials.

This method for olfactory assessment of natural and synthetic perfumery materials is based on comparison of a given material with its corresponding standard sample.

Terminology

For the purpose of this method the following definitions shall apply.

Top note. The initial and primary odour effect perceived by the olfactory nerves on smelling a strip freshly impregnated with the material being tested. The top note(s) is (are) usually of a short duration and may or may not be co-perceived along with the middle note.

Middle note. The secondary overall odour effect experienced by the olfactory nerves on smelling a strip impregnated with the material after the initial top note has evaporated. It lasts for a longer time on the strip than the top note.

Residual note (Dry-out note). The tertiary odour effect experienced by olfactory nerves on smelling a strip impregnated with a material after the top and the middle notes have disappeared. Besides indicating the lasting character and strength of the material, it may also reveal the nature of the lesser volatile materials.

By note. An odour effect, additional to the normal pattern of odours associated with the material, experienced by olfactory nerves on smelling an impregnated strip during any stage of evaporation. It is generally regarded as an index of foreign odour and/or undesirable adulterant and alien.

Odour description. Due to the absence of precise terms, descriptive words which are subjective in nature are commonly used to express the odour sensations perceived in the top, middle, residual and by-notes. Some of these terms are given below but the list is not intended to be exhaustive : acid, acrid, aldehydic, amber, animal, Balsamic, bitter, burnt, camphoraceous, choking, citrus, cloying, cool, dry, dull, earthy, exalting, faecal, fatty, fishy, floral, fungal, fresh, fruity, goaty, grassy, green, heavy, herbal, honey, intense, leafy, leathery, minty, mossy, mushroomy, musky, musty, nauseating, nutty, oriental, peppery, persistent, phenolic, piney, powdery, pungent, refreshing, sappy, sharp, sickly, smoky, sour, spicy, stem like, still odour, sulphuraceous, sultry, sweet, tarry, tart, woody.

Requirements. The following general precautions are required to be noted

Selection and training. Better results are obtained if individuals with a keen sense of smell and ability to distinguish between different odours are selected for training in olfactory assessment.

Fatigue. Continuous smelling causes olfactory fatigue and decreases critical odour perception. To avoid this, the number of samples assessed during a session should be limited as far as is practical. Further, during smelling, the body should be relaxed. Resting for an interval between smelling different samples is also advantageous. If the number of samples to be tested is fairly large, it is advisable to examine last those materials which are known to be pungent or strong in odour. It should be borne in mind that inability to correctly identify certain odours may arise from natural deficiencies such as specific anosmia. For instance, some people are unable to perceive musky odour.

Bias. The necessity of minimising all differences between samples other than that of odour in order to prevent the prejudicing of results is stressed. 'Blind' tests should be conducted by ensuring that the markings on the smelling strip do not dislodge the origin of the samples.

Time of olfactory assessment. The evidence relating to the most favourable time for conducting olfactory assessment is somewhat conflicting. However, the morning appears to be generally favoured. In general olfactory assessment should be done after a reasonable interval of time has elapsed after a meal or a beverage has been taken.

Freedom from contaminating odours. It is necessary to ensure that the hands, nose and smelling strips are free from contaminating odours as these are likely to vitiate the results. It is recommended that the individual responsible for assessing odour should wash his/her hands several times during a smelling session as well as clear his/her nose.

Material Requirements

The following materials, apparatus and environmental conditions are required :

Library of standard samples. For each essential oil, aromatic chemical or other perfumery material, there shall be a standard sample of approved odour value as agreed upon by the purchaser and the supplier. The standard samples shall be kept in well-stoppered, air-tight, neutral amber-coloured glass bottles and when not in use, they shall be stored in a refrigerator at about 5°C. The odour characteristics of standard samples are likely to change over a period of time however well they may be stored. Some materials improve in odour as a result of maturing while others deteriorate because of minute oxidative changes. An alteration in the odour characteristics of standard samples is not desirable and, in such cases, fresh standards should be adopted. Generally, all perfumery materials have recommended shelf life and the sample should be changed thereafter.

Ethyl alcohol. Perfumery grade.

Diethyl phthalate. Perfumery grade.

Smelling strips. These shall preferably be 1 cm wide and 15 cm long. They shall be made from odourless, thin, absorbent paper and shall be sufficiently stiff so that the strips do not bend under their own weight when held in a horizontal position. Smelling strips shall be packed in air-tight, odour-free containers and stored in a clean odour-free room. Those intended for daily use shall preferably be kept in a wide-mouthed glass bottle covered by a beaker.

Strip stand. A cruciform patterned 3-clip stand, approximately 21 cm high, or any other suitable device, to hold impregnated smelling strips as shown in Fig. 12.8.

Environment. A well-ventilated room, as free as possible from all outside disturbances. Ideally, the temperature and humidity suited are about 20°C and 80 per cent RH, respectively. The colouring of the room shall be sober and the furnishing restricted. The general environment shall have a restful rather than a distracting effect.

Procedure. One end of each smelling strip shall be clearly marked before use. Dip the unmarked end of one strip (about 0.5 to 1.0 cm) in the material under examination and of another strip to the same depth in the standard sample after it has attained room temperature. For certain perfumery materials, such as fatty aldehydes, absolutes and solids, use 1 to 10 per cent solutions in ethyl alcohol or diethyl phthalate for olfactory assessment.

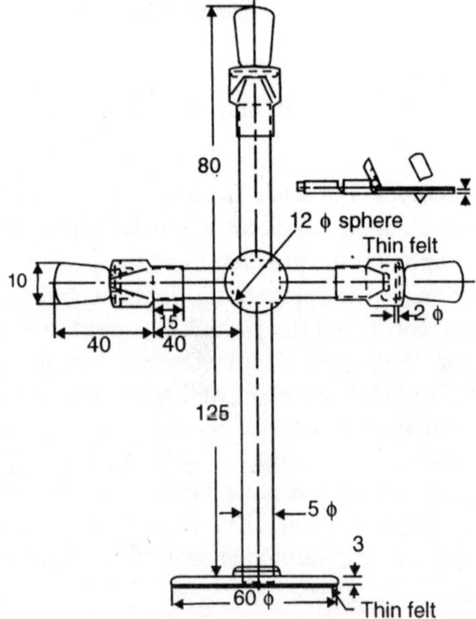

Fig. 12.8. Strip stand.

For semi-solid materials. The odour of semi-solid materials such as guaiacwood oil, oakmoss resinoid and absolute, labdanum resinoid and absolute, etc., should be taken on smelling strips but only after melting the contents completely under controlled temperature below 100°C preferably on water-bath.

For solid materials. The odour of solid materials, such as coumarin, musk, ambrette, etc., should be examined after pouring out some quantity (about 5 g) on odour-free non-absorbent paper (such as glassine). Solids should be smelt both as such, as well as after crushing to enable the occluded impurities to be perceived more easily.

For strong-smelling materials. In order to have a better perception, strong smelling substances irrespective of their physical appearance may also be smelt after dilution to about 1 to 10 per cent such as indole, fatty aldehydes, etc., using ethanol or diethylphthalate as a diluent.

Hold the strip impregnated with the standard sample at such a distance from the nose that there is incipient yet distinct perception of odour. While smelling, concentrate wholly on the sensations received and make mental observations. Repeat the procedure with the strip impregnated with the test sample. After about a minute's rest, repeat the comparison reversing the order of smelling the two strips. Finally, compare the two strips for their odour in a 'blind' test. If a difference in odour is observed, repeat the 'blind' test on the two strips five times. Record the observations of each 'blind' test.

It is important to note that although the room shall be well-ventilated, the strips kept under examination should not be exposed to a direct draught.

After this initial assessment for top notes, fix the two strips on a stand keeping them sufficiently apart to avoid inter-contamination. Examine the strips periodically by the 'blind' test and note the changes in quality and intensity of odour. Continue in this manner as long as the odour on each strip remains perceptible.

Report. Report the top, middle and residual odour assessment of the test sample as compared with the odour of the standard sample at corresponding stages of assessment.

Criterion for judgement of quality. The odour of the material under examination shall correspond to that of the standard at all stages of assessment. If it does not and the pattern of odour is considered to be inferior to that of the standard, the quality of the material shall be regarded as not satisfactory.

Referee test. In case of dispute, present the individual assessing odour with three suitably coded smelling strips, two of which have been dipped in the material under examination and the remaining one in the standard sample (or *vice-versa*). If the 'odd' sample is consistently picked five times in a 'blind' test, the material shall be deemed to have a pattern of odour different from that of the standard sample.

Determination of residue on evaporation

Apparatus

Evaporation basin, water-bath.

Preparation of sample

The quantity of the perfumery material shall be taken in accordance with Table 12.4.

Procedure

Weigh the dry basin. Weigh into the basin a suitable quantity of perfumery material as given in Table 12.4 and place the basin and contents on the vigorously boiling water-bath. Screen from draughts and heat continuously for 5 hours, or for 6 hours in case of bergamot oil. Remove the basin, wipe it and place it in a desiccator for 20 minutes.

Table 12.4. Mass of perfumery material to be taken for the determination of residue on evaporation.

Expected Residue on Evaporation, Per cent by Mass (1)	Mass of the Material, g (2)
Less than 5.0	4.8 to 5.2
5.0 to 8.0	2.8 to 3.2
More than 8.0	1.8 to 2.2

Calculation

Residue on evaporation, per cent by mass $= \dfrac{100\ M_2}{M_1}$

where

M_1 = mass in g of the sample taken, and

M_2 = mass in g of the residue.

Express the results to the first decimal place.

Chemical Analysis

Chemical Analysis is important from the viewpoint of quality of perfume. It provides composition of the perfume. The quantitative estimation of alcohol, water, acid value, ester value, carbonyl value, phenol and heavy metals is described.

Determination of content of alcohol insoluble solid impurities

Principle

Extraction of the perfumery materials is done with ethanol and weighing of any residue after filtration.

Reagent. Ethanol – 95 per cent (v/v).

Apparatus. Hot-air-oven.

Procedure

Weigh accurately about 20 g of the perfumery material into a conical flask. Add about 200 ml of ethanol, mix the contents of the flask thoroughly and filter through a sintered glass crucible (porosity 4 is suitable) which has previously been dried in the oven at 105°C and weighed. Wash any insoluble matter remaining in the flask into the crucible, suck dry and then dry the crucible and its contents in the oven at 105°C until the loss of mass on drying in the oven for an additional 30 minutes does not exceed 1 mg.

Calculation

Content of alcohol-insoluble solid impurities,

$$\text{Per cent by mass} = \frac{100\,(m_3 - m_2)}{m_1}$$

where

m_1 = mass in g of the sample of perfumery material taken,

m_2 = mass in g of the empty sintered glass crucible, and

m_3 = mass in g of the sintered glass crucible and residue.

Determination of water content

Principle

Distillation of the perfumery materials is done with a water-immiscible solvent.

Reagent

Heptane or toluene.

Apparatus

Dean and stark apparatus, electric heating mantle or oil-bath.

Procedure

Weigh 20 ± 0.5 g of the perfumery material into the distillation vessel. Half fill the vessel with solvent and add a few dry porcelain chips to regulate boiling. Assemble the apparatus. Maintain the contents of the flask at the boil using the heating mantle or oil-bath until the level of the water layer in the receiver remains constant for 30 minutes. Cool the apparatus to room temperature and detach any globules of water from the condenser wall by means of the spray tube.

Calculation

$$\text{Content of water, per cent by mass} = \frac{V \times 100d}{m}$$

where

V = volume in ml of water collected in the receiver,

m = mass in g of perfumery material taken, and

d = density of water at the air temperature (0.997 $\frac{1}{2}$ g/ml at 27°C).

Determination of acid value

Principle

Acid Value — The number of milligrams of potassium hydroxide required to neutralise the free acids contained in 1 g of the perfumery material.

Apparatus

Conical flask, measuring cylinder and burette

Reagents

Ethyl alcohol, potassium hydroxide solution (0.1 N), phenolphthalein (95%) and phenol red.

Procedure

Weigh about 2.5 g of dried material and dissolve in 20 ml of neutral rectified spirit. Add about 10 drops of indicator (except in the case of phenolic perfumery material) and titrate with potassium hydroxide solution until the solution remains faintly pink after 10 seconds of shaking. If the perfumery material under examination contains phenols or compound with phenolic groups, use phenol red as indicator instead of phenolphthalein.

Calculation

$$\text{Acid value} = \frac{56.1 \times V \times N}{M}$$

where

V = volume in ml of potassium hydroxide solution,

N = normality of the potassium hydroxide solution, and

M = mass in g of the material taken for the test.

Determination of ester value, content of esters and combined alcohols

Principle

The number of milligrams of potassium hydroxide required to neutralise the acids liberated by the hydrolysis of ester present in 1g of the perfumery materials.

Apparatus

Saponification flask, reflux condenser, burette, pipettes.

Reagents

Ethyl alcohol, (95 % *v/v*), alcoholic potassium hydroxide (0.5 N), hydrochloric acid (0.5 N), phenolphthalein indicator and phenol red alcoholic solution.

Prepared sample

The material shall be taken for the determination of ester value, content of esters and combined alcohols, with a suitable quantity as given in Table 12.5.

Table 12.5. Mass of perfumery material to be taken for determination of ester value, content of esters and combined alcohols.

Expected Ester Value (1)	*Mass of Perfumery Material* (2) g
Less than 50	4.5 to 5.0
50 to 70	3.5 to 4.0
70 to 90	2.5 to 3.0
90 to 110	2.0 to 2.5
110 to 140	1.5 to 2.0
140 to 180	1.2 to 1.5
180 to 220	1.0 to 1.2
220 to 280	0.9 to 1.0

Procedure

Blank determination. To a saponification flask containing a few fragment of pumice stone or porcelain, add 5 ml of ethyl alcohol and 25 ml of 0.5 N alcoholic potassium hydroxide solution and gently reflux on the heater for the period specified in specification for individual perfumery material. Then allow the liquid to cool, remove the reflux condenser, add five drops of phenolphthalein solution and neutralise with 0.5 N hydrochloric acid contained in the burette. Alternatively under similar conditions gently reflux the contents reserved from the acid value determination after the addition of 25 ml of the 0.5 N alcoholic potassium hydroxide solution and fragments of pumice-stone or porcelain. Then allow to cool and remove the reflux condenser adding five drops of phenolphthalein indicator except in case of phenolic perfumery material and neutralise the liquid with 0.5 N hydrochloric acid as indicated for the blank test carried out separately under identical conditions.

Calculation

Ester value. Calculate the ester value of the material as follows :

$$\text{Ester value} = \frac{56.1 \times N(V_1 - V_2)}{m}$$

where

N – normality of hydrochloric acid used for titration.

V_1 – volume in ml of hydrochloric acid used for the blank,

V_2 – volume in ml of hydrochloric acid used in determination to neutralise the excess alkali after hydrolysis, and

m – mass in g of the material taken.

Ester content. The percentage of esters present in the material may be calculated as given below :

$$\text{Ester \% (by mass)} = \frac{E \times M}{561}$$

where

E = the observed ester value; and

M = molecular mass of the ester.

Combined alcohols. The percentage of alcohols combined as esters in the material may be calculated as follows :

$$\text{Combined alcohols \% (by mass)} = \frac{E \times M}{561}$$

where

E = the observed ester value, and

M = molecular mass of the alcohols.

Saponification value. The saponification value is equivalent to ester value plus acid value.

$$\text{Saponification value} = \text{Ester value} + \text{Acid value}$$

Determination of ester value after acetylation and free alcohols

Method A - Acetic anhydride method

The perfumery material is acetylated by acetic anhydride in the presence of sodium acetate followed by alkaline hydrolysis. Excess alkali is titrated to determine the ester value after acetylation.

Apparatus

Acetylation flask and saponification flask.

Reagents

Acetic anhydride, Sodium acetate, Sodium carbonate (2 %), Sodium chloride solution (saturated), Magnesium sulphate or sodium sulphate, Alcoholic potassium hydroxide solution (0.1 N), Hydrochloric acid (0.5 N).

Procedure

Mix approximately 10 ml of the sample of perfumery material, 10 ml of acetic anhydride and 2 g of anhydrous sodium acetate in the acetylation flask. Add fragments of pumice-stone or porcelain and fit the flask with its reflux condenser. Heat the flask by means of any suitable heating device and gently reflux the liquid for 2 hours. At the end of this period, allow the liquid to cool; add 50 ml of water and heat at a temperature between 40 and 50°C for 15 minutes with shaking frequently. Cool to room temperature, remove the reflux tube and transfer the liquid to a separating funnel; wash the flask twice with 10 ml of water and add these washings to the contents of the separating funnel. Wait until separation of the liquid is complete, then reject the aqueous layer. Wash the oil layer by shaking successively with :

(1) 50 ml of sodium chloride solution;

(2) 50 ml of sodium carbonate;

(3) 50 ml of sodium chloride solution; and

(4) 20 ml of water.

Shake the acetylated perfumery material vigorously with the saturated solutions, and gently with water which, if the washings have been properly conducted, will be natural to litmus paper.

Run the acetylated material into a dry tube and shake intermittently several times with about 2 to 3 g of anhydrous magnesium sulphate or ahydrous sodium sulphate till it is dried. Filter the dried material through dried filter paper in a covered funnel.

In the saponification flask, about 2 g of the acetylated perfumery material, and add 2 ml of water and 0·5 ml of phenolphthalein indicator.

Neutralise the liquid with the ethanolic potassium hydroxide solution using the burette.

Determine the ester value adding 25 ml of 0.5 N ethanolic potassium hydroxide solution depending upon the mass of the sample taken.

Blank determination. Simultaneously carry out a blank determination of the ethanolic potassium hydroxide by the same procedure but omitting the perfumery material.

Calculation

$$\text{Ester value after acetylation} = \frac{56.1 \times N \times (V_1 - V_2)}{m}$$

where

N = normality of hydrochloric acid used for titration,

V_1 = volume in ml of hydrochloric acid required for the blank,

V_2 = volume in ml of hydrochloric acid required to neutralise the excess alkali after hydrolysis, and

m = mass in g of acetylated material.

$$\text{Content of free alcohols \% by mass} = \frac{M \times (E_1 - E_2)}{561 - 0.42\,E_1}$$

where

E_1 = ester value of the material after acetylation,

E_2 = ester value of the material before acetylation, and

M = molecular mass of the specified alcohol.

Method B — Dimethylaniline acetylchloride method (Fiore Method)

Principle

The perfumery material is acetylated by acetyl chloride in the presence of dimethylaniline and acetic anhydride followed by alkaline hydrolysis. Excess alkali is titrated to determine the ester value after acetylation.

Apparatus

Saponification flask, Graduated measuring cylinders, Burette.

Reagents

Dimethylaniline. Freshly distilled and free from methylaniline and aniline. Acetic anhydride, Acetyl chloride, Sodium sulphate solution – 10 per cent (m/v), Sulphuric acid – 2.5 per cent (m/v), Sodium bicarbonate solution – 50 per cent (m/v), Magnesium sulphate – anhydrous, neutral, freshly ignited and powdered., Potassium hydroxide solution – 0·1 N and 0.5 N solution in 95 per cent ethanol (v/v), Hydrochloric acid – 0.5 N, Phenolphthalein indicator solution – Dissolve 0.2 g phenolphthalein in 100 ml of 95 per cent ethanol.

Procedure

Introduce 10 ml of the sample of perfumery material into the 100 ml glass-stoppered conical flask. Add 20 ml of dimethylaniline. Stopper the flask and mix thoroughly. Cool in melting ice and add 8 ml of acetyl chloride followed by 5 ml of acetic anhydride. Keep the flask in the ice-bath throughout the addition and allow it to stand at room temperature for 30 minutes. Immerse the flask for 16 hours in the water-bath maintained at $40 \pm 1°C$. Cool to room temperature, transfer the contents of the flask to the separating funnel and wash by shaking successively with :

(1) Twice with 75 ml of sodium sulphate solution,

(2) Five times at least with 50 ml of sulphuric acid until the washings are free from dimethylaniline as demonstrated by freedom from turbidity when the last washing is made alkaline.

(3) Twice with 25 ml of sodium bicarbonate solution, and

(4) Twice with 25 ml of sodium sulphate solution.

Shake vigorously for 30 seconds with each solution and allow to separate. Check the neutrality of the last solution to litmus paper.

Dry the acetylated material with 2 g to 3 g of magnesium sulphate for at least 2 hours with intermittent shaking. Then, filter the material through a dry filter paper in a covered funnel.

Add 2 ml of water and 2 ml of phenolphthalein indicator solution. Neutralise the free acidity with potassium hydroxide solution using the burette. Not more than one drop should be required to give a pink colour. Take the first pink colour as the neutral point as certain esters, namely, linalyl acetate are easily hydrolysed.

Determine the ester value of the acetylated material as described earlier.

Blank determination. Simultaneously with the hydrolysis of the acetylated material by the procedure as described above. Carry out a blank determination on the ethanolic potassium hydroxide by the same procedure but omitting the acetylated material.

Calculation

Calculate the ester value after acetylation using the formula given above under ester value after acetylation.

Determination of carbonyl value and content of carbonyl compounds

Two methods, namely, hydroxylammonium chloride method and free hydroxylamine method are available for determining the carbonyl value of a perfumery material. Any one of these may be adopted for the determination of carbonyl value of a perfumery material, as specified in the individual standard or as may be agreed to between the purchaser and the supplier. The procedure for the two methods have been prescribed as under.

Hydroxylammonium chloride method — Method A

Principle

The carbonyl compounds to be determined are converted into oximes by reaction with hydroxylammonium chloride. The hydrochloric acid liberated during the reaction is determined by an ethanolic potassium hydroxide solution.

Apparatus

Conical flask, Graduated measuring cylinder, burette.

Reagents

Potassium hydroxide. Approximately 0.5 N solution in 95 per cent (*v/v*) ethanol.

Bromophenol blue indicator solution. Dissolve, while warming, 0.2 g of bromophenol blue in 3 ml of ethanolic potassium hydroxide, 0.1 N hydrochloric acid, using bromophenol blue as indicator and running the alkali into the acid.

Hydroxylammonium chloride, ethanolic solution. Dissolve 50 g of hydroxylammonium chloride in approximately 100 ml of water, and about 800 ml of 95 per cent (*v/v*) ethanol, then 10 ml of the ethanolic bromophenol blue solution and dilute to 1000 ml with 95 per cent (*v/v*) ethanol. Add the ethanolic potassium hydroxide solution until the solution is green, if the liquid is observed in a thin layer, or until red, if the layer is thick.

This solution is suitable for use if a lemon-yellow colour is obtained when 0.05 ml of the hydrochloric acid is added to 20 ml of the solution, and a red colour is obtained when 0.05 ml of the potassium hydroxide solution is added to another 20 ml of the solution. The solution is stable for 1 week.

Procedure

Weigh accurately into the conical flask the amount of perfumery material specified in the individual standard. Add from the graduated cylinder 25 ml of hydroxylammonium chloride solution and set aside for the time specified in individual standard. If additional heating is necessary, this will be specified in individual standard.

After the time prescribed, and if necessary after cooling to room temperature, titrate the contents of the flask with the ethanolic potassium hydroxide solution, taking care to avoid going beyond the greenish-yellow colour of the indicator, continue the titration with the potassium hydroxide solution until a bluish green colour persisting for 5 minutes is obtained.

Calculation

The carbonyl value and content of carbonyl compounds are calculated as follows :

$$\text{Carbonyl value} = \frac{56.1 \times V \times N}{m}$$

where

V = volume in the ethanolic potassium hydroxide solution used in the determination;

N = the normality of the ethanolic potassium hydroxide solution used; and

m = mass in g of the sample.

Content of carbonyl compounds (as aldehyde or ketone), per cent by mass :

$$= \frac{M \times V \times N}{10\,m}$$

where

M = molecular mass of the carbonyl compounds in which the results are to be expressed,

V = volume in ml of ethanolic potassium hydroxide solution used in the determination,

N = the normality of the ethanolic potassium hydroxide solution used, and

m = mass in g of the sample.

Free hydroxylamine method — Method B

Principle

The carbonyl compounds are converted into oxime by reaction with hydroxylamine freed through the action of a solution of potassium hydroxide on the hydroxylammonium chloride. The hydroxylamine remaining after the reaction is determined by titration with standard hydrochloric acid.

Apparatus

Alkali-resistant glass flask, Pipette, Conical flask, Burette.

Reagents

Hydrochloric acid – 0.5 N, Hydroxylammonium chloride solution, Bromophenol blue ethanolic solution, Potassium hydroxide solution 0.5 N in ethanol.

Procedure

Weigh accurately into a flask, a mass m of the perfumery material as prescribed in individual standard. Introduce by means of a pipette, 20 ml of hydroxylammonium chloride into a conical flask. Add 10 ml of potassium hydroxide solution measured from a burette. Mix it well.

Pour the liquid into the flask containing the perfumery material and reserve the conical flask without washing. In case of aldehydes, allow to stand for 15 minutes at room temperature while for ketones, reflux gently for one hour using a water condenser or an air condenser at least 75 cm long. Cool to room temperature. Where the individual standard specifies, add bromophenol blue indicator (as for dark coloured perfumery materials).

Neutralise with hydrochloric acid contained in the burette till a greenish-yellow colour is observed. Transfer the liquid to the conical flask used to mix the reagent and the alcoholic potassium hydroxide solution, and pour about half back again into the flask. Continue the neutralisation in one of the flask only till the greenish-yellow colour of the indicator changes to lemon-yellow. The colour obtained in both flasks, shall not alter further when two drops of 0.5 N hydrochloric acid are added.

Alternatively this titration may be carried out by a potentionmetric method to pH 3.5.

Blank determination. Simultaneously using above procedure, carry out a blank test with the same reagents and following the same procedure but omitting the perfumery material.

Calculation

The carbonyl value is calculated as follows :

$$\text{Carbonyl value} = \frac{56.1\,(V_1 - V_2)N}{m}$$

where

V_1 = volume in ml of standard hydrochloric acid used in the blank titration.

V_2 = volume in ml of standard hydrochloric acid used in the determination.

N = the normality of the hydrochloric acid used for titration, and

m = mass in g of the sample.

The content of carbonyl compounds expressed in terms of designated aldehydes and ketones is calculated as follows :

Content of carbonyl compounds (as aldehyde and ketone),

$$\% \text{ by mass} = \frac{M(V_1 - V_2)}{20\,m}$$

where

M = molecular mass of the aldehyde or ketones specified in the individual standard for the particular perfumery material,

V_1 = volume in ml of standard hydrochloric acid used in the blank titration,

V_2 = volume in ml of standard hydrochloric acid used in the determination, and

m = mass in g of the sample taken.

Determination of phenols

Principle

Phenols react with alkali hydroxides, giving rise to water soluble phenolates, which can be separated from the unabsorbed material. In the case of materials deficient in phenol content which are apt to yield troublesome stable emulsions, 2 ml of xylene may be added to the measured volume of the material to help in breaking the emulsions, and allowance made for the volume of added xylene in the final volume of unabsorbed residual materials.

Apparatus

Cassia flask with a graduated neck

Reagents

Tartaric acid, xylene, 0.1 N aqueous potassium hydroxide solution.

Procedure

Preparation of test sample. Remove as far as possible any visible water present in the sample by decanting, then add about 15 per cent by mass of neutral, freshly ignited and powdered anhydrous magnesium sulphate or anhydrous sodium sulphate and shake the mixture vigorously by from time to time during a period of 2 hours. However, before drying the material with magnesium sulphate or anhydrous sodium sulphate, shake vigorously a quantity of the material greater than 10 ml in a conical flask with 0·02 g of the tartaric acid per millilitre of the material.

Determination

Pipette out 10 ml of the perfumery material prepared with the pipette into the Cassia flask containing approximately 75 ml of potassium hydroxide solution. Shake the mixture six times at 5 minutes intervals at room temperature.

Raise the unabsorbed portion of the material into the neck of the Cassia flask by the addition of more of the potassium hydroxide solution. Facilitate the separation of the material drops attached to the walls by rotating the flask between the hands and gently tapping.

After allowing the flask to stand for few hours, read off the volume of the unabsorbed material if all of it is gathered into the neck. If a quantity of emulsion is observed, add 2 ml of xylene measured with the pipette, agitate the emulsified layer by means of a glass capillary tube and allow to stand. If the emulsion has disappeared, read the volume of the unabsorbed material. If, however, the emulsion persists, repeat the test with the addition of 2 ml of xylene to the portion before the initial shaking.

In the two latter cases, subtract from the reading 2 ml, corresponding to the volume of xylene added.

Calculation

Phenol content, per cent by volume = $(10 - V)$

where

V = volume in ml of the unabsorbed portion of the perfumery material.

Determination of heavy metals

Place in a test tube 10 ml of the perfumery material and add an equal volume of distilled water to which one drop of concentrated hydrochloric acid has been added. Shake thoroughly and then pass hydrogen sulphide through the mixture until it is saturated. Carry out simultaneously a blank determination to which no hydrogen sulphide is passed. In the absence of heavy metals, like copper or lead no darkening in colour in either the material or the water layer is produced. A comparison of the colours of the blank and of the determination will aid in establishing the absence of heavy metals, or the presence of traces. The formation of a scum at the surface between the material and the water layers is no indication of the presence of heavy metals, unless the scum is dark in colour. If the dark colour scum is obtained, follow the procedure of lead content.

Determination of Lead Content

Principle. Volatile components of the perfumery material are removed by boiling a suspension in dilute hydrochloric acid followed by extraction with carbon tetrachloride. The lead in the aqueous acid solution is then determined absorptiometrically as lead dithizonate.

Reagents. Carbon tetrachloride (Lead-free), Chloroform, Aqueous ammonia, Hydrochloric acid (Lead free), Nitric acid solution 1%, Ammonium citrate (Lead free), Sodium sulphite (4%), Ammonical sulphite-cyanide solution, dithiozone solution, Potassium cyanide 10% (Lead free) and Hydroxylammonium chloride solution (20%), Sodium hexametaphosphate solution (Lead free), Lead solution, Thymol blue indicator solution (0.1%).

Ammonium citrate solution. $(NH_4)_3C_6H_5O_7$, *lead-free, prepare as follows:*

Dissolve 125 g of ammonium citrate in 400 to 450 ml of water, make faintly alkaline to litmus paper with lead-free aqueous ammonia (5 M) and extract with chloroform and appropriate additions of the stock dithizone solution. Continue extraction until all metals have been removed and the extract is faintly green, then make the solution just acidic by adding lead-free hydrochloric acid (5 M), and extract with further portions of chloroform until the final extract is colourless.

Ammonical sulphite-cyanide solution. Mix 340 ml of the aqueous ammonia, 75 ml of the sodium sulphite solution, and 30 ml of the potassium cyanide solution and 605 ml of water.

Dithizone stock solution (0.1%)

Dithizone solution. Prepare by shaking 6 ml of the stock dithizone solution with 9 ml of water and one ml of ammonia. Separate and reject the lower layer. Centrifuge the aqueous layer until clear.

Sodium hexametaphosphate solution lead free. Dissolve 10 g of sodium hexametaphosphate in 90 ml of water, adjust to pH 9 with thymol blue indicator solution by adding (lead-free) aqueous ammonia and extract with dithizone in chloroform until free from lead. Make the solution just acidic and remove the dithizone traces by extraction with chloroform. Finally adjust to the maximum blue colour of the indicator.

Standard lead solution. Dissolve 1.60 g of lead nitrate [Pb $(NO_3)_2$] in water, add 10 ml of nitric acid, and dilute to 1000 ml. Further dilute one volume of this solution freshly as required to 100 volumes with water. One ml of this solution is equivalent to 10 μg of lead.

Apparatus

Separating funnel, spectrophotometer with cells.

Procedure

Weigh, approximately 5 g of the sample into a 250 ml beaker. Add 80 ml of water and 10 ml of hydrochloric acid. Boil gently with constant stirring for 10 minutes to volatilise most of the materials. Cool and transfer the acid solution to a 250-ml-separating funnel. Add 10 ml of the carbon tetrachloride to the beaker, boil, cool and transfer to the separating funnel. Repeat this procedure, then rinse the beaker twice with 10 ml of water, adding the washings to the separating funnel. Extract with 20 ml portions of carbon tetrachloride until no further colour is extracted. Filter the aqueous layer through a wetted filter paper into a 100-ml volumetric flask, rinsing out the separating funnel and filter with water until 100 ml of filtrate is obtained.

Transfer 10 ml of this solution to a small beaker, add 5 ml of ammonium citrate solution and 10 ml of sodium hexametaphosphate solution, a few drops of thymol blue indicator solution and sufficient aqueous ammonia to give a blue-green colour indicating pH 9·0 to 9·5. Cool, add one ml of potassium cyanide solution, one ml of hydroxylammonium chloride solution and adjust the alkalinity to pH 9·0 to 9·5, if necessary. Transfer to a 100 ml separating funnel containing 10 ml of the chloroform, washing in with a few millilitres of water. The volume of the aqueous layer at this stage should be about 50 ml. Add 0·5. ml of dithizone working solution, shake vigorously for one minute and allow to separate. If the lower layer is red, add dithizone working solution until, after shaking, a purple blue or green colour is obtained. Run the chloroform layer into a second separating funnel and rinse the first separating funnel through with one ml or 2 ml of chloroform. Add, to the liquid in the first separating funnel, 3 ml of chloroform and 0·2 ml of dithizone working solution, shake vigorously for 30 seconds, allow the chloroform layer to separate and add it to the main chloroform extract. The last chloroform extract should be green. If it is not, carry out further extractions with chloroform and dithizone working solution until the green colour of the final extract indicates that all the lead has been extracted. Alternatively, repeat the extractions using a smaller portion of the test solution. Reject the aqueous layer. Shake the chloroform extract vigorously for one minute with 10 ml of nitric acid solution. Separate, discarding the chloroform layer as completely as possible.

Add 30 ml of the ammoniacal sulphite-cyanide solution, exactly 10 ml of chloroform and 0.5 ml of dithizone working solution, and shake the mixture vigorously for one minute. Allow it to settle. If the supernatant liquid is not pronounced orange-yellow, add a further 0.5 ml of dithizone working solution, shake the mixture again for one minute and allow it to settle. Run off a few drops of the chloroform layer, insert a plug of cotton wool into the dry stem of the funnel and, after rejecting the first runnings, fill a one cm cell with the filtered chloroform solution. Measure the optical density of the solution against chloroform and determine the lead content from a standard curve obtained with 0, 10, 2.0, 3.0 and 4.0 ml of the standard lead solution, diluted to a volume of 10 ml with nitric acid solution, using the same procedure. Measure the optical density of the test solution against chloroform.

Reagent blank determination

Carry out a blank determination as prescribed above but omitting the perfumery material.

Expression of results

Deduct the value of the optical density for the reagent blank from that for the test solution and determine from the calibration graph the lead content of the perfumery material. Express the result as milligrams of lead per kilogram of the perfumery material.

Determination of Iron

Apparatus. Nessler cylinders – 100 ml capacity.

Reagents

Hydrochloric acid - 2 N, Ammonium persulphate, Potassium thiocyanate solution – approximately 5 per cent.

Standard iron solution

Dissolve 0.702 g of ferrous ammonium sulphate [$FeSO_4$ $(NH_4)_2$ SO_4. $6H_2O$] in 100 ml distilled water, add 10 ml of concentrated sulphuric acid and dilute with water to 1000 ml mark. Transfer 100 ml of this solution to a 1000 ml volumetric flask and dilute up to the mark. One millilitre of this solution is equivalent to 0.01 mg of iron (as Fe).

Procedure

Add 25 ml of 2 N hydrochloric acid to 25 ml of perfumery compound, shake and boil slowly for about 10 minutes. Cool and separate the aqueous layer. Add about 30 mg of ammonium persulphate and boil to oxidise the iron. Cool and transfer to a Nessler cylinder, add 2 ml of potassium thiocyanate solution and dilute to 100 ml mark with water. In another Nessler cylinder, take the same amounts of hydrochloric acid, ammonium persulphate and potassium thiocyanate solution and dilute to about 90 ml. From a burette add standard iron solution in small portions so that after dilution to 100 ml, the colour obtained matches with that obtained with the material. The iron content of the perfumery material is then equal to the amount of iron added to obtain equivalent colouration as produced by sample solution.

Detection of petroleum and mineral oils

The saturated paraffinic hydrocarbons, found in the petroleum oils, are chemically very inert; they are not destroyed by fuming sulphuric acid. Other compounds are attacked, giving rise to reaction products which are soluble in sulphuric acid. The material has to be added to the fuming sulphuric acid very cautiously since too rapid an addition of perfumery materials is apt to cause the liberated sulphur dioxide to carry part of the acid and material out of the flask. Since petroleum fractions often contain aromatic and unsaturated compounds as well as paraffins, the separation of the paraffinic portion described above does not usually represent the total amount of added petroleum. In general the portion thus separated is usually a small per centage of the adulterant. The test may be rendered more sensitive by preliminary fractionation of the material.

The addition of petroleum fractions to a material causes a lowering of the specific gravity, refractive index and optical rotation. The solubility of the material usually is also affected.

Apparatus

Cassia Flask.

Reagents

Fuming Sulphuric Acid —15 per cent oleum, Concentrated Sulphuric Acid.

Procedure

Place 20 ml of fuming sulphuric acid (15 per cent oleum) in a dry Cassia flask of approximately 150 ml capacity, and cool thoroughly in an ice-salt mixture. Cautiously add 5 ml of the material from a small burette, drop by drop, with frequent shaking and cooling in the ice-salt mixture. After the material has been added, again shake the flask and allow it to stand at room temperature for 10 minutes. Warm on a steam-bath for 5 minutes with frequent agitation, cool to room temperature, then fill to the mark with concentrated sulphuric acid and allow the flask to stand overnight, when the mineral oil will rise into the neck and separate as a colourless or straw-coloured liquid. To confirm the presence of mineral oil, remove a small amount of the separated mineral oil from the Cassia flask, using a glass tube drawn out to a small tip, and determine its refractive index. For petroleum products and mineral oils, the refractive index is expected to be less than 1.440 0.

Estimation of chlorine

Principle

Absence of even a transient green colour when the material is ignited on a copper gauze in a non-luminous flame is used for determining freedom from chlorine.

Apparatus

Copper wire and Bunsen burner.

Procedure

Place the copper strip in the non-luminous flame of the Bunsen burner until it glows without imparting a green colour. Cool the gauze and repeatedly ignite it until an oxide coating has formed. Cool the gauze and add 2 drops of the sample by means of a medicine dropper, permitting it to burn in air. Again cool and add 2 more drops of the test material and burn as before. Continue the procedure until 6 drops have been ignited. Hold the gauze in the outer edge of the non-luminous flame whose height has been adjusted to about 4 cm.

Result

The flame shall be free from even a transient green colour.

Determintaion of cineole content

Principle

The crystallising temperature is measured of a mixture of perfumery materials and o-cresol. This temperature depends on the cineole content of the perfumery materials.

Apparatus

Calibrated thermometer, test tube, stout-walled test tube and wide mouth jar or bottle.

Reagents

o-Cresol, Cineole, Cineole-o-Cresol complex.

Procedure

Prepare the sample as per the procedure described in phenol estimation.

Preliminary test. Weigh 3 g of the freshly prepared test sample in the dry test tube, and add 2.1 g of melted o-cresol. Place the tube in the apparatus and allow the mixture to crystallise by cooling, and stirring. When crystallisation takes place, there is a small increase in temperature t_1.

Determination. First assemble the apparatus as per Fig. 12.5. Remelt the mixture, at a temperature not exceeding t_1 by more than 5°C, using the water bath. Place the test tube into the apparatus maintained at a temperature 5°C below t_1. When crystallisation begins, or when the temperature of the mixture has fallen to a value 3°C below t_1, stir continuously. Induce the crystallisation by rubbing the wall of the test tube with the bath of the thermometer. Note the maximum temperature at which the mixture recrystallises, t_2.

Repeat the determination until the two highest values obtained for t_2 do not differ by more than 0·2°C. If supercooling occurs, induce the crystallisation by adding a small crystal of the complex.

If t_2 is below 27.4°C, repeat the determination after the addition of 5.1 g of the cineole-o-cresol complex.

Calculation

The content of cineole corresponding to the highest temperature observed t_2 is given in Table 12.6. If 5.1 g of the cineole-o-cresol complex has been added, the cineole content of the sample is calculated and given below :

Cineole content, per cent

by mass $= 2 (A - 50)$

where

A = percentage of cineole as given in Table 12.6.

The content of cineole, corresponding to the highest temperature observed t_2, is obtained, where necessary, by interpolation from the values as given in Table 12.6.

Table 12.6. Cineole content as a function of the crystallisation temperature of the perfumery materials-*o*-cresol mixture.

Crystallisation temperature 0°C	Cineole Contents, Per cent									
	0.0	0.1	0.2	0.3	0.4	0.5	0.6	0.7	0.8	0.9
24	45.6	45.7	45.9	46.0	46.1	46.3	46.4	46.5	46.6	46.8
25	46.9	47.0	47.2	47.3	47.4	47.6	47.7	47.8	47.9	48.1
26	48.2	48.3	48.5	48.6	48.7	48.9	49.0	49.1	49.2	49.4
27	49.5	49.6	49.8	49.9	50.0	50.2	50.3	50.4	50.5	50.7
28	50.8	50.9	51.1	51.2	51.3	51.5	51.6	51.7	51.8	52.0
29	52.1	52.2	52.4	52.5	52.6	52.8	52.9	53.0	53.1	53.3
30	53.4	53.5	53.7	53.8	53.9	54.1	54.2	54.3	54.4	54.6
31	54.7	54.8	55.0	55.1	55.2	55.4	55.5	55.6	55.7	55.9
32	56.0	56.1	56.3	56.4	56.5	56.7	56.8	56.9	57.0	57.2
33	57.3	57.4	57.6	57.7	57.8	58.0	58.1	58.2	58.3	58.5
34	58.6	58.7	58.9	59.0	59.1	59.3	59.4	59.5	59.6	59.8
35	59.9	60.0	60.2	60.3	60.4	60.6	60.7	60.8	60.9	61.1
36	61.2	61.3	61.5	61.6	61.7	61.9	62.0	62.1	62.2	62.4
37	62.5	62.6	62.8	62.9	63.0	63.2	63.3	63.4	63.5	63.7
38	63.8	63.9	64.1	64.2	64.4	64.5	64.6	64.8	64.9	65.1
39	65.2	65.4	65.5	65.7	65.8	66.0	66.2	66.3	66.5	66.6
40	66.8	67.0	67.2	67.3	67.5	67.7	67.9	68.1	68.2	68.4
41	68.6	68.8	69.0	69.2	69.4	69.6	69.7	69.9	70.1	70.3
42	70.5	70.7	70.9	71.0	71.2	71.4	71.6	71.8	71.9	72.1
43	72.3	72.5	72.7	72.9	73.1	73.3	73.4	73.6	73.8	74.0
44	74.2	74.4	74.6	74.8	75.0	75.2	75.3	75.5	75.7	75.9
45	76.1	76.3	76.5	76.7	76.9	77.1	77.2	77.4	77.6	77.8
46	78.0	78.2	78.4	78.6	78.8	79.0	79.2	79.4	79.6	79.8
47	80.0	80.2	80.4	80.6	80.8	81.1	81.3	81.5	81.7	81.9
48	82.1	82.3	82.5	82.7	82.9	83.2	83.4	83.6	83.8	84.0
49	84.2	84.4	84.6	84.8	85.0	85.3	85.5	85.7	85.9	86.1

(Cont'd)

Crystallisation temperature 0°C	Cineole Contents, Per cent									
50	86.3	86.6	86.8	87.1	87.3	87.6	87.8	88.1	88.3	88.6
51	88.8	89.1	89.3	89.6	89.8	90.1	90.3	90.6	90.8	91.1
52	91.3	91.6	91.8	92.1	92.3	92.6	92.8	93.1	93.3	93.6
53	93.8	94.1	94.3	94.6	94.8	95.1	95.3	95.6	95.8	96.1
54	96.3	96.6	96.9	97.2	97.5	97.8	98.1	98.4	98.7	99.0
55	99.3	99.7	100.0	—	—	—	—	—	—	—

GAS CHROMATOGRAPHIC ANALYSIS OF PERFUMERY MATERIALS

Principle

A small amount of the perfumery material is introduced into a gas liquid partition column. The various components that are volatile under the conditions of test are vapourised and transported through the column by a carrier gas. The separated components are measured in the effluent by a detector and recorded as a chromatogram. The chromatogram is interpreted by applying component attenuation and detector response factors to the peak areas, and the relative concentrations are determined by relating the individual peak responses to the total peak responses.

Apparatus

The system which enables GLC analysis, is a standard gas chromatography which consists of: (i) Oven; (ii) Injection Port; (iii) Columns; (iv) Detector; (v) Temperature Controllers; (vi) Carrier Gas and Flow Regulators; (vii) Recorder and Data Processor and; (viii) Power Supply.

Procedure

The gas chromatograph is prepared for perfumery chemicals or essential oils under the conditions indicated in the relevant standard of the chemical oil being tested.

After stabilisation of the desired temperature of the column, injection port and detector, a suitable amount of the sample is injected with microlitre hypodermic syringe. The amount of liquid samples would be 0.2 to 1.0 microlitre for packed columns and 0.1 to 0.2 microlitre of the material in a 30 to 50 per cent solution of pure analytical grade acetone or cyclohexane for capillary columns. The solid sample would also be dissolved in acetone or cyclohexane and the solution would be injected in appropriate amount.

On the chromatogram obtained, all peaks of interest should be of suitable dimensions and that except in case of attenuation, none should exceed 90 per cent of all the available recorder paper width. If required, sample is re-injected to get a better chromatogram.

Calculation

Peak areas are calculated either by the most commonly used triangular method or automated integration. The sum of the areas is equated to 100 per cent, it is tactly implied that all substances have come out of the column and all substances have the same response factor, however, this is not the case in truth due to response factor of detector. Yet this method of calculation is fairly accurate.

Chapter 13

Raw Materials for Cosmetics

INTRODUCTION

A wide range of chemical and natural materials are used in the manufacture of cosmetics. Some of the most commonly used materials include; water, surfactants, shampoos and bath additives, oil components, silicone oils, cream bases, oil-in-water (o/w) emulsifiers, water-in-oil (w/o) emulsifiers, humectants, aerosol propellants, antioxidants, colours and perfumes.

WATER

Water is highly reactive substance. In biochemical process, oxidation, reduction, condensation, and hydrolysis occurs. The water is also used as solvent for many ingredients of cosmetics. The water is always associated with many impurities hence it is purified by techniques such as, ion-exchange process, distillation, reverse osmosis, microbial technique, uv–radiation and filtration.

SURFACTANTS

Surfactants are surface active agents which reduces the surface tension of the liquids. Anionic, cationic and non-ionic surfactants are used in cosmetics and toiletries due to their balanced properties. They are excellent foamers. The basic or primary surfactants are the backbone of the cleansing products. They include alkyl ether sulphates, linear alkyl benzene and alkane sulphonates. Mild anionic surfactants are sulphosuccinates, cocoyl isethionates, acyl amides, alkyl ether carboxylates, and magnesium surfactants. Amphoteric surfactants include, alkyl betaines alkyl amide betaines, amine oxides, and ethoxylated products, alkyl polyglycosides are non-ionic surfactants. The list of cationic surfactants include monoalkyl, dialkyl, trialkyl, benzyl and ethoxylated quaternaries.

SHAMPOO AND BATH ADDITIVES

Thickeners

A high viscosity is often very important both for product stability and for handling of a cosmetic product. A different principle of thickening is obtained by the use of special, high molecular weight thickeners such as distearate, myristyl glycol methyl glucose dioleate. A side-effect of all these thickeners is that they modify the flow properties, leading to increased Newtonian flow. This contrasts with salt- or polymer-thickened systems, which show typically pseudoplastic flow behaviour. Another type of rheological modifier is the inorganic bentonites, which can be used to obtain a yield point.

Foam stabilisers

The stability of shampoo foam is reduced in the presence of oily soils such as sebum. The so-called foam boosters act as stabilisers and also modify the foam structure to give a richer, dense foam with small bubbles e.g. monoethanolamides, diethanolamides.

Pearlescent agents

Ethylene glycol mono- and distearates (EGMS, EGDS) are most often used as pearlescent agents in surfactant formulations. They have to be incorporated at high temperatures (approx. 70–75°C). Direct contact with perfume oils can result in coagulation of the polystyrol, and must be avoided.

Conditioning Agents

The addition of conditioning agents improves the wet-combability and reduction of static charge build-up. These are especially effective if the formulation also contains amphoteric surfactants such as betaines or amines oxides. Cationic surfactants used in hair rinses are normally incompatible with anionic surfactants and cannot be used in shampoo formulations. This problem can be overcome by the use of quaternised polymers.

Emollients

Emollients are added to overcome possible harsh effects of shower and foam baths on the skin. Real lipids like isopropyl myristate are difficult to incorporate due to their limited solubility in water-based formulations. Water-soluble lipids, which still have a certain lipophilic character can be obtained by ethoxylation to an HLB of approx. 6 to 8. They do not interfere with the foaming of the surfactants.

Sequestering Agents

Sequestering agents are used to prevent the formation and deposition of Ca and Mg soaps. Hard water along with these agents can be used for the production of shampoo and bath preparations. As a positive side-effect they can also give support to the preservation system.

Oil Components

Mineral oil

Mineral oils for cosmetic use are high-boiling fractions obtained from crude oil distribution that are purified and refined by treatment with sulphuric acid. Two types are of importance: (i) liquid paraffin (viscosity 110 to 230 mPas); and (ii) light liquid paraffin (viscosity 25 to 80 mPas). Mineral oils are colourless, clear and odourless liquids, insoluble in alcohol or water. They are excellent cosmetic emollients because they are inert and do not penetrate into the skin. They therefore have excellent skin compatibility and little or no comedogenic potential. Since they are not considered 'natural', mineral oils have been attacked repeatedly but are still the most widely used oil component in skin-care formulations.

The fatty character of mineral oils help to form a film on the skin which increases hydration by blocking the normal evaporation of water. Combinations with synthetic esters such as isopropyl stearate are recommended in order to open the film, and to guarantee the right balance for water evaporation.

Natural Oils

Triglycerides

Natural oils from vegetable sources are mainly glycerol esters based on mixtures of fatty acids. The most important are the saturated lauric, myristic, palmitic and stearic acids and the unsaturated oleic and linoleic acids. These fatty acids are derived from natural sources and typically have even numbers of carbon atoms in their chains.

Important characteristic values for triglycerides are the saponification number and the iodine number. The latter indicates the amount of unsaturated fatty acids present.

Triglycerides are relatively fatty and spread very little on the skin. They generally show moderate comedogenicity, exceptions being sunflower oil and safflower oil where no comedogenic effects have been found. Castor oil is used in lipsticks and also in alcoholic preparations.

Jojoba oil

It is a liquid, unsaturated wax from the esters of long carbon-chain fatty acids and long carbon-chain unsaturated alcohols. Jojoba oil has a very fatty character and shows excellent cosmetic properties.

Synthetic Oils

Synthetic oils are esters, usually obtained by direct reaction of fatty acids with alcohols. A wide range of combinations to produce different synthetic esters is possible, and the list of synthetic oils available to the cosmetic industry is long.

Isopropyl esters

The most important synthetic ester for cosmetics is isopropyl myristate. It has a very dry character and shows good spreading on the skin. It can open films of mineral oil or other fatty oils in order to let water evaporate, and it is an excellent solvent for perfumes, ultraviolet filters, etc. The isopropyl esters can cause difficulties when used in formulations that are filled into polystyrene or polyethylene containers since it is good solvent.

Ethylhexyl esters

Ethylhexyl esters of palmitic acid or stearic acid have become important products for the cosmetic industry. Compared with isopropyl esters, the molecular weight is higher, but spreading properties are similar.

Oleic acid esters

The two most important oleates are decyl oleate and oleyl oleate. The character of these esters is relatively fatty and, as with unsaturated esters, they can become rancid. The advantage of the oleates is that, even at high molecular weight, the esters are liquid.

Caprylic/capric acid esters

The most important caprylic/capric esters are the triglyceride and the propylene glycol diester. The glycerol ester is mainly used as a replacement for natural triglycerides, with the advantage that it is stable against oxidation.

N-Butyl stearate

N-butyl stearate is an excellent skin-compatible ester with very good emollience. The character is not very fatty compared with the isopropyl esters.

Isocetyl stearate

Isocetyl stearate is a saturated liquid ester of high molecular weight and good emollience. Water can evaporate easily through the ester and even small amounts can perforate a hydrophobic film of mineral oil.

Octyldodecanol

Octyldodecanol is not an ester but may be used as a cosmetic oil component. It combines a fatty character with good spreading properties.

Di-isopropyl adipate

Di-isopropyl adipate is an ester of a dicarbonic acid. Contrary to most other oils, it is soluble in alcohol and is therefore used, for example, as a refatting agent in aftershave formulations.

Pentaerythritol tetraisostearate

Pentaerythritol tetraisostearate is a high molecular weight, highly branched ester of pentaerythritol and isostearic acid. It has a very fatty character and shows no comedogenic effects. Consequently, it is an ideal ester for skin-protection formulations such as baby products and, due to its water-repellent properties, water-resistant suntan preparations.

WAXES

The word wax has two different meanings. From a chemical point of view it means an ester of a fatty acid and a fatty alcohol. Jojoba oil is therefore a liquid wax. In this section, however, the word wax is considered from a physical point of view, and means compounds having a high melting point (approx. 50–100°C).

Natural waxes

The most important wax, and a classical component for creams, is beeswax, which is the construction material of honeycombs. Chemically, it consists of mixed esters of long-chain alcohols plus fatty acids and hydroxy fatty acids of chain length of 16–26. At room temperature the beeswax is very hard. The melting point is around 61 to 66°C. Untreated beeswax is dark-yellow and is usually treated to improve the colour.

Beeswax is a very good consistency regulator in creams and ointments and is also used in stick formulations. The addition of approx. 6 per cent of borax, calculated on the beeswax, results in partial saponification and provides the beeswax with some emulsifying properties.

Two other natural waxes that are harder and therefore mainly used in stick formulations are carnauba wax (melting point approx. 85°C) and candelilla wax (melting point approx. 70°C).

Synthetic Waxes

These are often mixtures of different chemicals. An example of a beeswax replacement is the mixture of glyceryl hydroxystearate (and) cetyl palmitate (and) micro-crystalline wax (and) trihydroxy stearine. This product, which shows very similar behaviour compared with beeswax, is easy to emulsify and can be used as a general consistency regulator for cosmetic emulsions.

SILICONE OILS

The high molecular weight organo polysiloxanes (dimethicone) are hydrophobic oils with good skin protection and non-sticky skin-feel. They show very high spreading and are used in small amounts in stearate creams to avoid the soap-up effect. To prevent eye-sting, non-volatile oils should not be used in products that are applied to the eye area. Due to their water-repellent properties, silicone oils are important for waterproof sun products. They are usually classified by viscosity.

CREAM BASES

Fatty Alcohols

Fatty alcohols are important raw materials for surfactants, emollients and emulsifiers. Pure fatty alcohols, mainly cetyl alcohol and stearyl alcohol, are also used *per se* as consistency regulators and co-emulsifiers in creams, lotions and hair rinses. The so-called natural fatty alcohols are obtained by hydration of fatty acid methyl esters.

Polyol Esters

The most important polyol ester is glycerol monostearate (GMS). Generally, it is not a pure product but a mixture of mono- and diesters of stearic and palmitic acids. The main distinguishing feature is the content of monoester. Products with approx. 40% monoester are obtained by direct esterification of stearic acid and glycerol. Products with approx. 60% monoester are produced by glycerolysis of triglycerides with glycerol. The 90% material can be obtained by molecular distillation but is seldom used in cosmetic formulations because the mono/diesters provide the best applicational properties. Typical combinations are :

(1) GMS plus potassium stearate ('GMS self-emulsifying')

(2) GMS plus sodium lauryl sulphate

(3) GMS plus ethoxylated fatty alcohols ('GMS self-emulsifying, acid stable')

Fatty Acids

Fatty acids, like the fatty alcohols, are important raw materials for cosmetic ingredients such as surfactants, emulsifiers and emollients.They are seldom used in emulsions. They act as consistency regulators but tend to crystallise in the formulation.

Fatty acids especially stearic acids, are mainly used in the form of soaps. As the pH increases, they become truly anionic and act as oil-in-water (O/W) emulsifiers, forming a group of so-called stearate creams.

OIL-IN-WATER (O/W) EMULSIFIERS

Anionic O/W emulsifiers

Anionic O/W emulsifiers can be used to obtain very stable emulsions because they can build an electrical double layer around the droplets, which prevents the droplets from coalescing. On the other hand, anionic emulsifiers are sensitive to low pH and electrolytes.

The most important anionic O/W emulsifier is soap, which forms the stearate creams. The use of triethanolamine as a neutralising agent is more common. A disadvantage of stearic acid is the typical 'soap-up' effect, a type of foaming that occurs when the cream is rubbed into the skin. To prevent this, silicone oil is frequently added to the formulations.

Other very effective anionic emulsifiers are alkyl sulphates, particularly sodium cetearyl sulphate. In principle, all anionic surfactants can be used. However, due to the fact that the soaps are relatively alkaline and fatty alcohol sulphates are irritant to the skin, the mild surfactants are of more interest. These include sodium cocoyl isethionate as an extremely mild emulsifier, and phosphoric acid esters, like potassium cetyl phosphate which are very effective at low concentrations.

Cationic O/W Emulsifiers

Cationic emulsifiers have generally been avoided in skin-care products because they are often more irritant compared with anionic emulsifiers. They are important in hair-care formulations, where they act as conditioners and anti-static agents. Their advantage, particularly distearylammonium chloride, is that they produce emulsified products with excellent cushion feel on skin and mitigate the heavy feel imparted by glycerine.

Non-ionic O/W Emulsifiers

A wealth of non-ionic O/W emulsifiers is available. These are mainly PEG derivatives such as ethoxylates or PEG esters. Typical O/W emulsifiers are :

(1) Ethoxylated fatty alcohol

(2) PEG esters of fatty acids

(3) Ethoxylated sorbitan esters

(4) Ethoxylated monoglycerides

(5) Ethoxylated castor-oil derivatives

The degree of ethoxylation determines the primary properties of that part of the molecule that results in the HLB classification. Good O/W emulsifiers can be found in the HLB range between 8 and 18. However, the type of hydrophobic moiety is also very important for the stability of the emulsions. Contrary to anionic emulsifiers, the non-ionics are unaffected by changes in pH and, in the case of pure ethers, they can also be used at extreme pH values.

O/W Stabilisers

Carbomers are very efficient and widely used stabilisers for O/W emulsions. These are polyacrylate resins which have to be neutralised in order to form gels, usually by the use of triethanolamine (TEA). Amounts up to 0.5% are recommended for cosmetic emulsions. Too much can leave an unpleasant feel on the skin. The principle of stabilisation is the thickening of the outer phase. The use of preneutralised copolymers, e.g., acrylamide/sodium acrylate copolymer can make incorporation and handling easier.

WATER-IN-OIL (W/O) EMULSIFIERS

From the dermatological point of view, W/O emulsions are preferable to O/W emulsions. The natural lipid film on the skin is also a W/O emulsion. W/O emulsions can improve the hydration of the skin and are an ideal base for lipid-soluble active ingredients. A problematic point is that they usually leave a 'fatty-feeling' on the skin, and this is not particularly appreciated by customers. Another problem can be 'oiling-off' or separation of the oil phase on storage. Viscosity and stability are extremely process sensitive.

Single W/O Emulsifiers

A limited number of W/O emulsifiers are available. This is because ionic emulsifiers will not work in the case of W/O emulsions and, since very low HLB is required, variations using ethylene oxide are not possible. W/O emulsifiers have HLB values between 3 and 6. However, the type of the alkyl chain is also important. In practice, oleyl derivatives have shown good effects. The most widely used W/O emulsifiers are sorbitan monooleate, sorbitan sesquioleate and glycerol monooleate, and with increasing importance, polyglycerol esters.

Lanolin Derivatives

Lanolin or wool wax obtained from sheep wool is a mixture of different esters from higher alcohols, mainly cholesterol, with higher fatty acids. Since lanolin can cause allergies, lanolin alcohols extracted after saponification of the wool wax are now mainly used. Due to the high amount of cholesterol, they are better emulsifiers than lanolin itself.

Absorption Bases

Since it is relatively difficult to obtain stable W/O systems, pre-mixtures, which are able to bind high amounts of water are available. Relatively simple absorption bases are, for example, mineral oil (and) lanolin alcohol or petrolatum (and) lanolin alcohol. Very complex systems are also offered and include petrolatum (and) decyl oleate (and) sorbitan sesquioleate (and) beeswax (and) mineral oil (and) ceresin (and) aluminum stearate, or mineral oil (and) petrolatum (and) ozokerite (and) glycerol oleate (and) lanolin alcohol. Using these premixtures it is relatively simple to obtain stable emulsions, although the resulting emulsion often behaves more like an ointment than a cosmetic product.

W/O Stabilisers

The stable W/O emulsion is obtained using block copolymer like dodecyl glycol. These products, with molecular weights of 2000 to 4000, can improve the stability of W/O emulsions by protection of the droplets from coalescence. The hydrophilic middle part is responsible for a good anchorage in the water phase, while the highly branched ends give a steric hindering effect. The high molecular weight imparts excellent skin compatibility. These co-polymers typically reduce the viscosity. It is possible to obtain light, highly stable W/O lotions that are good bases for sun preparations and which give a skin-feel very close to emulsions.

HUMECTANTS

Humectants protect emulsions or toothpaste from 'drying-up'. The most important humectants are glycerol, obtained from saponification of triglycerides, and sorbitol, a hexa-alcohol. Sorbitol is a hygroscopic powder but is most often used in the form of a 70% aqueous solution. Glycerol is typically used at 99% with a density of 1.26 or at 85% with a density of 1.22. Both are non-toxic and have a sweet taste, which makes them ideal for the use in toothpastes.

Other humectants for emulsions are lactates, 1,3-butylene glycol and 1,2-propylene glycol. The latter should be of high quality as it has occasionally been implied as a skin irritant. A positive side-effect of humectants is a reduction of the freezing point. Humectants are also important in moisturising products.

AEROSOL PROPELLANTS

The fluoro-carbons were popular aerosol propellants however, it is banned due to reduction in ozone layer. Hydrocarbons and dimethyl ether are used now-a-days.

Hydrocarbons

The hydrocarbon propellants are propane (freezing point – 42.1), *n*-butane (freezing point – 0.5) and isobutane (freezing point – 11.7). Commercial butane is always a mixture of *n*-butane and isobutane. Propane mixtures with different ratios are usually used to adjust the pressure. Hydrocarbons are cheap, stable and of low toxicity, but are highly flammable.

Dimethyl ether

Dimethyl ether (DME) has been known as a propellant and, it is a gas (freezing point – 24.8) with a vapour pressure of 4.2 bar at 20°C. It shows very low toxicity and does not damage the ozone layer. Like hydrocarbons it is inflammable, but unlike hydrocarbons it is miscible with water. This can reduce or prevent the risk of flammability of DME aerosol sprays, since 6%

water can be solved in DME, or 34% DME can be solved in water. Alcohol/ DME mixtures are good solvents for PVP/PVA resins for hair sprays, whereas hydrocarbon propellants are difficult to use in this field.

ANTIOXIDANTS

Cosmetic preparations containing fats and oils particularly those have unsaturated linkage are susceptible to oxidative deterioration. Even a very small quantity of highly unsaturated ingredient is sufficient to start a typical oxidative deterioration. Some of the most commonly used antioxidants are :

(1) Phenolic Type

(2) Quinone Type

(3) Amine Type

(4) Organic acids, alcohols and esters

(5) Inorganic acids and their salts

Selection of Antioxidant

The selection of antioxidant depends on the following :

(1) The Nature of the fat/or oil present as ingredient of cosmetics

(2) The physical form of cosmetic

(3) pH of cosmetics

(4) Intended use of cosmetics

(5) Expected shell life of cosmetics

(6) Conditions of storage of cosmetics

An ideal antioxidant would be one which is stable and effective over a wide range of pH and is soluble in its oxidised form. The reaction products of antioxidant should be colourless and odourless.

COLOURS

Colours have been used in cosmetics from the very beginning. Basically a desire to buy a cosmetic product is controlled by three senses, namely, sight, touch and smell. As such colour is an important ingredient of cosmetic formulation.

Colours which can be used in cosmetics can be classified into three classes: (i) Natural Colours; (ii) Inorganic Colours; and (ii) Coal tar Colours.

Natural Colours

Some of the important natural colours used are :

Cochineal

Cochineal is a red dye stuff.

Saffron

Saffron is dried stigma of flowers of plant.

Chlorophyll

Chlorophyll occurs abundantly in nature. Green colour to leaves is imparted by chlorophyll. Chlorophyll can be extracted from plant leaves.

Inorganic Colours

A number of inorganic colour pigments which are used in cosmetics are : Iron oxides, chromium oxides, ultramarines, carbon black, titanium dioxide, zinc oxide, aluminium hydrate.

PERFUMES

There may be only few types of cosmetics which may not contain perfume. A perfumer blends a number of natural and synthetic materials to make a perfume. This mixture is usually known as perfume compound. These blends may contain two or three materials (which is very rare) or may contain many (about dozen or more) materials. Raw materials of perfume compound can be classified as : Plant oil, Animal Secretions, Chemical Substances.

Plant Oil

Plant oil may be essential oils, flower oils or resins, gums etc.

Animal Secretion

Many animals secrete odourous materials. Some of the materials of importance are: Musk, Civet, Ambergris and Castoreum.

Chemical Substances

The chemical substances used in perfumes and cosmetics may be isolates from plants. There are certain chemical substances which are derived from either isolates or essential oils by chemical reactions. Examples of such substances are : esters, geraniol, terpineol. The other synthetic materials may be those which occur in nature, e.g., phenylalcohol, (found in jasmine oil) or those which do not occur in nature.

Chapter 14

Surface-active Agents

INTRODUCTION

The fundamental phenomenon of surface activity is *adsorption*. It leads to lowering of one or more of the boundary tensions at interfaces in the system, and/or stabilisation of one or more of the interfaces by the formation of adsorbed layers.

A surface-active agent (surfactant) is a material which, has the property of altering the surface energy of a surface with which it comes into contact. This lowering of surface energy can easily be observed in, foaming, the enhanced spreading of a liquid on a solid, the enhanced suspension of solid particles in a liquid medium and the formation of emulsions.

The use of surfactants is well established in cosmetics and toilet products and falls into five main areas depending on the surface-active properties required :

(1) Detergent: Where the main problem involves the removal of soiling matter, for example in shampoos and toilet soaps.

(2) Wetting: In products where good contact is required between a solution, for example in the application of hair colourants and permanent waving lotions.

(3) Foaming: Some products need to have a high level of foam in use, for example in shampoos and foam baths.

(4) Emulsification: In products where the formation and stability of an emulsion is a vital feature, for example in skin and hair creams.

(5) Solubilisation: Products in which it is necessary to solubilise an insoluble component need a surface-active agent with the appropriate properties, for example the solubilisation of perfumes and flavours.

These qualities are not mutually exclusive; they are shared to some degree by all surface-active agents.

All surface-active agents have one structural feature in common: they are all amphipathic molecules; that is, the molecule has two distinct parts—a hydrophobic unit and a hydrophilic unit.

Hydrophobic units are usually hydrocarbon chains or rings or a mixture of the two. Hydrophilic units are usually polar groups such as carboxylic, sulphate or sulphonate groups, or, in non-ionic surfactants, a number of hydroxyl or ether groups. The dual nature of these molecules allows them to adsorb at interfaces and this accounts for their characteristic behaviour.

Classification of Surfactants

Surfactants may be classified on the basis of the uses to which they may be put, on the basis of their physical properties or on the basis of chemical structure. None of these is entirely satisfactory, but probably the most logical is to classify them according to their ionic behaviour in aqueous solution. Using this procedure there are four types of surfactant—anionic, cationic, non-ionic and ampholytic surfactants. In addition, the different structures of the hydrophobic and hydrophilic groups have to be considered.

Anionic surfactants

Anionic surfactants are those molecules in which the surface-active ion is negatively charged in solution. The classic example is soap: $C_{17}H_{33}COO^-$ Na^+ (sodium oleate). The anionic surfactants are further sub-divided according to the manner in which the anionic group is attached to the hydrophobic part of the molecule.

Cationic surfactants

The positively charged ion in aqueous solution is referred as cationic surfactant.

Non-ionic surfactants

Non-ionic surfactants are characterised by the fact that the hydrophilic part of the molecule is usually made up from a multiplicity of small uncharged polar groups, for example hydroxyl groups or the ether linkages in ethylene oxide chains. The same linkages are used to reinforce the hydrophilic character in certain anionic surfactants, for example alkyl ether sulphates.

Ampholytic surfactants

Ampholytic surfactants are characterised by their ability to form a surface-active ion with both positive and negative charges.

Properties of Surface-active Agents

The characteristic of surface active molecule is the change in surface properties as concentration of surfactant increases in aqueous solution. For example, as concentration rises the surface tension of an aqueous solution of, say, sodium dodecyl sulphate ($C_{12}H_{25}OSO_3Na$) falls rapidly, with corresponding changes in the physical properties such as interfacial tension, electrical conductivity, etc. At a certain concentration level a discontinuity occurs and surface tension and other properties no longer fall. The concentration at which this discontinuity occurs is called the *critical micelle concentration* (CMC).

It has been found that surface tension fell as the concentration of single anions increased (for instance $C_{12}H_{25}OSO_3^-$ in the example given) until at the CMC the single ions began to associate into groups which is called micelles. These micelles may be in the form of spheres of molecular size, in which the hydrophobic tails of the anions are oriented to the centre of the sphere, while the hydrophilic heads are at the outer surface. Thus a spherical micelle of sodium dodecyl sulphate would consist of a group of $C_{12}H_{25}$ tails pointing towards the centre of the sphere, with OSO_3^- heads at the surface. This micelle would correspond very roughly to a droplet of dodecane of molecular size. In fact, micelles do have the property of dissolving water-insoluble organic matter. This phenomenon is called solubilisation and is one of the characteristics of surface-active agents important to the cosmetic chemist.

Selection and Use of Surface-active Agents

Detergency

Detergency is a complex process which involves the wetting of a substrate (hair or skin), the removal of greasy soiling matter, the emulsification of the removed grease and the stabilisation of the emulsion.

For skin cleansing, soap is still an excellent detergent. Custom dictates that a high level of foam is necessary, though it performs no function.

Increased foaming can easily be achieved by superfatting with long-chain fatty acids (as in shaving soaps).

Hair washing is more complex and here foam volume does appear to play some part. Sodium lauryl ether sulphate (SLES) is a common component of shampoos, and foaming is frequently enhanced by the addition of alkanolamides. Ampholytic surface-active agents are used for specialised shampoos.

Wetting

All surface-active agents have some wetting properties. Short chain (C_{12}) alkyl sulphates, alkyl ether sulphates and alkyl aryl sulphonates are all commonly used.

Foaming

High foam volume and stable foams are generally achieved by the use of SLES reinforced by an alkanolamide.

Emulsification

Usually a good emulsifying agent requires a slightly longer hydrophobic unit than does a wetting agent. Soap is still used as an emulsifying agent in cosmetic products, very often because of ease of preparation. If a fatty acid is incorporated in the oil phase and the alkali in the aqueous phase, then stable oil-in-water emulsions are easily formed *in situ* by simple mixing. Water-in-oil emulsions (such as in certain hair creams) are frequently stabilised by calcium soaps. Non-ionic surface-active agents are also of value in emulsions.

Solubilisation

All surface-active agents above the CMC have solubilising properties. This is important when it is required to incorporate a perfume or an insoluble organic component into a clear product, for example a shampoo. Soaps, alkyl ether sulphates and indeed most surface-active agents have been used for this purpose. It is of course necessary to use high concentrations to give good solubilisation.

All the above properties may be modified by the presence of electrolytes. In general electrolytes tend to lower the CMC and this should improve solubilisation. They may also tend to break emulsions and in general electrolytes should not be added to cosmetic products containing surface-active agents until their effects on surface-active properties have been fully checked.

Other Properties

All cationic products adsorb strongly at protein and other negatively charged substrates. They are thus used to modify the surface of a substrate, for example to improve the feel and appearance of hair. Cationics have some anti-microbial properties also and may be used as components of special shampoos and of mouthwashes (for example chlorhexidine).

Sodium N-lauryl sarcosinate is known to inhibit the enzyme *hexokinase* (which is involved in the glycolytic breakdown of sugars in the mouth) and has been used in toothpastes.

Anionics can have an effect on each other; for example, the foam produced by SLES can easily be destroyed by soap (both anionic). This property is made use of in the formulation of low foam detergents.

Biological Properties of Surface-active Agents

As surface-active agents are adsorbed at the surface, hence they may have biological effects. All cosmetic products containing surface-active agents should be checked rigorously to ensure that they do not have harmful effects on users.

Dermatological effects

Surface-active agents wet the skin and may remove grease from the surface of the skin. When wrongly used they may create chapping, cracking and dryness of the skin. The C_{12} moiety seems particularly active in this respect and the C_{12} sulphate, for example, is used to create chapping artificially. Fortunately the effects are easily reduced by mixture with sulphates of other chain length, by addition of ethylene oxide and by other means.

In general, all cosmetic products containing surface-active agents should be patch tested (and if appropriate eye-tested) to ensure that they produce no adverse reaction.

Biodegradation

It is good practice to use only biodegradable surface-active agents. In some countries this is compulsory. Branched-chain alkyl aryl sulphonates are not biodegradable, but the corresponding straight-chain compounds are, as they are all soaps and alkyl sulphates.

Toxicological effects

Surface-active agents are not, compounds of high toxicity. Nevertheless, since they may be ingested either accidentally or from a toothpaste or mouthwash, it is wise to check oral toxicity of cosmetic products containing them.

Cationics are the most toxic and have LD_{50} values of the magnitude of 50–500 mg per kg body weight; anionics are roughly in the range 2–8 g per kg and nonionics range upwards from about 5 g per kg[12]. Cosmetic products containing surface-active agents should therefore be reasonably safe from toxic hazards.

The various surface-active agents used in cosmetics are as under :

Basic Surfactants

Alkyl ether sulphates

Alkyl ether sulphates are the most widely used surfactants for cosmetics and toiletries due to their well-balanced properties. They are excellent foamers, although the foam structure is relatively coarse. Foaming is not affected by hard water and this, together with their low critical micelle concentration, makes alkyl ether sulphates ideal for use in foam-bath preparations, shampoos and shower baths.

Alkyl sulphates are excellent foamers, producing a rich and creamy foam, but are unstable in hard water. This is not a problem for shampoo or shower applications, but means that they cannot be used as the main component in foam baths. Instead, they must be combined with surfactants such as olefin sulphonates or sulphosuccinates, which are stable in hard water.

A very important application of alkyl sulphates is their use in toothpastes. Sodium lauryl sulphate is available in very pure form with low amounts of unsulphated matter. This results in a very neutral taste compared to the bitter taste of other surfactants.

α-Olefin sulphonates are a mixture of alkene sulphonates and hydroxy alkane sulphonates prepared by sulphonation of α-olefins followed by alkaline hydrolysis.

They are stable in hard water and show hydrotropic properties. This results in low cloud points on cooling, and high solubilising power for superfatting agents. The difficulty of thickening can be overcome by combination with other surfactants. For example, a mixture of 60% olefin sulphonate and 40% sulphosuccinate, which are individually very difficult to thicken, shows a thickening behaviour comparable to alkyl ether sulphates. Contrary to the alkyl sulphates and the alkyl ether sulphates, olefin sulphonates are stable at both acidic and alkaline pH values.

Other basic surfactants

Linear alkyl benzene sulphonate and alkane sulphonate are very powerful surfactants with good detergency and foaming properties. However, due to their strong defatting action, they leave a harsh and dry feel on skin and hair. For very cheap formulations, small amounts can be used to reduce the costs of raw materials. Their main use is in dish-washing liquids and all-purpose cleaners.

Mild Anionic Surfactants

Compared to the basic surfactants, a much wider group of mild anionic surfactants is used in cosmetics and toiletries. The purpose of mild or secondary surfactants is to improve skin and eye compatibility of the formulation. On the other hand, mild surfactants usually show reduced foaming and cleansing compared with basic surfactants.

Sulphosuccinates

The sulphosuccinates have good skin compatibilities compared with the basic surfactants. In general, the amide-based sulphosuccinates are better than the alcohol-based types. Sulphosuccinates are good foamers and are relatively cheap. A disadvantage is their instability at both low and high pH values.

Cocoyl isethionates

Cocoyl isethionate is mainly used as a surfactant in syndet bars. In normal soaps it improves skin compatibility and acts as a lime soap dispersant.

Acyl amides

The two most important acyl amides used in cosmetics and toiletries are the sarcosinates and the protein fatty acid condensates. They are prepared by reaction of fatty acid chloride with N-methylglycine or with oligopeptides. The acyl amides can be considered as modified soaps with good water solubility. Due to their enlarged polar ends, they are stable in hard water.

Alkyl ether carboxylates

Alkyl ether carboxylates are prepared by reaction of chloroacetic acid with fatty alcohol ethoxylates.

Unlike alkyl ether sulphates, alkyl ether carboxylates have no ester linkages and are stable at low pH-values. They are available as 100% active material in the acidic form, or as neutralised solutions.

Magnesium surfactants

The significant applicational property of magnesium surfactants is that they are practically unchanged compared with the sodium forms. From this point of view they belong to the group of basic surfactants showing high foaming and excellent cleansing properties.

Alkyl phosphates

These are used for mild shampoos and shower-bath preparations. Both the alkyl phosphates based on fatty alcohols, and the alkyl phosphates based on low ethoxylated fatty alcohols.

Amphoteric Surfactants

Amphoteric surfactants are surfactants where the charge changes as a function of the pH value of the formulation in which they are used. They are generally regarded as mild surfactants but this is not a simple matter and may not always be true. Amphoteric surfactants build complexes in combination with anionic surfactants and these complexes are milder than the individual surfactants.

Alkyl betaines

Alkyl betaines are prepared by condensation of an alkyl dimethyl amine with sodium chloroacetate. Betaines can improve the foaming of a formulation, particularly the structure of the foam, which becomes finer and more creamy.

Alkylamido betaines

Alkylamido betaines are prepared by condensation of fatty acids with dimethylamino propyl amino (DMAPA), followed by reaction with sodium chloroacetate. The application properties of alkylamido betaines are generally similar to those of the alkyl betaines. The coco-type is by far the most important.

Amine oxides

Unlike betaines, amine oxides are never anionic. However, they do show cationic or non-ionic behaviour depending on the pH, and therefore behave quite similarly. As much, they are excellent foamers. In combination with anionics, small amounts can act as foam boosters and can improve the foam structure. Like betaines, they are good thickeners for anionic surfactants. In addition, amine oxides are good conditioning agents in hair rinses.

Non-ionic Surfactants

Non-ionic surfactants are having poor foaming properties. These surfactants are rarely used as shampoo surfactants, although they are often added as solubilisers for perfumes.

Ethoxylated products

Ethoxylation allows the production of a wide range of products. Variables are the hydrophobic part of the molecule, and the number of ethylene oxide units added.

Alkyl polyglycosides

A relatively new range of commercially available surfactants is the alkyl polyglycosides (APG). Based on carbohydrates, APGs are mixtures of different isomers, (i) stereoisomers, (ii) binding isomers and (iii) ring isomers.

Surface active APGs are prepared by the glycosylation of starch or monomer glucose with fatty alcohols. This can be done directly or by transglycosylation using butylpolyglucosides as intermediates in a two-step process.

Cationic Surfactants

Unlike the anionic or amphoteric surfactants, cationics are seldom used for cleansing applications. These surfactants show a high substantivity to surfaces such as skin and hair due to conditioning. They can act as emollients for the skin, or as conditioning agents on the hair.

Monoalkyl quaternaries

The monoalkyl quaternaries are still of major importance in conditioning. The most important is cetyl trimethyl ammonium chloride (CTMAC) which shows a light to moderate conditioning intensity combined with excellent water solubility. The substantivity and conditioning properties are improved as the length of the carbon chain increases.

Dialkyl quaternaries

The dialkyl quaternaries are well known as relatively strong conditioning agents. It exhibits strong conditioning properties, good detangling and combing, and adds manageability to unruly or damaged hair.

Trialkyl quaternaries

An increase in the number of alkyl chains leads to considerable improvement in properties. A relatively new product is tricetyl methyl ammonium chloride (TCMAC), which shows superior detangling, static control, and excellent combing properties on both wet and dry hair.

Benzyl quaternaries

Benzyl quaternaries are obtained by reaction of alkyl dimethyl amines with benzyl chloride. Based on short-chain amines, quaternaries with good anti-microbial properties are obtained.

Ethoxylated quaternaries

They are more hydrophilic than normal quaternaries, but show relatively good conditioning properties regarding both wet-combing and, particularly, static control. As in the case of betaines and aminoxides, the ethoquats are compatible with anionic surfactants.

Chapter 15

Humectants and Antiseptics

INTRODUCTION

The property of absorbing water vapour from moist air until a certain degree of dilution is attained is termed as humectants. The degree of dilution depends on the character of the humectant used and the relative humidity of the surrounding air. Equally, aqueous solutions of humectants can reduce the rate of loss of moisture to the surrounding air until equilibrium is attained.

The addition of humectants to cosmetic creams specifically oil-in-water type reduce drying out when such cream is exposed to air. As the humectants are hygroscopic in nature, on application they influence the texture and condition of the skin. They also control in use by reducing the rate at which water disappear and viscosity decreases. It is believed to minimise 'balling' and 'rolling' of a product in use.

The terms 'antiseptic' and 'germicidal' are predominantly used to describe preparations applied to living tissue to prevent infection although 'germicidal' is often applied to antibacterial soap bars. Although the term 'disinfectant' is more correctly used to describe preparations for treatment of inanimate objects such as floors, toilets, drains and so on, the term 'skin disinfectant' is often applied to products used by medical and other personnel to prevent transmission of infection in hospitals, the food industry and other high risk areas.

Use of antiseptics in toilet preparations should be distinguished from use of preservatives, in that the former are expected to render the product active against micro-organisms present on the skin or scalp or in the mouth, whereas the function of preservatives (often the same anti-bacterial agent) is to maintain the product in a satisfactory condition during its shelf-life and use.

DRYING OUT

The cosmetic product may dry out at any time between manufacture and final use by the consumer. It is determined by the temperature of the product, its degree of exposure to air and the relative humidity of the air to which it is exposed. It is essentially a rate process which proceeds towards the equilibrium state in which the water vapour pressure of the product is equal to that of the surrounding air.

The nature of the container and the means of closure, play keyrole in preventing drying out on storage. With an efficient closure, the humectant is of less importance since there is only a small space above the product to be saturated with water vapour.

In the case of emulsion products the type of emulsion is critical. Water-in-oil emulsions lose water at a much lower rate than oil-in-water emulsions, because of the lower water content and the fact that the external phase is oil. Oil-in-water creams are very difficult to maintain in a factory-fresh state even with a screw cap and compressible wad.

Toothpaste packed in a metal tube and with a screw cap presents a slightly different problem. The product will not normally dry out if the cap is kept in position, but if the cap is left off after use the drying out at the nozzle can cause blocking of the orifice. This can be serious in pressure-packed products where there is no easy means of clearing the obstruction. Fortunately, toothpastes can tolerate high concentrations of humectant and a level of 30 per cent of glycerol is not unknown.

Humectants certainly do reduce drying out, but their effect should not be exaggerated. The concentration of humectant in the water phase of a typical cosmetic product is normally much too low for it to be in equilibrium with average atmospheric humidity. All that the humectant can do is to reduce the rate of water loss to the atmosphere and this effect can and should be reinforced with an effective pack closure. The ideal humectants should have the following specifications :

(1) The product must absorb moisture from the atmosphere and retain it under normal conditions of atmospheric humidity.

(2) Within the normal r.h. range, change of water content should be small in relation to r.h. changes.

(3) A low-viscosity humectant is easily mixed into a product, but conversely a high-viscosity helps to prevent creaming or separation of emulsions, or settling of suspensions.

(4) The viscosity-temperature curve should be relatively flat.

(5) The humectant should be compatible with a wide range of raw materials: solvent or solubilising properties are desirable.

(6) Good colour, odour and taste are essential.

(7) The humectant should be non-toxic and non-irritant.

(8) The humectant should be non-corrosive to normal packing materials.

(9) The humectant should be non-volatile and should not solidify nor deposit crystals under normal temperature conditions.

(10) The humectant should preferably be neutral in reaction.

(11) Humectants should be freely available and should be as inexpensive as possible.

Types of Humectant

There are three general types of humectant: inorganic, organo-metallic and organic.

Inorganic humectants

Calcium chloride is typical of inorganic humectants, which is quite efficient but which fail badly on corrosion and compatibility. They find only limited use in cosmetic products.

Organo-metallic humectants

The principal organo-metallic humectant is sodium lactate which has, in fact, greater hygroscopic powers than glycerine. However, it is incompatible with some raw materials, can be corrosive, has a pronounced taste and may discolour. It has not been widely used in cosmetics but has been recommended for use in skin creams, particularly because lactates occur naturally in the body and there is no risk of toxicity or dermatitis. The problem of pH can be overcome by admixture with lactic acid which is also fairly hygroscopic. Buffered solutions can be obtained between pH 7.1 and pH 2.2 at 5 per cent sodium lactate/lactic acid.

Organic humectants

Organic humectants are usually polyhydric alcohols, their esters and ethers. The simple unit is ethylene glycol and by progression up the series the most common products are: Glycerol (trihydroxypropane), Sorbitol (hexahydrohexane).

A series can be built up by the addition of ethylene oxide to a basic unit or just to itself. This produces, for example, polyethylene glycols of varying molecular weight which often have useful cosmetic properties of their own. The multiple ether linkages reduce hygroscopic properties which depend primarily on the ratio of –OH groups to C atoms.

The soap industry inevitably produces glycerol as a by-product and this can also be synthesised from petroleum building blocks. Glycerol is probably the most popular humectant used in cosmetics though this accounts for only a small percentage of total use.

Sorbitol (in the form of 70 per cent syrup) has recently replaced or partially replaced glycerol in many cosmetic products. The replacement increases the water content of the final product. This is not important in most cosmetic products, but it may be vital in toothpastes which normally have a low water content.

Thus the compounds most generally used in cosmetic products for hygroscopic purposes are : Ethylene glycol, Propylene glycol, Glycerol, Sorbitol, Polyethylene glycol.

Hygroscopicity

Hygroscopicity is determined by constructing a curve of relative humidity of atmosphere against humectant concentration of equilibrium. This is done by exposing small weighed amounts of solutions of known composition in atmospheres of controlled humidity and weighing periodically. The controlled humidities can be achieved in small desiccators charged with crystals wetted with their own saturated solutions (Table 15.1). Other humidities in the lower range are best achieved over sulphuric acid solutions of known concentration. The humidities are given in standard tables.

Table 15.1. Suitable crystals for establishing atmospheres of controlled humidity.

	Relative humidity *(%)*
$K_2Cr_2O_7$	98
$Na_2SO_4.10H_2O$	94
$BaCl_2.2H_2O$	88
$NaCl$	75
KI	71
$NaNO_2$	66
$NaBr.2H_2O$	58
$NaHSO_4.H_2O$	52
$Na_2Cr_2O_7.2H_2O$	52
$KCNS$	47

(Cont'd)

	Relative humidity (%)
$CaCl_2.6H_2O$	33
CH_3COOK	20

Blends of humectants do not always show hygroscopicities exactly in accord with the arithmetic average of their individual hygroscopicities. It has been investigated that a humectant could show different efficiencies at different concentrations reported a maximum efficiency in the region of 1 per cent humectant in a typical vanishing cream formula. This is because to surface-active effects which predominate at very low concentrations but become insignificant at higher concentrations where hygroscopic effects predominate.

Stability of Emulsions

The humectant will affect stability through its viscosity and in addition by chemical nature. It is possible that in some complex systems, especially those containing monoglycerides, glycerine can positively promote stability.

Safety

Ethylene glycol is not considered safe, since it is oxidised in the body to oxalic acid and any absorption through the skin might lead to renal calculus; for the same reason diethylene glycol is considered toxic. The mono-ethyl ether of diethylene glycol (Carbitol) has been widely used in cosmetic and toilet preparations.

Glycerine, in particular, has been questioned because of the hygroscopicity of pure glycerol which had been considered to be capable of drawing water from the skin.

Skin Moisturising

It would appear that the presence of a humectant may be expected to stabilise the water content of the residual film from a cream on the skin and prevent excessive drying out. The relative transfer rates of water between the atmosphere and the film and the skin depends on changes in the ambient humidity, and whether the skin is dried or moistened. On the whole it is justified to use humectants in skin products for moisturising in container and on the skin.

MICROBIAL FLORA OF THE BODY

The normal flora of the body surface comprises two distinct groups of organisms — the resident flora and the transient flora.

It should be noted that the population of bacteria varies considerably on different parts of the body; the hair, face, axilla and groin harbour the greatest numbers of organisms, whereas colonisation of more exposed and dry areas such as legs, arms and hands is relatively less extensive. Most resident organisms are found on the superficial skin surface but 10–20 per cent of the total flora is concentrated in hair follicles, sebaceous glands and so on where lipid and superficial cornified epithelium make their removal difficult.

Various areas of the body (although mainly the hands) also contain, in addition to the resident flora, transient flora consisting of contaminants picked up continuously from the environment and other body areas such as the nasal mucosa and gastrointestinal tract. This flora may contain any number of different organisms including pathogenic strains of *Pseudomonas, Enterobacter, Salmonella, Shigella* and *Escherichia coli.* In general, however, these transient contaminants survive for only relatively short periods owing to insufficient moisture and the presence of bactericidal substances such as fatty acids on the skin surface. In contrast to the resident flora, these organisms are only loosely attached to the skin and may be removed in substantial numbers by washing and bathing.

EFFECTS OF ANTIBACTERIAL AGENTS ON BODY FLORA

It has become increasingly apparent that the effectiveness of antiseptic preparations depends not only on the properties of the antimicrobial agent but also on the nature of the formulation, which may be a soap bar, emulsion, liquid soap or detergent formulation.

In the development of effective antiseptic toilet preparations, it is necessary that products are formulated according to the desired effect (that is, reduction in resident or transient flora) or the need for immediate or progressive and prolonged reduction in bacterial flora. Medicated toilet preparations used for routine washing and bathing are intended largely to protect the individual against minor skin infections by both resident and transient bacteria. They are used to control the conditions such as body odour and halitosis. The routine handwashing associated with toilet visits, food hygiene and the handling of newborn infants and sick persons is intended for removal of transient organisms from the skin to prevent transmission of infection.

Antibacterial Soap Bars and Other Skin Degerming Preparations

It is suggested that the sustained effects of medicated soaps are due to the substantive properties of antibacterial agents which remain on the skin after handwashing. It is possible that routine use of antibacterial soap bars, medicated shampoos, etc., may assist in their control of skin infections in healthy family. Antiseptic preparations are generally used to prevent body surface and particularly the hands play an important part in transmission of infection in the community.

Antiseptic preparations made of iodophors or chlorhexidine increases removal of transient skin flora. They produce rapid and immediate bactericidal action. The skin washing is vital in controlling dispersal of infection, disinfectant usage is probably not justified except in certain 'high risk' hospital areas and that the substantial removal of transient contamination achieved by soap and water washing is adequate for most purposes of food and toilet hygiene.

Antiseptics in first aid

Antiseptic creams or other skin disinfectants are used in first aid treatment of cuts, burns and other wounds to prevent infection during healing. They are also used for treatment of minor skin infections, particularly those associated with the face. For first aid purposes, rapid acting compounds such as iodophors, chlorhexidine and chloroxylenol products are recommended.

ANTIMICROBIAL AGENTS COMMONLY USED IN ANTISEPTIC PRODUCTS

Phenols and Cresols, Bisphenols, Hexachlorophene, Dichlorophene (G4) (2,2'-methylene-*bis*-(4-chlorophenol)), Irgasan, Fentichlor (*Bis*-(2-hydroxy-5-chlorophenyl) sulphide), Salicylanilides and Carbanilides, Cationic Surface-active Antibacterials, Quaternary Ammonium Compounds, 1-(3-Chloroallyl)-3-5-7-triaza-1-azoniaadamantane chloride, Chlorhexidine, Amphoteric Surface-active Compounds, Miscellaneous Antimicrobial Agents, *n*-Trichloromethylthio-4-cyclohexene-1,2-dicarboximide, Dioxin, Imidazonidyl Urea, Halogens, Mercury Compounds, Pyridine N-oxides and Tetramethylthiuram Disulphide.

Chapter 16

Preservatives and Antioxidants

INTRODUCTION

Preservatives are added to products for two reasons: firstly to prevent spoilage, that is, to prolong the shelf-life of the product, and secondly to protect the consumer from the possibility of infection. It is recognised that products may require protection from contamination during manufacture, although preservation must never be used to hide bad manufacturing procedures. It is also recognised that cosmetic products, are liable to consumer abuse. Although products cannot be protected against extremes of abuse, such as the use of saliva in the application of eye make-up, the manufacturer should anticipate misuse when is formulated.

The role of microbiologist during the formulation of a product from its development stage is very important. One can not assume that a manufacturer has taken care that the product in question leaves his premises in a microbiologically satisfactory condition. In cosmetics the effects of oxidation are normally deleterious and can lead to complete spoilage.

Two of the problems associated with an understanding of the general oxidation reactions have been the very wide spectrum of organic materials which are subject to this type of decomposition and, secondly, the large number of factors which can effect both the rate and course of the reactions. Amongst these latter may be numbered the effects of humidity, oxygen concentration, temperature, uv irradiation and the presence or absence of anti- and pro-oxidants.

MICRO-ORGANISM

Micro-organisms grow and multiply by utilising the materials in their immediate environment. In considering spoilage problems created by micro-

300

organisms, the variety of chemical reactions they can carry out and the rate at which these can occur must be taken into account. Table 16.1 lists some of the genera of micro-organisms.

Table 16.1. Some micro-organisms isolated from toilet preparations.

Fungi	Bacteria	Yeasts
Absidia	Acinetobacter	Candida*
Alternaria*	Alcaligenes	Monilia
Aspergillus*	Bacillus*	Torula
Citromyces	Diphtheroids	Zygosaccharomyces
Cladosporium*	Enterobacter	
Dematium	Enterococcus	
Fusarium	Escherichia	
Geotrichum	Klebsiella*	
Helminthosporium	Micrococcus	
Hormodendrum	Proteus	
Mucor*	Pseudomonas*	
Paecilomyces	Sarcinia	
Penicillium*	Serratia	
Phoma	Staphylococcus*	
Pullularia	Streptococcus*	
Rhizophus*		
Stemphylium		
Thamnidium		
Trichothecium		
Verticillium		

* Several different species of these genera have been reported.

Bacteria and fungi are widely distributed in nature and there are few places on or near the surface of the earth that are free from them. In growing, bacteria and fungi can cause rapid and profound changes in their immediate environment and in the synthesis of new protoplasm many complex chemical reactions are accomplished within a remarkably short period of time. The organisms carry out these reactions by means of enzymes and some of the basic reactions that can occur are: hydrolysis, dehydration, oxidation, reduction, decarboxylation, deamination, phosphorylation and dephosphorylation.

Preservative Requirements

The essential requirements of a preservative are :

(1) Freedom from toxic, irritant or sensitising effects at the concentrations used on the skin, mucous membranes and, in the case of orally administered products, on the gastro-intestinal system.

(2) Stability to heat and prolonged storage.

(3) Freedom from gross incompatibility with other ingredients in the formula and with the packaging material, which could result in loss of antimicrobial action.

Other requirements are that the preservative should be active at low concentration; should retain its effectiveness over a wide range of pH; should be effective against a wide range of micro-organisms; should be readily soluble at its effective concentration; should have no odour or colour; should be non-volatile; should retain its activity in the presence of metallic salts of aluminium, zinc and iron; should be non-corrosive to collapsible metal tubes and non-injurious to rubber. Table 16.2. lists some of the preservatives used in cosmetics and toilet preparations.

Table 16.2. Some preservatives used in cosmetics and toilet preparations.

p-Hydroxybenzoic acid	Phenol
Benzoic acid	Cresol
Sorbic acid	Chlorothymol
Dehydroacetic acid	Methylchlorothymol
Formic acid	Chlorbutanol
Salicylic acid	*o*-Phenylphenol
Boric acid	Dichlorophene
Vanillic acid	Hexachlorophene
p-Chlorobenzoic acid	Parachlormetaxylenol
o-Chlorobenzoic acid	Parachlormetacresol
Propionic acid	Dichlormetaxylenol
Sulphurous acid	*p*-Chlorphenylpropanediol
Trichlorphenylacetic acid	β-Phenoxyethylalcohol
	β-*p*-Chlorphenoxyethylalcohol
Methyl *p*-hydroxybenzoate	β-Phenoxypropylalcohol
Ethyl *p*-hydroxybenzoate	Potassium hydroxyquinoline sulphate
Propyl *p*-hydroxybenzoate	8-Hydroxyquinoline
Butyl *p*-hydroxybenzoate	*p*-Chlorphenylglyceryl ether
Benzyl *p*-hydroxybenzoate	Formaldehyde
	Hexamine
Benzethonium chloride	Monomethylol dimethyl hydantoin
Benzalkonium chloride	2-Bromo-2-nitro-1,3-propanediol
Cetyltrimethyl ammonium bromide	1,6-*Bis*-*p*-chlorophenyl diguanidohexane
Cetylpyridinium chloride	Phenyl mercury acetate

(Cont'd)

Dimethyldidodecenyl ammonium chloride	Phenyl mercury borate
β-Phenoxy-ethyl-dimethyl-dodecyl ammonium bromide	Phenyl mercury nitrate
Tetramethylthiuramdisulphide	Sodium ethyl mercurithiosalicylate
1-(3-Chloroallyl)-3,5,7-triazonia- adamantane chloride	Tetrachlorsalicylanilide
5-Bromo-5-nitro-1,3-dioxan	Trichlorsalicylanilide
6-Acetoxy-2,4-dimethyl-*m*-dioxan	Trichlorcarbanilide
Imidazolidinyl urea	
Vanillin	
Ethyl vanillin	

Factors Influencing the Effectiveness of Preservatives

Dissociation and pH

Formulations of cosmetics and toiletries encompass a wide pH range and since micro-organisms of one sort or another are capable of growing between pH 2 and pH 11, ideally a preservative should be effective over this range. In practice many preservatives are pH-dependent, the majority of them being more active in the acidic than alkaline range. Some preservatives with a wide pH profile have the disadvantage of being chemically highly reactive compounds (for example formaldehyde and formaldehyde donors) which react with other components of the formulation. pH may also have an effect on the microbial cell surface and may affect the partitioning of an antimicrobial agent between the cell and the product.

Selection of a Preservative

While selecting a preservative due consideration should be given to a theoretical basis the factors that are likely to influence the preservation of a new product. This approach, coupled with simple laboratory tests on various combinations of the formula components, while not a substitute for thorough microbiological testing of the finished formula, can save time and frustration.

The complexity of modern formulae often means that there is a variety of materials present, some of which will act in favour of good keeping qualities while others will act against whatever preservative is chosen; the relative hostility of micro-organisms or the nutritive value of the formula itself is also of importance.

Steps for selecting preservatives are :

(1) Check ingredients for the likelihood of contamination (for example water, materials of natural origin, packaging, etc.).

(2) Consider which materials might provide sources of energy for microbial growth (for example glycerine, sorbitol, etc., at concentrations below 5 per cent; non-ionic surfactants at almost any useful concentration; soaps and anionic surfactants at concentrations below about 15 per cent, proteins, carbohydrates, cellulose derivations and natural gums).

(3) Determine the pH of the aqueous phase of the product before attempting to use any of the preservatives. Consider changing the pH to provide enhanced antimicrobial activity.

(4) Depending on the ratios of water and oil present in the formula, estimate whether certain preservatives will be partitioned between the two phases, possibly leaving insufficient in solution in the aqueous phase to be effective. The change in partition coefficient will affect the effectiveness of preservative. Certain agents may alter CMC, e.g. urea increases the CMC of non-ionic surfactants, thus reducing the number of miscelles and the degree of preservative inactivation.

(5) As a guide, estimate the approximate ratio of total to free preservative in the presence of macromolecules in the formulation, and multiply the normally effective concentration by the appropriate factor. Equations relating preservative capacity to surfactant concentration and the interaction between surfactant and preservative may also give useful data.

(6) The preservative selected should be less toxic so that it provides long shelf life.

In any new formulation first consider the well known and tested preservatives. The several other materials listed below also can be tried in system :

(1) *Bronopol* (2-bromo-2-nitro-1,3-propanediol) which, is active at low concentrations against *Pseudomonas* species, is only slightly reduced in activity by nonionics and has low toxicity.

(2) *Chlorhexidine* (*bis*(*p*-chlorophenyl-diguanido)hexane), is a wide-spectrum antimicrobial agent with a good record of safety.

(3) *Dehydroacetic acid*, which is suitable for formulae of low pH, is relatively unaffected by the presence of high levels of non-ionic emulsifiers and appears to be safe for use on the skin.

(4) *Imidazolidinyl urea*, which is not pH-dependent, has high water solubility, is non-toxic, non-irritating and non-sensitising. It is active against Gram-positive and Gram-negative bacteria, but selectively active against yeasts and moulds. It retains its activity in the presence of many cosmetic ingredients, including surfactants, and frequently acts synergistically with other preservatives, for example Parabens.

Mixtures of preservatives that are effective against different micro-organisms are also often useful. For example β-p-phenoxy-ethyl alcohol, which is highly active against Gram-negative bacteria and fungi, can be used with a quaternary ammonium compound such as benzalkonium chloride which acts against Gram-positive bacteria at very high dilutions. In addition there are advantages in using combinations which are not only active against a wide range of organisms but also act synergistically, for example imidazolidinyl urea and Parabens.

Safety Aspects

Preservatives are commonly expensive ingredients and it is always advisable to use the lowest effective concentration. The cost, however, is secondary to the more important question of safety to the consumer.

To achieve the desired antimicrobial action the concentration of preservative in case low, it is necessary to consider the toxicity of preservative. The ratio of bound to free preservative is unlikely to remain unchanged when the product is actually in use. Thus from a toxicity point of view the *total* amount is important rather than only that fraction which is acting as a preservative in the particular vehicle. Application of the product to the skin, for example, will disturb the original preservative equilibrium between the various phases of the product and will almost certair'y result in liberation of the preservative previously bound. Evaporation of water will increase the concentration of the preservative available to the skin and may result in primary irritation or, in some cases, sensitisation.

A reasonably continuous spectrum of toxicity, ranging from the very low concentrations to which a few people may show an adverse reaction, to the high levels where both primary irritance and allergic responses will be more numerous. The toxicology of the p-hydroxybenzoate esters has been thoroughly studied and no primary irritation following their use at concentrations up to about 0.3 per cent has been reported. Levels of between 5 and 10 per cent have been used in powders, ointments and solutions to treat athlete's foot and, even at these levels, adverse reactions have not been numerous.

Sorbic and benzoic acid have also been used in products at concentrations far in excess of those required for normal preservation. Benzoic acid appears to have a reasonably clear bill of health but sorbic acid has caused primary irritation characterised by erythema and itching at concentrations below 0.5 per cent.

The organic mercury compounds are, of course, recognised poisons. Although they have for a number of years apparently been safety used at concentrations below 0.01 per cent, they present a toxicity hazard to those who have to handle them in concentrated form in factories. Several leading scientists have advised against the use of phenyl-mercuric acetate, borate, and nitrate and also methiolate, on the grounds of their ability to penetrate the skin and endanger the liver and kidneys.

The quaternary ammonium compounds have been extensively tested for skin irritation and sensitising properties. At concentrations below 0.1 per cent most of those commonly used as preservatives appear to cause little or no irritation; higher concentrations can cause erythema and drying of the skin. Cases of sensitisation to cetrimide at concentrations of about 1 per cent have been reported. Their substantivity to the human skin has caused concern about plant safety.

Formaldehyde is well known to be a skin irritant, and for this reason, and for reasons of volatility and odour, it has not been used extensively as a preservative.

The toxic thresholds of preservatives will depend not only upon the concentrations at which they are used but also upon the vehicle. A certain concentration of a particular preservative may be quite harmless in one system while the same level might evoke adverse skin responses in another because of the presence of substances which increase its penetration through the skin.

Tests for Preservative Effectiveness

Initial screening tests

A rough indication of whether a particular preservative is likely to be effective or not is found by agar plate test.

The test normally involves the use of agar plates seeded with a variety of micro-organisms; wells are cut in the seed agar and small amounts of the product under test are placed in the wells before incubation of the plates. Gram-negative and Gram-positive bacteria, together with fungi typical of those which frequently contaminate toilet preparations and confirmatory tests for preservation should be carried out using organisms of these or similar kinds.

In this type of test some of the preservative will inevitably diffuse into the agar leaving a lower concentration than was originally present in the product in the well. Although this kind of test can give a rapid indication of whether the preservative shows any likelihood of being effective in the product, it is not by any means definitive and should not be used as a substitute for longer-term and more rigorous evaluation. It is necessary to check compatibility of the preservative in the new system by carrying out microbiological test before formulation.

Inoculation tests

Methods by which known numbers of bacteria or fungi are introduced into the product and samples taken at intervals to estimate survival are by far the most reliable. Various test procedures have been proposed. In general, ascertaining the resistance of a product to bacterial contamination involves inoculating a suitable size sample (for example 10 g) of the product with the test organism to give a final concentration of 10^5–10^7 organisms g^{-1}. The number of survivors in the sample is determined at intervals after storage at room temperature. The standard which must be met in order that a product may be considered effectively preserved varies depending on its intended use. Differences in opinion regarding interpretation of results are reflected by the different standards set by BIS, the Society of Cosmetic Chemists test and the Toilet Goods Association test.

Longer-terms test, in which smaller numbers of micro-organisms are used and the samples are observed for changes in their physical characteristics over several months, may be more meaningful. Products inoculated with spore-bearing organisms can only be observed for physical changes, as microbiological sampling is unrealistic since spores which might remain dormant in the product will germinate when transferred to nutrient culture medium. In samples inoculated with vegetative organisms, a gradual diminution in numbers can be traced over a period of time if the preservative is effective, but with fungi and spore-bearing organisms one can only wait for the appearance of visible spoilage and this sometimes takes several months to occur.

Test organisms should be chosen to represent the types of organism that are known to be frequent product contaminants, for example *Pseudomonas* species, and those with which the product is likely to come into contact, for example *Staphylococcus* species. In addition, organisms isolated from the manufacturing environment, raw materials and, where possible, from contaminated products (preferably of the same or similar formula) should be used. In the testing of shampoos, for example, tap water provides a suitable source of test organisms.

The maintenance of test organisms is crucial since their resistance is influenced markedly by the medium on which they are grown. To ensure suitable resistance, test organisms may be grown in a medium containing the preservative or product in low concentration. Product contaminants may be kept in unpreserved or inadequately preserved preparations.

Mixed cultures may be used initially to reduce the amount of testing required to assess the adequacy of the preservative system, while pure culture challenge may be employed to give more detailed information about preservative adequacy against specific organisms.

Most tests run for a minimum of 28 days, the product being sampled for viable organisms at various intervals, depending on the probable frequency of usage during this period. Slow adaptation of micro-organisms to their environment makes it essential to test for long enough to determine whether inoculated bacteria and fungi will grow after a dormant period, and in some instances tests lasting as long as six months may be too short.

Tests should also be performed on products that have been stored for specified time intervals at temperatures and humidities which the product is likely to meet during use, in order to ensure that adequate preservative activity is retained throughout the shelf-life. A concurrent chemical assay of the preservative gives additional valuable information.

Products may be tested using a single inoculation or using an inoculation-sampling cycle. The latter method, in which the sample is subjected to more than one challenge, has been advocated by several workers since it is considered to be more representative of in-use conditions and has the advantage of indicating at what point the preservative system will fail. A criticism of the capacity test is that it may lead to excessive preservation, with consequential use of dermatologically unsafe preservative concentrations. The balance between over-preserving and under-preserving a product will depend on the number of challenges, which can only be chosen after a period of experimentation.

ANTIOXIDANTS

The role of antioxidants is to suppress oxidation either by preventing formation of free radical or it may react with free radical formed thereby obstructing build-up of reaction chains. The formation of free radicals cannot be wholly prevented and therefore substances which behave as free radical acceptors—antioxidants—are important.

Special problems also exist in the choice and relative efficiency of antioxidants in emulsified and solubilised preparations. With emulsions it is an advantage for the antioxidant to be present at the interface between the oil drop and the continuous aqueous phase. Therefore the antioxidant should exhibit a suitable balance between lyophilic and lyophobic groupings. If it exists in both phases, then in the aqueous phase it will decompose to give free radicals which may initiate oxidation of oil.

Synergism

Synergism is said to occur when two or more antioxidants present in a system show a greater overall effect than can be accounted for by a simple addition of their individual effectiveness. The phenomenon is associated with two separate systems: (i) mixed free radical acceptors; and (ii) the metal chelating agents.

Mixed free radical acceptors

In a synergistic system involving a material such as ascorbic acid (BH) which has a low stearic factor and hydroquinone (AH) in which steric factors would not be important, Uri has suggested that the following reactions will take place :

$$RO_2 + AH \longrightarrow RO_2H + A$$

$$A + BH \longrightarrow B + AH$$

The possible disappearance of A by reaction with oxygen is thus eliminated and the effective antioxidant AH is regenerated. On its own BH would not produce any significant antioxidant effect as the reaction :

$$RO_2 + BH \longrightarrow RO_2H + B$$

would be prevented by steric factors.

Metal chelating agents

The normal effect of metal chelating agents is to bond with pro-oxidant metallic ions and thus prevent their catalytic effect on the normal oxidation chain reaction. Metallic pro-oxidants that are already present as part of complex organic structures are not usually affected by chelating agents. Stabilisation has been achieved by reaction of the metal with organic acids of the tartaric or citric acid type or with materials such as ethylenediaminetetra-acetic acid (EDTA).

Typical antioxidants and synergistic systems used in cosmetics are given in Table 16.3.

Table 16.3. Antioxidants for use in cosmetic systems.

Aqueous systems

Sodium sulphite	Ascorbic acid
Sodium metabisulphite	Isoascorbic acid
Sodium bisulphite	Thioglycerol
Sodium thiosulphate	Thiosorbitol
Sodium formaldehyde sulphoxylate	Thioglycollic acid
Acetone sodium metabisulphite	Cysteine hydrochloride

Non-aqueous systems

Ascorbyl palmitate	Butylated hydroxyanisole
Hydroquinone	α-Tocopherol
Propyl gallate	Phenyl α-naphthylamine
Nordihydroguaiaretic acid	Lecithin
Butylated hydroxytoluene	

Synergistic systems

Antioxidant	*per cent*	*Synergists*
Propyl gallate	0.005–0.15	Citric and phosphoric acid
α-Tocopherols	0.01–0.1	Citric and phosphoric acid
Nordihydroguaiaretic acid (NDGA)	0.001–0.01	Ascorbic, phosphoric, citric acids (25.50% NDGA content) and BHA
Hydroquinone	0.05–0.1	Lecithin, citric acid and phosphoric acid, BHA, BHT
Butylated hydroxyanisole (BHA)	0.005–0.01	Citric and phosphoric acids, lecithin, BHT, NDGA
Butylated hydroxytoluene (BHT)	0.01	Citric and phosphoric acids up to double the weight of BHT and BHA

Choice of Antioxidant

The ideal antioxidant should be stable and effective over a wide pH range and be soluble in its oxidised form, and its reaction compounds should be colourless and odourless. It should be non-toxic, stable and compatible with the ingredients in the products and their packages. The list of effective antioxidants permitted for use in foodstuffs includes the materials shown below:

Guiaacum resin	Ascorbic acid
Tocopherols	Ascorbyl palmitate
Lecithin	Monoisopropyl citrate
Propyl gallate	Thiopropionic acid
Butylated hydroxyanisole (BHA)	Dilauryl thiodipropionate
Butylated hydroxytoluene (BHT)	
Trihydroxybutyrophenone	

Phenolic Antioxidants

Guaiacum resin

Guaiacum resin is largely phenolic in character, but is a less effective antioxidant than most of the other phenolics mentioned above. It is more effective in animal than in vegetable oils and possesses an advantage over some other antioxidants in that it is equally effective in the presence and absence of water and it is not seriously affected by heating.

Nordihydroguaiaretic acid

A synergistic effect occurred with 0.003 per cent NDGA and 0.75 per cent citric acid. This, of course, arises from the sequestering effect of citric acid on heavy metals.

Tocopherols

They have some antioxidant effect with animal fats such as tallow and with distilled fatty acids, particularly in the presence of a synergist such as citric acid, lecithin, or phosphoric acid, but are of little value for the preservation of vegetable oils.

Gallates

Gallates constitute one of the most important classes of antioxidants. The propyl ester is the only one permitted in foodstuffs in most countries but methyl, ethyl, propyl, octyl and dodecyl gallates are commonly used in cosmetics. Gallic acid itself is a powerful antioxidant, but tends to turn blue in the presence of traces of iron.

Butylated hydroxyanisole (BHA)

BHA consists mainly of two isomers, 2- and 3-*tert*-butyl-hydroxyanisole. It is seldom used alone, as its activity in most systems is less than that of propyl gallate, but it forms a number of very useful synergistic mixtures with the gallate esters. Thus a mixture of 20 per cent BHA, 6 per cent propyl gallate, 4 per cent citric acid, and 70 per cent propylene glycol is commonly used in both the food and cosmetic industries. If such a mixture is used at levels of about 0.025 per cent total antioxidant, most animal and vegetable oils can be protected, as can the fatty esters such as methyl oleate.

Butylated hydroxytoluene (BHT)

BHT is widely used as an antioxidant for fatty acids and vegetable oils, and possesses several advantages over the other phenolic antioxidants in its freedom from any phenolic smell, its stability towards heating, and its low toxicity.

Auto-oxidation of fatty materials takes place with a logarithmic velocity coefficient, so that it is important to stop such oxidation as early as possible in the life of a material.

Trihydroxybutyrophenone

The 2,4,5-trihydroxyphenones, especially the butyrophenones, have outstanding effects with lard, groundnut oils, and tallow. Although the product is a recognised food additive in the USA, it does not appear to be widely used in the cosmetics industry.

Non-phenolic Antioxidants

Many non-phenolic antioxidants are chelating agents. Ascorbic acid and ascorbyl palmitate appear to act by stopping the free-radical oxidation process. Ascorbyl esters are particularly effective in vegetable oils, and make an excellent synergistic mixture with phospholipids such as lecithin and tocopherol.

Among the sequestering agents, the thiodipropionates are widely used, usually in conjunction with phenolic antioxidants. The esters of fatty alcohols have greater solubility in oils.

Lecithin is an effective synergist for many phenolic antioxidants, mainly because it is an oil-soluble phosphate with excellent sequestering properties. Members of another class of oil-soluble sequestering agents are MECSA (mono-octadecyl ester of carboxymethylmercapto-succinic acid) and METSA (mono-octadecyl ester of thiodisuccinic acid). Under some conditions these materials can function as effective antioxidants in concentrations as low as 0.005 per cent. They possess the disadvantage that they decompose on heating, and must therefore be added, like perfume, during the cooling phase of manufacture.

Photo-deterioration

Photo-deterioration generally manifests itself as fading of the colour of the product or development of off-colours caused by uv-visible light.

Packaging in opaque containers or wrappers so as to exclude all light is, of course, an obvious way to avoid this type of deterioration. It is frequently possible to wrap or pack in transparent material suitably coloured or containing a uv absorber to screen out the offending portions of the spectrum. In some cases where uv energy is causing the deterioration, the uv absorber can be incorporated in the product.

Uv-induced deterioration frequently involves the presence of traces of metals, particularly, iron. In such case the screening agent may be reinforced, or in some cases replaced, by a chelating agent such as ethylene-diamine tetra-acetic acid (EDTA). The permeability of the cell walls of some bacteria, notably the Gram-negative *Pseudomonas aeruginosa*, is altered by EDTA. It has been found that the addition of concentrations in the region of 0.05 per cent greatly enhances the antibacterial potency of phenolic antiseptics. Small amounts of this material might thus serve two useful purposes in protecting systems prone to deterioration by uv and susceptible to the omnipresent *Pseudomonas* species.

Temperature

The susceptibility to microbial attack will vary with the temperature of storage, so that a cosmetic kept at room temperature will be liable to spoilage by different organisms from those that flourish in a product kept in a hot environment (for example one left in the sun or in a hot car). Bacteria generally prefer temperatures of 30°–37°C and fungi and yeasts 20°–25°C.

Chapter 17

Emulsions

INTRODUCTION

Every chemist working in the area of cosmetics has knowledge of emulsions which are mixtures of oils, fats and water. They are made by mixing oil-soluble and water-soluble substances together in the presence of an emulsifying agent.

Emulsions—creams and lotions—form a very important part of the cosmetics market. Clearly, however, no cosmetics chemist can consider himself competent until he understands how to formulate emulsions of his own and how to incorporate into them certain desired characteristics. In order to do this, he must learn the fundamentals of emulsion technology.

BASIC PRINCIPLES

Few substances show an 'affinity' for each other and others do not, e.g. water and ethanol are completely miscible. These two materials show an 'affinity' for each other which is obviously not shared by (say) mineral oil and water. This idea of 'affinity' plays an important part in emulsion technology. Under normal circumstances each molecule is in turn attracted by many others around it in all directions, any two molecules might be pulled together. This phenomenon is called 'cohesion' and the force of cohesion between molecules is attributed to their 'cohesive energy'. The magnitude of these cohesive forces depends upon the size of the molecules taking part and their chemical make-up. The basic principle is that 'like attracts like'. Related molecules such as water and ethanol show no tendency to separate because the cohesive forces between water and ethanol molecules are similar in magnitude to those between water and water or ethanol and ethanol. When mineral oil is introduced into water, however, the cohesive forces between water and mineral oil are negligible and hence separation rapidly occurs.

'Affinity' indicate solubility as well as the concept of the 'phase'. When two or more materials in contact with each other co-exist as overtly different and separate entities, each is referred to as a 'phase' (Table 17.1). In two-phase systems one phase may be distributed as a large number of distinct and separate entities in the other. Under these circumstances, the former is known variously as the 'internal', 'disperse' or 'discontinuous' phase and the latter as the 'external' or 'continuous' phase. When one material is dispersed in a finely divided state within another in this way, the area of contact between the two phases is exceedingly large. The characteristics exhibited by two phases depend primarily on the chemical or physical natures of the two surfaces and the interaction between them. This is certainly true of cosmetic emulsions.

Table 17.1. Some common two-phase systems.

System	Continuous phase	Disperse phase
Smoke	Gas	Solid
Gas	Gas	Liquid
Foam	Liquid	Gas
Dispersion	Liquid	Solid
Emulsion	Liquid	Liquid
Foam	Solid	Gas

Principles of Emulsion Stability

An oil and water mixture can be emulsified by shaking. The more vigorous the shaking, the finer the size of the droplets of dispersed phase. Sooner or later, however, the droplets of dispersed phase become noticeably larger as they coalesce until the initial phase separation occur. Emulsion can be explained by mechanical model and thermodynamics.

Mechanical Model of Emulsification Coalescence

When a quiescent mixture of oil and water is shaken, large volumes of one phase inevitably get isolated and trapped within the other phase. The fate of these isolated globules depends partially upon the turbulence which they encounter in their immediate surroundings. If the size of the local eddy currents is smaller than that of the globule, the latter will break up into a number of smaller drops under the influence of the shear force exerted by the eddy. This shear force is resisted by the surface tension at the interface between the drop and the liquid. As the droplet size becomes reduced it must find smaller, more powerful eddys—and therefore greater turbulence—to become even smaller. Thus the final droplet size depends almost exclusively upon the surface tension at the interface and the degree of turbulence set up in the continuous phase.

Thermodynamic Description of Emulsification and Coalescence

When the surface area of a liquid is increased (for example, by agitation), molecules from the interior rise to the surface. They do so against the force of attraction of neighbouring molecules and hence some mechanical work or energy is always required to increase the surface area. The surface also tends to become cooled and thus heat flows into it from the surroundings Hence there is an increase in surface energy equivalent to the sum of mechanical energy expended and the heat energy absorbed. The relationship between the increase in surface energy, ΔS, associated with an increase in surface area, ΔA, is as follows :

$$\Delta S = T \cdot \Delta A$$

where T is the interfacial surface tension between the liquid and its surroundings. Thus it can be seen that surface tension is no more than the increase in surface energy associated with a unit increase of surface area.

It is a well-known principle in mechanics that an object is in stable equilibrium when its potential energy is at a minimum. Given the opportunity, therefore, the emulsion will lose its considerable energy excess to its surroundings in the form of heat by coalescence of the droplets of internal phase and phase separation.

STABILISATION OF COSMETIC EMULSIONS

The following recommendations can be made to prevent thermodynamically unstable system from separating into two layers :

(1) By increasing the viscosity of the external phase, the mobility of internal phase droplets will decrease making it more difficult for them to collide with each other.

(2) By ensuring that the internal phase is of the smallest and most uniform drop size possible. The likelihood of adhesion between two drops will decrease when internal phase is smallest and uniform drop size.

(3) By increasing the mechanical strength of the interface, will make this less susceptible to rupture with the resulting coalescence of adhering drops.

(4) The thermodynamic 'driving force' for coalescence will be decreased
   by decreasing surface tension.

It should be noted that the increase in stability which results from the
formation of internal phase droplets of very small size represents an
apparent anomaly. It has already been shown that decreasing the droplet size
causes a rapid increase in surface area and also that a large surface area can
only be achieved, in a given system, by a larger energy input. Such a
system should therefore possess a high excess energy content—which
seems to be in conflict with the rule about high energy systems being less
stable than those of low energy content. Apparently, therefore, the
stabilising effect of a low probability of adhesion between internal phase
droplets far outweighs the influence of excess free surface energy in
bringing about coalescence.

Classification of Emulsifiers

The classification of emulsifier is based on nature of hydrophilic end of
molecule. Hence they are classified as: (i) anionic; (ii) cationic; and (iii)
amphoteric or non-ionic.

FACTORS AFFECTING THE STABILITY OF EMULSIONS

The stability of emulsions is influenced due to interfacial tension, electrical
charge, viscosity, ratio of oil to water phase, temperature and concentration
of hydrogen ion.

Interfacial Tension

The relative lowering of the surface tension on each side of the surfactant
interface helps to determine the nature of the emulsion and the ease of
emulsification. It does not, however, determine the stability of the
emulsion—this is a point which deserves great emphasis. The essential
factors governing the integrity of the interfacial film and its resistance to
rupture are its extension, its compactness and its electrical charge. A
lowering of interfacial tension is not vital for the stability of an emulsion.

Electrical Charge

The mobile ions are present in the external phase of an emulsion, they are
attracted by the charged droplets of the internal phase (if these have an
opposite charge) giving rise to the formation of an electrical double layer.
The nature and effect of this double layer are markedly different in oil-
in-water emulsions. The thickness of the double layer around oil droplets

in oil-in-water emulsions amounts to only 10^{-3} to 10^{-2} μm. Electrical repulsion therefore occurs at very short inter-globular distances and this results in a very considerable electrical barrier which must be overcome before two droplets can coalesce. On the other hand, the electrical double layers around water droplets in oil are very diffuse (several μm in size) and the electrical potentials of adjacent droplets overlap, lowering the potential barrier. The stability of water-in-oil emulsions cannot, therefore, be attributed to electrical repulsion of charged droplets.

Viscosity of the Continuous Phase

Viscosity is an important parameter, because it can be readily varied— usually by the addition of a thickening or gelling agent (provided that these are compatible with the emulsifier system). Although a large number of such agents is available, the formulator is limited in his choice because the rheological behaviour of most emulsions is almost entirely determined by that of the external or continuous phase. Thus the viscosity, thixotropy and the 'feel' on application of the total emulsion may be affected by the thickener chosen for the continuous phase.

Ratio of Oil Phase to Water Phase

Although the proportion of oil phase to water phase has a marked effect on such parameters as the 'feel' and overall viscosity and appearance of the emulsion, it can also influence the stability. The higher the proportion of internal phase, the greater the number of droplets. The chances of collision are thereby increased and the average distance which one droplet must travel to collide with another (the 'mean free path') is reduced. All this increases the likelihood of coalescence.

Temperature

Solubility is temperature dependent phenomena. It is unlikely that as the temperature of an emulsion changes, the relative solubilities of both ends of all its emulsifier system will change in strict proportion. In other words, hydrophilic-lipophilic balance is to some extent a temperature-dependent property itself. Variation of temperature can therefore decrease the stability of an emulsion. This is obviously something which must be borne in mind when formulating products for differing climates.

Hydrogen Ion Concentration

Nevertheless, the influence of hydrogen ion concentration in oil-in-water emulsions is dramatic whenever an ionisable emulsifier system is used because of the change of species which can be brought about. Anionic emulsifiers are converted to non-ionisable salts in acidic media and the reverse is true of cationic emulsifiers. In both instances, water solubility and therefore all emulsifier activity can be lost. The effect of pH on amphoteric emulsifiers is less dramatic but obviously dictates whether the anionic or cationic form predominates.

PRACTICAL ASPECTS OF EMULSIFIER CHOICE

Probably the first question that needs to be settled is the chemical classification of emulsifier system to be used. This, in turn, can depend upon the content of the other two phases. If the product is to be alkaline, cationics should not be considered. It would be equally unwise to use an anionic emulsifier in an emulsion of low pH and, if electrolyte concentration in the aqueous phase is to be high, non-ionics are the best choice. On the whole, members of this latter group are probably least affected by incompatibilities with the remainder of the formulation—apart from the well known ability of polyoxyethylene chains to deactivate certain preservatives. Unfortunately, complete guidance on the choice of emulsifier type is not possible because of the many and varied factors involved; the formulator must experiment for himself to gain good practical experience in order to be able to come to a quick decision.

Determination of Required HLB

The optimum or 'required' HLB value of the emulsifier system for a given composition of oil and water phases provides a useful starting point in the selection of emulsifiers which will give an emulsion of good stability. The determination of the optimum HLB value is based upon a series of practical experiments in which a set of emulsions is produced, identical in every way except for variation in the ratio of a pair of emulsifiers. These emulsifiers are a matched pair, one lipophilic and one hydrophilic, of known HLB values. For example, sorbitan monostearate (HLB 4.7) and polyoxyethylene sorbitan monostearate (HLB 14.9). These are mixed in ratio to give combined HLB values according to the formula :

$$HLB = xA + (1 - x)B$$

where x is the proportion of a surfactant having an HLB value of A and the other surfactant has a value of B. This is a straight-line relationship and can therefore be computed graphically. The HLB values of the series of mixtures are chosen to differ by an increment of 2 throughout the range bounded by values of the two emulsifiers chosen. In the example above, a range of 4.7, 6, 8, 10, 12, 14.9 might be considered. For each of the test emulsions, an excess of emulsifier (approximately 10 per cent of the weight of the oil phase) is used and all the emulsions are made in precisely the same way. Usually, one or more of the emulsions will give better stability than the others. The trial-and-error process will enable to arrive at an idea of the optimum HLB value for this system.

Selection of the Best Chemical Type

Selections can be made on the basis of the simple rule 'like attracts like'. For example, if the oil phase is to contain a high proportion of unsaturated or highly branched molecules, then a choice of emulsifier based on oleates or 'iso' esters might be appropriate. Eventually, however, the final choice must depend upon trial-and-error. It should be remembered that best stability will be obtained with a mixed emulsifier system of the optimum HLB value.

ASSESSMENT OF EMULSION STABILITY

Although all emulsions will eventually lose their excess energy by breaking down, it is obviously important that any commercial product should maintain its integrity throughout its useful life. Because no emulsion can be separated from its environment, the influence of such factors as temperature variation, light, mechanical vibration, atmospheric oxygen and microbiological contamination cannot be ignored in any assessment of stability. For this reason, almost every cosmetic emulsion is certain to be subjected to one or more of the following accelerated aging processes at some time during its development.

(1) Storage at ambient temperature for up to nine months in glass or plastic containers.

(2) Storage at 35°–40°C for up to three months in glass or plastic containers.

(3) Storage in partially filled containers at ambient or elevated temperature.

(4) Storage at low temperatures (–5°C to + 5°C) for up to three months.

(5) Storage in freeze-thaw cycle cabinets (–5°C to + 30°C, two cycles per 24 hours).

(6) Centrifugation tests.

(7) Microbiological challenge tests.

Generally, these storage and monitoring techniques can be valuable help in formulation, in the development of manufacturing procedures and in production control.

CHARACTERISTICS OF EMULSIONS

Of prime importance when considering cosmetic emulsions is their appearance, since this can help to determine their customer appeal. Emulsions may vary tremendously in appearance from glossy opaque whiteness through a grey translucence to sparkling clarity. If, however, the refractive indices of both phases are identical, or nearly so, no such reflections and refractions take place; light travels unhindered through the emulsion which has a sparkling clear appearance. As the particle size of the internal phase diminishes, the familiar milky-whiteness appears: as the size reduction continues, the colour takes on a blueish hue, becoming grey, semitransparent and finally transparent. These changes in appearance occur as the particle size of the droplets approaches that of the wavelength of light itself. The probability that a light ray will collide with (and be reflected by) a tiny particle is enormously reduced once the particles become so small that they are comparable in size to the wavelength of light. Under these circumstances the majority of rays pass through the emulsion without being reflected or refracted and the emulsion appears to be transparent.

As the droplet size approaches that of the wavelengths at the red end of the spectrum, the reflected or refracted light is made up increasingly of smaller wavelengths at the blue end of the spectrum until, eventually, the droplets become too small for interaction at all.

In practice, it is difficult to formulate emulsions in which both phases have similar refractive indices—micro-emulsions are far more frequently found, although even these are not common (Table 17.2).

Table 17.2. Effect of particle size of internal phase on emulsion appearance.

Internal phase droplet size	Emulsion appearance
$\geq$ 0.5 mm	Globules clearly visible
0.5 mm to 1 μm	Milky-white
1 μm to 0.1 μm	Blue-white
0.1 μm to 0.05 μm	Grey, semi-transparent
< 0.05 μm	Translucent or transparent

The gloss of the emulsion is a function of the microscopic smoothness of its surface. For ultimate smoothness and gloss, the internal phase particles must be relatively small and even in distribution and there must be no inclusions in the external phase such as large crystallites of stearic acid or inorganic matter of large particle size.

Rheological Properties

The rheological behaviour of emulsions is an important subject, not only because of its influence on the 'feel' and acceptability to the consumer but also because of its impact on the manufacturing process.

Application Properties

The in-use properties of cosmetic emulsions can be thought of as those which are apparent during its application to the skin or hair (the 'feel') and the after-effects once the product has been applied. Both types of property are important since even the most effective products will not appeal to the consumer if the 'feel' on initial application is unpleasant.

The initial feel of an emulsion is largely dependent on that of the external phase; thus an oil-in-water emulsion will feel like water, whatever is dispersed in the aqueous phase. Water-dispersible thickeners and additives such as glycerine, sorbitol and glycols will all exert some effect. Water-in-oil emulsions will feel oily—but whether or not they are sticky, for example, depends upon the choice of oil-phase ingredients. Viscosity also plays an important part in the initial impact of an emulsion: high viscosities tend to give a cream 'richness'.

During the application, some emulsifiers tend to promote the appearance of a foam-like whitening, often referring to as 'soaping'. Anionic emulsifiers are particularly prone to this happens, while being easy to detect, is somewhat difficult to describe in words.

Finally, the after-effects are determined by the choice of oil-phase ingredients (which may be greasy or non-greasy) and any non-volatile water-phase ingredients.

DETERMINATION OF EMULSION TYPE

A combination of all three of the following methods may be expected to give a reliable indication of the orientation of an emulsion:

(1) The emulsion is subjected to an electric voltage. If no current flows, the external phase is non-conducting (that is, oily). If an appreciable current flows, the external phase is conducting (that is, water). If a small current flows, this may indicate a dual emulsion or a gradual inversion.

(2) Oil-in-water emulsions will disperse easily in water, water-in-oil emulsions will disperse easily in oil.

(3) Water-soluble dyes will spread through oil-in-water emulsions, oil-soluble dyes through water-in-oil emulsions.

QUALITY CONTROL AND EMULSION ANALYSIS

The properties of emulsions most commonly examined for the of purpose quality control are: Colour, odour and general appearance; weight per millilitre; apparent viscosity; hydrogen ion concentration; water content; volatile content; stability; and chemical identity of separated phases.

Chapter 18

Equipment for Manufacture of Cosmetics

INTRODUCTION

Cosmetics manufacture is concerned with a very broad range of processes, there are enough common elements to allow a relatively simple overall view of the subject; this helps considerably in a study of the basic principles of cosmetic production technology. The manufacture of cosmetics is conveniently divided into two parts: (i) bulk manufacture; and (ii) Unit manufacture. The bulk manufacture is carried out in three steps. They are: (i) Mixing; (ii) Pumping; and (iii) Filtering. The most important step is mixing.

MANUFACTURE OF BULK PRODUCT

A convenient way of classifying the mixing processes is represented in table 18.1. Every single cosmetics manufacturing process contains minimum one mixing operation.

Table 18.1. Scope of mixing operations within the cosmetics industry.

Type of mixing	Examples
1. Solid/Solid	
(a) Segregating	None
(b) Cohesive	Face powders, eye shadows and all dry mixing
2. Solid/Liquid	(i) Dissolution (of water-soluble dyes, preservatives, powder surfactants, etc.)
	(ii) Suspensions and dispersions (pigments in castor oil and in other liquids)

(Cont'd)

Type of mixing	Examples
3. Liquid/Liquid	
(a) Miscible	(i) Chemical reactions (formation of soaps from acid and base)
	(ii) pH control
	(iii) Blending (spirituous preparations, clear lip gloss products)
(b) Immiscible	(i) Extraction (none)
	(ii) Dispersion (emulsions)
4. Gas/Liquid	(i) Absorption (none)
	(ii) Dispersion (aeration and de-aeration)
5. Distributive	
(a) Fluid motion	Heat transfer (during emulsion and other manufacture)
(b) Limited flow	Pumping (pastes and other highly viscous products)

e.g. The manufacture of pigmented emulsion-based cream include four mixing steps. They are:

(1) Preliminary dry blending of pigments and excipient (type lb).

(2) Dissolution of oil-soluble and water-soluble materials separately in their appropriate phase (type 2 example i and type 3a).

(3) Dispersion or suspension of pigments in the oil or water phase (type 2, example ii).

(4) Mixing of the two phases to form an emulsion, possibly with the formation *in situ* of a soap as part of the emulsifier (types 3a and 3b).

(5) Adjustment of pH (type 3a).

(6) De-aeration of the bulk (type 4).

(7) Cooling to ambient temperature and pumping into a storage vessel (type 5a).

The subject of pumping is not clearly separated from that of mixing since pumping implies the forced flow of product. Any flow will naturally introduce an element of mixing if the product is not already homogeneous. Further, since flow is a common element of both processes, the same product characteristics (for example, rheological behaviour) must be taken into account.

Filtering is not usually a unit operation of major importance in cosmetics manufacture except in the production of spirituous preparations (colognes, aftershave and perfumes). It is possible to regard filtering as un-mixing and certainly the flow characteristics of the filtered product are again of prime importance.

Unit Manufacture

Most cosmetic products are filled from bulk in machines specifically designed to handle the units of a particular product type. While it is true that great care must be taken in the choice and the setting up of such machines, the main problems encountered are often concerned with the characteristics of the machines themselves rather than with the manufacture or processing of the product. There are at least two areas, however, where special understanding of the product units and their characteristics are essential for the achievement of efficient production: these are the moulding processes (lipsticks, wax-based sticks, alcohol-stearate gels) and compression processes (compressed eyeshadow, blushers and face powders). A description of unit manufacture could include all filling and packaging operations.

THE MANUFACTURE OF BULK COSMETIC PRODUCTS

Mixing

The object of a mixing operation is to reduce the inhomogeneities in the material being mixed. As Table 18.1, shows, inhomogeneity may be of physical or chemical identity or of heat. Further, in the processes demanded by cosmetic manufacture, the mixing is designed to be permanent—or as permanent as it is possible to make it.

The reduction in inhomogeneity depends on the efficiency of the mixing apparatus used and also on the physical characteristics of the materials constituting the mixture. For miscible liquids, homogeneity can be produced at a molecular level whereas for mixture of powders homogeneity is limited to the sizes of the powder particles themselves. When examining a mixture for quality, therefore, the *scale of scrutiny*—the magnification at which the mixture is examined—must vary from product to product. At an acceptable scale of scrutiny, *perfect mixing* implies that all samples removed from the mixture will have exactly the same composition. This is rarely achievable. *Random mixing* is achieved if the probability of finding a particle of a given component in a sample is the same as the proportion of that component in the whole mixture. Random mixing is the aim of all industrial mixing operations and whereas samples removed from such a mixture will not be identical, the variations should be very small. If the scale of scrutiny is reduced sufficiently, however, this may no longer be true.

Mixing can only occur by relative movement between the particles of the constituent components of the mixture. Three basic mechanisms for achieving this relative movement have been identified: bulk flow, convective mixing and diffusive mixing. *Bulk flow* (which includes shear mixing, cutting, folding and tumbling) occurs in pastes and solids, when relatively large volumes of mixture are first separated and then redistributed to another part of the mixing vessel. *Convective mixing* involves the establishment of circulation patterns within the mixture. Finally, *diffusive mixing* occurs by particle collisions and deviation from a straight line. In miscible liquids of sufficiently low viscosity, the thermal energy which is possessed by the constituent molecules may be enough to achieve a good mixture quality by thermal diffusion without additional energy being applied, although this process is usually too slow for industrial purposes.

Many mixing problems arise from the tendency of mixture particles to segregate during attempts to mix them. *Segregation* is defined as the preference of the particles of one component to be in one or more places in a mixer rather than in other places. The size of the non-uniformities in an imperfect mixture is sometimes referred to as the 'scale of segregation' and the difference in composition between neighbouring lumps of volumes is the 'intensity of segregation'. Segregation is not, fortunately, a major problem in cosmetics manufacture although it does manifest itself occasionally (as, for example, in the flotation of pigments during lipstick processing).

SOLID-SOLID MIXING

Solid-solid mixing is of two types: (i) Segregating; and (ii) Cohesive. Free-flowing powders exhibit many process advantages (such as easy storage, easy flow from hoppers, smooth flow of product), but have the disadvantage that they tend to segregate unless all the constituent particles are of very similar shape and size. Cohesive powder, on the other hand, lacks mobility, and individual particles are bonded together and move as clumps or aggregates. Although segregation does not appear to be a problem (except, as will be seen, at very small scales of scutiny), cohesive powders are difficult to store and do not easily flow from hoppers.

MANUFACTURE OF PIGMENTED POWDER PRODUCTS

Powder eyeshadows, face powders and powder blushers are commonly composed of the following types of material : Talc, Pigments, Pearl agents, Liquid binder and Preservative.

The order in which these ingredients are mixed and the process by which the mixing is carried out depend largely upon the type of equipment that is available.

Hammer Mill

The processing of bulk pigmented powder products is carried out using the hammer mill Fig. 18.1. The hammer mill was designed as a comminution machine. It consists of a fast rotating shaft fitted with freely swinging hammers mounted in a cage which is equipped with a breaker plate against which the feed is disintegrated, chiefly by impact from the hammer. The very high speed at which the hammers move (60–100 ms^{-1}) increases the chance of a hammer making contact with each particle and the dwell-time of particles within the chamber is increased by the placement of a variable size screen over the exit.

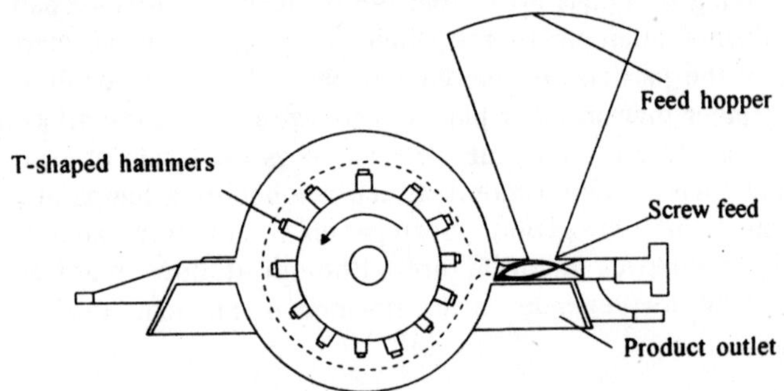

Fig. 18.1. Hammer mill.

Hammer mills are very efficient in the comminution of brittle particles in the range of 1500–50 μm but below this size their efficiency (the probability of direct impact) falls off rapidly. The very high rotational speed of the hammers and the air flow within the chamber ensure that there are enough weak secondary impacts (particle-wall and particle-particle) to break the much weaker pigment agglomerates—which may be up to 50 μm in diameter. The disintegrated agglomerate fractions then stabilise by becoming coated on to larger talc particles and should not be further changed by subsequent passes through the mill.

Hammer mill, has certain disadvantages. For example, from the viewpoint of energy consumption, a hammer mill used is very inefficient. The feed-rate and therefore, the processing time for all but the smallest batch sizes of powder is very slow. Increasing the residence time of powder within the grinding chamber by decreasing this mesh size can cause the screen to become blocked with compacted powder, resulting in overheating and damage to machine and product. The hammer mill is a continuous processing device being used for batch processing. For this reason, it must be fed with a powder mixture which has already been effectively mixed, otherwise the colour of the milled product changes as each section of unmixed bulk passes through. This preliminary mixing must be efficient although it is not necessary for any extension to be achieved at this stage.

The most widely used is the 'ribbon blender' which comprises a horizontal drum containing a rotating axial shaft which carried ribbon-like paddles. In such a device, the pre-mix can take anything between 20 and 60 minutes. Other mixers are now available which utilise higher energy input but are quicker. Table 18.2, summarises the properties of some of the more conventional powder mixers. Since it is relatively easy to achieve good mixture quality (at a large scale of scrutiny) in cohesive powders, any mixing device will eventually produce a satisfactory even distribution of components provided that it contains no dead spots where mixing does not take place.

Table. 18.2. Conventional powder mixers.

Type of mixer	Batch/ continuous	Main mixing mechanism	Speed of mixing	Ease of cleaning	Energy consumption	Quality of extension
Horizontal drum	B	Diffusive	Poor	Good	Low	Poor
Löedige-type	B	Convective	Good	Fair	Medium	Fair
Ribbon blender	B or C	Convective	Poor	Fair	Low	Poor
Nauta mixer	B	Convective	Good	Poor	Low	Poor
V-mixer (with cutters)	B	Diffusive	Poor	Good	Medium	Unknown
Airmix	B	Convective	Good	Fair	Low	Unknown

It is usual to add the liquid binder during this preliminary mixing stage. The binder may be poured into a suitable orifice in the mixer although many production chemists prefer to spray it into the mixer cavity as an aerosol through a venturi or similar device. This procedure helps to distribute the liquid more evenly and avoids the formation of wet, lumpy areas in the powder body. The separation of large agglomerates which takes place subsequently in the mill normally assures the completion of the wetting process provided only that the binder is correctly chosen. Should the binder still appear to be unevenly distributed after the passage of the powder through the mill, the product can often be rescued by passing it through as fine a mesh sieve as possible.

Batch Colour Correction

It is not unusual for the bulk powder product, even though it has been correctly processed, to require colour correction in order to obtain a satisfactory match to the standard. Since any addition of pigment or talc needs to be extended, a passage through the mill is necessary. A common procedure is as follows. After the preliminary coarse mix has been completed, a small amount of the bulk (usually about 5 kg), which is assumed to be representative of the whole, is passed through the mill. This is examined in the laboratory, and if necessary, a pigment addition specified. This correction is added to the 5 kg of milled product, mixing in roughly by hand and the 5 kg is re-milled. The twice-milled sample is returned to the remainder of the bulk and re-mixed in the original mixer. A further 5 kg is then removed and the process is repeated until a match is obtained. The use of pigments previously extended on talc and stored as such. This has the merit of speeding up the correction process.

When pearl agents are part of the formulation, unless an un-pearlised standard is provide, the pearl must be added in the correct proportion to the laboratory sample before colour can be assessed. Pearl is only added to the bulk in the last stage of the manufacturing procedure. The ideal equipment would probably have the following properties :

(1) It would be capable of breaking up weak particles in the size range 50–0.5 μm without damaging talc or mica particles of similar diameter.

(2) It would be a low energy device, consuming little power itself without heating the powder mixture excessively.

(3) It would be a batch processing device capable of mixing and extending in one operation.

(4) It would be rapid: processing times of less than 10 minutes would be acceptable.

(5) It would not cause excessive aeration of the powder (since this causes further processing problems in later processing).

(6) It would be easy to clean.

(7) Its efficiency would not vary with the cohesiveness of the powder; it would not be affected by poor flow characteristics.

(8) It would be quiet and clean in operation.

Other comminution devices have been shown to produce extension, particularly pin mills and fluid energy mills, yet none seems to work as efficiently as the hammer mill. Recently two types of mixers have been developed. They are: (i) Vertical vortex mixer; (ii) Plough-shear.

Vertical vortex mixer

The powder mixture is placed in a vertical, cylindrical chamber and is then accelerated outwards and upwards into a fluidised vortex motion. The motion may be produced by compressed air blasted sequentially from a series of nozzles contained in a lower cone-shaped section; alternatively, a propeller-shaped tool of 'poor aerodynamic' design may be used which rotates rapidly in the dished base of the mixing bowl. Mixing and dispersion occur at the point of conversion of the powder particles (in the upper point of the mixing bowl) by particle-particle collisions.

Plough-Shear

The high-speed mixer is often referred to as a 'plough-shear' device, because of the unusual shape of the mixing paddles which rotate on an axial shaft in a cylindrical horizontal mixing chamber. These paddles cause the powder from all parts of the chamber to be thrown about in such a way that it all passes rapidly through a zone occupied by a series of rapidly revolving blades on a separate shaft, referred to as a 'chopper'. The chopper is largely responsible for the powder extension and may be switched on or off independently of the main axial drive.

Both types of mixer have been used as partial or complete replacement for the traditional blender-hammer mill combination. The plough-shear type may also be used for wet-processing.

MIXING PROCESSES INVOLVING FLUIDS

Although there are similarities between the flow of powders and the flow of liquids it is obviously easier to set up and sustain flow patterns in the latter. On the whole this makes the mixing processes easier to perform and a much larger variety of equipment is consequently available to choose from.

General Principles of Fluid Mixing

Some mixing operations can be thought of as simple blending – for example, the blending of colour solutions into miscible bulk liquids and the blending of oils, alcohol and water in perfumes and colognes. On the other hand, the formation of an emulsion, the suspending of a gelling agent and the distribution of pigment agglomerates in a viscous liquid all involve the breaking up of one of the constituents of the mixture into finer particles during the mixing process. For this reason, it is referred to as 'dispersive' mixing to distinguish it from simple blending.

On the industrial scale, mixing occurs as the result of forced bulk flow within the mixing vessel. Two types of flow can be distinguished, laminar and turbulent. Laminar flow occurs when the fluid particles move along streamlines parallel to the direction of flow. The only mode of mass transfer is by molecular diffusion between adjacent layers of fluid (Brownian motion). In turbulent flow, the fluid elements move not only in the parallel paths but also on erratic and random paths, thus producing eddies which transfer matter from one layer to another. For this reason, turbulent mixing is rapid compared with other mixing mechanisms.

When a quiescent liquid is slowly stirred the flow is laminar but as the velocity increases it may become turbulent; thus the velocity is a significant factor in determining the type of flow set up in the mixing vessel. A valuable aid in describing the critical point at which laminar flow becomes turbulent is due to Reynolds who, in 1883, first demonstrated turbulence. The dimensionless number which bears his name, Re, can be calculated for agitated vessels as follows:

$$\mathrm{Re} = \frac{D^2 N \rho}{\eta} \qquad \qquad \dots 18.1$$

where D is the diameter of the impeller, N the impeller speed (rpm), ρ the density of the mixture and η its viscosity.

Although little is known about the mechanism of turbulence, experience has shown that in agitated tanks the onset of turbulence occurs at Reynolds numbers of about 2×10^3. For fully developed turbulence, Reynolds numbers greater than 10^4 are required and are found in many cosmetic mixing processes. The majority of products exhibit non-ideal (non-Newtonian) behaviour which can often be more appropriately described by the expression

$$F = (\eta_{\mathrm{app}})^n A \times \text{velocity gradient} \qquad \dots 18.2$$

In this case η_{app} is termed as the 'apparent viscosity' and n usually has a value between 0 and 1. The name given to this type of behaviour is 'pseudoplastic' and the basic difference between materials exhibiting this property and ideal or 'Newtonian' fluids is illustrated in Figure. 18.2. As can be seen, pseudoplasticity is manifested by a fall in viscosity with increasing shear rate at constant temperature. Many cosmetic liquids exhibit this behaviour – especially emulsions and suspensions of particles of the order of 1 μm or less in size.

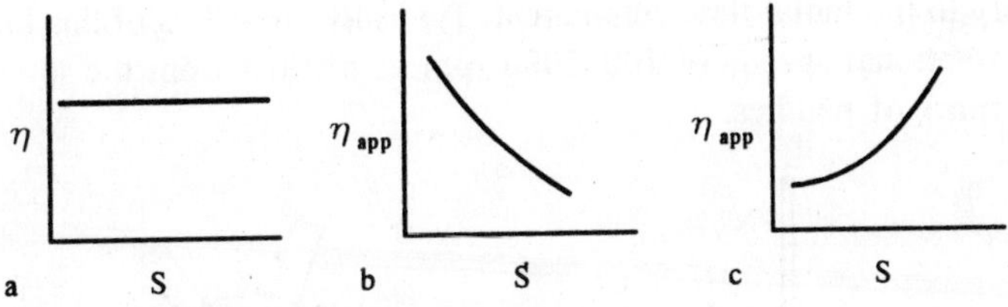

Fig. 18.2. Rate of shear (S) plotted against viscosity (η) or apparent viscosity (η_{app}).

Mixing Equipment for Fluids

In the mixing of fluids, all three mixing mechanisms — bulk flow, turbulent diffusion and molecular diffusion — are usually present. As viscosity increases, however, and turbulence becomes correspondingly more difficult to establish, the parts played by turbulent and molecular diffusion become less important. Mixing equipment can therefore be divided into two categories, laminar shear and turbulent mixers.

Laminar shear/distributive mixers	*Turbulent mixers*
Helical screw/ribbon blenders	Turbine-agitated vessels
Two-blade mixers	Pipes
Kneaders	Jet mixers
Extrusion devices	Sparged systems
Calenders	High-speed shear mixers
Static mixers: low Re	Static mixers: high Re

Paddle mixers

Paddle mixers are simple and cheap but very inefficient for all but very low viscosity liquids. They produce mainly tangential flow and are usually mounted centrally because of their large diameter compared with that of the tank.

Turbines

Turbines are probably the most common impeller type used in cosmetics processing since they can cope with a wide range of viscosities and densities. For liquids of low viscosity, the flat-blade impeller is sometimes used Figs. 18.3 (a) and (b). For very viscous materials the blades may be curved backwards in the direction opposite to the rotation, since these require a lower starting torque and seem to give better energy transfer from impeller to liquid Fig 18.3 (c).

Fig. 18.3 (d) illustrates a fixed-pitch *axial flow impeller*. Used without baffles, however, the axial component generated by such turbines remains secondary to the radial flow component. Typically, impellers of this kind are used at rotational speeds of 100–2000 rpm as distinct from the low speed (15–50 rpm) of paddles.

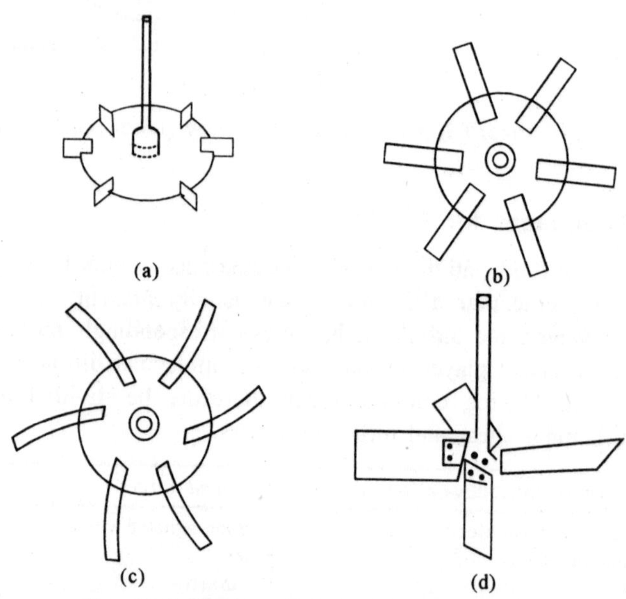

Fig. 18.3. Various designs of turbine impeller.

Propeller mixers

Propeller mixers are restricted to use with low viscosity fluids. They have pitched blades of which the blade angle varies along the length from centre to tip. Flow patterns developed by propeller mixers have a high axial component and the rate of circulation is high. They are usually of relatively small diameter, typically three-bladed, and are used at speeds between 450–2500 rpm. Such stirrers are used extensively in the cosmetics industry for simple blending operations but are not suitable for the suspension of particles which settle rapidly or for the dissolution of sparingly soluble heavier materials.

Many portable mixers are of the propeller type. If the mixer is mounted centrally in the mixing tank Fig. 18.4 (a), because of the entrainment of liquid above the impeller, the surface becomes depressed and a vortex is formed. Generally, vortices are to be avoided because of the low order of turbulence and the air-entrapment which they cause. When they are mounted eccentrically, however Fig. 18.4 (b) turbulence is increased and vortices avoided.

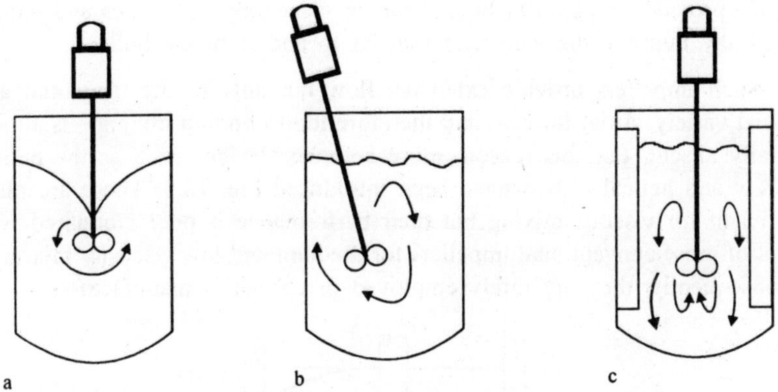

Fig. 18.4. Portable mixers.

Mixing in non-Newtonian liquids of low or medium viscosity

Many liquids — perhaps the majority — encountered in cosmetics processing are of the shear-thinning and/or elastic rheological type. Naturally, if the liquid is already of low viscosity, the effect of shear-thinning may not be noticeable. On the other hand, more viscous liquids showing these characteristics present considerable problems to the cosmetics processor. The fluid close to the rotating impeller of a mixer is sheared at a high rate and so becomes relatively mobile, but as this is pumped away from the impeller it encounters regions of less intense flow and hence of much higher viscosity. Turbulence is therefore rapidly damped out, decreasing the turnover in the vessel and slowing down the mixing process. Moreover any elasticity shown by the liquid results in the absorption of energy by deformation of a recoverable variety, thus damping out turbulence even further.

Impeller types and mixers for high viscosity fluids

Propellers and turbines, as already mentioned, work best under turbulent conditions at relatively high rotational speeds. In viscous products, given that such speeds are attainable at all, flow is confined to the regions very close to the impeller, and large stagnant regions in the mixer exist where no mixing can occur without the employment of some secondary mechanism. To eliminate these stagnant regions, large impellers such as paddles, gates, anchors and leaf impellers may be used; these sweep a much greater proportion of the vessel and produce more extensive flow. Usually such impellers are designed to have close clearances with walls, giving a degree of wall-scraping. This helps to eliminate build-up of unmixed materials at walls. provides a region of high shear for dispersing agg..gates and lumps, and may improve the wall heat transfer to and from the bulk.

Such impellers provide extensive flow but only of the tangential and radial variety. Axial flow — and therefore top-to-bottom mixing—is almost totally absent. For this reason, more complex designs such as the helical screw and helical ribbon have been introduced Fig. 18.5. These are more efficient for viscous mixing but their performance is poor compared with that of more conventional impellers for medium and low viscosity mixtures. Consequently, they are rarely employed in cosmetics manufacture.

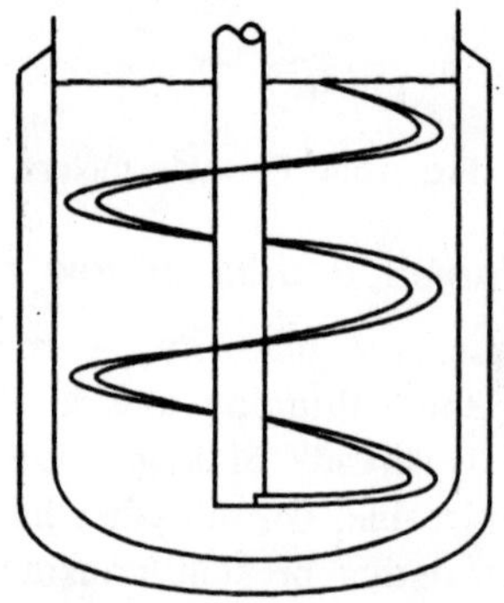

Fig. 18.5. Helical impeller.

Axial flow cannot be achieved by the introduction of baffles as when mixing lower viscosity fluids, but some success has been achieved by the use of impeller-draught tube combinations. As the name implies a draught tube is a tubular, axially orientated enclosed space within the main mixing chamber containing an impeller or some other means of forcing the flow of mixture along it. Small impellers or helical screws designed to fill most of the cross-section of such a tube have been successfully used to promote axial flow in most liquids, even those of very high viscosity.

An alternative approach to the problem created by lack of flow in viscous media is the use of impellers which progressively sweep the whole contents of the vessel while the mixture remains stationary. Examples of this include the 'Nauta'-type mixer in which a helical screw sweeps the wall of a conical mixing chamber.

For even more viscous products such as mascara and very thick pastes, equipment which exhibits a greater degree of distributive mixing may be utilised. Such mixers are designed to produce bulk flow and laminar shear by spatial redistribution of elements of the mixture. Perhaps the most commonly encountered mixers of this type are of the single or double action planetary type or the two-blade 'dough' mixer. Their essential feature involves the cutting and folding of a volume of the mixture and the physical replacement of it into another part of the mixer where it is cut and folded again. An example of this distributive mechanism is illustrated in Fig. 18.6, in which for clarity a volume of a mixture has been isolated and divided into six equal segments, one of which consists of a black minor component. The cube is compressed to one quarter of its initial height, cut and reassembled as shown. Redistribution of the minor component has been achieved which, were the process to be repeated often enough, would eventually achieve the desired level of homogeneity.

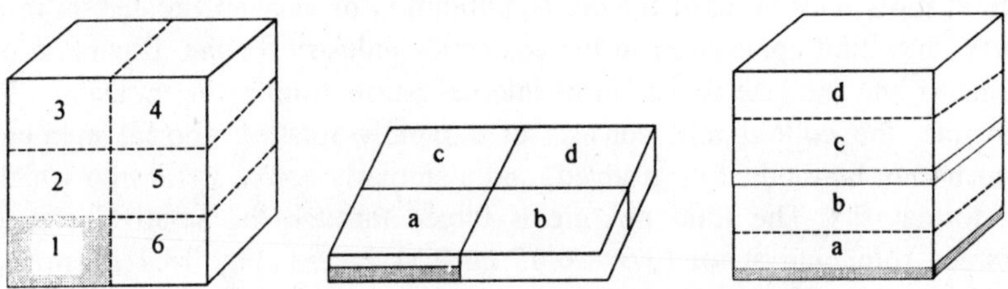

Fig. 18.6. Distributive mixing mechanism.

A more recent innovation is the so-called static mixer, of which several designs are now commercially available. Static mixers are essentially in-line mixing devices in which mixtures flowing through a pipe are cut and folded by a series of helical elements in a circular tube Fig. 18.6. These elements (which do not move — hence the name 'static') turn the flowing mixture through an angle of 180°. Since alternative elements have opposite pitch and are displaced 90° to each other, this causes the bulk flow to reverse direction at each junction, and thus the leading edge of each element becomes a cutting device, splitting and re-folding the mixture in on itself.

Finally, mention must be made of extruders, in which a helical screw forces the bulk mixture to flow down a tube. Here, the pressure generated can be enormous, as in soap-plodding, and such energy can cause materials with the viscosity of toilet soap to undergo laminar flow. The actual flow pattern produced is complex, being a combination of pressure and drag flow within the tube.

High shear mixers and dispersion equipment

Generally, a high shear rotor-stator mixer may be used either for batch processing all-enveloping chamber. Used as a batch mixer, it is capable of generating considerable turbulence because of the great velocity with which fluid is pumped out of the mixing head. As with other devices, however, this high energy is increasingly converted into heat with increasing viscosity of the mixture. A serious disadvantage for certain processes is the tendency of the mixer to cause aeration when used in the top-entry mode. For this reason, such devices are often incorporated into the bottom of processing vessels.

Another high shear rotor-stator device in common use is the colloid or stone mill. Such equipment is commonly thought of as a comminution device; this serves to illustrate the fineness of the dividing line between mixing and comminution with high shear equipment. While it is true that colloid mills may be used for the comminution of very soft materials in a slurry, they find application in the cosmetics industry for the dispersion of pigments and the size reduction of internal phase droplets in emulsions. In principle, the colloid mill consists of a rapidly rotating conical member (which may be toothed or grooved) and a similarly coned stator into which the former fits. The fluid mixture is forced through the small clearance between rotor and stator (0.5 – 0.05 mm) as before. Fig. 18.7 illustrates the design of colloid mills in greater detail.

Colloid mills are used exclusively as an 'in-line' or continuous device. They may be water-cooled and can be adjusted as the moving parts wear down.

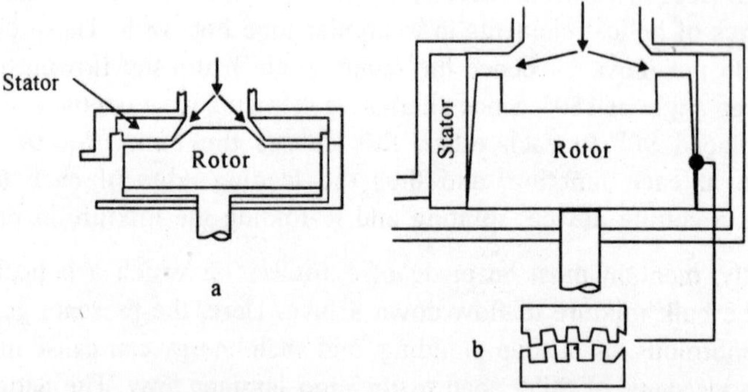

Fig. 18.7. Colloid mills (a) Stone (b) Toothed.

For more viscous products, an alternative device is the triple roll mill Fig. 18.8. The device consists of three steel rollers which rotate in the directions indicated in the diagram. Each roller is water-cooled and machined to great accuracy so that the gaps between each pair of rollers can be set very fine. The product is applied at the top of roller A, passes between A and B and round the underside of B between rollers B and C.

As it traverses each gap, the product is subjected to enormous compression forces which are particularly effective in breaking down pigment agglomerates. Strictly speaking, therefore, the triple roll mill is not a high shear device but a compression device. It is included here, however, as a real alternative to the rotor-stator mixers in bringing about effective pigment dispersion in liquids — particularly in viscous liquids.

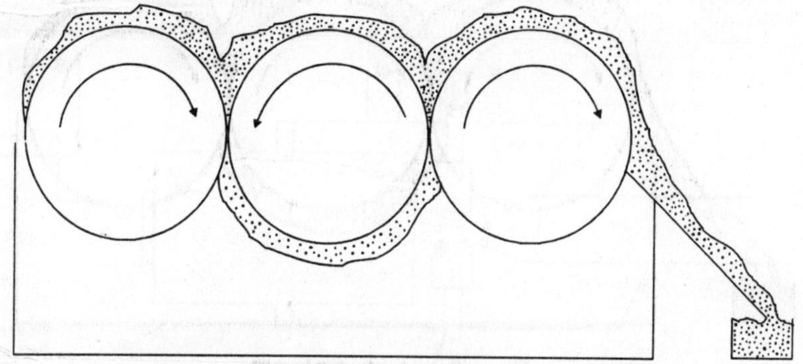

Fig. 18.8. Triple roller mill.

Perhaps the highest shear stress of all is generated by a valve homogeniser, which is still extensively used in the production of emulsions with very fine internal phase droplets.

An interesting alternative to the valve homogeniser, and one which is finding increasing use in cosmetics manufacture, is the ultrasonic homogeniser. When high intensity ultrasonic energy is applied to liquids, a phenomenon known as cavitation occurs. Cavitation is complex and not fully understood. As ultrasonic waves are propagated through the fluid, areas of compression and rarefaction are formed and cavities are produced in these rarefied areas. When the wave passes on, these cavities collapse and change to an area of compression and it has been demonstrated that the pressure in these cavities just before their collapse can be as much as several thousand atmospheres. Most of the effects of ultrasonic radiation in liquids are attributed to the powerful shock waves produced immediately following the collapse of such cavities. One design of ultrasonic homogeniser is illustrated in Fig. 18.9.

No description of dispersing equipment would be complete without mention of ball mills and sand mills. In both these devices the breakdown of agglomerates is achieved by attrition between rapidly moving grinding elements which take the form of pebbles, balls or (in the case of sandmills) finer sand-like particles < 1 mm diameter. The movement of these grinding particles may be achieved in a number of ways. In a tumbling mill, the elements tumble over each other as a horizontal drum is rotated on trunnions Figure 18.10. Sand mills may be horizontal or vertical cylinders in which the grinding medium is stirred by a rotating agitator (Fig. 18.11), whereas in vibration mills movement of the whole chamber may be caused by eccentric cams, by out-of-balance weights on a drive shaft, or electrically.

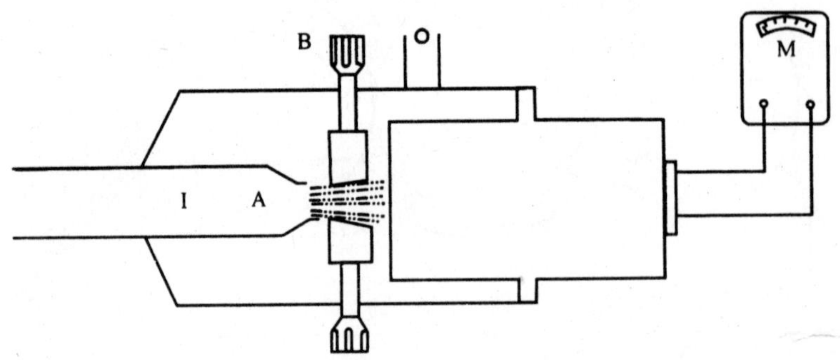

Fig. 18.9. Ultrasonic homogeniser.

The roughly premixed product enters at I and, on passing through orifice A, is subjected to intense ultrasonic energy by vibrating blade B. The treated product leaves via O. The meter, M, and 'tuning' devices at B combine to allow the operator to achieve maximum effect with each different product type

Small ball mills and sand mills are used extensively in the dispersion of pigments into liquids (as, for example, in the production of castor oil lipstick pastes) and for the dispersion of bentone into nail varnish media. Although they are very effective, their chief disadvantage is the extremely protracted cleaning time required when changing from one colour to another. For this reason many users prefer to keep separate sets of grinding media for each different colour they wish to produce.

The basic mechanism by which mills of this kind produce their effect is attrition between grinding elements. Very little shear is developed.

SOLID-LIQUID MIXING

The production of a cosmetic product often involves the incorporation of a powdered solid material into a liquid. The objective may be to dissolve the powder completely (as with salt or preservatives in water), to effect a colloidal dispersion of water-swellable particles (as with bentones and other gelling agents) or simply the dispersion of insoluble materials such as pigments. To do this efficiently with a wide range of powder particle size and surface characteristics and with a range of different liquids of varying viscosity, quite a variety of mixing equipment is available.

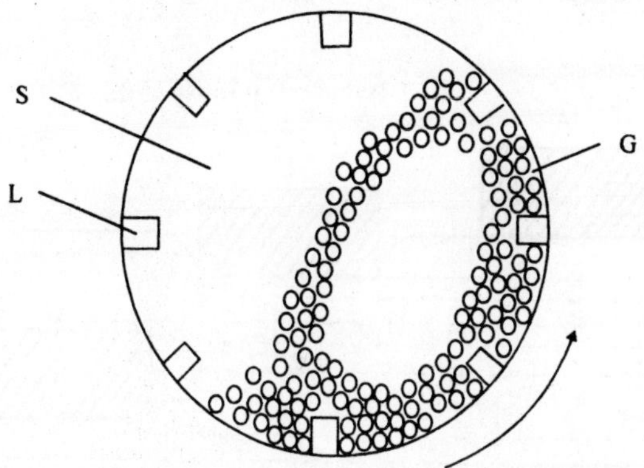

Fig. 18.10. Ball mill showing ball pattern in rotating drum. (G–Grinding medium; L–Lifter elements; S–Slurry).

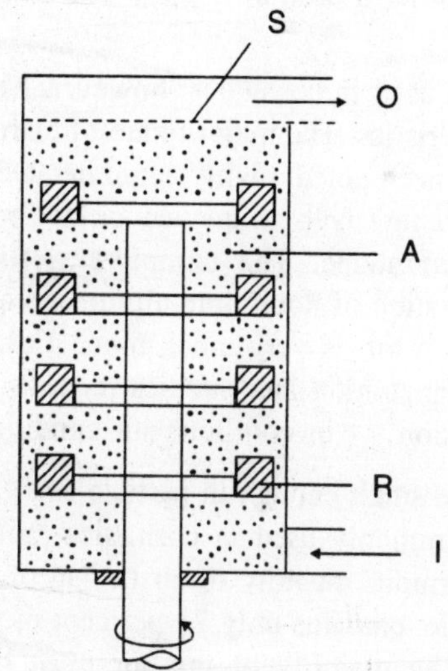

Fig. 18.11. Sand mill. (I–Slurry in; O–Slurry Out; S–Screen; R–Rotating agitator; A–Glass beads, sand or shot).

Perhaps the easiest of these processes to carry out is the dissolution of a fairly large size, smooth-faced solid, such as salt. The initial incorporation of each crystal into the liquid involves the complete replacement of the air-solid surface with a liquid-solid surface. This may be considered to be a three-stage process of adhesion, immersion and spreading Fig. 18.12.

Immersion is complete when all the air has been displaced and the surface of the crystal has been completely wetted by the liquid. This process is aided by a low surface tension and low contact angle between the liquid and solid.

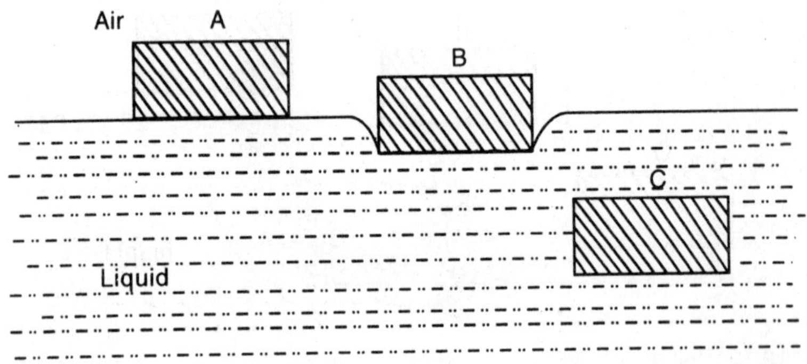

Fig. 18.12. Immersion of a solid in a liquid. ((A) Adhesion (B) Immersion (C) Spreading).

Not all powders used in cosmetics, however, have such favourable size and surface characteristics. The majority are of extremely small particle size and, as has already been noted, highly agglomerated. Each agglomerate will have a complex structure with an uneven surface and will be perforated by cavities of irregular shape. The complete wetting of such structures, involving the penetration of liquid into all the crevices and cavities together with the expulsion of air, is very much more difficult. It should be noted, for example, that penetration into cavities requires a low contact angle but a high surface tension — in conflict with conditions for easy wetting.

Powders that are small enough in particle size to form agglomerates are by far the most commonly used in cosmetics processing. In the dry state they entrap an enormous quantity of air (a bag of cosmetic-grade titanium dioxide, for example, contains only 25 per cent of powder together with 75 per cent of air). The majority of this air must be expelled if a smooth uniform mixture is to be obtained.

Immersion is only the first stage in the production of cosmetic quality dispersions. Even if the agglomerates were evenly distributed, the larger of them would give rise to 'grittiness'. Further, for pigments maximum colour can only be developed when these agglomerates are broken up and the maximum possible surface area of pigment is exposed. Disagglomeration is therefore the next step in the production process.

In cosmetic processes, the disagglomeration of solid particles in liquid media can be brought about by a variety of machines. In lipstick processing, for example, pigments are 'ground into' castor oil by preparing a coarse mixture which is then passed over a triple roll mill, or further ground in a colloid mill, ball mill or sand mill. These machines are used specifically because they will deal effectively with media of lipstick-paste viscosity.

For less viscous media (for example, the dispersion of pigments into the aqueous phase of an emulsion), a high shear device of the rotor-stator type is frequently used. In this case, processing time can be shortened by ensuring that the whole contents of the vessel are brought into the catchment area of the shearing head by secondary stirring. As with all disagglomeration, shear stress is largely responsible for the partial disintegration of the agglomerate.

For soluble powders, the enormous increase in the solid-liquid interface brought about by immersion and disagglomeration will ensure that the actual process of dissolution can proceed at the maximum possible rate. For insoluble powders, however, there remains the problem of maintaining a good stable dispersion.

Disagglomeration is usually a reversible phenomenon and it can usually be assumed that the opposite process—'flocculation'—will be simultaneously taking place.

It was noted when considering powder-powder dispersion that stabilisation could be achieved by the introduction of particles of larger size to which the disintegrated agglomerates could adhere. In some cases this can be applied to solid-liquid systems — for example, by pre-extending pigments onto talc before adding them to a liquid foundation base — but in many instances all the solid particles are of too small a size for this to be done. Under these circumstances rules similar to those used in emulsion technology can be applied. Thus the rate of flocculation can be slowed by some or all of the following means :

(1) The use of surface-active agents (sometimes as polymer coating of the powdered solid) to inhibit flocculation by steric hindrance.

(2) The manipulation of electrostatic charges on the surfaces of the powder particles.

(3) The manipulation of the viscosity of the dispersion.

Surface-active agents have a part to play at two stages in the process of manufacturing a stable dispersion. It has already been seen that the lowering of the solid-liquid contact angle speeds up the wetting process. In practice, the best results are often achieved not with a surfactant, which measurably lowers the surface tension of the liquid, but with what is sometimes described as a 'surface activator' which reduces the interfacial tension between solid and liquid. These surface activators (which are also described as 'dispersants' or 'wetting agents') can, if correctly chosen, cause an immediate improvement in the quality of the dispersion which is made manifest by a sudden increase in the colour intensity.

Suspension of Solids in Agitated Tanks

If a particulate solid is dispersed in a liquid in which it does not dissolve and the suspension so formed is allowed to stand undisturbed in a vessel, provided that the densities of the two components are dissimilar some degree of settlement or flotation will eventually take place. Where the particles are present in sufficiently low concentration to have a negligible effect on the viscosity of the suspension, re-suspension can be achieved by the establishment in the liquid of flow patterns of sufficient turbulence. The suspension of solids in agitated tanks is frequently encountered in cosmetics processing, as an aid to dissolution or as a means of obtaining a good dispersion of particles prior to a change of viscosity of the liquid medium by gelling or cooling.

Three conditions can be recognised during the production of a suspension, namely complete suspension, homogeneous suspension and the formation of bottom or corner fillets.

Complete suspension

Complete suspension exists when all particles are in motion and no particle remains stationary on the bottom or surface for more than a short period. Under these conditions, the whole surface of the particles is presented to the fluid, thereby ensuring the maximum area for dissolution or chemical reaction.

Homogeneous suspension

Homogeneous suspension exists when the particle concentration and (for a range of sizes) the size distribution are the same throughout the tank. The homogeneous suspension speed is always considerably higher than 'complete suspension speed' and more difficult to achieve and to measure. Nevertheless, homogeneous suspension is very desirable for certain types of cosmetics applications and particularly so for continuous processing. In practice, for such processes the requirement is only that the particle size distribution and concentration in the discharge and the vessel are the same.

Sometimes heavier particles are allowed to collect in corners or on the bottom of the vessel in relatively stagnant regions to form *fillets*. This may have the practical advantage of very large saving in power consumption compared with the energy that may be needed to achieve complete suspension (provided, of course, that this saving offsets the loss of active solids in the fillets).

In general, it may be said that propeller or 45°-angle turbines offer the best advantage for rapid suspension for low power consumption—particularly if draught tubes are introduced. If, on the other hand, radial flow agitators need to be used, these should be of relatively large length-to-diameter ratio, be placed close to the bottom of the tank and have turbine blades extending to the shaft to prevent problems with central stagnant regions.

LIQUID-LIQUID MIXING

As indicated by Table 18.1, it is convenient to consider separately the case in which the liquid components are all mutually soluble and the case in which some or all of them can coexist as separate phases (that is to say, they are sparingly or partly soluble in each other).

Miscible Liquids

The mixing of miscible liquids (blending) represents perhaps the simplest mixing operation in cosmetics manufacture. Several examples have already been cited and no further elaboration is needed except to reiterate that it is important to choose the mixing apparatus best suited to the viscosities of various components in order to carry out the operation efficiently.

Immiscible Liquids

Practically the only representatives of this category of mixing operation are emulsions. All cosmetic emulsions consist of two major immiscible liquids, one dispersed as fine droplets in the other and separated by a layer of surface-active agent at each liquid-liquid boundary.

The emulsification process

The two major phases (these will be referred to as 'oil' and 'water') together with the emulsifier are brought together under turbulent conditions. Depending on the prevailing conditions, one major phase is broken up into droplets (predominantly by the action of shear stress imparted by turbulent eddies) and distributed throughout the other major ('continuous') phase.

While the droplets remain larger than the eddies, they will continue to break up into ever smaller droplets. Eventually a point is reached in this process when the available power giving rise to the turbulence cannot provide the shear stress necessary to reduce the droplet size any further. All this stage there exists an emulsion containing droplets of a certain mean diameter but over a range d_{min} to d_{max}. Provided that it is correctly chosen, the emulsifier prevents the rapid coalescence of these droplets and a stable emulsion is formed.

Orientation of phases

In any emulsion the orientation of the phases (that is, whether the oil or the water phase is continuous) is determined principally by the choice of emulsifier and the volume ratio of oil to water. Usually, however, there is a range of volume ratio over which either phase may be dispersed, depending upon the method of manufacture. If a quiescent mixture of two phases coexisting as two simple layers (one upon the other) is agitated, the phase into which the agitator is placed is most likely to form the continuous phase in the resulting emulsion. In other words, drops are drawn into the phase in which the impeller is placed. If initially only one phase is present in the mixing vessel containing the impeller, an added second phase will inevitably form the disperse or discontinuous phase. If, however, the continued addition of the second phase combined with the choice of emulsifier eventually leads to a volume ratio at which the system is more stable with the added phase being continuous, then the emulsion will spontaneously invert to achieve this end result. When inversion takes place, it is very often accompanied by a change in droplet size. Where this change is a decrease, then inversion leads to a more stable emulsion and gives rise to a valuable method of manufacture.

Addition of surfactant

In a batch manufacturing process for emulsions, there are four possible methods of adding the emulsifier. The first of these involves dissolving (or dispersing) the emulsifying agent in water, to which the oil is added. An oil-in-water emulsion is initially produced but inversion to water-in-oil may take place if more oil is needed.

Alternatively, the emulsifier may be added to the oil phase; the mixture may then be added directly to water to form an oil-in-water emulsion or water may be added to the mixture to form a water-in-oil emulsion. Many emulsions, on the other hand, are stabilised by soaps which are formed at the interface between the two phases. In this case, the fatty acid end of the soap is dissolved in the oil and the alkaline component is dissolved in the water. The two phases can be brought together in any order.

Finally, a less used method is one in which water and oil are added alternately to the emulsifying agent. Usually, the improvement in product quality obtained by the use of this method does not warrant the complication it causes in the manufacturing procedure.

Batch processing equipment

It will be evident from the discussion so far that there are at least two important elements of emulsion processing, namely shear (for the emulsification and particle size reduction process) and flow (in order to bring the whole contents of the vessel through the region of the high shear). Flow is also important in the heating and cooling of the emulsion. Most emulsion processing vessels are equipped with a jacket through which steam or hot water can be circulated to heat the contents and cold water circulated to cool them. Evidently then, to be effective, the mixing mechanism must be able to provide adequate flow to and from the vessel walls.

For these reasons, most emulsion batch processing vessels contain a high shear turbine or rotor-stator device (typically bottom or side entry rather than top entry, to decrease the likelihood of air entrapment) and a high flow, low shear mixing device which may be driven by a separate motor. This high flow device is of variable design, the most popular being a gate stirrer in which the arms are inclined at about 45° to the horizontal so as to give an element of axial flow. In more complicated designs, a central shaft carries more blades which sweep the area between the first set, the sets of blades rotating in opposite directions. Whatever the design, the frame holding the outer blades normally carries spring-loaded plastic scraper blades to prevent the build-up of product on the inner vessel wall, which would interfere with efficient heat exchange across the surface (Fig. 18.13).

The main motor may be driven by electricity or by air (up to about 100 psi or 9 bar). The advantages of air-driven motors are that they are infinitely variable in speed, torque-sensitive (and therefore less likely to become damaged when subjected to sudden loads), they do not constitute a hazard in the processing of low flash-point materials and generally they require less maintenance. Electric motors can be built to match some of these advantages (with slipping clutches and flameproofing), but only at considerable expense.

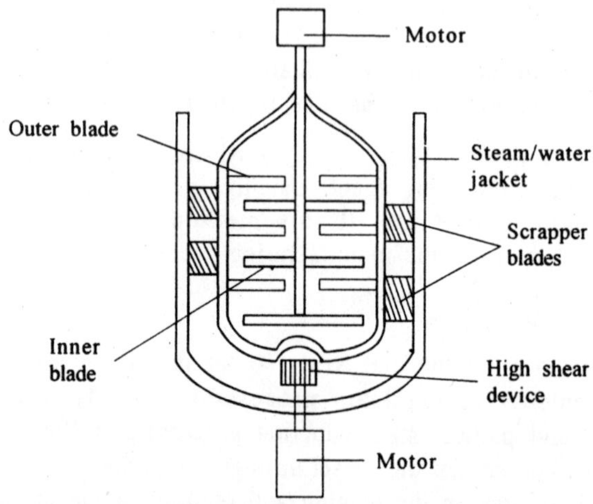

Fig. 18.13. Batch emulsion processing plant.

This traditional gate-type impeller system suffers from the grave disadvantage of limited axial flow. This is not noticeable in smaller vessels (below 600 litres capacity) but becomes a major problem in large tanks. One approach to the problem is to provide top-to-bottom transfer of the contents by means of a pump and an external pipe. A more satisfactory arrangement for the manufacture of emulsions of medium and low viscosity is to replace the gate stirrer by one or more axial flow impellers mounted centrally on a single central shaft. Although it becomes more difficult to provide wall-scrapers, the excellent flow around the vessel walls makes scraping less necessary.

The problem of ensuring that all the product passes through the region of high shear has led to the idea of passing the batch through an external circuit containing an in-line homogeniser; this may be a rotor-stator device, colloid mill or valve homogeniser.

Continuous processing

In view of the difficulties encountered in manufacturing large batches of emulsion, a logical extension of an external circuit with in-line homogeniser is a continuous processing plant. A simple form of such a plant is illustrated diagrammatically in Fig. 18.14. Such a plant is more correctly referred to as 'batch-continuous' since in essence a single batch is manufactured at a time. For long-run products, a truly continuous plant would be suitable, such as that illustrated in Fig. 18.15. In this case, the addition of second vessels A' and B' (which are exact duplicates of A and B respectively) together with the three-way valves V_1 and V_2 means that a second batch of each phase can be prepared while the first is being used. In this way, a continuous supply of each phase is assured by the turning of a valve. In reality, continuous plants tend to be slightly more complex than is illustrated in the diagrams with the inclusion of take-off points for sampling and other sophisticated features. Nevertheless, continuous manufacture is a very practical and, for some applications, extremely economical method of processing emulsions.

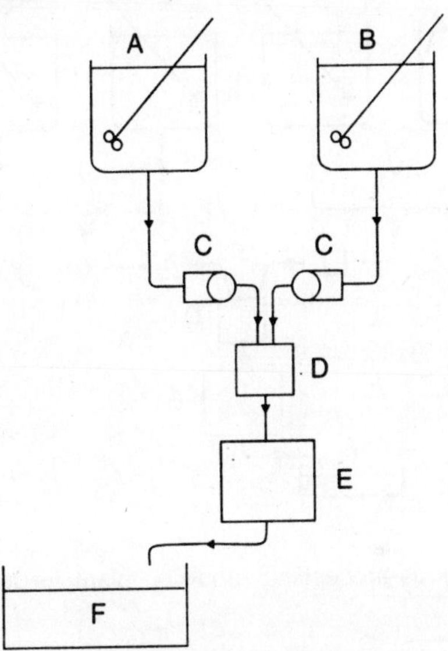

Fig. 18.14. Simple continuous-processing emulsion plant. (The two phases are prepared separately in tanks A and B, then pumped in correct proportions via metering pumps, C, into an in-line premixer, D (such as a static mixer), and then through a homogeniser, E. Finally, the formed emulsion is pumped into F, which may be a storage tank or the hopper of a filling vessel. A heat exchanger can be incorporated between E and F for rapid cooling.)

Emulsion temperature

The primary reason for raising the temperature of the phases during emulsion manufacture is to ensure that both are in the liquid state. In particular, the oil phase may contain fats and waxes that are solid at room temperature; there is very little point in raising the temperature of the oil phase much above that at which these liquefy. Excessive heating of the phases during manufacture prolongs the manufacturing time and wastes energy.

If the water phase is liquid at room temperature, it is customarily heated to approximately 5°C above the temperature chosen for the oil phase (so as not to cause the sudden solidification of the latter on blending). There is, however, an interesting alternative — namely, emulsification between hot oil phase and cold water phase. The plant for this procedure is illustrated in Fig. 18.16, which shows that mixing of the phases and homogenisation take place simultaneously. The obvious advantage of such a method is the saving of time and energy in not having to heat the aqueous phase.

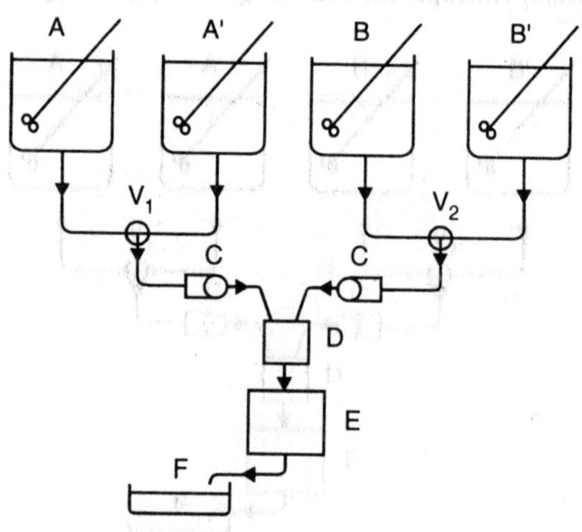

Fig. 18.15. Continuous-processing emulsion plant for long-run products.

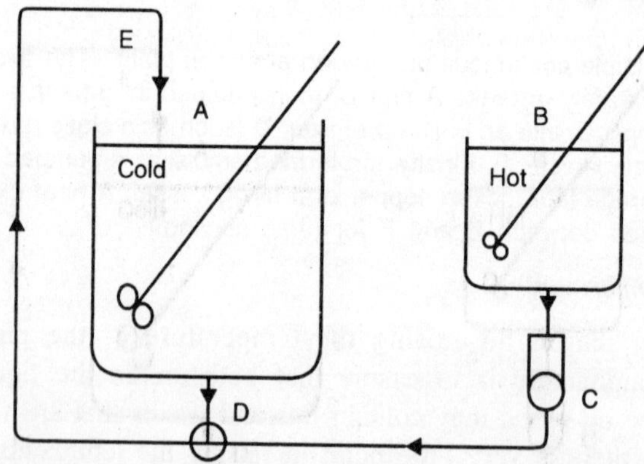

Fig. 18.16. Hot/Cold processing emulsion plant. (The hot oil phase from tank A and the cold water phase from tank B are pumped into an in-line homogeniser at D and thence into the main tank B at E).

Storage of Cosmetic Powders

Two factors have an important effect on stored cosmetic powders: moisture and pressure. It is not always appreciated that a small increase in relative humidity can give rise to sufficient moisture in the stored powder to change the main mechanism of particle-particle bonding, increasing the bond strength of agglomerates by a factor of 2 or more. Such an increase in cohesiveness can make the handling and flow problems already inherent in cosmetic powders perceptibly worse and can change the processing characteristics of (say) an eyeshadow to the point where all the pressing machine settings may have to be altered to compensate.

In the same way, powder bulk that has been stored in large vertical containers exhibits increasingly difficult flow characteristics as the container gradually empties. The lower layers, having been compressed by the weight of powder above them, become increasingly cohesive as the bottom is approached. For this reason it is far better to store powder in a large number of small well-sealed containers than in loosely covered large bins.

Chapter 19

Control of Microbial Contamination

INTRODUCTION

Products of high microbiological quality do not just happen; they are so designed from the earliest stages of manufacture. Frequent occurrence of low-level microbial contamination of the finished product is a warning that the equipment on which it was made may not have been adequately sterilised. Appropriate precautions should be taken against product contamination risks of all kinds. Insanitary processing and filling equipment can also be detrimental to the finished item. Toilet and washing facilities must be appropriately located, designed and equipped.

The presence of micro-organisms in large numbers in cosmetic preparations is undesirable because spoilage of the product may result. The product can change in colour, odour or consistency, or manifest visible growth. Furthermore, the presence of microbial contaminants constitutes a potential hazard to public health, although reports of infections traced to contaminated cosmetics are rare.

SOURCES OF CONTAMINATION

Environment

Control of the manufacturing environment significantly reduces the risk of product contamination. Floors and wall surfaces should be impervious and resistant to antimicrobial agents. Walls should be washed periodically and floors should be washed each night. Puddles, dirt and debris should be removed as soon as practical. Drains on the floor should be kept covered and should receive particular attention as they can easily become sources of contamination. Air flow through production areas should be minimised.

Air ducts, light fittings and other piping can create problems unless they are routinely cleaned. In general, an acceptable and effectively controlled production environment requires all exposed surfaces to be kept clean. Other potential hazards of contamination are cleaning utensils — buckets, mops, brushes, etc., should themselves be kept in a clinically clean condition. Sinks and areas where equipment is washed should be kept clean at all times. Traffic through production areas must be restricted to essential personnel and necessary equipment.

Factory Buildings

Buildings should be constructed to protect as far as possible against the entrance and harbouring of vermin, birds and other pests. The buildings should be effectively lit and ventilated with air control facilities appropriate both to the operations undertaken within them and the external environment. Certain areas should not be used as a general right of way for materials or personnel passing through to other parts of the factory. Laboratories, processing areas, stores and the factory in general should be maintained in a clean, neat and tidy condition, and free from accumulated waste. Waste material should be collected into suitable receptacles for removal to collection points outside the buildings. It should be then disposed off at regular and frequent intervals. The operations carried out in any particular area of the premises should be such as to minimise the contamination of one product by another.

Floors and Walls

Floors should be made of impervious materials, laid to an even surface, and free from cracks and open joints. They should be maintained in a good state of repair. Dust, dirt and other contaminants that may be carried into manufacturing and filling areas on the soles of shoes and on trolley wheels can be collected on special plastic screens that pick up such material on their surface, have good non-slip properties and are easy to maintain.

Walls and ceilings should be finished with a smooth, impervious and washable surface and maintained in a good state of repair. Pipe work should not contain uncleanable recesses and should be effectively sealed into walls and partitions through which it passes. Extraction fans should be sited to avoid cross-contamination hazards caused by intake or exhaust.

Equipment

Hygienic considerations are often neglected in equipment design. Cooperation between production managers and microbiologists is essential and this applies even to minor modifications to existing equipment. Equipment should be made of materials capable of withstanding conventional cleaning methods, such as the use of steam, detergents and chemical antimicrobial agents; the most efficient and widely used material is stainless steel. Other essential components of production machinery such as gaskets, fittings, tubing and rails should also receive proper attention. Hoses that are old, cracked and decaying and equipment with crevices and corners are nearly impossible to clean and sanitise properly. Pumps that are old or frequently taken apart can be a major source of contamination. Such hazards can be avoided by establishing routine cleaning and maintenance procedures. Discretion and a general awareness of all aspects of sanitary hygiene on the part of the microbiologist and the engineering personnel enable appropriate equipment choices to be made and routine maintenance procedures to be established, so that unnecessary clean-up and prolonged stoppages in processing and filling are avoided.

The Manufacturing Process

Potential sources of microbial contamination during the manufacturing process can be summarised as follows :

	Cause of contamination
Storage	Airborne micro-organisms
	Inadequate cleaning of vessel/tank
Filling	Airborne micro-organisms
	Transfer via filling machine
	Transfer via container
	Transfer via operator

Unsuspected sources of contamination

Where all reasonable precautions have been taken but sporadic and serious outbreaks of infection still occur, the source is likely to be an unsuspected reservoir of contamination. The underlying cause is probably among the following :

(1) Poor communication between management and staff.

(2) Poor supervision, especially early in the morning.

(3) Poor hygienic design of equipment or layout.

(4) Changes in cleaning/sterilising procedure, introduced to reduce costs.

(5) Rapid staff turnover.

(6) Assumptions made without verification by laboratory test.

CLEANING AND DISINFECTION

Marketing of elegant cosmetics can be defeated by failure to exercise the same strict control over the cleaning and sanitation of buildings, plant and equipment and over the conditions of bulk manufacture, filling and storage of products. The decisions to be made relate to the labour requirement, the cleaning equipment, the detergents and the standard of cleanliness required. The following important considerations must be taken into account :

(1) The quality and therefore the value of the end product depends essentially on the cleanliness of the production plant.

(2) The sterilisation capacity of the product is a function of the initial microbial count.

(3) The shelf life of creams, lotions and powders that are made from sterilised ingredients is affected by reinfection during manufacture and filling.

(4) Production of pathogenic micro-organisms such as staphylococci, *Pseudomonas*, coliforms, clostridia and *Candida* must be prevented.

Cleaning schedules can only be established by careful study of the problem, an appreciation of what is required, and a knowledge of how cleaning and sterilisation are best carried out.

The Cleaning Logbook

The unexpected problems and unusual situations may occur despite the sterilisation of equipment. The supervisor or manager responsible can quickly make himself familiar with any variation from the expected pattern and take remedial action. When poor microbiological results are obtained from manufactured bulk or filled finished product, reference to the cleaning logbook can help to identify the cause if it lies in a variation in cleaning procedure.

Cleaning Staff

Staff of suitable quality should be engaged, their role in the business should be explained to them and they should be instructed in the type of plant employed in the factory, and the cleaning equipment, procedures and materials to be used. Cleaning staff should be supplied with overalls, rubber gloves, rubber boots and hats. Arrangements should be made to replace this equipment where necessary with adequately clean items in order to minimise the possibility of cross-infection from dirty clothing. The cleaner should be provided with a composite cleaning kit comprising brushes, abrasives, detergents and sterilants. Also advantageous would be an equipment cupboard with glass sliding doors for stocking cleaning equipment; cleaning personnel should be given the responsibility for maintaining it in a scrupulously clean condition.

Many of the cleaning and disinfecting materials are potentially hazardous, so it is essential to instruct and train cleaning personnel to handle them in a completely safe manner. Supervisors must know about the toxic properties and dangers of the chemicals used and be aware of suitable treatments for injuries arising from mishandling.

Equipment Cleaning

It is very important that the cleaning must be effective. For efficient cleaning following points must be taken into account :

(1) The likely contamination occurring on the plant and equipment, and failure to remove it.

(2) The cleaning process should be planned.

(3) Cleaning equipment and its use should be proper.

(4) Use optimum concentration of correct detergent and sterilant concentration.

The method to be employed to clean a specific item of equipment will vary according to the nature of the product being processed and the quality of the surfaces, but certain general principles can be established Fig. 19.1. The equipment should first be dismantled and all product residues removed. Any part of the equipment that comes in contact with the product should be removed and cleaned: all processing vessels, pumps, hoses, valves, storage cans, various hand utensils, and the filling equipment should undergo this cleaning process. In acessible parts of this equipment should be given extra attention. This preparation may take a long time, but it is well worth the effort if a serious contamination problem is thereby avoided.

Detergents

Detergents employed in cleaning plant and equipment must be able: to wet the surface to be cleaned thoroughly, removes the residual product from the surface, holds the removed product in suspension, resists deposition of the residual product on dilution, that is, have good rinsability.

The factors to be considered in selecting a suitable detergent, such as: Prevention of corrosion, emulsfication of dried cosmetic product, prevention of scale formation, quick and complete solubility and, economy in use.

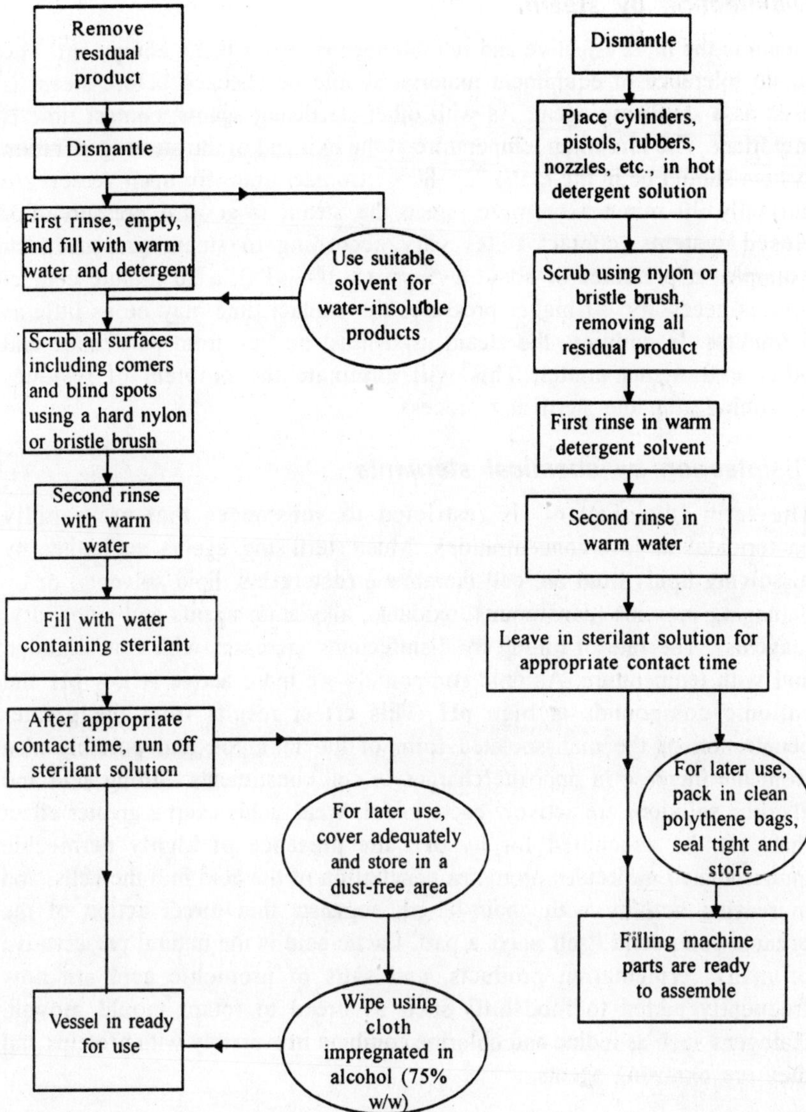

Fig. 19.1. Sequence of equipment cleaning and sterilisation operations.

Equipment Disinfection

Once equipment is thoroughly cleaned, the actual disinfection process can be initiated, using steam or chemical sterilants.

Disinfection by steam

Steam is the most effective and reliable means of sterilising equipment. The steam tolerance of equipment material should be checked before steam is used as a sterilising agent. As with other sterilising agents, contact time is important. The minimum temperature at the exit end of the steam generating system should be in the range 72°–80°C. Contact times for open vessels are normally 30 minutes or more, since the steam is at zero pressure. For closed systems, contact times vary according to steam pressure; for example, at pressures of about 1–5 psi (5–100 kPa), a 20 minute contact time is necessary. At higher pressures the contact time may be as little as 5 minutes. In addition, the steam used must be free from particulate and other extraneous matter. This will eliminate the problem of residues remaining after the sterilising process.

Disinfection by chemical sterilants

The term 'disinfectant' is restricted to substances that are rapidly bactericidal at low concentrations. Most sterilising agents act either by dissolving lipids from the cell membrane (detergents, lipid solvents) or by damaging proteins (denaturants, oxidants, alkylating agents and sulphydryl reagents). The rate of killing by disinfectants increases with concentration and with temperature. Anionic compounds are more active at low pH and cationic compounds at high pH. This effect results from the greater penetration of the undissociated form of the inhibitor, and possibly also from the increase in opposite charges in cell constituents. Strong acid and alkaline solutions are actively bactericidal. Weak acids exert a greater effect than can be accounted for by pH: the presence of highly permeable undissociated molecules promotes penetration of the acid into the cells, and increasing activity with chain-length suggests that direct action of the organic compound itself plays a part. Lactic acid is the natural preservative of many fermentation products, and salts of propionic acid are now frequently added to foodstuffs such as bread to retard mould growth. Halogens such as iodine and chlorine combine irreversibly with proteins and they are oxidising agents.

Alkylating agents

Suitable dilutions of formaldehyde and of ethylene oxide in carbon dioxide (in appropriate gas-tight enclosures) replace the labile H atoms on $-NH_2$ and $-OH$ groups which abound in proteins and nucleic acids, and also on $-COOH$ and $-SH$ groups of proteins. The reactions of formaldehyde are in part reversible, but the high-energy epoxide bridge of ethylene oxide leads to irreversible reactions. These alkylating agents, in contrast to other disinfectants, are nearly as active against spores as against vegetative bacterial cells, presumably because they can penetrate easily (being small and uncharged) and do not require water for their action.

Phenols

Phenol itself is both an effective denaturant of proteins and a detergent. Its bactericidal action involves cell lysis. The antibacterial activity of phenol is increased by halogen or alkyl substitutes on the ring, which increase polarity of the phenolic OH group and also make the rest of the molecule more hydrophobic. The molecule becomes more surface-active and its antibacterial potency may be increased a hundred-fold or more. Phenols are more active when mixed with soaps, which increase their solubility and promote penetration. With increasing chain-length the potency of phenols first increases and then decreases, presumably because of low solubility. With Gram-negative organisms the maximum is reached at a relative short chain-length.

Alcohols

The aliphatic alcohols upto 8–10 carbon atoms having good solubility in water are sterilant-disinfectant. Isopropyl alcohol has the advantages of being less volatile and slightly more potent than ethyl alcohol. The sterilant-disinfectant action of alcohols, like the denaturing effect on proteins, involves the participation of water. Ethyl alcohol is most effective in 50–70 per cent aqueous solution; at 100 per cent it is a poor sterilant, in which anthrax spores have been reported to survive for as long as 50 days, and its bactericidal action is negligible at concentrations below 10–20 per cent. Some organic disinfectants such as formaldehyde and phenol are less effective in alcohol than in water because of the lower affinity of the disinfectant for the bacteria relative to the solvent. On the other hand, alcohol removes lipid layers that may protect skin organisms from some other disinfectants.

Other Chemical Sterilants

Ether, benzene, acetone and chloroform also kill bacteria but are not reliable disinfectants. Glycerol is bacteriostatic at concentrations exceeding 50 per cent and is used as a preservative for vaccines and other biologicals, since it is not irritating to tissues.

Propylene glycol and diethylene glycol reduce the bacterial count in air when dispersed in fine droplets at concentrations non-toxic to man. Their activity is unfortunately highly sensitive to humidity. At high humidities the glycol droplets take up water and became too dilute, while at low humidities the desiccated bacteria no longer attract glycols. Moreover, the glycols do not disinfect surfaces such as floors, walls or working surfaces, from which aerial contamination is renewed. The chemical sterilants mentioned above are widely used in the cosmetics industry. They are generally used on equipment that cannot tolerate steam or to which steam cannot be applied.

Chlorine is commonly used in the food and cosmetics industries. Hypochlorites or chloramines, and has a relatively broad spectrum of activity. It is fairly inexpensive but it is corrosive and is not a cleaning agent. Equipment must be thoroughly cleaned or the chlorine will be inactivated.

Iodine-containing sterilants are useful and can be formulated with detergents in order both to clean and to sterilise. The disadvantages of materials containing iodine are their poor rinsability and their less than satisfactory effect against certain micro-organisms. Staining of equipment components may also be a problem.

Quaternary ammonium compounds are odourless and less corrosive than some of the other chemical agents. Their antimicrobial effectiveness is slightly less than that of alternative materials, but when used at appropriate concentrations and elevated temperatures they can be satisfactory. Quaternary ammonium compounds have the advantage of being non-toxic and are therefore more easily handled than other sterilising agents.

Detergent sterilisers work more efficiently at higher temperatures, and at lower concentration. At lower temperatures, a higher concentration and increased contact time is needed. Different types of surface may require different contact times with the sterilant; smooth, non-porous surfaces require less time than moving parts. Processing systems containing moving parts that are not always in contact with the sterilant will require frequent cycling of the parts during the exposure time to give adequate contact time. Properties of some major classes of antimicrobial chemicals are listed in Table 19.1.

The first step for selection of appropriate sterilant is to review the technical literature and set out its advantages and disadvantages. In the determination of minimum inhibitory concentrations, should be conducted using a wide variety of possible contaminants encountered in cosmetic products. The results will give information on both the spectrum of effectiveness and the potential range of concentration required for the sterilant. From further investigations into temperatures and contact times, a degree of sterilant effectiveness may be established.

Table 19.1. Properties of some major classes of antimicrobial chemicals.

Property	Inorganic hypochlorites	Detergent quaternaries	Detergent iodophors
Broad microbial spectrum	Yes	No	Yes
Activity against pseudomonas	Yes	No	Yes
Product stability	Limited	Yes	Yes
Interference by water hardness	No	Yes	Yes
Interference by alkalinity	Yes	No	Yes
Odour (use dilution)	Yes	No	Slight
Visual indicator of activity	No	No	Yes

Microbial Adoptation

Regular use of a single type of sterilant can result in the development of strains of organisms resistant to the antimocrobial agent in use. Over long periods of time organisms that were once eliminated are found to have adapted and to survive. This adaptation of equipment-borne micro-organisms can be prevented by the rotation of sterilants. Regular use of a different sterilant will eliminate the occurrence of resistant organisms. For instance, a quaternary sterilant may be used for a given period of time, followed by a chlorine-based sterilant for a further period of time. However, if steam is used as the sterilant, rotation is not normally necessary. In order to avoid serious equipment-borne contamination particular attention should be given to the phenomenon of microbial adaptation.

CONTROL OF CONTAMINATION

Hazards from Personnel

Eating, drinking and smoking should be forbidden in all production areas. The high standards of personal hygiene should be observed by all persons concerned in production processes. The direct contact between materials and operators, hands should be avoided by the use of correct and suitable equipment. All operators should wear protective garments appropriate to the process being carried out. The garments should be regularly and frequently laundered. Persons not regularly employed in a production area, whether employees of the firm or not, should wear protective garments where appropriate and necessary. No person known to be suffering from a disease in communicable form, or to be the carrier of such disease, and no person with open lesions or skin infection on the exposed surface of the body, should be employed on production processes.

The bacteria that live on human skin are adapted to the skin habitat and use skin secretions such as sebum and sweat as food; furthermore, pores and cavities function as shelter. Skin bacteria such as *Staphylococcus epidermidis* and *Corynebacterium acnes* adhere to epithelial layers that form the cornified skin surface, and extend between the squames and down the mouths of the hair follicles and glands opening onto the skin surface. These bacteria can only be reduced in numbers, but never eliminated, by scrubbing and washing. They produce odouriferous substances by metabolising secretions of the apocrine sweat glands, giving the body a smell that modern man, at least, finds offensive.

The ability of micro-organisms to spread from one host to another is of great importance for bacterial distribution and their eventual survival. Clearly if micro-organisms do not spread from individual to individual they will die with the host and will be unable to persist in nature. A successful parasitic micro-organism is one that lives on or in the individual host, multiplies, spreads to fresh individuals, leave descendants, and from an evolutionary point of view avoids its extinction and that of its host. On the other hand, if an infection is too lethal or crippling, there will obviously be a reduction in numbers of the host species and thus in the numbers of the micro-organisms. Although a few micro-organisms cause disease in the majority of those infected, it is to be expected that most are comparatively harmless, causing even no disease, or disease in only a small proportion of those infected. A few microbial parasites achieve the supreme success of causing zero damage, thus failing to be recognised as parasites by the host. It is the virulence and pathogenicity of micro-organisms, their ability to kill and damage the host, that makes them important to the medical scientist and to the microbiologist.

Many micro-organisms are effectively transmitted from faeces to mouth after contamination of water used for drinking. All these organisms are potentially pathogenic. They can very easily be transferred and cause contamination and infection. Microbial transmission by the respiratory route depends on the production of aerosols containing micro-organisms. These are air-borne particles produced to some extent in the larynx, mouth, and throat during speech and normal breathing. Harmless commensal bacteria are shed and more pathogenic streptococci, meningococci and other micro-organisms are also spread in this way, especially when people are crowded in small rooms. Micro-organisms in the mouth, throat, larynx and lungs are expelled to the exterior with much great efficiency during coughing; a cough will project bacteria into the air.

Dust, dandruff, hair, skin particles, dried sputum and droplets from coughing and sneezing have all been shown to harbour all types of bacteria associated with the human body, and these can obviously be transferred to products.

Washrooms and Toilets

Stringent procedures should be set up and adhered to for the cleaning and disinfection of sinks, toilets, showers, tools, lockers and floors. The use of tablet soap and roll towels for handwashing should be discouraged and discontinued. Instead, antiseptic liquid soap and disposable paper towel dispensers should be installed. Work clothing should be provided to employees and freshly laundered garments should be made available on a regular and frequent basis. Dirty clothing cannot be overlooked as a source of contamination and therefore every effort should be made to eliminate it.

Bacteria originating from faeces are often found in large numbers on all surfaces even after cleaning of washrooms and toilets daily. These areas readily come into contact with people. One third by weight of faeces is bacteria, many of which are alive. The areas likely to cause cross-infection are the toilet seat, wash basin, overflow tap, handles, inside handle of the entrance door. Splashing of the water in the toilet pedestal during defecation is another potential source of cross-infection; this varies with cistern height and bowl design. Thus simple cleaning and the maintenance of a good flushing cistern are the main principles of toilet care. Flush handles can be a source of transmission of contamination from one individual to another and foot-operated pedals for flushing are recommended.

Raw Materials

Ingredients used in cosmetic formulae should be prevented from infection until their quality is determined. Those found to be acceptable should be protected from contamination and should be examined microbiologically on a routine basis. A major raw material in cosmetic products is water. If contaminated, the entire water system should be shut down, emptied, thoroughly cleaned, and disinfected by disinfectant solution circulating throughout the entire system, including all feed lines, tanks, and resin beds. After disinfection the system should be drained and flushed with clean water until all traces of the disinfectant solution have been rinsed away. On the other hand if a reduced bacterial count is detected on routine testing, and found to be increasing, it may be sufficient to inject a small quantity of disinfectant solution as a prophylactic measure in order to control the bacterial count. The water supply may be a major source of *Pseudomonas* contamination, and where ambient temperatures exceed 18°C few bacterial cells should be allowed to reach over 10^6 bacteria per gram. Raw materials of natural origin, such as the natural gums tragacanth and acacia, may be very heavily contaminated and so may kaolin, chalk and starch. Significant microbial contamination has been reported in bentonite, Quaternium-18, hectorite, sodium magnesium silicate and aluminium powder.

There are a number of methods for taking samples aseptically for microbiological examination of raw materials. A sample is scooped or pipetted out from the raw material container, placed in a sterilised sample container and returned to the laboratory for immediate testing. The raw material container should immediately be sealed.

If, after completion of all microbiological tests, the material is accepted, it is then removed from the holding quarantine area and placed in stock. The excessive humidity and large fluctuations in temperature from ambient may effect the physical, chemical and microbial properties of raw materials. Raw materials that are retained for long periods of time should be retested at specific intervals (at least every six months). Raw materials that are purchased pre-sterilised to assure the microbiological quality should also be tested to verify: (i) that the sterilisation procedure was effective; and (ii) that post-sterilisation contamination has not occurred.

Storage Areas

All raw materials should be stored in such a way that the tested degree of microbial purity is maintained. Storage areas for raw materials should be maintained in as clean a condition as bulk product manufacturing and product filling areas. The goods such as raw materials, packing materials etc. should be stored off the floor. This will enable them to maintain clean, dry and orderly condition.

Product Packaging

The product filled into container is protected if it is closed efficiently. The seal should be secured and tight. This protection may be described as a criterion of the physical preservation of the product. Packaging components such as jars, bottles, tubes, container closures and closure liners should not be ignored as sources of contamination. Before use if possible they should be air-blown with dry, clean air to rid them of any extraneous particles and stored in clean, dry areas with efficient packing. It is recommended strongly that air-blowing is carried out under an extraction hood with vacuum facilities so that the particles blown off are immediately extracted from the filling area. If possible, the supplier should be persuaded to supply the packaging components in sealed plastic bags which should be opened just prior to use. This will obviously avoid the complications of air rinsing.

Microbiological Standards

Microbiological standards serve several quite distinct purposes :

(1) Control of the danger from pathogenic organisms.

(2) Assurance that the cosmetic product has never been grossly contaminated.

(3) Confirmation of a reasonable expected life in storage—that is, an estimate of perishability.

The usual guidance given to the industry is to exclude 'named pathogens' from cosmetics. With few exceptions, it is doubtful whether a single bacterial cell ever did anyone any harm. It appears to be necessary for growth that at least a few organisms should establish themselves and adapt to their environment.

The need to ensure that the cosmetic product has never been seriously contaminated emphasises the importance of quantitative tests in microbiological control. An attempt should always be made to assess the number of organisms present, even though the error may be large. The reason for this is that the onset of an infection, or the development of a microbiological defect in a product, is dependent on the number of organisms originally present.

Microbial specifications within the cosmetics industry are essential aids in maintaining the sanitary quality and stability of the end product, when used to control raw materials, processes and factory hygiene.

If equipment has been properly cleaned and sterilised the number of bacteria left will not exceed one per cm^2 by a swab test or one per ml by a rinse test. These tests are therefore quite adequate to assess the efficiency of cleaning in a general sense. It can be assumed under ordinary working conditions that if the results are satisfactory (that is, less than 1 colony per cm^2 or ml) then all pathogens have been killed or removed. It is also unlikely that micro-organisms will have survived in sufficient numbers to cause trouble.

The cleanest factory conditions can easily be invalidated if unhygienic methods are practised; microbial contamination will only be avoided by careful attention to all aspects of production and control. Disregard for any of the following points results in weaknesses and, eventually, costly complications.

(1) Check the quality of raw materials.

(2) Unpack raw materials in a separate building, especially if embedded in sawdust, cotton waste, straw, etc.

(3) Apply a biocidal treatment where necessary if this is practicable.

(4) Control the quality of water used in the factory.

(5) Check the bacterial purity of the air near fillers, etc.

(6) Check the hygienic condition of all containers.

(7) Remove all residue, broken or split containers, etc. as soon as possible.

(8) Do not use cloths for mopping up spillages, unless these are maintained in a clean condition; paper towels are much better.

(9) Thoroughly clean equipment immediately after use.

(10) Sterilise or cleanse all equipment, as may by necessary, immediately before use.

Common Fallacies in Hygiene

It is not true that :

(1) Splashing disinfectant over floors, etc., 'solves the hygiene problem'.

(2) Forcing steam round a circuit, with the production of great clouds of 'steam' and considerable noise, necessarily sterilises the equipment.

(3) If equipment looks clean, then it must be clean—this is not true microbiologically.

'Window-dressing' devices, such as making people wear white coats and caps, provision of glaring uv lamps, use of disinfectant aerosols, may serve some useful purpose and undoubtedly exert a psychological influence, but in real terms their value is not great.

The following phenomena are common in insufficiently preserved cosmetics: creams and lotions can exhibit visible mould colonies; products can separate with or without the production of a rancid or putrid odour; discolouration can occur; clean preparations can become turbid owing to precipitation; fermentation can produce gas, sometimes causing tubes to swell or glass bottles to burst. The result is that a product manufactured at high cost and believed to have received all possible care has become unsaleable.

The sanitary quality of a cosmetic product is normally the summation of the level of sanitation in the production of raw material ingredients, finished raw materials and cosmetic product manufacturing plant. The definition of the term cosmetic is 'any substance or preparation intended to be placed in contact with various superficial parts of the human body (epidermis, hair system, nails, lips and external genital organs) or with the teeth and mucous membrane of the oral cavity with a view exclusively or principally to perfume them, clean them, protect them and keep them in good condition, to change their appearance or to correct body odours'. As such, cosmetics demonstrating poor microbiological quality have a potential for being 'injurious to health' with unpredictable frequencies and at unpredictable times.

Chapter 20

Packaging of Cosmetics

INTRODUCTION

The packaging of any product and cosmetics have same principle. However, in case of cosmetics for successful marketing more importance is given to package design and development.

Packaging utilises a wide variety of materials such as plastics, glass, paper, board, metal and wood combined with wide range of technologies including printing, machinery design and tool making.

Packaging has been defined as the means of ensuring the safe delivery of a product to the ultimate consumer, in sound condition at the minimum overall cost. Other definitions are :

(1) Packaging is the art or science of, and the operations involved in, the preparations of articles or commodities for carriage, storage and delivery to the customer.

(2) Packaging sells what it protects and protects what it sells.

PRINCIPLES OF PACKAGING

Packaging must contain the product, restrain the product, protect the product, identify the product, sell the product and give information about the product, and do this within a cost related to the marketing, profit margin, selling price and image of the product.

Marketing and Packaging

The package projects the style and image, not only of the product, but often of the company which markets the brand. The pack must therefore project

368

the image that it has been designed for, not only to the customer through advertising and point of sale but also to the retailer and wholesaler chain.

Packaging is particularly important in the self-service retail trade. The package designer has a responsibility to ensure not only that the pack has the type of appeal that will make the customer pick it up and be encouraged to purchase on impulse, but also to ensure that the pack will stack on self-service shelves and give the retailer the maximum profit per linear unit of shelf space.

Advertising has ensured that the package is now more widely seen than ever before. With the predominance of colour in advertising — in television, cinema and press advertisements and on posters — the package must be made of materials that have good aesthetic appeal and which will take the hold colour.

TECHNOLOGY AND COMPONENTS

Plastics

The use of plastics for producing primary components and point-of-sale material now dominates packaging technology. Two main groups are used—thermoplastic resins and thermosetting resins. Thermoplastics can be extruded at their melt temperature and then blow moulded or injection moulded. After cooling, the resin can be remelted by heating to the limits of thermal fatigue and oxidation. Thermosetting resins, by contrast, are moulded using an irreversible chemical reaction and the resins tend to be rigid, hard, insoluble and unaffected by heat up to decomposition temperature.

Metals

Metals are suited in cosmetics packaging are aerosol containers, powder dispensers, shallow tins and collapsible tubes. Tin plate is the most commonly used metal for the rigid packs, although aluminium also finds widespread use. The impervious nature of the metal gives the collapsible tube the great advantages of reduced risk of contamination of the product and reduced losses of volatile materials from the contents.

Laminates

The various requirements of packages for cosmetics and toiletries (such as attractive appearance; impermeability to water and volatile oils) are not always available from a single material. In such case the use of composite material in laminar form can be made. Laminates have found particular application in the production of sachets and of collapsible tubes as alternatives to pure metal tubes for toothpastes.

Laminates are used for *flat sachets* that are heat sealed around the periphery (as distinct from the fatter *pillow sachets* which are made from PVC tube and sealed ultrasonically). The laminate must be able to withstand the pressure of the contained product and provide leak-proof seals.

Glass

Glass containers are still used widely in the toiletries industry by virtue of the basic packaging characteristics of glass. Glass is chemically very inert and generally will not react with or contaminate high quality cosmetic and perfume products. Glass is manufactured in many different formulations but the most common in packaging is soda lime glass.

Paper and Board

Practically every cosmetic and toiletry products uses paper or paperboard in some form. Many grades of paper and board are available. Uses of paper and board in cosmetics packaging include labels, leaflets, corrugated cases, printed cartons and soap wraps.

Printing and Decoration

All packaging components can be printed to give a wide range of decorative effects. Different processes are used depending on the application. The five main processes used in the printing of packaging components are screen printing, letter press, flexographics, offset lithography and gravure printing.

PACKAGE DEVELOPMENT AND DESIGN

Package development has the aim of increased sales and profit through the correct design of the package. Packaging must be considered as early as possible in the development of a new product to allow time to ensure that pack and product are compatible. The development process begins with a detailed analysis of the product so that a pack can be designed to give protection. Finally, in the development of the package the environmental aspects of the pack should be considered in terms of disposability and litter. It should also consider the re-use of scarce raw materials.

Technical Aspects of Design

The packaging material must be sufficiently strong to survive any treatment that it is liable to receive from the time to first delivery at a factory, through to filling, distribution, sale and actual use. It is essential that failure rate due to accident must be very low. Hence very strict testing and quality control of packs is essential in a factory producing toiletries or cosmetics.

A pack must be attractive to the consumer and contain the product in an efficient manner and render the product available as soon as the consumer desires to use it.

Closures

Unless it has an efficient closure, container has no value. Ideally the closure should be easy to remove and replace. However, it should give a seal that prevents the diffusion of gases and vapours and the seepage of liquids. Many bottle and tube closures consist of a screw cap containing a compressible wad. But wadless closures—achieved by new designs of thermoplastics screw and snap-on—are now extensively used for liquid products such as shampoos and hair conditioners.

PACKAGE TESTING AND COMPATIBILITY

Testing

The testing of packaging material and finished pack is required :

(1) For effective material selection;

(2) For assessing performance of material; and

(3) For checking quality.

The packaging industry examines the properties like, mechanical, physical chemical and compatibility.

Mechanical properties for example, compression, tensile, flexural and impact strengths. *Physical properties* for example, water absorption, moisture vapour transmission rates, accelerated aging, flammability and thermal conductivity. *Chemical properties* for example, resistance to the product or the chemical environment, and corrosion testing. *Compatibility testing* is performed when the final product formulation and packaging system have been decided. Samples of the product should ideally be taken from trial batches, and the complete packaging system should be assembled using actual samples or pilot tooling samples representing the final component.

The general compatibility of the pack and product needs to be checked by storage testing which will enable an assessment to be made of the effect of the pack on the product as well as that of the product on the pack. It is important to remember that the effect of spillage on the outside of the pack is important, and should certainly be included in any testing schedule.

Shelf-life testing is also necessary in order to determine the rate at which volatiles may be lost and this includes not only water or alcohol but the perfume. The assessment of loss of perfume can only sensibly be done by nose, although weighing will obviously allow determination to be made of the loss of solvents.

Finally, the convenience of any cosmetics pack must be tested by in-use tests as well as by laboratory tests and for this purpose it is necessary to take into account both the local customs and the climate of the country in which the product is to be marketed: for instance, bathrooms in the UK tend to be cold, and tests for pourability should reflect this; similarly, products for tropical countries should be tested in conditions of high temperature and humidity.

Chapter 21

Skin Creams, Astringents and Skin Tonics

INTRODUCTION

The human body is protected from the surrounding the skin. The entry of radiations, bacteria, fungi and water is controlled by the dead outer layer of the skin. The skin also absorbs oxygen (it does breathe), eliminates waste, secretes protective lubricants, grows hair, and manufactures new skin cells at a furious rate. No simple organism could do all this.

The layer of dead corneal cells does not feel like a rough, splintered board because it is coated with an oily liquid secreted by underlying sebaceous glands and by moisture from sweat glands. This film coats the scales, smoothing and sealing them. The fatty secretions and sweat combine to create an "acid mantle" of pH 4–6 on the skin. This mantle— or film – keeps the skin's surface supple and regulates the moisture content of the deeper as well as the outer skin strata. Most dermatologists believe that the mantle also inhibits the growth of bacteria and fungi on the skin's surface, most of which flourish at higher pH (6.0–7.5) and moisture levels than those of the corneal layer. This may explain why people whose skin cannot maintain normal pH and moisture ranges are prone to skin problems.

Age and Skin

As the body ages, its functions change and its needs alter. Some cosmetics can meet the skin's changing needs and cover ageing effects. In no case do they "cure."

In old age the output of sebum from the sebaceous glands drops and the skin dries out. The skin becomes flaky and itchy, particularly on the lower legs. Household cleaning products, which can strip sebum off the skin, can cause a similar condition in younger people. Furthermore, since many household cleansers are alkaline, they can swell and damage the keratin of the corneal cells.

Wrinkles are another change. Wrinkles occur when the connections between the epidermis and the dermis loosen. No known cosmetic can eliminate wrinkles by filling it in or by swelling, pinching, or oiling the skin. However it makes unnoticeable.

Cosmetics and Skin

Most skin cosmetics fall into two categories : those with a purely decorative effect, and those which have subsurface effect. The decorative cosmetics, such as eye make-up, face make-up, and fragrances, should affect only the surface of the skin. A good product should not irritate or sensitise the skin, or cause infections, or harm the health of the user in any other way. On the other hand, no special benefits to the skin can be expected from even the best of decorative skin cosmetics. Cosmetics with a subsurface effect, such as moisturisers and cleansers, are usually intended to alleviate or change some skin condition. Since judgements concerning the effectiveness of make-up or fragrances are largely subjective the chapters on decorative cosmetics will focus on alerting consumers to potential health problems associated with the use of such products. In our treatment of subsurface beauty preparations, we will show how they act upon the skin to help users evaluate their efficacy and safety.

SKIN CREAMS

In the context of cosmetics, the term 'cream' usually signifies a solid or semi-solid emulsion. It is also applied to non-aqueous products such as wax-solvent-based mascaras, liquid eyeshadows and ointments.

Classification of Skin Creams

Traditionally, cosmetic creams have been marketed and sold on the basis of their 'function'. Thus customers have come to learn what type of emulsion they can expect from a jar marked 'cold cream' or 'night cream'. The customer, therefore, is likely to make her own judgement on the basis of subjective features, using the manufacturers' functional labels as a guide to end-use and quality. The characteristics of various skin creams are given in Table 21.1.

Table 21.1. Characteristics of skin creams.

Functional	Physico-chemical	Subjective
Cleansing creams	Medium-to-high oil content	Oily
Cold creams	Oil-in-water or water-in-oil	Difficult to 'rub in'
Massage creams	Low slip-point oil phase	May be stiff and 'rich'
Night creams	Neutral pH may contain surfactants to improve penetration and suspension properties	Also popular as lotions
Moisturising creams	Low oil content	Easily spreadable and 'rub in' quickly
Foundation creams	Usually oil-in-water	
Vanishing creams	Low slip-point oil phase Neutral to slightly acidic pH may contain emollients and special moisturising ingredients	Available as creams or lotions
Hand and body protective	Low-to-medium oil content Usually oil-in-water	Easily spreadable but do not 'rub in' with the ease of vanishing creams
	Medium slip-point oil phase May have slightly alkaline or acidic pH may contain 'protective factors', especially silicones and lanolin	Very popular in lotion form
All-purpose creams	Medium oil content Oil-in-water or water-in-oil	Very often slightly oily but should be easy to spread

Cleansing creams

The regular cleaning removes grime, sebum, and other secretions and keeps skin healthy. Water is a very cheap and effective cleansing agent for certain types of facial soil but fails to remove oils. By the process of emulsification, soaps and other detergents are able to improve the cleansing properties of water dramatically. However, this combination suffers from disadvantages: it is inconvenient to use outside the bathroom, and it may remove too much oil from the surface, leaving it feeling dry and rough. The alkalinity of soap, may cause the outermost cells to lift and separate from their neighbours.

Cleansing creams and lotions can, by a combination of water and the solvent action of oils, effect the cleansing of the skin surface efficiently and pleasantly. Moreover, if they are properly formulated they can accomplish this without completely degreasing the skin (indeed, by leaving behind a very thin emollient layer of clean oil, they may give a healthy supple feel to the skin surface).

A cleansing cream or lotion is spread onto the skin, using the finger-tips, and massaged onto the surface. This action serves to loosen and suspend the grime and soil in the emulsion. A subsequent wipe with a tissue or cotton wool pad removes the majority of the applied cleansing emulsion, and with it the skin soil, grime or make-up. The cream should have a medium-to-high percentage oil phase and easily spreadable, should not 'rub in' and should not irritate the skin. In addition, it can also leave a residual emollient film on the skin, so much the better.

Today, beeswax–borax emulsions are still popular and commercially viable, although the development of secondary or alternate emulsifiers has enabled the formulator to produce a wider range of emulsions around the beeswax–borax theme.

Beeswax itself suffers from two disadvantages as an ingredient in skin creams. It has a distinctive smell which usually has to be masked in the final product. The beeswax odour is not unpleasant, but it is not compatible with the sophisticated skin-care image which many manufacturers now-a-days build into their products. Nevertheless, the qualities of neutralised beeswax as an emulsifier are such that it will continue to be used in cleansing and cold creams for the foreseeable future.

The sodium salts of wax acids are formed when borax solution is mixed with molten bees-wax at the oil water interface. A stable and textured cream is obtained using borax quantity less than theoretical value, i.e. 5–6%. The amount of borax–neutralised beeswax in a cold cream can vary from 5 to 16 per cent. The lower levels produce softer creams which can be stiffened (if required) by incorporating other waxes. Formulae 1, 2 and 3 are all water-in-oil emulsions.

| *Formulae* | *1* | *2* | *3* |
Constituents	*per cent*	*per cent*	*per cent*
Beeswax	5.0	16.0	12.0
Mineral oil	45.0	50.0	–
Borax	0.2	0.8	0.5
Micro-crystalline wax	7.0	–	–
Spermaceti	–	–	12.5
Sesame oil	–	–	40.0
Water	32.8	33.2	35.0
Perfume, preservative	q.s.	q.s.	q.s.
Paraffin wax	10.0	–	–

Alternatives to waxes as thickeners for a continuous oil phase are bentones (quaternary hectorites and related chemical species); formula 4 is a product of this type :

Formula 4

Constituents	*per cent*
Beeswax	12.0
Mineral oil	53.0
Quaternium-18 hectorite	0.7
Borax	0.7
Water	33.2
Isopropanol	0.4

Procedure: Mill the bentone with the isopropanol and some of the mineral oil. Heat the resultant gel with the beeswax and the remainder of the oil to 75°C. Dissolve the borax in the water, bring to 70°C and slowly pour into the oil phase with stirring. Continue stirring to 45°C, adding perfume at a late stage.

It is a peculiarity of the beeswax–borax system that both water-in-oil and oil-in-water creams may be produced without the aid of secondary emulsifiers. The ratio of oil to water, the proportion of the beeswax that is saponified, the constituents of the cream (which will affect the HLB requirement) and the temperature have influence on the type of emulsion.

It has been found that preparation at high temperature tends to produce cold creams of the water-in-oil type and phase inversion may occur during processing. Phase inversion also occurs on the skin when an oil-in-water emulsion is spread onto the skin surface and the water phase begins to evaporate.

Non-ionic emulsifiers like sorbitan fatty acid esters can be used to increased flexibility and stability to the emulsion. Formulae 5 and 6 illustrate the point; formula 5 is a water-in-oil emulsion whereas 6 is oil-in-water.

Formulae	*5*	*6*
Constituents	*per cent*	*per cent*
Beeswax	10.0	10.0
Mineral oil	50.0	20.0
Lanolin	3.1	3.0
Borax	0.7	0.7
Hydrogenated vegetable oil	–	25.0
Antioxidant	–	0.5

Sorbitan sesquioleate	1.0	–
Sorbitan stearate	–	5.0
Polysorbate	–	2.0
Water	35.2	33.8
Perfume, preservative	q.s	q.s

If required, the external phase of oil-in-water cleansing creams may be thickened by the use of cellulose derivative alginates and other hydrocolloids formula 7.

Formula 7	
Constituents	*per cent*
Beeswax	8.0
Mineral oil	49.0
Paraffin wax	7.0
Cetyl alcohol	1.0
PEG-15 Cocamine	1.0
Borax	0.4
Carbomer-934	0.2
Water	33.4
Perfume, preservative	q.s.

A number of beeswax derivatives are manufactured with modified emulsifier properties, for example, a range of ethoxylated beeswax derivatives available with HLB values ranging from 5 to 9. Although they still have the beeswax odour, it is claimed that creams made from them are softer, liquefy readily, allow the incorporation of larger amounts of water, are neutral and are stable at 50°C. Formulae 8 and 9 are both oil-in-water creams.

Formulae	*8*	*9*
Constituents	*per cent*	*per cent*
Mineral oil	50.0	50.0
Beeswax	–	7.0
PEG-8 Sorbitan beeswax	12.0	–
PEG-20 Sorbitan beeswax	3.0	8.0
Polysorbate-40	–	2.0
Water	35.0	33.0
Perfume, preservative	q.s.	q.s.

Night and massage creams

Night or massage creams are designed to be left on the skin for several hours or to remain mobile on the skin even after vigorous rubbing. They are composed with a substantial oil phase which will spread easily without disappearing but also without rubbing off onto clothing or bed linen in use. Such creams tend to be high-oil-content, water-in-oil, soft solid or viscous liquid creams.

A moisturising effect is resulted due to formation of the occlusive layer on the skin surface thereby reducing the rate of trans-epidermal water loss. Hence the skin surface feel smooth by lubricating action and allowing any 'saw tooth' cells in the outer layer of the stratum corneum to be smoothed down.

Massage has a valuable part to play in skin care. It is well-known that vigorous rubbing of the skin helps to prevent the build-up of excessive numbers of dead surface cells and keeps the epidermal blood supply in good condition.

The term 'moisturising' has also been applied to water-in-oil creams of this type. The recent research has broadened the concept of moisturising from the simple occlusive skin barrier principle. Many modern moisturising creams are comparatively light and easy to rub in compared with those of the overnight and massage types, although there still remains a market for the heavier moisturising creams.

Vitamins in skin creams

It has been found that fat-soluble as well as water-soluble vitamins are capable of being taken up through the skin. The use of stabilised vitamins in cosmetic preparations for external application is justified.

Pantothenic acid, a part of the water-soluble vitamin B complex. Its precursor and the related materials panthenol, pantethine and pangamic acid have all been quoted as having a beneficial action on the skin and being useful in skin and/or hair preparations. Although there is no certain proof that they penetrate the skin and reach the location where they might exert an influence, vitamin B complex, panthenol and vitamin B_6-pyridoxin are used in some cosmetics.

Vitamin D, like vitamin A, is oil-soluble and is essential for skin health, but deficiencies are best corrected by oral administration to achieve a systemic effect. However, vitamins D_2 and D_3 (calciferol) are used, sometimes in conjunction with vitamin A. A mixture of vitamins A, E and D_3 has been claimed to be synergistic.

Vitamin E is said to enhance percutaneous resorption, and vitamin H is claimed to help fat and cholesterol synthesis.

Other vitamins having some use in topical preparations include the so-called vitamin F, now known as essential (unsaturated) fatty acids (EFA), which can undoubtedly heal the skin symptoms characteristic of rats brought to a chronic state of EFA deficiency. The relevant point here is that it would be virtually impossible to bring the human to a corresponding EFA-deficient state.

Oil phase constituents

The predominant oil phase constituents in massage and night creams are petrolatum, mineral oil, lanolin and low-melting-point waxes such as beeswax and low-melting mineral waxes (ceresins and paraffin). Esters such as isopropyl palmitate, iso-propyl myristate and purcellin oil are reserved for lighter 'vanishing cream' types of product.

Moisturising, vanishing and foundation creams

As the term 'vanishing' implies, creams and lotions falling within this category are designed to spread easily and to seem to disappear rapidly when they are rubbed into the skin.

Moisturisers

Of all the beneficial properties claimed for cosmetic creams, 'moisturising' is possibly the most widely used. Water is the only material which will plasticise the outer dead layers of the epidermis to give the much desired attribute we call 'soft, smooth skin', hence the turn moisturiser.

If water is lost more rapidly from the stratum corneum than receiving from epidermis, the skin dehydrates. Hence loses flexibility which can not restored by oil alone.

Dry skin is of two types. (i) It is due to prolonged exposure to low humidity and air movement, which modifies the normal hydration gradient of the stratum corneum; and (ii) It is due to ageing, continual degreasing. Ageing occurs due to influence of ultra-violet rays.

The dry skin is hydrated using three different routes—occlusion, humectancy, and restoration of deficient materials—which may be (and often are) combined.

Occlusion

Occlusion reduces the rate of transepidermal water loss through old or damaged skin or in protecting otherwise healthy skin from the effect of a severely drying environment. It instantly decreases the rate of water loss through the epidermis. This results the stratum corneum to become more hydrated, making it softer and more supple. However, this extra hydration increases the diffusion coefficient of water across the epidermis. After application of petroleum jelly within three hours rate of water loss actually increases to a value higher than the pre-treatment value. (This, of course, in no way detracts from the usefulness of this approach to moisturisation, since it achieves the desired hydration of the stratum corneum.) Examples of occlusive are : mineral and vegetable oils, lanolin and silicones. Their effect is increased by the use of mixtures of lipids and other fatty chemicals which have been designed to imitate the composition of the skin's natural oily secretions.

More recently, skin substantive barrier materials (mainly based on quaternary ammonium complexes) have become available which seem to be able to influence the rate of transepidermal water loss without putting an occlusive or greasy barrier layer on the surface. These materials can be shown to be substantive to skin (and hair) and act not only as moisturisers, but as emollients and skin conditioning agents. Examples are quaternium, a hydroxyethyl-cellulose derivative.

Humectants

A second approach to the moisturising problem is the use of humectants to attract water from the atmosphere, so supplementing the skin's water content. Although popular in use, such a concept is, to say the least, somewhat doubtful from the physiological viewpoint. It is, after all, easy to demonstrate that externally applied water will not increase the flexibility of the stratum corneum — in fact, it can have precisely the opposite effect.

The humectants most frequently used as moisturisers are glycerol, ethylene glycol, propylene glycol and sorbitol. They can be used alone or in admixture at various levels. Whether or not they penetrate the skin surface is a moot point, but at least they will attract moisture to the skin.

Moisturisation

The moisturisation of skin is to determine the precise mechanism of the natural moisturisation process, to assess what has gone wrong with it in the case of dry skin and to replace any materials in which such research has shown damaged skin to be deficient.

Emollients

'Emollience' is another ill-defined term often used in connection with skin creams. Emmollient means imparting of a smoothness and general sense of well-being to the skin, as determined by touch. (In a sense, therefore, water is an emollient).

Every liquid, semi-solid or low-melting-point solid of a bland nature and cosmetic quality has been used as emollients. Among the most popular water-soluble emollients are glycerine, sorbitol, propylene glycol, and various ethoxylated derivatives of lipids. Oil-soluble emollients include hydrocarbon oils and waxes, silicone oils, vegetable oils and fats, alkyl esters, fatty acids and alcohols, together with ethers of fatty alcohols (including polyhydric alcohols). The choice is determined by personal preference, data on potential skin irritation, the degree of 'greasiness' and apparent residual film on the skin, cheapness and availability.

Mineral oils and silicone oils do not 'disappear' from the skin very readily when used in any quantity and are therefore useful, in cleansing and night creams. Propylene glycol is an efficient preservative against certain micro-organisms at concentrations of more than 8 per cent, but it is a potential sensitiser.

The alkyl esters represent a range of interesting emollients ranging, as they do, through lactates, oleates, myristates, adipates, linoleates with the possibility of straight-chained, branch-chained, unsaturated, or saturated precursors. Some are almost water-thin liquids which rub quickly into the skin (decyl and isodecyl oleates, isopropyl myristate) and others are waxy solids which melt near body temperature and give 'body' to creams. Lanolin was considered once to be an extremely desirable emollient and the claim 'contains lanolin' was felt to be a product 'plus'.

Vanishing and foundation creams

In order to achieve their rapid 'rub-in' effect, vanishing creams are composed, in the oil phase. A low percentage oil phase is usually chosen of emollient esters which leave apparent film on the skin.

Foundation creams possess many of the same properties. These creams are for day-time use to protect and 'condition' the cleansed skin. They must therefore leave the skin surface non-greasy and preferably matt so that other make-up can easily be applied over it. Modern foundation creams are of excellent appearance and stability. They contain emollients and moisturisers. Sun-screen agent present helps to protect the consumer's skin from the harmful, ageing effects of short-wave solar radiation. A typical 'simple' vanishing cream formula is as under :

Formula 10

Constituent	*per cent*
Stearic acid	15.0
Potassium hydroxide	0.7
Glycerine	8.0
Water	76.3
Perfume, preservative	q.s.

Pigmented foundation creams

Pigmented foundation creams can contain from 3 to 25 per cent of pigments. Those with between 3 and 10 per cent can form a suitable substrate for the subsequent use of powder, whereas those with higher pigment concentrations can be used as complete make-up and are often termed powder creams. They can be water-continuous or oil-continuous systems in liquid or solid form. The difficulties encountered in the preparations are: (i) The preferential absorption of emulsifier in the high surface of the pigment may sometime cause inversion of emulsion; (ii) the inadequate dispersion of the pigment for reproducible colours. Pigments can be suspended by the use of cellulose derivatives or inorganic silicates such as bentonite or hydrated magnesium silicate. The basic types of tinted foundation cream are illustrated by the following examples.

Formulae (Water-in-oil creams)	*11*	*12*
Constituents	*Solid* *per cent*	*Liquid* *per cent*
Light mineral oil	4.0	30.0
Isopropyl myristate	8.0	–
Lanolin	–	8.0
Ceresin	19.2	–
Micro-crystalline wax	–	1.0
Sorbitan sesquioleate	2.8	2.3
Polysorbate	–	0.1
Powder base	q.s.	8.0
Titanium dioxide	3.0	–
Glycerine	–	5.0
Water, perfume, preservative to	100.00	100.0

Formulae (Oil-in-water *solid creams)*	*13*	*14*	*15*
Constituents	*per cent*	*per cent*	*per cent*
Mineral oil	30.0	–	–
Stearing acid	3.0	8.0	12.0

Isopropyl palmitate	–	–	1.0
Glyceryl stearate	3.0	–	–
Sorbitan stearate	–	–	2.0
Polysorbate-60	–	–	1.0
Cetyl alcohol	2.0	–	–
Triethanolamine	1.0	–	–
Glycerine	–	10.0	–
Sorbitol	–	–	2.5
Propylene glycol	–	–	12.0
Lenette wax	–	8.0	–
Pigment and powder base	5.0	10.0	11.0
Water, perfume, preservative, etc.to	100.00	100.00	100.00

Formula 16 (Oil-in-water liquid make-up)

Constituents	per cent
Mineral oil	20.0
Cetyl alcohol	1.0
Spremaceti	1.0
Sodium lauryl sulphate	0.5
Glyceryl stearate	1.0
Bentonite	2.5
Powder base and colour	8.0
Water, perfume, preservative, etc.	to 100.0

The foundation preparations which do not contain water, for example :

Formula 17

Constituents	per cent
Sesame oil	64
Zinc oxide	11
Oxycholesterol	2
Triglyceryl stearate (polyglyceryl-3-stearate)	1
Perfume and colouring	6
Titanium dioxide	16
Preservative	q.s.

Hand creams and hand-and-body creams

The hands are the main unprotected area of the body other than the face. It is important that the skin which covers them should remain soft and smooth. The main features of good hand creams or lotions are therefore that they should be easy and quick to apply without leaving a tacky film, They should also soften the hands and perhaps help them to heal without interfering with normal hand perspiration. They are usually coloured and are lightly perfumed to make their use pleasant.

The most popular healing agent is allantoin (2,5-dioxo-4-imidazolidinyl-urea) one weak metal complexes such as aluminium dihydroxy alltoinate. Additionally, some of the newer, skin-substantive quaternary salts have also been shown to have healing and soothing effects—for example, Quaternium, a hydroxyethylcellulose derivative.

Further examples of formulations suitable for the starting-point of a hand lotion or cream are given below.

Formula 18

Constituents	*per cent*
Glyceryl stearate SE	2.7
Cetyl alcohol	1.5
Dimethicone	1.5
Lanolin oil	2.0
Squalene	3.0
Sodium lauryl sulphate	0.3
Water, perfume, preservative	q.s.

Formula 19

Constituents	*per cent*
DEA-oleth-3 phosphate	0.5
Lanolin alcohol	1.0
Mineral oil	4.0
Stearic acid	1.0
Glycerine	3.0
Triethanolamine	0.5
Carbomer 941	0.1
Dimethicone	1.0
Water, perfume, preservative	q.s.

Formula 20

Constituents	*per cent*
Stearic acid	7.0
Lanolin	0.5
Sorbitan oleate	0.5
Polysorbate-60	0.5
Sorbitol	10.0
Water, perfume, preservative	q.s.

Formula 21

Constituents	per cent
Cetrimonium bromide	1.5
Isopropyl myristate	3.0
Cetyl alcohol	2.5
Lanolin	2.0
Glycerine	8.0
Water, perfume, preservative	q.s.

All-purpose creams

All-purpose creams should comply with the following requirements :

(1) It must provide a satisfactory foundation base for make-up without being too greasy as foundation cream.

(2) It should be oily in nature, liquefy readily and not readily absorbed by skin as a cleansing cream.

(3) It should be emollient yet not leave a greasy or sticky film on the skin as a hand cream.

(4) It should leave a continuous but non-occlusive oil film on the skin as a protective and emollient cream.

Some suggested starting formulae for all-purpose creams are given in formulae 22-24.

Formula 22

Constituents	per cent
Trieolyl phosphate	3.0
Petrolatum	18.0
Glyceryl stearate	5.0
Isopropyl palmitate	4.0
Cetyl alcohol	2.0
Stearyl heptanoate	0.5
Cetearyl octanoate	0.5
Sorbitol	5.0
Water, perfume, preservative	q.s.

Formulas	23	24
Constituents	*per cent*	*per cent*
Stearic acid	15.0	15.0
Lanolin	4.0	2.0
Beeswax	2.0	2.0

Mineral oil	23.0	24.0
Polysorbate-85	1.0	–
Sorbitan trioleate	1.0	–
PEG-40 stearate	–	5.0
Sorbitol	12.0	10.0
Water, perfume, preservative	qs.	q.s.

PROTECTIVE CREAMS AND HAND CLEANSERS

The chemicals that are in everyday use and are capable of inflicting damage on unprotected skin. The list include domestic materials such as detergents, floor and metal polishes, bleaches, oven cleaners, paints and varnishes. Industrial hazards include acids, alkalies, organic solvents, resins, dyestuffs, weedkillers, insecticides, lubricants and many more.

The types of hazard from which the skin is to be protected are :

(1) Dry solids, dust and dirt.

(2) Aqueous solutions or suspensions.

(3) Non-aqueous materials, including oils, fats and solvents.

(4) Emulsions.

(5) Physical hazards, such as heat, cold, uv radiation and abrasion.

Barrier Materials—Protective Creams and Gels

The use of barrier material in protective creams and gels provide protection from water borne hazards. There are many hydrophobic substances that can be spread upon the skin in a continuous film to form a water-repellent occlusive film. These include petrolatum, paraffin, waxes, vegetable oils, lanolin, silicones and occlusive esters. The water-repellent materials added are capable of modifying the film-forming agent to improve its aesthetic or functional qualities. These include alumina, zinc oxide, zinc stearate, talc, titanium dioxide, kaolin and stearic acid. Oil-repellent films can be formed from water-swellable polymers such as alginates, cellulose derivatives, bentonites and natural clays. The combination of water-repellent and oil-repellent can form the basis of a general purpose barrier cream.

Formulae — creams and lotions

The barrier properties of lanolin, petrolatum and kaolin are used in formula 25. It also contains a small amount of sodium stearate to facilitate easy wash-off. Such a cream might best be used to protect the skin against dust and dry powders.

Formula 25

Constituents	per cent
Stearic acid	6.00
Cetyl alcohol	3.00
Lanolin	3.00
Petrolatum	2.00
Sodium hydroxide	0.65
Water	67.35
Kaolin	18.00
Colour, preservative, perfume	q.s.

The presence of stearic acid thickens water-in-oil emulsion of non-ionic emulsifier. The barrier properties of silicon oil are used in formula 26 and the water-repellent properties of zinc stearate together with a film-forming material, methylcellulose are used in formula 27. The presence of sorbitol in each case prevents 'rolling' on application and provides an additional emollient effect. (Formula 26 can be used as an aerosol if packaged in a nitrogen-pressurised container.)

Formulae (Protective hand creams)	26	27
Constituents	per cent	per cent
Stearic acid	20.00	15.00
Dimethicone	5.00	–
Zinc stearate	–	5.00
Isopropyl myristate	2.00	–
Sorbitan stearate	1.50	1.50
Polysorbate	3.50	2.00
Sorbitol	20.00	6.00
Methylcellulose (4% aqueous)	–	25.00
Water	48.00	45.50
Perfume, colour, preservative	q.s.	q.s.

The barrier properties of silicones are used in formulae 28 and 29. The emulsifier system consists of a mixture of polyethylene glycol esters of cetyl alcohol and the viscosity is controlled by the presence of another film-forming agent, Carbomer (neutralised with triethanolamine).

Formulae (Protective hand lotion)	28	29
Constituents	per cent	per cent
Dimethicone	10.0	10.0
Catech	7.5	4.0

Triethanolamine	0.2	0.2
Carbomer	0.2	0.2
Water	82.1	82.0
Perfume, colour, preservative	q.s.	q.s.

A simpler cream containing silicone, mineral oil and sodium magnesium sulphate is represented by formula 30.

Formula 30 (Barrier cream)

Constituents	*per cent*
Dimethicone	19.0
Mineral oil	19.0
Glyceryl oleate	2.0
Sodium magnesium silicate	2.0
Water	58.0
Colour, preservative, perfume	q.s.

There have been attempts, however, to utilise film-forming materials of a type completely impervious to and unaffected by water. A method of using such polymeric films is to apply the material in a water-soluble form and then to convert it, *in situ*, into an insoluble analogue. The use of acid polymers that are soluble as alkaline salts but insoluble in the free acid form could be considered. If the substance were to be presented as an ammonium salt, the volatile alkaline end of the molecule would detach itself leaving the polymer behind in the free acid form. Similarly, soluble sodium salts of certain film-formers (for example, alginates) can be converted into an insoluble form by the subsequent application of calcium or other suitable ions.

Non-aqueous barrier products

The protection against water-soluble irritants can best be given by film-forming materials applied from a non-aqueous medium. The non-aqueous protective product of all is the zinc and castor oil ointment which has been smeared onto the more vulnerable skin surfaces of countless babies and young children. The following anhydrous formulation is suggested as protection against flash burns in formula 31.

Formula 31 (Protective base)

Constituents	*per cent*
Dimethicone	50.0
Titanium dioxide	30.0
Magnesium stearate	18.0
Iron oxide	2.0

A much more pleasant and sophisticated product is the gelled oil formulation given in formula 32.

Formula 32 (Protective gel)

Constituents	*per cent*
Lanolin oil	12.00
Lanolin wax	2.50
Mineral oil	50.48
Olive oil BP	20.00
Sorbitan oleate	5.00
BHT	0.02
Silica	10.00

Testing of protective preparation

Although laboratory screening tests to determine the relative efficacy of protective creams have been described. No test can predict the behaviour of the protective cream. Some of them give some useful indications, particularly in the negative aspect, which can eliminate unsuitable preparations.

The simplest test is described here. A film of the barrier or protective cream is applied to a series of clean glass microscope slides. After drying for specified period slides are immersed in solvents like alcohol, acetone, water and oils to obtain resistance of the product. The slide may be subjected to standardised conditions of agitation, and the integrity. of the barrier film after such treatment examined, preferably against a control preparation.

An apparatus for measuring the permeability of films of barriers cream was devised. Here the passage of the test solution through the barrier could be measured by reading the meniscus from time to time in a glass tube of fine bore, graduated in hundredths of a millilitre.

Barrier creams will usually remain effective for at least 4 hours, and they are usually applied twice a day. However, when dealing with particularly corrosive substances or whenever the cream is likely to be rubbed off more regularly, more frequent applications of the barrier cream are indicated in order to provide full protection for the hands.

All protective creams should be properly labelled. The substances against which protection will be conferred should be clearly indicated. Hence the user may avoid contact with irritants against which a particular cream will not protect him.

Hand Cleansers

Skin cleansers of various types are now considered to be a valuable part of the skin care regime along with toners and moisturisers. Such cleansers are formulated to remove everyday grime, secretions and make-up. They may not be equally effective against the heavy stains, greases, resins, adhesives, oils, paint, tar, and dyestuffs with which the skin (particularly the hands) may get covered in the modern home, garage or workplace. Heavy-duty cleansers offer the possibility of removing many of these contaminants with little risk of permanent damage to the skin.

The first heavy-duty skin cleaners other than soap and water are sulphonated oils. These being particularly valuable in the removal of oils and solvents where the habitual and frequent use of soap had caused skin irritation. Sulphonated oils have now been largely superseded by the so-called waterless hand cleansers. The term 'waterless' is misleading because many of them contain water in the formulation, and 'waterless' refers to the fact that they can be used without the use of additional water (although a final rinse-off in water is often recommended by the manufacturer).

Waterless hand cleansers can be formulated as pastes, creams, gels, lotions or clear liquid and consist of a cleansing agent, a thickener, an emulsifier and (usually) water.

The heavy-duty cleansers are not water soluble. Aliphatic solvents are used as cleansing agents. These solvents are effective, cheap, innocuous and readily available. Thus the majority of formulae given below contain odourless kerosene, mineral spirits or mineral oils.

Any agent which will thicken either the water phase or oil phase of the product may be used as a thickener. The soaps used as emulsifiers to produce gels may be the sodium, triethanolamide or monoethanolamide salts of stearic or oleic acids or a mixture of both.

Emollients are added to improve the application properties and to prevent the defatting of the skin. The emollient used in the following formulae are lanolin (formula 33), ethoxylated lanolin (formula 34), myristyl myristate (formula 41) and propylene glycol (formula 34) :

Formulae (Waterless hand gels)	*33*	*34*
Constituents	*per cent*	*per cent*
Deodorised kerosene	35.00	25.00
Lanolin	10.00	–
Cocamide	4.00	–

Stearic acid	2.43	6.00
Oleic acid	3.64	8.00
Sodium hydroxide	0.38	0.80
Amphoteric	–	2.00
PEG Lanolin	–	0.50
Water	44.55	57.70
Perfume, colour, preservative	q.s.	q.s.

Formula 35 (Waterless hand gel)

Constituents	per cent
Deodorised kerosene	20.00
Alkylaryl sulphonate (amine neutralised)	5.00
Cocamide	2.00
Oleic acid	8.00
Monoethanolamide (20%)	8.00
Water	57.00
Perfume, colour, preservative	q.s.

Formulae (Waterless hand cleansers)	36	37
Constituents	per cent	per cent
Deodorised kerosene	20.00	55.00
Mineral oil	20.00	–
Glyceryl stearate	3.00	–
Stearic acid	5.00	–
Oleic acid	–	4.50
Stearamide-stearate	–	6.00
Propylene glycol	5.00	–
Triethanolamine	1.50	1.50
PEG-Cocoate	–	3.00
Water	45.50	30.00
Perfume, colour, preservative	q.s.	q.s.

Formula 38 (Skin cleansing gel)

Constituents	per cent
Mineral oil	15.50
Stearic acid	4.40
Triclosan	0.10
Sodium magnesium silicate	2.00
Triethanolamine	1.60
Water	76.40
Perfume, colour, preservative	q.s.

Formula 38 illustrates the possibility of using antiseptic agents (such as triclosan) and other additives to provide additional benefits to the product.

Formula 39 is soapless, utilising non-ionic emulsifiers to produce a cream. Such a formulation might be favoured by users who find that alkaline products are irritating to their skin. Formula 40 is a much milder form of cleanser in simple liquid form :

Formula 39 (Waterless hand cleansing cream)

Constituents	*per cent*
Magnesium aluminium silicate	2.50
Sorbitan stearate	2.00
Polysorbate	8.00
Deordorised kerosene	35.00
Methylcellulose	0.50
Water	52.00
Perfume, colour, preservative	q.s.

Formula 40 (Liquid hand cleanser)

Constituents	*per cent*
PVM/MA polymer	0.40
PEG	5.00
Octoxynol	5.00
Water	89.60
Potassium hydroxide	to pH 7
Perfume, colour, preservative	q.s.

The formula is 41 unusual. It comprises a base which may be turned into a number of different products of varying consistency by the addition of solvent together with variable amounts of oleic acid :

Formula 41 (Waterless hand cleanser)

A	*Base (smooth paste)*	*per cent*
	Cocamidopropylamine oxide	18.00
	Cocamide	38.00
	Dioctyl sulphosuccinate	15.00
	Myristyl myristate	15.00
	Oleic acid	14.00

Mix until smooth at 45–55°C.

B	*Finished products*	Gel *per cent*	Cream *per cent*	Lotion *per cent*
	Base	15.00	13.90	12.95

Odourless kerosene	30.00	27.70	25.90
Oleic acid	1.50	1.40	1.40
Water	53.50	57.00	59.75
Perfume, colour, preservative	q.s.	q.s.	q.s.

All the ingradients are stirred and heated except water at 55–65°C. Gradually add water at 55°C with stirring till homogeneous and smooth product is formed.

ASTRINGENTS AND SKIN TONICS

Astringents are a class of materials which are identified by their local effect on skin when applied topically. These effects may include all or some of the following (not all astringents are equally active) : the erection of hairs, the tightening of skin (or at least the sensation of tightening), the temporary reduction of pore size, antiperspirancy, the mitigation of 'oily skin', the rapid coagulation of blood from a fresh wound, skin healing, the promotion of tissue growth and other, more subjective sensations such as a refreshing or invigorating feeling. While these beneficial properties ensure that astringents are regarded as important and valuable cosmetic materials, not all the claims made for them can be substantiated by careful experiment.

Types of Astringent

Metal salt astringents

The astringent effects of many metal ions have long been known. The list of active metals includes iron, chromium, aluminium, zinc, lead, mercury, tin, copper, silver and zirconium, although they vary in their degree of astringency. Of these only useful are the salts of zinc, aluminium and zirconium.

The effect of the anion in metal salt astringents is not entirely passive. It can be shown, for example, that the astringency of a metal ion is partially dependent upon the identity of the anion. Furthermore, the anion will also help to determine the solubility of the salt in various cosmetic media.

Organic acids

Low-molecular-weight organic acids with an ionisable proton show astringent properties. Most commonly encountered are lactic and citric acids.

Alcohols

Both ethanol and (less frequently) isopropanol are used as astringents, usually as aqueous solutions of strength up to 60 per cent w/v. Solutions of ethanol greater in strength than 20 per cent may cause stinging when first applied although this may be regarded as beneficial in certain types of product.

Auxiliary additives

The materials added to astringent preparations to assist or enhance their effect are referred as auxiliary additives. Menthol included to produce a cooling effect in an after-shave, an antibacterial in a styptic stick, and rose water as a refreshing adjunct to a skin toner are various agents added to tonics to promote soothing and the healing of damaged skin.

Astringent Products

Many products contain raw materials with astringent properties. It is possible to categorise these various products according to the particular astringent feature which they utilise. Antiperspirants, for example, exploit the property which some zinc, aluminium and zirconium compounds have of causing 'anhidrosis' or decrease in activity of the sweat glands.

Astringent lotions may be subdivided into a number of related products. These include pre-shave and after-shave lotions, skin tonics, toners, colognes and fresheners.

Astringent emulsion include antiperspirant creams as well as plain astringent creams and milks together with emulsion colognes and after-shaves.

Stick astringents are generally formed from sodium stearate – alcohol gels. Among the products presented in this form are antiperspirants, deodorants, colognes, after-shaves and styptic pencils. Some astringents are made in gel form, including some cleansing products and face masks.

Aqueous and aqueous—alcoholic lotions

Toners

Toners have become an accepted part of the facial skin treatment regime. They are normally recommended for use after cleansing and before the application of a moisturising cream. The primary purpose of such a product is to tighten the skin, reduce the pore size and reduce any tendency of areas of the face and neck to oiliness. Formulae 42-44 are shown below. In some product ranges, toners are offered in a series of increasing alcoholic strength for dry, normal and oily skins (formulae 45, 46 and 47).

Formulae (Skin toners—aqueous)	44	45	46
Constituents	per cent	per cent	per cent
Potassium alum	1.0	–	4.0
Zinc sulphate	0.3	1.0	–
Glycerine	5.0	–	6.0
Zinc phenolsulphonate	–	2.0	–
Rose water	50.0	57.0	35.0
Orange flower water	–	–	35.0
Water	43.7	40.0	20.0
Perfume, preservative	q.s.	q.s.	q.s.

There is a variety of water-soluble or alcohol-soluble emollients that can be used to offset the drying effect of ethanol. In formulae 45, 46 and 47 use is made of an ethoxylated lanolin derivative, together with propylene glycol.

Formulae (Skin toners—aqueous—alcoholic)	45	46	47
Constituents	per cent	per cent	per cent
Denatured ethanol	20.0	35.0	50.0
Water	72.0	58.0	42.0
Propylene glycol	5.0	5.0	2.0
Laneth-10 acetate	3.0	3.0	1.0
Perfume, preservative	q.s.	q.s.	q.s.

Skin tonics

Skin tonics can be made by the addition of auxiliary additives and perhaps menthol or camphor to produce a slightly 'medicated' fragrance. Possibly the most commonly used skin healing and soothing additive is allantoin, a chemical of the purine group. Allantoin has been shown to have regenerative, healing, softening, soothing and keratolytic properties. More recently, two allantoin combination compounds, chlorohydroxyaluminium allantoinate and dihydroxyaluminium allantoinate have become available. These possess mild astringent qualities in addition to the healing and soothing properties of allantoin itself and their use has been specifically recommended in astringent lotions of the 'tonic type'.

Azulene and its derivatives, particularly guaiazulene, have long been accredited with healing and soothing properties. Newer soothing ingredients to become available include a cationic cellulose polymer. No doubt other skin-substantive cationic polymers will be found to have similar desirable properties.

The various benefits to be derived from the use of such additives make them logical ingredients of pre-shave and after-shave products. Pre-shave lotions — especially those designed to be used before electric shaving — make use of the ability of astringents to make facial hairs stand erect. After-shave lotions may be designed to soothe and cool the skin. Also to tighten it, to close the pores, to sterilise and stem the flow of blood from any inadvertent nicks or cuts. Formulations of skin tonic/after shave are given in formulae 48 and 49.

Formulae (Skin tonic/after-shave)	*48*	*49*
Constituents	*per cent*	*per cent*
Alcohol	40.00	20.00
Water	55.30	77.75
Allantoin	0.20	–
Polysorbate	1.50	–
Sorbitol	1.50	–
Glycerine	1.50	2.00
Quaternium	–	0.25
Perfume, colour, preservative	q.s.	q.s.

The normal pH of skin is slightly acid; body lotions and shaving preparations often make use of acids to 'restore' this acidic condition as well as for reasons of astringency, as in formulae 50 and 51.

Formulae (Skin tonic/shaving lotions)	*50*	*51*
Constituents	*per cent*	*per cent*
Alcohol	10.00	45.00
Zinc sulphate	0.40	–
Citric acid	1.00	–
Sorbitol	6.00	5.00
Water	82.50	48.90
Lactic acid	–	1.00
Menthol	–	0.10
Perfume, colour, preservative	q.s.	q.s.

Astringent lotions such as these may be thickened with cellulose ethers (formulae 52 and 53) or carbomer (formulae 54 and 55) or any other suitable agent, such as an alginate.

Formula 52 (Body lotion)

Constituents	*per cent*
Alcohol	43.00
Aluminium chlorhydroxyallantoinate	0.20
Propylene glycol	3.00
Menthol	0.05
Aluminium chlorohydrate (50%)	5.00
Hydroxypropylmethylcellulose (3%)	47.75
Mica (and) titanium dioxide	1.00
Perfume, colour, preservative	q.s.

Formula 53 (Body lotion)

Constituents	*per cent*
Magnesium aluminium silicate	1.50
Hydroxypropylmethylcellulose	0.75
Water	80.05
Acetylated monoglyceride	3.00
Di-isopropyl adipate	5.00
Camphor	0.40
Menthol	0.80
Methyl salicylate	6.50
Zinc phenolsulphonate	1.00
Mica (and) titanium dioxide	1.00
Perfume, colour, preservative	q.s.

Formulae 52 and 53 contain a small quantity of titania coated micas to give a 'pearly' effect, which is slightly unusual in products of this kind.

Formula 54 (Skin freshener)

Constituents	*per cent*
Carbomer 940	2.00
Alcohol	73.00
Propylene glycol dicaprate	1.00

PEG-25 castor oil	5.00
Di-(2-ethyl hexyl)amine	2.00
Water	17.00
Perfume, colour, preservative	q.s.

Formula 55 (Astringent gel)

Constituents	*per cent*
Alcohol	50.25
Water	47.26
Carbomer 940	0.70
Menthol	0.04
Di-isopropylamine	0.50
Octoxynol-9	1.25
Perfume, colour, preservative	q.s.

Astringent emulsions

Astringent creams or lotions (of the emulsion variety) serve many purposes in the cosmetic field. Specifically as antiperspirants, shaving products and perfumed creams. The following formulae (56 and 57) illustrate a few of the ways in which astringents can be used in emulsions. Formula 58 is non-ionic and formula 59 is anionic. Formulae 60 and 61 shows the formulation of astringent lotion and creams.

Formula 56 (Alcoholic astringent cream {anionic})

Constituents	*per cent*
Sodium magnesium silicate	2.00
Isopropyl myristate	5.00
Triethanolamine	0.80
Glycerine	2.00
Stearic acid	3.00
Cetyl alcohol	0.50
Triclosan	0.10
Alcohol	30.00
Water	56.60
Colour, perfume	q.s.

Formula 57 (Alcoholic after-shave lotion {(non-ionic})

Constituents	*per cent*
Laneth-40	0.5
Oleth-10	0.5
Mineral oil	1.0

Alcohol	15.0
Water	72.5
Glycerine	1.0
Di-isopropylamine (10% aqueous)	1.5
Carbomer 941 (91% dispersion)	8.0
Perfume, colour, preservative	q.s.

Formulae (Astringent/antiperspirant creams)	*58*	*59*
Constituents	*per cent*	*per cent*
Stearic acid	14.00	10.60
Mineral oil	1.00	1.00
Beeswax	2.00	1.00
Sorbitan stearate	5.00	–
Polysorbate 60	5.00	–
Aluminium chlorohydrate (50% aqueous)	40.00	32.00
Water	33.00	42.70
Glyceryl stearate	–	6.40
Propylene glycol	–	5.00
Sodium lauryl sulphate	–	1.30
Colour, perfume, preservative	q.s.	q.s.

Formula 60 (Astringent lotion non-ionic/cationic)

Constituents	*per cent*
Glyceryl stearate	5.00
Quaternium-7	5.00
Aluminium chlorohydrate	15.00
Water	67.00
PEG-20 stearate	3.00
PEG-8	5.00
Colour, perfume, preservative	q.s.

Formula 61 (Astringent cream with witch hazel)

Constituents	*per cent*
Water	69.00
Witch hazel extract	10.00
Cetyl alcohol	3.00
Propylene glycol stearate	12.00
Sorbitol	5.00
Isopropyl myristate	1.00
Colour, perfume, preservative	q.s.

Chapter 22

Bath Preparations

INTRODUCTION

In recent years range of bath product are available. They have undergone considerable change in terms of volume. The product range include, bath oils, shower gels, after bath body lotions and newer hydro-alcoholic products like bath satins.

FOAM BATHS

Foam baths are most popular bath preparations, which enjoyed a very healthy growth in recent years. Generally, they are available in liquid, gel and powder forms. Body cleansing, the primary function of a bath, is performed very well by a bubble bath. In addition it offers the opportunity to apply many desirable health and beauty ingredients to the skin, although not necessarily so efficiently as when these aids are used individually. Functioning as a body cleanser, a bubble bath soaks off and suspends dirt, grime and body oils. It prevents the formation of the 'bath-tub ring' that usually results from the use of soap. Therefore a bubble bath is superior to the traditional soap and water bath. Additionally, a well formulated bubble bath will condition the skin, deodorise, perfume the body and the bathroom, stimulate the senses, yet promote relaxation.

A good foam bath has following characteristics :

(1) It should provide copious foam at minimal detergent concentration.

(2) The foam should be stable, particularly in the presence of soap and soil and within wide limits of temperature. The foam stability should be reasonable and not excessive. It is possible to prevent premature breakdown of the foam, by late use of soap. This will satisfy the aesthetic requirements of the bather and facilitating the subsequent removal of the dirty water from the bath.

401

(3) It should prevent bath-tub ring formation.

(4) It must be non-irritant to the eyes, skin and mucous membranes. Bubble baths have been claimed to produce symptoms of irritation in the lower urinary tract and it is essential to check the irritation potential of all these products before marketing.

(5) It should have adequate detergent power so that it will cleanse the body efficiently. In order to counteract any excessive harshness to the skin it is advisable to include a low level of skin emollient.

Formulation of Foam Baths

As already mentioned, foam baths are available in several physical forms and clearly the choice of raw materials is highly dependent on the final product form. Before discussing the product types in detail, it is of interest to examine the raw materials available.

Foaming agents

The foaming agent is the most important ingredient in all foaming bath products. While selecting a surfactant it is important to bear in mind the properties that good foam baths should exhibit. The anionics are the most widely used; both non-ionics and amphoterics are also of considerable interest. Due to eye irritancy and incompatibility with soaps and anionic, cationics are not advisable to use in bath formulations.

The anionic surfactants used are the sodium, ammonium and alkanolamine salts of fatty alcohol sulphates, fatty alcohol ether sulphates and (sometimes) alkyl benzene sulphonates. Soaps are not suitable surfactants for use in foam baths, since in hard water calcium and magnesium soaps precipitate as a dirty hard scum.

Perhaps the most popular surfactants used in foam baths are the fatty alcohol ether sulphates, especially the sodium salts and, in particular, those based on lauryl-myristyl alcohols and containing 2–3 moles of ethylene oxide per mole of the alkyl ether sulphate. Copious foaming in hard water, reasonable foam stability in the presence of soap, fragrances, oil additives and body debris, together with a good skin compatibility, make these materials an obvious choice for bath products. A high degree of proficiency as a lime soap dispersant prevents bath-tub ring even in very hard water. Other advantages of these surfactants are good colour, which permits the use of very delicate pastel colours if required, a good viscosity response to electrolyte and unusual solubilising powers for perfumes.

Primarily for reasons of cost, alkyl benzene sulphonates (mainly branched-chain dodecyl benzene sulphonates) soon found their way from household detergent uses into toiletry items such as bubble baths. Of all the synthetic surfactants, the sodium alkyl benzene sulphonates are probably the most suitable for spray drying and until recently most of the commercial products on the market in bead form consisted of this material and various salt builders and extenders. When biodegradability considerations became critical, the importance of linear alkyl benzene sulphonates grew.

The alpha-olefin sulphonates are commercially attractive and have been used as alternatives to linear alkyl benzene sulphonates in powder foam baths. Lauryl sulphoacetates are sometimes used in high-priced powder or granular bubble baths but limited solubility has greatly restricted their use in liquid products.

The sulphosuccinates, particularly the monoesters such as the di-sodium lauryl alcohol polyglycol ether sulphosuccinate, are considered to be very mild detergents with good foaming properties. They are free from any tendency to irritate the skin and mucous membranes. Furthermore, they are also claimed to increase the tolerance of the skin to other detergents such as fatty alcohol ether sulphates and alkyl benzene sulphonates.

The paraffin sulphonates produced by the sulphoxidation of n-paraffins are relatively cheap, in particular, the secondary alkane sulphonates. This material, in addition to being biodegradable and exhibiting good physiological properties, is also a good foamer with high solubility in water. The use of alkane sulphonate with alkyl ether sulphates prevent excessive degreasing of the skin and easy thickening of finished product.

The non-ionic components of bubble baths are used to stabilise foam, to enhance the viscosity of the product, or to solubilise skin care ingredients and fragrances. The non-ionic surfactants used in bubble baths are alkanolamide and amine oxides. In addition, ethoxylated derivatives such as ethoxylated fatty alcohols, ethoxylated fatty acids, alkyl phenol ethoxylates, ethoxylated alkanolamides, ethoxylated propylene oxide condensates (Pluronics) and ethoxylated sorbitan fatty acid condensates are occasionally utilised.

Emollients

Special ingredients known as emollients are often added to the formulation to achieve and maintain a healthy and attractive skin. The value of many of these conditioners has been challenged but they continue to be used with apparent success. Although it has been argued that the bather stays in the bath for too short a time to receive any real benefit, equally strong opinions have been expressed that, even in this short time, the skin absorbs active substances. Many such ingredients are available. The following list gives some of the most popular in common use: branched chain esters, for example isopropyl myristate; decyl oleate; ethoxylated partial glyceride fatty acid esters; protein derivatives; lanolin derivates; and fatty alcohol ethoxylates; etc.

Perfumes

There is no question that the perfume used in a foam bath is extremely important. Most of the larger companies marketing foam baths spend a great deal of time and money in selecting the perfumes for their products. A good perfume must of course, convey the marketing image of the brand and it should also fulfil the following requirements :

(1) It should be acceptable when sniffed in the bottle.

(2) In use it should be fresh and have sufficient volatility to give a strong impact.

(3) It should linger on the skin to give a feeling of freshness and wellbeing.

(4) It must have an acceptable shelf-life in the product.

The level of perfume used will vary between 1 and 5 per cent, depending on the cost. The nature of the perfume ingredients may require the use of additional solubilisers; the most commonly used are non-ionics such as ethoxylated fatty alcohols, ethoxylated fatty acid esters, ethoxylated sorbitan fatty acid esters and ethoxylated propylene oxide condensates (for example, pluronics).

Perfumes often cause problems of product instability. They not only affect odour stability, but can cause discolouration, upset preservation systems and cause instability in clear, opacified and emulsion products. The need to test the shelf life of all new products adequately cannot be overstressed. The ingredients used in perfumes should also be adequately tested before use for safety.

Herbal extracts are used in bath preparations, usually to help convey the brand image and to justify therapeutic claims with respect to minor skin disorders. Herbal extracts of most living plants can be obtained if required, but their medicinal nature is open to speculation.

Viscosity controllers

The problem of achieving the required viscosity of liquid products depends on the choice and level of surfactant and foam booster. Even certain perfumes have been found to have significant effect on viscosity. Generally, inorganic salts such as sodium and potassium chloride are used to thicken the product where possible, whereas alcohol, hexylene glycol, propylene glycol and polyethylene glycols are used to lower the viscosity. Certain thickening problems, however, can only be solved by the use of natural gums such as tragacanth and gum acacia, or synthetic gums such as methyl cellulose and hydroxyethyl cellulose.

Colour

Colour of foam baths is clearly important in marketing and care should be taken to select colours that are stable in the chosen product. Obviously, if the products are marketed in clear bottles adequate light testing should be carried out. Interesting colour effects in the bath can be achieved by the use of indicator colours and fluorescein. The colour should be selected as per the legislative requirement of the country.

Preservatives

Foam baths are easily attacked by moulds and bacterias, particularly *Pseudomonas*. Hence, they should have sufficient amount of preservative. Bacterial attack can produce opacity in products that are intended to be clear, separation in emulsified and pearlescent products and can cause changes in both perfume and colour system. Suitable preservatives include ethanol; methyl, propyl and butyl hydroxy benzoate; phenylmercuric nitrate; formaldehyde; Bronopol; and many others. The best preservative for a particular foam bath can only be determined by properly designed microbiological testing. Good housekeeping in the manufacturing unit is, however, just as important as choice of preservative if product contamination is to be avoided.

Opacifying agents

When an opaque liquid foam bath is required, an opacifier is needed. Higher alcohols such as stearyl or cetyl alcohols; ethylene glycol mono- and distearates; glyceryl and propylene glycol stearates and palmitates; and the magnesium, calcium and zinc salts of stearic acid are opacifier used.

Clearly, viscosity is an important factor in the stability of such systems. The manufacturing technique used is vitally important in achieving maximum stability. Ideally, when using the opacifiers mentioned above, all the ingredients (except perfume) should be heated to 65°–70°C and allowed to cool slowly to ambient temperature with gentle mixing, when the pearl will develop. Rapid cooling will produce less pearly products that may be unstable. It is possible, however, to obtain blended opacifier – detergent concentrates of some of these opacifiers which eliminate the need to heat the batch.

Types of Product

Liquids

Liquids can be further divided into clear, translucent, opaque, pearlescent and multi-layer products. The formulation possibilities for a medium-priced clear liquid foam bath are endless. A typical and very simple product formula is shown below :

Formula 1

Constituents	*per cent*
Sodium lauryl ether sulphate (28% active)	50
Coconut diethanolamide	3
Perfume	1–2
Citric acid	q.s. to pH 7
Colour, preservative, emollients, solubiliser	q.s.
Sodium chloride	q.s. to required viscosity
Water	to 100

More expensive but milder formulations can be achieved by replacing part of the sodium lauryl ether sulphate by a coconut imidazoline betaine.

Translucent and pearlescent products can be created by the addition of insoluble stearates. These are readily available from all major surfactant manufacturers and the depth of opacity is governed by the level of incorporation, which is usually between 1 and 5 per cent. Opaque non-pearlescent products are achieved by the use of polymeric materials.

Multi-layer products can be achieved and a typical formulation is as follows :

Formula 2

Constituents	*per cent*
Sodium lauryl ether sulphate (28%)	50.0
Coconut diethanolamide	9.0
Hexylene glycol	14.0
Neutral monoethanolamine citrate	13.0
Citric acid	3.0
Perfume, colour, preservative, water	to 100.0

Gels

Basically gels are very similar to liquid products, having much higher
viscosity. This is achieved by increasing the level of detergent, foam
stabiliser or electrolyte content, depending on the particular formulation.
Shower gels are virtually identical in formulation to the highly viscous liquid
bath products; however, because they are sold for direct application to the
body they must be very mild. It is usual to adopt for shower gels the types
of formulation, already mentioned, that have a milder action on the skin and
eyes.

Dry bubble baths

Dry bubble baths, of considerably less importance than liquid bubble baths.
Basically they consist of a mixture of one or more foaming agents, fillers
and water softeners to add bulk or act as carriers, perfume, colour and
free-flow agents.

The major surfactant ingredients are usually dry products, sometimes
fortified with liquid surfactants, for example sodium lauryl ether sulphate,
to give good flash foam. The principal surfactants employed are benzene
sulphonates. Other surfactants that have been used include sodium lauryl
sulphate, sodium lauryl sulphoacetate and isethionate derivatives.

Inorganic fillers such as sodium chloride and sodium sulphate are often
used in the more inexpensive products. These can, however, be replaced by
functional fillers also have water-softening properties. These include sodium
hexametaphosphate, sodium sesquicarbonate and tetrasodium pyrophosphate.
Free-flowing and anti-caking is obtained by tri-calcium phosphate, calcium
silicate, or sodium silica aluminate to powdered bubble bath. Bentonite or
starch is usually employed to absorb the perfume, in order to disperse it
throughout the product. The colour is usually incorporated by premixing
with one of the fillers. Formulae 3 and 4 are given as under :

Formula 3

Constituents	*per cent*
Alpha-olefin sulphonate (40% active spray-dried beads)	20
Lauric isopropanolamide	3
Sodium sesquicarbonate (low density)	60
Sodium chloride	14
Perfume	3
Colour	q.s.

Formula 4

Constituents	per cent
Sodium lauryl sulphate	30
Sodium lauryl sulphoacetate	10
Lauric isopropanolamide	3
Sodium sesquicarbonate (low density)	50
Calcium silicate	4
Perfume	3
Colour	q.s.

Product Assessment

A large consumer panel is required for complete assessment of foam bath products. As it involves complete interpretation of the initial and residual feel of skin, efficiency of special ingredients, fragrance, amount of texture of foam. The laboratory testing is limited to the assessment of foam volume and stability.

Of the stabilisers examined on alkylamido betaine was shown to be the best stabilising agent against soap, with the N-alkyldimethyl betaine second.

BATH SALTS

Ingredients and Formulation

Salts

Sodium sesquicarbonate ($Na_2CO_3.NaHCO_3.2H_2O$), which is a mixed salt, is probably the most popular material used in the preparation of bath salts. It is available in uniformly sized, elongated, attractive translucent crystals which are extremely stable, non-caking and free-flowing. It dissolves rapidly and completely in water, is easy to colour and perfume. It is an excellent water softener and quite mild to the skin, having a pH value of about 9.8 for a 1 per cent solution.

Other carbonates that have been used are sodium carbonate decahydrate, $Na_2CO_3.10H_2O$ (washing soda), and the monohydrate, $Na_2CO_3.H_2O$. The decahydrate, is a good water softener and consists of large attractive crystals that are readily soluble in warm water. Unfortunately, it has several serious disadvantages including a low melting point of 35°C, at which temperature it dissolves in its own water of crystallisation. This clearly precludes its use in hot climates. Even under ordinary storage conditions it tends to effloresce to become powdery and unsightly, although this can be overcome by coating the crystals with a film of humectant such as glycerine. It is also more highly alkaline than the sesquicarbonate. The monohydrate is the most stable form of sodium carbonate and is available in attractive crystal agglomerates with excellent stability. Its main disadvantages are its slow rate of dissolution and the fact that it is more alkaline than the sesquicarbonate.

Phosphates like the hexametaphosphate (Calgon), tetrasodium pyrophosphate and sodium tripolyphosphate are used to improve softening. Trisodium phosphate is a good water softener but because of its highly alkaline nature it cannot be used without buffering. However, it can be used in combination with either sodium sesquicarbonate or borax, both of which buffer it quite effectively.

Borax ($Na_2B_4O_7.10H_2O$) is less alkaline than the carbonates and possesses a mild detergent action. Although it is less effective as a water softener than the carbonates and is slow to dissolve.

Rock salt, NaCl, is used in bath salts of the type that provide fragrance only. It is very stable and its large crystals are attractive and easily coloured. It is of course non-alkaline and mild to the skin. Its disadvantage are that it has no water-softening properties interfere with soap lathering the large crystals do not dissolve easily.

For effervescent systems, sodium bicarbonate and tartaric or citric acid are included.

Fragrance

The instability of fragrance is attributed to its distribution in a thin film on the surface of an inorganic salt which can be quite alkaline. Hence, the perfume must have good fixative properties in addition to adequate stability to alkalies, light and oxidation. The deposition, retention and stability of the fragrance and they may also improve the flow characteristics of the product is achieved by using calcium silicate or fumed silica.

Colour

As with fragrance, the choice of colour is limited by stability to both alkalies and light. Colour stability may also be affected by perfume. Insoluble colours have been recommended because of their better stability to alkalies and light. These are best dispersed at a concentration of about 1 per cent in a suitable medium such as a glycol or liquid non-ionic surfactant.

Example formulations

Formula 5 (Bath salts—fragrance only)

Constituents	per cent
Sodium chloride	95–99
Perfume	1–5
Colour	q.s.

Formula 6 (Water-softening bath salts)

Constituents	per cent
Sodium sesquicarbonate	95–99
Perfume	1–5
Colour	q.s.

Formula 7 (Effervescent bath salts)

Constituents	per cent
Sodium sesquicarbonate	25
Sodium bicarbonate	50
Tartaric acid	20
Perfume, colour	5

The manufacture of bath salts is a straightforward process and can be carried out in most types of dry powder blender. Colouring is done either by spraying with colour solution, mixing and drying or by immersing the salts in the colour solution, followed by drying. Colour solutions should be hydro-alcoholic or, if possible, alcoholic. Alcohol, in addition to reducing the drying time, helps to prevent solution of the bath salts. Fragrance is added either by spraying an alcoholic solution or by dispersing the perfume in the colour solution. Perfume can also be mixed with oil-absorbent powders. Clearly, in colouring effervescent salts no water can be used.

Bath cubes and tablets

Bath cubes and tablets employ the same materials as bath salts, but in powder form. The powdered bath salts are first granulated with starch and a small amount of alcohol-soluble gum binder is normally added. The granules are then compressed into tablets or cubes. A suitable disintegrating material like sodium lauryl sulphate or starch should be included to aid the dissolving process before compressing.

BATH OILS

The primary function of a bath is to cleanse the body. The function of bath oil is not only cleansing but of skin lubrication. In addition, they are used to impart fragrance to the body.

The oil bath has emerged as the simplest and most effective method of lubrication for generalised skin dryness. Dry skin affects both young and old and. in its mildest form, appears as a slight roughening and scaling of the skin. Severe dryness can result in disturbing itching and both the incidence and serverity of itching becomes progressively worse as the individual becomes older. Clinical efforts to overcome skin dryness are based on the concept that a surface oil or lipid film on the skin retards water loss through evaporation — hence the importance of oil baths in combating dry skin.

Various investigations have been conducted, aiming to quantify the absorption of different oils by skin. Recently, the first objective attempt to quantify all deposition on skin using the technique of arm emersion in a bath of oil stipulates that product based on mineral oil adhere to skin better than vegetable oil formulation. Further, an oilated oatmeal preparation (colloidal oatmeal combined with mineral oil and lanolin) can be adsorbed very poorly by the skin. The likely explanation for this being that oatmeal is a better adsorptive substrate for oils than is the skin. It is also observed that adsorption increased as the temperature of the bath was raised and as the concentration of oil increased. Soaking for longer than 20 minutes, however, did not cause a significant increase in oil adsorption.

Classification of Bath Oil

Bath oils can be classified into four main categories: floating or spreading bath oils which are water-immiscible; a dispersible or blooming type which turns milky on addition to water; a soluble type which forms a clear dispersion in water; and a foaming type similar to the foam bath.

Floating or spreading oils

Floating or spreading bath oils are hydrophobic in nature. They have low specific gravity. Hence, they float on top of the bath water, which covers the bather's skin with an oily film on emergence from the water. It provides the bather with a luxurious emollient coat. This type of product is ideal for augmenting the aesthetic nature of the bath with pleasant fragrance to the bathroom, since the oil layer on top of the hot bath water allows the fragrance to distil readily into the atmosphere. One problem with this type of product is the occurrence of an unsightly 'ring' around the bath caused by the oil deposit. This deposit is compounded by a soap scum if the bather also uses a true soap in conjunction with hard water. Furthermore, since oils are natural foam depressants, the floating oil layer is likely to impede the lathering performance of the soap. Ideally, a floating bath oil should fully cover the surface of the water and be deposited on the skin in a very thin film, covering as much of the skin surface as possible. A lubricating bath oil should not be deposited in a heavy greasy layer, which is unattractive to the user, nor should it leave in the bath a heavy oily film that is difficult to remove.

Perfume is obviously an important ingredient. Other ingredients sometimes included are antioxidants, colours and sunscreens.

Typical formulae for floating bath oils are given below:

Formula 8

Constituents	*per cent*
Arlamol	49
Arlatone	1
Light mineral oil	45
Perfume	5

Formula 9

Constituents	*per cent*
Light mineral oil	46
Isopropyl myristate	48
Arlatone	1
Perfume	5

Dispersible or blooming oils

Dispersible bath oils consist of emollient oils, perfume oils and contain a surfactant. To emulsify the oils in the water instead of making them spread on the surface surfactant is used. When poured into the bath water they bloom into a milky cloud. They are sometimes preferred to the floating type because the oils are dispersed uniformly throughout the bath water, providing thorough contact with the body during the bath. When properly formulated they leave very little oily stickness or ring in the tub after the water is drained. The emollient oils used tend to be similar to the ones used in floating bath oils, with perfume levels between 5 and 10 per cent. A typical formula for a dispersible bath oil would be as follows (formula 10).

Formula 10

Constituents	*per cent*
Mineral oil	65
Isopropyl myristate	20
Brij	10
Perfume	5

Soluble oils

Soluble bath oils contain large quantities of surfactants to solubilise the high-fragrance oil concentrations. And also to dissolve or disperse these oils readily in the bath water. They leave no residue in the bath and have no emollient effect on the skin. Soluble bath oils are either anhydrous concentrates consisting of perfume and surfactant or solubilised products consisting of perfume, surfactant and water. They contain 5 to 20 per cent perfume oil, which can usually be solubilised by means of a hydrophilic surfactant. A typical formulation is shown in formula 11.

Formula 11

Constituents	*per cent*
Perfume	5
Polyoxyethylene	
Sorbitan monolaurate	5–25
Preservative	q.s.
Water	to 100

The quantity of tween (polyoxyethylene sorbitan monolaurate) is clearly dependent on the type of perfume used.

Foaming oils

Foaming bath oils can be considered either as bubble baths with high fragrance levels or as soluble bath oils with foaming agents and stabilisers. These products provide both fragrance and foaming action. They also serve to eliminate the bath-tub ring. Like the soluble bath oils they generally have no emolliency properties. The foaming agents and stabilisers discussed under foam baths are used. Polyoxyethylene sorbitan monoalaurate is often used to solubilise the fragrance in the foaming bath oil. Thickeners such as carboxymethyl cellulose, methyl cellulose and other gums are commonly added, as are sometimes sequestrants. A typical formulation is shown in formula 12.

Formula 12

Constituents	*per cent*
Perfume	5
Polyoxyethylene	
Sorbitan monolaurate	20
Sodium lauryl ether sulphate (28% active)	40
Coconut diethanolamide	2
Preservative	q.s.
Water	to 100

Very occasionally emollient oils are included in these products at significant levels but most of these have such a foam-depressing action that the final product can barely be considered to be a foam bath.

AFTER-BATH PRODUCTS

After-bath products are body or dusting powders and the various lotions.

Body or Dusting Powders

Body or dusting powders are also known as body talcs or talcum powders. They have a wide appeal because of the smooth feeling and cooling effect which they impart while they temporarily absorb moisture. The extra heat loss from the large surface area of the talc particles creates cooling effect.

Talc used in these formulations, must have good slip characteristics, covering power, body adhesion and absorbency. The slip and texture properties are essentially based on the talc. It is essential, therefore, that grit-free, alkali-free high quality cosmetic talc is used. Talc should be free of bacteria and sterilised grades should always be used. In order to improve adhesion properties, metallic stearates such as zinc or magnesium stearate and kaolin are incorporated. Whereas magnesium carbonate, starch, kaolin and precipitated chalk all improve absorbency. Zinc and titanium oxides at low levels along with earth colours can be incorporated for tinting purposes when required. Perfume oils are easily incorporated and should be sufficiently powerful to cover the base odour yet not interfere with other perfumes that may be used. Other ingredients sometimes included are boric acid to act as a skin buffering agent and fumed silica to give a powder of lower density. A typical formulation (Formula 13) is as follows :

Formula 13

Constituents	*per cent*
Talc	75
Kaolin	10
Fumed silica	2
Magnesium carbonate	6
Zinc stearate	6
Perfume	1

After-bath Emollients

After-bath emollients or moisturisers are applied to the body to replace natural skin liquids removed during bathing. Their function, therefore, is to prevent the occurrence of dry skin. Basically four types of product exist, anhydrous oil-based systems, water-in-oil emulsions; oil-in-water emulsions; and hydro-alcoholic emulsions. The first two types are less popular due to their greasy nature. Oil-in-water emulsions are perhaps most widely used. Generally, they are sold in lotion form and are usually formulated to give good 'rub-in' and feel properties. A typical formulation (Formula 14) is shown below :

Formula 14

Constituents	*per cent*
Crodafos N-3 acid	0.5
Emollient oil	5.0
Carbopol 9.41	0.4
Denatured ethanol (95%)	40.0
Triethanolamine	q.s. to pH 6.5
Perfume	1–5
EDTA	q.s.
Water	to 100.0

Such a product could be made without heat by hydrating the Carbopol in the water with efficient mechanical agitation, followed by the addition of a mixture of emollient oil, perfume and Crodafos. Finally, the emulsion is completed by the addition of the alcoholic solution of triethanolamine. EDTA or other suitable sequestrant is included to stabilise the Carbopol gel, since metal ions can depolymerise Carbopol and this can lead to loss in viscosity and hence emulsion instability.

The higher concentration of alcohol, above 50% should be avoided as it may destabilise due to coagulation of the carbopol. It may destabilise the emulsion unless the levels of emulsifier and gelling agent are increased at the same time. Increasing the level of alcohol serves to provide greater perfume lift and quicker drying but produces a thinner, less opaque and sometimes a more unstable emulsion. The rise in the level of emollient oil increases emolliency which produces more opaque product.

Chapter 23

Skin Products for Babies and Young People

INTRODUCTION

The skin of the child undergoes extensive change and development — during the very earliest weeks after birth. Therefore, the skin of the young child differs from that of the adult and from that of the older child. The very young skin is very thin, less cornified, less hairy and contains a relatively high proportion of water in comparison with the adult. Sebaceous glands are not only present in the newborn skin, but begin to function very early. Apparently, transpidermal water loss at this time is lower than for adults (at least for some body areas).

Typically, the skin of babies and very young caucasian children is pink, very soft and smooth to the touch. The majority of skin products purchased by young people in this age group are concerned with the treatment, prevention or camouflage of blemishes or of oily skin.

FUNCTIONAL REQUIREMENTS OF BABY PRODUCTS

The baby skin care products need to protect the skin from a hostile environment, to cleanse the skin thoroughly from sebum, grime and excreta and to keep the skin surface as dry as possible. Within few days of birth most baby skins come into contact with soap and water. Although there is no evidence that ordinary soap has a bad effect on the baby, most baby soaps are white and free from perfume.

Zinc and castor oil creams or ointment provides the protection to baby skin. Zinc oxide is mildly antiseptic, astringent and anti-inflammatory. This accounts for its use in protective products, in the concentration range 2–10 per cent. Petrolatum, castor oil, beeswax, lanolin, silicone oil and polyethylene wax gives protective, occlusive barrier. These may be used as anhydrous preparations or as the oil phase of protective baby creams. Inorganic salts of stearic acid and oleic acid are used to improve the water-repellent effect of creams while stabilising the emulsion.

Baby powder is another traditional and valuable toiletry product. It provides a dry and lubricated surface to skin that has been cleaned and protected with oils or lotions. The main constituent of baby powder is talc, but since this lacks the absorbency of other powders it is often blended with such materials as kaolin, hydrated aluminium silicate, magnesium and calcium carbonates, starches and pyrogenic silica. The grades of materials used are naturally important — particularly talc, which should be the very purest available and devoid of any fibrous materials. The adhesive power of baby powders as well as their water repellency can be improved by the incorporation of aluminium, zinc and magnesium stearates; cetyl and stearyl alcohols and zinc oxide perform a similar function.

One of the main problems associated with the use of talc is its susceptibility to contamination by micro-organisms. Various methods are available for the sterilisation of talc, some being more suitable and successful than others. Ethylene oxide treatment may leave irritating residues, while heat treatment is not always sufficient to sterilise the material completely.

The use of boric acid and borates in baby powders as a mild antiseptic and as a neutralising buffer, although once popular, has now largely ceased because of the potentially toxic nature of these substances.

Safety of Baby Products

It is extremely important to ensure that all baby products are free from bacteria. They contain adequate preservative systems to prevent accidental contamination during use. Such principles should apply equally to all cosmetic and toiletry products.

The logical conclusions to be drawn from these considerations is that the raw materials used in baby products should, wherever possible, be chosen for their low toxicity and their non-irritating character when applied topically.

Example Formulations

The first two formulae for baby creams and lotions illustrate the use of a comparatively new type of mild, non-toxic emulsifier based on sucrose esters of palmitic and stearic acids. Formula 1 is a lotion and formula 2 a cream.

Formulae	*1*	*2*
Constituents	*per cent*	*per cent*
Mineral oil	25.00	35.00
Cetearyl alcohol	–	0.50
Petrolatum	–	4.20
Lanolin alcohol	–	1.25
Mono-ester	3.00	–
Di-ester	0.50	–
Tri-ester	–	3.00
Hydroxyethylcellulose	0.20	–
Water	71.30	55.05
Glycerine	–	1.00
Perfume, preservative	q.s.	q.s.

Baby creams and lotions are based upon the triethanolamine stearate (anionic) emulsifier system, of which formulae 3 and 4 being representative.

Formulae	*3*	*4*
Constituents	*per cent*	*per cent*
Mineral oil	26.00	15.00
Lanolin	1.04	5.00
Stearic acid	0.94	2.00
Triethanolamine	0.52	1.00
Water	69.68	52.00
Stearyl alcohol	0.94	–
Cetyl alcohol	0.52	–
Sodium alginate	0.36	–
Isopropyl palmitate	–	2.00
Beeswax	–	8.00
Propylene glycol	–	5.00
Stearate	–	10.00
Perfume, preservative	q.s.	q.s.

The use of some of the non-ionic emulsifiers based upon sorbitol in baby creams and lotions is shown in formulae, 5 to 7 :

Formula 5	
Constituents	*per cent*
Cetearyl alcohol	1.00
Mineral oil	4.00
Polysorbate	1.70
Sorbitan isostearate	1.00

Glyceryl stearate	1.00
Liquid lanolin	0.25
Water	83.35
Hydroxyethylcellulose	0.20
Glycerine	7.50
Perfume, preservative	q.s.

Formula 6

Constituents	per cent
Mineral oil	35.50
Lanolin	1.00
Cetyl alcohol	1.00
Sorbitan oleate	2.10
Polysorbate	4.90
Dimethicone	5.00
Water	50.50
Perfume, preservative	q.s.

Formula 7

Constituents	per cent
Pertrolatum	20.00
Sorbitan isostearate	2.10
Micro-crystalline wax	3.34
Mineral oil	10.55
Glycerine	3.14
Water	60.87
Perfume, preservative	q.s.

Formulae 5 and 6 are of oil-in-water creams, whereas formula 7 is water-in-oil. The last cream formulation (formula 8) illustrates the use of polyoxyethylene sorbitan lanolin derivatives, which are also thought to be fairly mild.

Formula 8

Constituents	per cent
Mineral oil	15.00
Stearic acid	15.00
Beeswax	2.00
Lanolin	1.00
PEG-20 sorbitan lanolate	5.00
PEG-40 sorbitan lanolate	1.00
Sorbitol	10.00
Water	51.00
Perfume, preservative	q.s.

In view of its potentially irritating nature, the use of lanolin itself in baby products should be carefully considered.

In the transition from emulsions to baby oils, the use of anhydrous ointments (of which the zinc and castor oil cream is an example) should be considered. The following formulation (formula 9) is a little less sticky and more pleasant to use than zinc and castor oil, while still affording excellent barrier protection against excreta.

Formula 9

Constituents	*per cent*
Mineral oil	83.50
Acetylated lanolin alcohol	1.50
Silica	5.00
Zinc oxide	10.00

Baby oils are composed predominantly of a very pure grade of mineral oil. Small amounts of fatty acid esters, vegetable oils, lanolin derivatives and other compatible materials should be included only after careful consideration of safety and irritation potential.

Baby powders function is to lubricate, dry and perhaps to impart a slight perfume to the skin. The absorbency and adhesiveness of talc can be improved by the blending-in of other materials is shown by formulae 10-13.

Formulae *Constituents*	*10* *per cent*	*11* *per cent*	*12* *per cent*	*13* *per cent*
Sterilised talc	80.00	74.00	95.00	90.50
Magnesium stearate	10.00	4.00	–	2.50
Calcium carbonate	10.00	–	–	–
Kaolin	–	20.00	–	5.00
Glyceryl stearate	–	1.00	–	–
Cetyl alcohol	–	1.00	–	–
Starch	–	–	5.00	–
Zinc oxide	–	–	–	2.00
Perfume	q.s.	q.s.	q.s.	q.s.

Chapter 24

Antiperspirants and Deodorants

INTRODUCTION

Antiperspirants are used primarily to reduce (axillary) wetness. Deodorants (except soaps) are employed to reduce axillary odour. Since this is considered a non-therapeutic purpose and a function of the body is not considered to be altered, they are classed as cosmetics.

Several metal salts have astringent properties including those of aluminium, zirconium, zinc, iron, chromium, lead, mercury and several rarer metals.

MECHANISM OF DEODORANTS AND DEODORANT INGREDIENTS

Since axillary odour is largely produced by the action of bacteria on nutrients present in apocrine secretion, any compound which inhibits the growth of those micro-organisms found in the axillae will, in theory, exhibit deodourant properties.

· A formulation (Formula 1) of a deodorant composition is as follows :

Formula 1

Constituents	per cent
Hexamethylenetetramine	14–20
Zinc oxide	16–23
Starch	16–23
Petroleum jelly	38–43
Perfume	0.5–1.2

The effect of this deodorant remains for 15 days.

Sodium bicarbonate (baking soda) is used as a deodorant. It is an acid salt which can act chemically as either a mild alkali or a mild acid. Underarm odours are largely caused by volatile acidic compounds which are absorbed by baking soda to form stable odourless salts.

Evaluation of Antiperspirants

Efficacy of Antiperspirants

The range of effectiveness (average sweat reduction) in laboratory hot-room tests of OTC antiperspirants submitted to the Antiperspirant Review Panel is given in Table 24.1.

Table 24.1. Range of average sweat reduction — US FDA OTC antiperspirant review panel.

Dosage form	Average reduction %
Aerosols	20–33
Creams	35–47
Roll-ons	14–70
Lotions	28–62
Liquids	15–54
Sticks	35–40

The efficacy of an antiperspirant is best defined as the percentage reduction in the rate of sweating in the axilla that may be achieved after a realistic application or series of applications of the test product. The preferred methods for the determination of efficacy are gravimetry or the use of electronic hygrometers.

Product Formulation—Antiperspirants

Aerosols

The antiperspirant in powder suspension, are prepared using micronised powdered aluminium carbohydrate and oil base 3-4 % active ingredient. Many combinations of raw materials are available for the formulation of aerosol antiperspirants. In addition, formulations must provide maximum antiperspirant and deodorant effectiveness, maximum safety, cosmetic elegance and minimum staining. A typical powder-in-oil formulation (Formula 2) is given below :

Formula 2

Constituents	per cent
Aluminium chlorhydrate (micronised)	4.50
Isopropyl myristate	3.70
Fumed silica	0.15
Perfume	q.s.
Propellants (65:35)	to 100.00

The emollient or carrier for the aluminium chlorhydrate is used to produce a smooth feel on the skin and to help the powder to adhere. Commonly an ester such as isopropyl myristate is used although some products contain volatile silicones to reduce staining of clothes.

A suspending agent is added to prevent agglomeration of the aluminium chlorhydrate which could lead to valve blockage and leakage. Typical suspending agents are silica and 'Bentone' derivatives. Fumed silica has an extremely fine particle size and forms a coating over the powder particles to prevent the development of hard caking.

Antiperspirant sticks

The antiperspirant stick usually consists of a wax-like matrix which serves as a carrier for aluminium chlorhydrate powder and volatile silicone. A low-melting matrix and a high-boiling volatile silicone are necessary in order to conserve the latter during processing. For this reason stearyl alcohol is preferred over stearic acid because of its lower melting point (58.5°C vs 69.9°C).

Typical formulations are as follow (Formulae 3 and 4) :

Formulae	*3*	*4*
Constituents	per cent	per cent
Volatile silicone	46	46
Aluminium chlorhydrate powder	20	20
Stearyl alcohol	24	24
Polyethylene glycol distearate	6	6
Carbowax	4	4

Procedure : Heat the stearyl alcohol, carbowax and polyethylene glycol distearate 6000 to 80°C. When melted, add the aluminium chlorhydrate and mix thoroughly. Cool to 70°C and rapidly mix in the volatile silicone 7158. When mixing is complete, pour the mixture into a stick container. Allow the mixture to cool undisturbed for 24 hours.

Antiperspirant creams

While oil-in-water emulsions provide a convenient vehicle for storing and delivering antiperspirant actives, compositions of this type tend to produce an undesirable wet, cold and/or sticky sensation when they are applied to and rubbed into the skin. This can be minimised somewhat by utilising compositions in anhydrous form. The formulation (Formula 5) is given as under :

Formula 5

Constituents	*per cent*
Isopropyl myristate	32.0
Bentone 38 (thickening/suspending agent)	7.0
Ethyl alcohol (gel-promoting agent)	3.0
Zirconium hydroxychloride/aluminium chlorhydroxide/glycine complex	47.0
Silicone (antisyneresis agent)	10.0
Perfume	1.0

This is a substantially anhydrous antiperspirant composition in the form of a cream and is resistant to syneresis. A normal oil-in-water cream antiperspirant formulation (Formula 6) is given as under :

Formula 6

Constituents		*parts by weight*
A	Neo-fat	10.6
	Mineral oil	1.0
	Beeswax	1.0
	Glyceryl monostearate (pure)	6.4
B	Chlorhydrol (50% solution of aluminium chlorhydrate)	32.0
	Perfume	q.s.
C	Propylene glycol	5.0
	Sodium lauryl sulphate	1.3
	De-ionised water	to 100.0

Procedure

Heat A to 70°–80°C. Add C with agitation and cool to 35° – 40°C. Add B and mix thoroughly.

Roll-on antiperspirants

They are generally either emulsion products or aqueous alcoholic solutions thickened with cellulose gums. The aqueous alcoholic products generally dry quicker and are less sticky than the emulsion products. The viscosity of the final product is important to avoid leakage around the roll-ball.

The volatile silicone reduces the sticking of the roll-ball due to the drying out of the aluminium chlorhydrate (Formula 7) is given as under :

Formula 7

	Constituents	per cent
A	Magnesium aluminium silicate	1.0
	De-ionised water	49.0
B	Glyceryl monostearate (acid stable)	8.0
C	Aluminium chlorhydrate (50% soln.)	40.0
D	Volatile silicone	2.0
E	Fragrance	q.s.

Product Formulation—Deodorants

Deodorant soaps

Among the deodorants, the toilet soap market is important market in the world. The most frequently used antimicrobial agents in soaps at the present time are trichlocarban (TCC), cloflucarban (CF_3) and triclosan. Hexachlorophene and tribromsalan (TBS) occupied the key spots in this application before they are ban. All but triclosan are active only against Gram-positive organisms in the presence of soap. Triclosan is active against both gram-positive and many gram-negative organisms.

Deodorant sticks

A typical formulation (formula 9) is given below :

Formula 9

Constituents	per cent
Sodium stearate	8.0
Ethyl alcohol	74.8
Propylene glycol	10.0
Isopropyl myristate	5.0
Triclosan	0.2
Perfume	2.0

The alcoholic sticks shrink if the packaging is poor. Hence, to avoid shrinkage non-alcoholic deodorants sticks are prepared. The formulation (Formula 10) is given below :

Formula 10

Constituents	*per cent*
Sodium stearate	8.0
Propylene glycol	10.0
Perfume	1.0
Coconut diethanolamide	5.0
PPG-3-myristyl ether	68.8
Triclosan	0.2
Water	7.0

Aerosol Deodorants

The alcoholic solution of bactericide is the base of aerosol deodorants. A product called a 'deo-cologne' is based on alcoholic solution of perfume. It is used as body spray. A typical formulation (Formula 11) for an aerosol deodorant is given as under (Lacquered monobloc aluminium or tinplate containers can be used):

Formula 11

Constituents	*per cent*
Triclosan	0.05
Propylene glycol	2.00
Alcohol (99% v/v)	57.45
Perfume	0.50
Propellant	40.0

Chapter 25

Depilatories

INTRODUCTION

Since many years the formulation for removal of unwanted hair is known. The depilators are those preparations applied for the removal of superfluous hair, specifically on the face, legs and axilla with no injury. A distinction must be drawn between the mechanical removal of hair by either plucking it with tweezers or by embedding it in an adherent material which can then be pulled away from the skin bringing the hair with it (a process referred to as epilation), destruction of hair by electrolysis, and the removal of hair after it has been sufficiently degraded by chemical means.

EPILATION

Epilatory preparation are based on mixtures consisting essential oil of rosin and beeswax, modified in some instances by the addition of mineral oil and/ or waxes. The following formulations (Formulae 1 and 2) are given as under:

Formula 1

Constituents	*per cent*
Rosin	75.0
Beeswax	25.0

Formula 2

Constituents	*per cent*
Light coloured rosin	52.0
Yellow beeswax	25.0
Paraffin wax	17.0
Petrolatum	5.0
Perfume	1.0

427

In addition to rosin and waxes, mineral or vegetable oil may be included (for example at a level of about 15 per cent). The discomfort experienced when the hair is pulled off is reduced by addition of camphor. A local anaesthetic, for example benzocaine, to enhance this effect and an antibacterial compound will reduce the chance of infecting the skin after damage or exposure.

CHEMICAL DEPILATION

The term 'depilatory' refers to preparations intended for the chemical breakdown of superfluous hair without injury to the skin. The essential requirements of a depilatory are :

(1) Non-toxic and non-irritant to the skin;

(2) Efficient in action, removing hair rapidly (prefefably in 4 – 6 minutes);

(3) Preferably odourless;

(4) Stable on storage;

(5) Harmless to clothing; and

(6) Preferably cosmetically elegant.

Depilatory preparations usually contain as their active component an alkaline reducing agent. The latter will cause the hair fibres to swell and produce a cleavage of the cystine bridges between adjacent polypeptide chains as a preliminary to the complete degradation of the hair.

Sulphides

Compositions based on alkali and alkaline earth sulphides are capable of producing rapid depilation, when used with a suspension of lime.

Strontium sulphide is a much milder depilatory, but must be used at a higher concentration than sodium sulphide to produce an equivalent dehairing action. Preparations containing strontium sulphide. Although largely replaced by those based on thioglycollates, are still available. They are very effective and work within 3–5 minutes after application.

In addition to the active agent, a depilatory preparation may contain a humectant such as glycerine or sorbitol. A thickening agent, methyl cellulose incorporated, so as to thicken the solution sufficiently to allow it to remain in contact with the hair as long as necessary. For a sulphide depilatory, the following formula (3) will be found effective :

Formula 3
Constituents

Constituents	per cent
Strontium sulphide	20.0
Talc	20.0
Methyl cellulose	3.0
Glycerine	15.0
Water	42.0

This is prepared using an emulsion base for smoothness and stability. The formulation of depilatories depends upon very careful adherence to detail; slight departures from formulation in the process of manufacturing can produce remarkable differences between the efficacies of different batches of supposedly the same product. For this reason any formulation a general guide.

Despite their disadvantages, sulphide-based depilatories are preferred by many black-skinned men for removing facial hair because of their comparatively rapid action.

Stannites

The stannites have no appreciable odour, they suffer from instability, forming stannates in the presence of water. The triethanolamine, dextrine sugar, can be used as stabilisers of water-soluble organic compounds having three or more carboxyl group. They are also used for soluble silicates.

Substituted Mercaptans

The majority of depilatories available today are substituted mercaptans. They are used in the presence of alkaline reacting materials, For example calcium thioglycollate, in conjunction with calcium hydroxide. These preparations possess less odour than the sulphide but slow to act. They are safer on the skin than sulphides. They are therefore used on the face — an area where superfluous hair can cause great distress and where women have a strong psychological aversion to using a razor. In general, thioglycollate preparations are more attractive than the sulphide type. However, their slowness in attacking the coarse resistant hair of the underarm has left a market open for sulphide depilatories used for this purpose only.

It is often said that depilatories can be used for smoothing the legs, but so much of the product is required to cover each leg that it becomes uneconomic for most users. In any case, the legs are easy to shave.

Thioglycollates

Thioglycollate-based preparations are non-toxic and stable at concentrations between 2.5 and 4 per cent. At these concentrations they may produce depilation in 5–15 minutes at pH 10.0 and 12.5 to produce depilation within a fairly short time and without irritating the skin. Basic formula 4 for preparing cream, semi-fluid and powder depilatories is given below :

Formula 4 (Depilatory cream)

Constituents	*per cent*
Evanol	6.5
Calcium thioglycollate	5.4
Calcium hydroxide	7.0
Duponol paste	0.02
Sodium silicate	3.43
Perfume	q.s.
Distilled water	to 100.0

Procedure: Add with stirring Duponol and Evanol to hot water (70°C) till it melts and disperse. After cooling add calcium hydroxide, perfume and calcium thioglycollate until uniform solution obtained. Formulae 5 and 6 are given below :

Formulae (Semi-fluid depilatories)

Constituents	5 *per cent*	6 *per cent*
Cream base		
Distilled water	60.0	60.0
Cetyl alcohol	6.0	6.0
Brij		
Final product		
Distilled water	17.3	17.2
Calcium thioglycollate	5.4	5.4
Calcium hydroxide	6.6	10.4
Strontium hydroxide	3.7	–
Perfume	q.s.	q.s.
Cream base (as above)	67.0	67.0

Procedure: Prepare the cream base at 70°C and allow to cool to room temperature. Add the calcium thioglycollate to the bulk of the water and mix well; add the calcium hydroxide slowly with stirring, followed by the strontium hydroxide and any remaining water. Combine the two parts and stir well, adding the perfume at this point. Formula 7 for powder depilatory is given below :

Formula 7 (Powder depilatory)

Constituents	Per cent
Calcium thioglycollate	20.0
Calcium hydroxide	23.1
Strontium hydroxide	8.9
Sodium lauryl sulphate powder	1.5
Cellosize	1.0
Magnesium carbonate	45.2
Perfume	0.3

Procedure: Mix the calcium thioglycollate, calcium hydroxide, strontium hydroxide, sodium lauryl sulphate and sellosize. Blend the perfume thoroughly with the magnesium carbonate. Add the latter to the former and blend thoroughly.

Other 'Thio' Compounds

A non-irritant, rapid-acting depilatory 'soap' bar formula 8 is given below:

Formula 8

Constituents	per cent
Thiolactic acid	30.5
Urea-sorbitol complex	25.5
Strontium hydroxide	4.5
Calcium hydroxide	4.5
Sodium lauryl sulphate	35.0

Enzymes

Depilatory preparations based on the enzyme keratinase have also been developed. They have the unpleasant odour of sulphide or even thioglycollate depilatories, and are non-irritant, but are not quite as effective.

Keratinase is used in depilatory preparations in a purified form, buffered to a pH within the range of 7–8, with an activity of 200k units per mg. The k unit is defined as the amount of enzyme which will digest wool keratin so as to produce an increase in optical density of 0.04 at 280 nm.

Facial Depilatories for Black Skin

Conventional thioglycollate-based depilatories take 15–20 minutes to remove beard hair which is regarded as far too long. Effectiveness outweighs cosmetic elegance. The tendency is to use a powder depilatory which must be mixed with water before use but which effects adequate hair removal in 3–7 minutes.

The active ingredients commonly used in powder depilatories are barium sulphide and calcium thioglycollate. The former is the most popular because of its effectiveness and despite its offensive odour.

However, powders based on calcium thioglycollate are said to be gaining favour since they are less odourous and therefore more easily perfumed, though less effective.

EVALUATION OF DEPILATORY EFFICACY

The procedure for determining the efficacy of depilatories, involves the measurement of the cross-sectional diameter and the length of a hair immersed in a solution of a depilatory, and observing the time of maximum hair swelling. Sigmoid curves are obtained when both the length and width of swelling hair are plotted against time.

The slope maxima of these sigmoid curves may be used to define an index of depilatory effectiveness *in vitro*.

Table 25.1. Relative efficacy of 'thio' compounds at 5 per cent concentration, adjusted to pH 11.0 – 12.0.

'Thio' compound	Average depilation time* (min)
2-Mercapto-ethanol	4.0
Thioglycerol	6.5
Thioglycollic acid	7.5
3-Mercaptopropionic acid	8.0
2-Mercaptopropionic acid	11.0
Thiodiglycol	15.0
Thiomalic acid	15.0

* Average of times for all alkalies used for neutralisation. Some combinations are found to be more effective than others.

Table 25.1 summarises the relative efficacy of various 'thio' compounds at a concentration of 5 per cent pH 11.0 – 12.0. It has been found that a concentration of 5 per cent 'thio' compound is sufficient.

The increase in concentrations do not increase the speed of depilation appreciably. Table 25.2 highlights the activities of thio-glycollic acid and thio-lactic acid neutralised with various alkalies.

Table 25.2. Depilatory activity of thio-glycollic acid and thio-lactic acid neutralised with various alkalies (mercapto–acid concentration 0.4M, pH 12.5)

Cation	Thio-glycollic acid		Thio-lactic acid	
	Depilation time (min)	Effect on skin	Depilation time (min)	Effect on skin
Calcium	7	0	10	0
Barium	12	0	7	0
Strontium	5	0	7	0
Sodium	4	+	5½	+
Potassium	3½	+	4	++
Lithium	3½	0	5	0

Chapter 26

Shaving Preparations

INTRODUCTION

A wide range of preparations are now available which prepare beard or face for shaving. These preparations increase speed and comfort during and after-shave. The shaving preparations has been divided into three types, wet-shaving, dry-shaving and after-shave.

WET SHAVING PREPARATIONS

Wet-shaving preparations are to soften the beard, to lubricate the passage of the razor over the face and to support the beard hair. The preparation should be non-irritating, assist in removing shaving debris from the face, should be stable over a range of temperatures, resistant to rapid drying out and collapse, non-corrosive to the razor blade and easily rinsed from the razor and face. There is good evidence for the hair softening and lubrication functions of the shaving preparation, but little has been reported on the hair-supportive role.

Lather Shaving Cream

Criteria for a good lather shaving cream

Reasons for successful shaving preparations are economic use, supply of water to beard, and maintaining hair in fully water saturated condition. The requirements of a good lather shaving are, it must produce rich copious lather, composed of small bubbles, non-irritant, good wetting properties, smooth, soft, adhere readily and retain texture at all temperatures. The following points evaluate shaving preparations.

(1) Ease of transfer to and spreading on the face.

(2) Wetting and drainage properties of foam.

(3) Comfort and closeness of shave.

(4) Foam texture, rigidity rheology and stability.

(5) Easy removal of the other shaving debris from the razor basin.

(6) Acceptability on perfume and life of razor blade.

Formulation

Lather shaving creams are concentrated dispersions of alkali metal soaps in glycerol and water. To maintain the desired level of foamability, consistency and product stability, careful control of the manufacturing process is essential. Even the slightest change to the formulation or manufacturing procedure can result in a disastrous phase separation of the cream at slightly elevated temperatures.

Lather shaving creams normally contain 30 to 50 per cent soaps. To produce voluminous lather, it is usual to add some coconut oil fatty acids or stearic acid. The satisfactory ratio of stearic acid to coconut oil is 75:25. A mixture of sodium and potassium hydroxides is used to saponify the fatty acids. It has been suggested that a 5:1 ratio of potassium hydroxide to sodium hydroxide with 3–5 per cent free fatty acids will give shaving creams of the correct degree of plasticity. A cream containing a high level of sodium soaps tends to be thick and stringy, from which it is often difficult to produce a good lather. Lather creams can be made with potassium soaps alone, but these tend to be less stable. Formulae 1 and 2 given below show the formulation of lather shaving cream :

Formulae	*1*	*2*
Constituents	*per cent*	*per cent*
Stearic acid	30.0	36.0
Coconut oil	10.0	9.0
Palm kernel oil	5.0	–
Potassium hydroxide	7.0	8.0
Sodium hydroxide	1.5	1.0
Glycerine	10.0	–
Water	36.5	43.0
Perfume	q.s.	3.0
Sorbitol (70% solution)	–	3.0

Saponification of mixture of half of stearic acid with oil at 75 – 80°C using alkali water and glycerine with agitation. Later on add left out stearic acid and perfume at 35°C. In second process saponify coconut oil using alkali at 75–80°C. Then add melted stearic acid, sorbitol, preservative and water. Add perfume after cooling at 35°C.

Lather Shaving Stick

A lather shaving stick can be prepared from a mixture containing 80 per cent fatty acid soaps, 5 – 10 per cent glycerol and 8 – 10 per cent water. The ratio of the fatty acids and the ratio of potassium to sodium soaps should be similar to those described under lather shaving creams. After mixing, the composition is chipped, dried and milled with perfume, colour or an opacifier. The soap flakes are packed to the desired shape using a soap plodder.

Aerosol Shaving Foams

Aerosol shaving foams are oil-in-water emulsions in which propellant droplets, liquefied under pressure, form a substantial part of the oil phase. When the emulsion is discharged to the atmosphere, the dispersed propellant droplets vapourise, producing a foam consisting of propellant vapour bubbles surrounded by an aqueous surfactant phase. The formula 3 of foam contain from 4 to 15 per cent soaps, mainly triethanolamine stearate with minor proportions of potassium and sodium stearates.

Formula 3

Constituents	*per cent*
Triethanolamine stearate	8.0
Sodium stearate	1.0
Potassium stearate	4.6
Water	72.5
Perfume	0.9
Borax	0.5
Propellant (fluorocarbon)	12.5

The composition of shaving foam concentrate is given in formula 4 :

Formula 4

Constituents	*per cent*
Potassium soap from stearic acid /coconut oil fatty acids (80:20)	1.5
Potassium Polyacrylate (polyacrylic acid mol, wt, 100 000-200 000)	1.0
Polyvinylpyrrolidone	0.5
Stearic acid /coconut oil fatty acids (80:20)	3.0
Castor oil	3.0
Lauric acid diethanolamide	0.5
Polyoxyethylene sorbitan monolaurate	0.5
Perfume	0.5
Water	89.5

Guidance on formulation

The hydrocarbon propellant base and scum-free aerosol shaving form (formulae 5,6) is prepared by heating parts A and B separately at 75°C. Then adding A to B at 35°C with stirring, also add perfume. At room temperature charge the aerosol container. The process for the preparation of soap free aerosol shaving foam (formula 7) except adjustment of pH of concentrate using citric acid remain same.

Formula 5

	Constituents	per cent
A	Palmitic acid	5.0
	Lauric acid	1.0
B	Sodium lauryl sulphate	1.0
	Polyethylene glycol (400) monolaurate	0.5
	Polyacrylic acid (40% aq.) mol.wt. 100 000	1.5
	Triethanolamine	2.0
	Potassium hydroxide	0.8
	Glycerol	5.0
	Water (deionised)	83.2
	Perfume	q.s.
	Concentrate	96.9
	Propellants, isobutane/ Propane	3.1

Formula 6

	Constituents	per cent
A.	Palmitic acid	1.95
	Myristic acid	0.62
	Myristyl alcohol	2.10
B.	Polyoxyethylene (20) cetyl ether	5.23
	Lauric diethanolamide	5.23
	Propylene glycol	0.82
	Glycerol	3.54
	Triethanolamine	1.54
	Water (deionised)	78.97
	Perfume	q.s.
Or	Concentrate	91.5
	Propellants 12/114(40:60)	8.5
Or	Concentrate	97.0
	Propellants (Butane 48)	3.0

Formula 7

	Constituents	*per cent*
A.	Myristyl Alcohol	2.1
B.	Dicarboxylated lauric imidazoline 40% solution	5.1
	Polyethylene glycol (1000) monolaurate	5.5
	Propoxylated polyol	0.7
	Glycerol	5.0
	Water (Deionised)	81.6
	Perfume	q.s
	Citric acid	q.s.
	Concentrate	97.0
	Propellant (Butane 48)	3.0

Brushless or Non-lathering Cream

Brushless or non-lathering shaving creams are oil-in-water emulsions. They contain components similar to those in vanishing creams. The main difference being that the concentration of oils and emulsifying agents tends to be higher in the shaving preparations. Ideally, the cream should vanish on completion of the shave, leaving the face free from irritation and with a matt appearance. Since a too rapid disappearance of the cream would be deleterious to the comfort and closeness of the shave, it should be possible at least to rub any remaining cream into the skin after the shave.

The composition of a typical brusheless shaving cream is given in formulae 8-10. The preparation is done by heating the oil and stearic acid at 75°C. Then disperse the carbopol in cold water and add the glycerol, surfactant and triethanolamine. Add the oil phase to the aqueous phase at 75°C with vigorous stirring. Cool the mixture rapidly, and add the the perfume.

Formulae	*8*	*9*	*10*
Constituents	*per cent*	*per cent*	*per cent*
Mineral oil	9.0	–	5.0
Lanolin	0.5	4.0	–
Stearic acid	14.5	18.0	18.0
Carbopol	0.5	–	–
Triethanolamine	2.5	1.0	1.0
Triethanolamine lauryl sulphate	1.0	–	–
Glycerol	5.0	2.0	–
Water	67.0	66.8	63.0
Preservative, perfume	9.5	0.2	9.5
Propylene glycol monostearate	–	4.0	–

Isopropyl palmitate	–	4.0	–
Silicone fluid	–	–	1.0
Polyoxyethylene	–	–	5.0
Sorbitan monostearate			
Borax	–	–	2.0
Sorbitol (70%)	–	–	5.0

Brushless Shaving Stick

Brushless shaving stick can be applied directly to the face. The continuous thin smear left on wetted skin provides adequate lubrication for the shaving operation. The stick is composed of fatty or waxy materials to which hydrophilic properties have been imparted soap or a partial fatty acid ester of a polyhydric alcohol. This ensures that the product is readily wetted. It is not more sparingly soluble in, water. Pigment, dyestuff or opacifier is incorporated to indicate the presence of the composition on the face formulae 11 and 12 are given as under :

Formula 11

Constituents	*per cent*
Sesame oil	35.35
Spermaceti	45.80
Stearin	7.60
Soap	5.00
Monoglycerides of coconut oil fatty acids	3.00
Titanium dioxide	2.00
Perfume	1.25

Formula 12

Constituents	*per cent*
Mineral oil	95.97
Dioctyl sodium sulphosuccinate (Aerosol OT)	1.44
Lanolin	0.96
Silicone fluid	0.67
Octadecanol	0.48
Fragrance	0.48
Preservative	q.s.

DRY SHAVING PREPARATIONS

It is generally recognised that electric shavers do not cut beard as close to the skin surface as a razor blade. Both electric and blade shaving result in there removal of skin, the amount removed for an individual being dependent on the pressure applied to the face. Generally, the closer the shave the greater the amount of skin damage. It has been suggested that pre-electric shave preparations may not increase the quality of the shave but may assist in reducing skin damage.

Lotions of the oily type aim to deposit on the face a film of lubricant which reduces the drag of the cutting head against the skin. It has been shown that a film of silicone oil substantially reduces the frictional force between skin and a smooth steel probe formulae 13-15 are given below :

Formula 13

Constituents	*per cent*
Ethanol	45.0
Sorbitol	5.0
Lactic acid	1.0
Water	49.0

Formula 14

Constituents	*per cent*
Zinc phenolsulphonate	1.0
Distilled extract of witch hazel	40.0
Ethanol	40.0
Water	18.8
Menthol	0.1
Camphor	0.1

Formula 15

Constituents	*per cent*
Aluminium chlorhydroxide (50%)	5.0
Isopropyl myristate	5.0
Ethanol	80.0
Perfume	q.s.
Colour	q.s.
Water	to 100.0

AFTER-SHAVE PREPARATIONS

The purpose of an after-shave preparation is to relieve the slight irritation or 'after-glow' and confer a pleasant of comfort and well-being after shaving. This is achieved by giving a slight coolness, anaesthesia, mild astringency or emolliency to the skin. At the same time, the preparation should be antiseptic and help to keep the skin free from bacterial infection during the short time it takes to recover from to the slight degree of injury inflicted during the shaving operation.

After-shave Lotion

An after-shave lotion is a clear aqueous ethyl alcohol solution containing a perfume. They contain 50-70 per cent by weight ethyl alcohol. The controlled ratio of ethyl alcohol to water produces balanced mild astringency and coolness. Perfume or solubiliser is also used. Non-ionic surfactants with a hydrophile-lipophile balance (HLB) number in the range 15–18 are often found to be the most effective solubilisers, although anionic surfactants have also been used. Humectants and emollients are at levels not exceeding 5 per cent. Polyols such as glycerol, sorbitol and propylene glycol help to maintain the water content of the skin. Glycerol has the best humectant properties of the group, but propylene glycol is often preferred because of its greater solvent power, lower viscosity and higher volatility. The feel of the skin can be improved by the addition of long-chain fatty esters, for example, isopropyl myristate or lanolin. Quantities are often limited by their low solubility in aqueous alcohol solutions. Water-soluble lanolin derivatives can be used at higher levels to provide emolliency and to assist in the solubilisation of the perfume oil.

The level of menthol should be kept below 0.1 per cent because of its lachrymatory properties and because its odour can upset the balance of the perfume. Odourless cooling agents may be more appropriate for this type of product. Menthol is also said to cause slight surface anaesthesia of skin; however, it is preferable to achieve this effect with lignocaine at a level of 0.025-0.05 per cent. A basic after-shave lotion, composition is given in formula 16.

Formula 16

Constituents	*per cent*
Ethyl alcohol, specially denatured	60
Propylene glycol	3
Water, demineralised	36
Perfume	1

It is prepared by dissolving the perfume and propylene glycol in the alcohol and add the water slowly with stirring. Allow the solution to stand or several hours at about 4°C, then filter. Formulae 17-19 given below show the formulation of antiseptic after-shave lotion :

Formula 17

Constituents	*per cent*
Hyamine	0.250
Ethyl alcohol	40.000

Menthol	0.005
Benzocaine	0.025
Water	59.720
Perfume	q.s.

Formula 18

Constituents	per cent
Witch hazel extract	15.00
Ethyl alcohol	10.00
Alum	0.50
Menthol	0.50
Ethyl p-amino benzoate	0.05
Glycerol	5.00
Water	69.40

Formula 19

Constituents	per cent
Hyxadecyl alcohol	0.8
Ethyl alcohol	53.7
Distilled water	33.0
Polawax	2.0
Perfume	0.5
Propellant	10.0

Some perfumes are unstable to ultra-violet light. Therefore after exposure to direct sunlight for a few months, an after-shave lotion in clear glass bottles develops a characteristic 'bottle-odour', quite unlike the original fragrance. Accelerated testing of the ultra-violet light stability of the packaged product should therefore be included in the evaluation of after-shave preparations. Many manufacturers circumvent the problem by using opaque or chromium-containing glass bottles to package the lotion.

After-shave Gel

An aqueous alcohol gel can be formed by neutralising a carboxyvinyl polymer with a base. The amount of gelling agent (usually less than 1 per cent) and the degree of neutralisation controls the stiffness of the gel . The gel may optionally contain a physiological cooling agent an emollient which is soluble in the alcoholic solution. It is prepared by dissolving perfume and menthol in alcohol and stirred in water to get clear solution. Disperse the Carbopol 940 in the aqueous alcoholic solution. Reduce the speed of the mixer and slowly add the di-isopropylamine dissolved in the small quantity of retained water. The resulting product should be a crystal-clear gel, to pack it in coloured or opaque containers. The formula 20 is given below :

Formula 20
Constituents	*per cent*
Ethanol	45.1
Water	53.0
Carbopol	1.0
Menthol	0.1
Di-isopropylamine	0.8
Perfume oil	q.s

After-shave Cream and Balm

The astringency of after-shave preparations containing more than 50 per cent by weight alcohol can be irritating to skin which has been excessively damaged by shaving or over-exposure to sun and wind. Formula 21 gives soap-free after-shave cream.

Formula 21 (Soap-free after-shave cream)
Constituents		*per cent*
A.	Glycerol monostearate	10.0
	Mineral oil	10.0
	Petrolatum	6.0
	Tegiloxan	0.5
	Lanolin	3.0
	Cetyl alcohol	3.0
B.	Glycerol	3.0
	Citric acid	0.2
	Potassium aluminium sulphate	0.1
	Water	64.2
	Perfumem	q.s.

Formula 22 (Opaque hydro-alcoholic balm)
Constituents		*per cent*
A.	Amerchol	5.0
	Isopropyl lanolate	10
	Poly-thylene glycol	3.0
B.	Carbopol	0.5
	Water	60.0
	Triethanolamine	0.5
	Ethyl alcohol	30.0
	Perfume	q.s.

Procedure: Add the Carbopol slowly to the water at room temperature with rapid agitation until cloudy dispersion is obtained. Heat the carbopol solution and the oil phase separately to 75°C. Add the carbopol solution to the oil phase and stir for 5 minutes before adding the triethanolamine. Cool with stirring to 38°C. Add the alcohol and perfume.

Chapter 27

Foot Preparations

INTRODUCTION

Foot discomfort is common among a large population in different form. Although foot fatique affects adversely which sharply reduces individuals effort, very little attention is paid to it. It is really amazing how little personal attention is lavished upon the much maligned human foot, although foot fatigue affects adversely both the physical and mental well-being of the sufferer, and leads to a sharp decline in the individual effort. Man-made materials impermeable to moisture vapour and thus to sweat have considerable influence on foot health. Wear trials have shown that it is the properties of the upper material which have by far the greatest effect on sweat accumulation in shoes.

Sweat accumulation in footwear has a direct effect :

(1) on mechanical properties of foot skin;

(2) on physicochemical properties of the shoe material; and

(3) on encouragement of micro-biological growth in the skin and the shoe materials.

Foot Malodours and Foot Ailments

A musty or mouldy odour to the shoe and foot is due to fungi and breakdown of perspiration due to bacteria. However, other odours can arise by interactions with the material of the shoe. This will lead to penetration of odour of sweaty feet, burning and itching sensation between the toes; painful, tired and swollen feet; softening of the toe-nail bed; moist skin irritation, creating ideal conditions for fungal infections.

445

Foot Infections, Care & Hygiene

The feet spend long periods covered with socks or stockings and encased in footwear. This results in persistently warm and moist conditions which form an ideal environment for microbial proliferation and activity. The use of communal washing and bathing facilities is a major cause of spread of such infections. The supreme importance of hygienic cleansing of communal bathrooms, showers, etc., is clear; regular cleansing with germicidal products to remove skin debris and microbial contaminants is essential. The most prevalent kinds of foot infection are caused by fungi—mycosis. The well known mycosis 'athlete's foot' has almost become an international disease.

The feet should be washed at least once a day with soap and water. After washing it should be dried thoroughly, especially between the toes, and dusted with a talcum or foot powder. If infection is suspected or likely to have been encountered, both the soap and the powder should certainly contain an antiseptic agent.

There are many types of foot preparation in existence. Some are used to provide relief for tired and aching feet, others to soften cornified skin or to combat foot perspiration, and there are those used for alleviating skin irritations, eruptions and infections and for providing an anti-bacterial or anti-fungal effect.

Bathing the Feet

The reduction of bacterial infection, stimulation of blood circulation, and elimination of sweat secretions can be done by bathing foot in hot water. If the bath is alkaline it softens the hard keratinous layer of the skin, corns, calluses, etc. Artificial sea-salt powders are used in foot baths (formula 1). One part of artificial powder of prepared by mixing is enough for 20 parts of water for foot bath.

Formula 1

Constituents	*parts*
Potassium iodide	1
Potassium bromide	2
Magnesium chloride	250
Calcium chloride	125
Magnesium sulphate	250
Sodium sulphate	500
Sodium chloride	1500
Colour, perfume	q.s.

A bubbling foot bath tablet releases oxygen and carbon dioxide. The composition of Deo-foam bath for the feet is shown in formula 2.

Formula 2

Constituents	*per cent*	
Lathanol	26.0	
Tartaric acid powder	26.0	
Salicylic acid	1.0	
Sodium bicarbonate powder	25.0	
Sodium sesquicarbonate powder	9.0	
Sodium hexametaphosphate	4.0	
Sodium perborate	5.9	
Perfume	1.5	or q.s.
Calflo	1.5	
Colour: FD & C Blue No. 1	0.1	or q.s.

Formula 3

Constituents	*per cent*
Texapon extract N25	55
Comperlan KD	6
Salicylic acid 20% solution*	25
Foromycen F10	2
Water	7
Neo-PCL water-soluble	1
Perfume oil	4
Salicyclic acid 20% solution:	
Borax	8
Salicylic acid	20
Water	72

(Heat to near boiling point to obtain a clear solution)

FOOT POWDERS

Foot powders keep the feet dry. If they contain a fungicidal or fungistatic agent a better opportunity is presented for a more continuous application as perspiration dissolves the medicament. In shoes they help to prevent reinfection (formula 4). They are recommended in the Indian Army.

Formula 4

Constituents	*per cent*
Thymol	1
Boric acid	10
Zinc oxide	20
Talc	69

Another simple fungicidal dusting powder for the feet is given below in formula 5 :

Formula 5
Constituents   *per cent*

Constituent	per cent
Zinc undecenoate	10.00
Undecenoic acid	2.08
Pine oil	0.47
Starch	50.00
Light kaolin	37.45

The main constituents of aerosol spray form foot powders are deodorant and antiperspirant. The different compositions are given in formulae 6 and 7.

Formula 6 (Aerosol foot powder)
Constituents   *per cent*

Constituent	per cent
Talc	92.0
Zinc stearate	1.0
Irgasan	0.2
Santocel	2.8
Span	2.0
Isopropyl myristate	1.0
Perfume oil	1.0
Powder base	15
Propellants	85

Formula 7 (Women's foot spray)
Constituents   *per cent*

Constituent	per cent
Fragrance	0.35
Menthol crystals	0.10
Talc 'lo micron'	8.35
Acetol	1.00
Chloroxylenol	0.20
Propellants 12/11 (40 : 60)	90.00

FOOT SPRAYS

Foot sprays are intended principally to cool and refresh the feet. They contain anti-microbial agents, and an antiperspirant material, an absorbent powder. They may be sprayed through socks or stockings and therefore are suitable for 'emergency' treatment.

The following three formulae, two for aerosol sprays respectively and one for a non-aerosol pump spray, illustrate modern products. The formula 9 is soothing and cooling spray. It provides relief for hot tired feet and leaves them feeling soft and smooth. 'A clear liquid solution for packing in a manually operated spray dispenser which dries quickly and tack-free leaving the feet cool and refreshed is formula 10 :

Formula 8 (Antiperspirant foot spray)

Constituents	*per cent*
Talc	34.0
Cab-o-sil	2.0
Microdry	5.0
Procetyl	`4.0`
PVP-VA Co-polymer	4.0
Menthol	0.5
Ethanol anhydrous	50.5
Concentrate	15.0
Propellant	85.0

Formula 9 (Aerosol foot deodorant spray)

Constituents	*per cent*
Ammonyx	0.125
Menthol (racemic)	0.125
2-Ethyl-1,3-hexanediol	2.25
Laneto	0.25
Ethanol anhydrous	27.25
Perfume	q.s.
Propellants	70.00

Formula 10 (Foot deodorant spray solution for pump spray)

Constituents	*per cent*
Ethanol 96%	28.00
Versene	0.03
Perfume	q.s.
Hyamine	0.25
Laneto	2.00
Deionised water	68.62
Silicone fluid	0.10
Camphor	0.50
Menthol (racemic)	0.50

FOOT CREAMS

Foot massage is extremely relaxing. Foot creams are a suitable adjunct to massage and may contain antimicrobials, antiperspirants, mildly keratolytic agents, vasodilators to stimulate circulation, cooling agents, as well as providing emolliency and skin-softening properties. A simple emollient and deodourant foot cream may be formulated (Formula 11) as given below :

Formula 11

Constituents	per cent
Glyceryl monostearate	15.0
Lanolin	1.0
Sorbitol syrup 70%	2.5
Glycerine	2.5
Antimicrobial agent	0.25–0.50
Water	to 100.0

The foundation creams formulae may be applicable to foot creams. Many of the formulae given under foundation creams may be adapted for foot creams. Some spermaceti substitute or high-melting wax, or fatty acid such as stearic acid, may be included to give a waxy film rather than a greasy layer on the foot. Camphor methyl salicylate and menthol are commonly included to give a cooling sensation. The formula 12 is given below :

Formula 12

Constituents	per cent
Glyceryl monostearate	12.0
Mineral oil	2.0
Glycerine	5.0
Spermaceti substitute	5.0
Camphor	1.0
Methyl salicylate	1.0
Water	73.9
Preservative	0.1

Formula 13 (Foot cream)

Constituents		per cent
A	Ottasept extra	1.00
	Lexemul AR	16.00
	Cetyl alcohol	1.00
	Amerchol L-101	3.00
	Solulan 98	0.50
	Mineral oil	3.00

B	Alcloxa	0.25
	Propylene glycol	5.00
	Distilled water	69.95
C	Perfume	0.30

Formula 14 (Oil-in-water foot massage emulsion)

	Constituents	*per cent*
A	Self-emulsifying	25.0
	Propyleneglycol dipelargonatc	3.0
	Hostaphat	0.5
B	Water	65.9
	Propylene glycol	5.0
	Preservative	0.3
	Borax	0.1
C	Perfume oil	0.2

Corn and Callus Preparations

Corn cures or paints contain salicylic acid in collodion either with or without cannabis extract. Other preparations include lactic acid, trichloracetic acid, glacial acetic acid is as given in formulae 15 and 16 :

Formula 15

Constituents	*per cent*
Salicylic acid	10
Lactic acid	10
Flexible collodion	80

Formula 16

Constituents	*per cent*
Salicylic acid	10
Cannabis extract	10
Flexible collodion	80

Five per cent castor oil may be added to the above if desired, for its plasticising effect. For callus softening, alkaline products are used, such as given formulae 17 and 18 :

Formula 17

Constituents	*per cent*
Tribasic sodium phosphate	8
Triethanolamine	12
Water	80
Perfume (alkali stable)	q.s.

Formula 18
Constituents — *per cent*

Constituents	per cent
Potassium hydroxide	0.5
Glycerine	15.0
Water	84.5

Formula 18 should be used with care and must not be left in contact with the skin as it is caustic. The composition of corn remover in gel form is shown in formula 19 :

Formula 19 (Corn remover)

Constituents	per cent
Salicylic acid	12
Benzoic acid	6
Pluronic	47
Water	35
Perfume, colour	q.s.

The product is effective on clean skin but the presence of oil or grease may cause it to 'smear' rather than 'roll'. The emulsion can be made pressure-sensitive by using minimal quantities of the correct type of emulsifier. The following formula (formula 20) will serve as a basis.

Formula 20

Constituents	per cent
Stearic acid	1.0
Beeswax	4.0
Cetyl alcohol	0.5
Paraffin wax	12.0
Triethanolamine	0.6
Sorbitol syrup	5.0
Veegum	0.5
Water	76.4

Chilblain Preparations

Chilblains are a condition resulting from poor circulation and an inadequate supply of blood to the hands and feet. The constituents for ointment (formula 21) is given below :

Formula 21

Constituents	per cent
Phenol	1.0
Camphor	6.0
Balsam of Peru	2.0
Soft paraffin	25.0

| Hard paraffin | 7.5 |
| Anhydrous lanolin | 58.5 |

Athlete's Foot Preparations

Since the introduction of aerosol sprays and powders the number of preparations for the treatment of athlete's foot has been increased. The antiseptics that are mainly present in aerosol foot preparations are selected primarily for the control of fungi, but they have also an important function as deodourising agents to inhibit the growth of odour-producing bacteria.

Salts of certain fatty acids such as propionic and undecylenic acids, often used in conjunction with the free acids, have been claimed to give good results in the treatment of *Trichophyton* and other dermatophytes. The formulae 22 and 23 are given for foot ointments.

Formula 22 (Propionate ointment)

Constituents	per cent
Sodium propionate	16.4
Propionic acid	3.6
Propylene glycol	5.0
n-propyl alcohol	10.0
Carbowax	35.0
Zinc stearate	5.0
Water	25.0

Formula 23 (Undecylenate ointment)

Constituents	per cent
Undecylenic acid	10.0
Triethanolamine	6.0
Propylene glycol	14.0
Carbowax 1500	10.00
Carbowax 4000	40.0
Water	20.0

These acids rarely cause any adverse reaction on the skin. Undecylenic acid is often used in conjunction with zinc undecylenate, which is similar to zinc stearate. It would be more suitable for use in aerosol powders than in presssurised lotions.

The antifungal foot powder based on a mixture of salts of propionic and caprylic acids is illustrated by the following formula 24 and gel form by formula 25.

Formula 24

Constituents	*per cent*
Calcium propionate	15.00
Zinc propionate	5.00
Zinc caprylate	5.00
Propionic acid	0.25
Talc	74.75

Formula 25

Constituents	*per cent*
Ethanol 96%	72.0
Water	21.0
Undecylenic acid	5.0
Carbopol	1.0
Di-isopropanolamine	1.0

Other Developments

Among newer developments are 'odour-destroying insoles' with absorbent and deodorant properties, a foot care kit with 'everything you need' for the care of the feet and a synthetic fibre 'foot sponge' for use with soap to remove thick hard skin. A recent patent describes a deodoriser for the foot, etc., based upon an ion-exchange material such as ion-exchange cotton.

Chapter 28

Sunscreen, Suntan and Anti-sunburn Products

INTRODUCTION

The sunlight has both beneficial and harmful effects on the human body, depending on the length and the frequency of exposure, the intensity of the sunlight and the sensitivity of the individual concerned. Erythema of the skin is obvious effect of sun rays. It is followed by the formation of tan. Factually tan is a protective reaction of the human body to minimise any damaging effect of solar irradiation. The intensity of erythema (reddening) produced depends on the amount of uv energy absorbed by the skin. Erythema usually starts to develop after a latent period of 2–3 hours and reaches its maximum intensity within 10 to 24 hours after exposure.

SUNLIGHT AND THE HUMAN BODY

Tanning

The tanning ability of an individual is genetically predetermined and depends on his capacity to produce melanin pigment within the melanocytes. The energy from 300 and 660 nm rays stimulates tanning immediately. Its maximum efficacy lies between 340 nm and 360 nm. It entails the immediate darkening of unoxidised melanin granules present in the epidermal layer of the skin, near its surface. It reaches a maximum about one hour after exposure to radiation and begins to fade within 2–3 hours after exposure.

The erythemogenic radiation (wavelength 295 to 320 nm) stimulates delayed and true tanning. True tanning starts about two days after exposure and reaches a maximum about two to three weeks later.

Beneficial and Adverse Effects of Sunlight

Beneficial effects

The moderate exposure of human body to sunshine gives sense of fitness, peace of mind and well-being. It stimulates blood circulation, increases the formation of haemoglobin, and may also promote a reduction in blood pressure. It plays a vital part in the prevention and treatment of rickets by producing—through the activation of 7-dehydrocholesterol (pro-vitamin D_3) present in the epidermis. This enhances the absorption of calcium from the intestine.

It has been used in the treatment of certain types of tuberculosis, such as the tuberculosis of glands and bones, and the treatment of certain skin diseases such as psoriasis. A beneficial influence on the autonomous nervous system and reduction in the susceptibility of individuals to various infections is observed. Finally, by producing melanin and causing thickening of the skin, it plays an essential part in the formation of the body's natural protective mechanism against sunburn.

Adverse Effects of Sunlight

Solar irradiation can have both short-term and long-term adverse effects.

Sunburn

The known symptoms of sunburn are temporary damage of the epidermis. These may range in severity from a slight erythema to painful burns and blistering accompanied in more severe cases, when large amounts of the skin have been affected, by shivering, fever and nausea, and sometimes pruritus. The symptoms of sunburn are the direct result of damage or destruction of cells in the prickle cell layer of the skin, possibly through denaturing of its protein constituents. The four degrees of sunburn are observed with exposure to mid-day sunshine. They are minimal perceptive erythema, vivid erythema, painful burn and burn and blistering burn.

Chronic exposure

It may also produce degenerative changes in the connective tissue of the corium and result in the so-called premature ageing of the skin. This is evidenced by the thickening of the skin, the loss of natural elasticity and the appearance of wrinkles, all resulting from the loss of the skin's water-binding capacity. There is also an increased tendency to form skin blemishes.

SUNSCREENS AND SUNTAN PREPARATIONS

The sunscreens and suntan are prepared to prevent or minimise the harmful effects of solar radiation or to assist in tanning the skin without any painful effects. Sunscreens are classified on the basis of applications as :

(1) Sunburn preventive agents are defined as sunscreens which absorb 95 per cent or more of uv radiation within wavelengths 290-320 nm.

(2) Suntanning agents absorb at least 85 per cent of uv radiation within the wavelength range from 290 nm to 320 nm but which transmit uv light at wavelengths longer than 320nm and produce a light transient tan. These agents will produce some erythema but without pain. Both agents are defined as sunscreens. These are chemical sunscreens absorbing a specific range of uv radiation. In some instances the same sunscreen may be employed in both product classes but at different concentrations (lower in a suntan product).

(3) Opaque sunblock agents aim to provide maximum protection in the form of a physical barrier e.g. Titanium dioxide and zinc oxide. Titanium dioxide reflects and scatters practically all radiation in the uv and visible range (290–777 nm), thereby preventing or minimising both sunburn and suntan.

Sunscreen Agents

Sunscreens should either scatter the incident light effectively, or they should absorb the erythemal portion of the sun's radiant energy. Opaque powdered materials, when applied to the skin either in the dry state or with suitable vehicles, will serve to scatter the ultra-violet light falling upon them.

The most important class of sunscreens are those which operate by absorbing the erythemal ultra-violet radiation. Sunscreen must have following properties :

(1) It must be effective in absorbing erythmogenic radiation in the 290–320 nm range without breakdown would reduce its efficiency or give rise to toxic or irritant compounds.

(2) It must allow full transmission in the 300–400 nm range to permit the maximum tanning effect.

(3) It must be non-violet and resistant to water and perspiration.

(4) It must possess suitable solubility characteristics to allow the formulation of a suitable cosmetic vehicle to accommodate the requisite amount of sunscreening.

(5) It must be non-odourous or at least sufficiently mild to be acceptable to the user and be satisfactory in other relevant physical characteristics such as stickiness, etc.

(6) It must be non-toxic, non-irritant and non-sensitising.

(7) It must be capable of retaining its protective capacity for several hours and stable under conditions of use.

(8) It must not stain clothing.

A number of proprietary products have been based upon menthyl salicylate. It is non-irritant, odourless and initially acts as a satisfactory sunscreen at concentrations of about 10 per cent.

The level of five sunscreen agents recommended by the US department of health, education and welfare to use in a base like mineral oil, cold cream or ethyl alcohol is, isobutyl *p*-amino benzoate, 5%, propylene glycol *p*-aminobenzoate 4.0%, monoglyceryl *p*-aminobenzoate 2.5%, diagalloyl trioleate 5.0% and benzyl salicylate and benzyl cinnamate 2% each.

Evaluation of sunscreen Preparations

All sunscreen preparations are necessarily assessed for their efficiency. The absorption characteristics measured in terms of concentration, thickness and wavelength which decides efficiency of products. In other words absorption is called as optical density. It is the logarithm of the ratio of the intensities of the radiation before and after passage through the solution. It is directly proportional to the concentration and to the thickness of the material, hence calculation is simple. These three categories of ingredients of sunscreen with concentration are tabulated in Table 28.1.

Table 28.1. Classification of sunscreen ingredients.

I Safe, Effective Compound	Discount limit (%)
p-Aminobenzoic acid	5–15
2-Ethoxyethyl-*p*-methoxy cinnamate	1–3
Diethanolamine-*p*-methoxy cinnamate	8–10
Digalloyl trioleate	1–5
2,2-Dihydroxy-4-methoxybenzophenone	3
Ethyl-4-bis-(hydroxypropyl) aminobenzoate	1–5
2, Ethylhexyl 2-cyano-3, 3-diphenyl acrylate	7–10
Ethylhexyl-*p*-methoxy cinnamate	2–7.5
2-Ethylhexyl salicylate	3–5

(Cont'd)

I Safe, Effective Compound	Discount limit (%)
Glyceryl aminobenzoate	2–3
3, 3, 5-Trimethylcyclohexyl salicylate	4–15
Lawsone with dihydroxyacetone Lawsone	0.25
(DHA)	3.0
Menthyl anthranilate	3.5–5
2-Hydroxy-4-menthoxy benzophenone	2–6
Amyl-*p*-dimethylaminobenzoate	1–5
2-Ethylhyexyl-p-dimethylamino benzoate	1.4–8
2-Phenyl benzimidazole-5-sulphonic acid	1–4
Red petrolatum	30–100
2-Hydroxy-4-methoxybenzophenone-5 sulphonic acid	5–10
Titanium dioxide	2–25
Triethanolamine salicylate	5–12

II Not safe, not effective

2-Ethylhexyl-4-phenyl benzophenone-2'-carboxylic acid

3-(4-methylbenzylidene)-camphor

Sodium 3, 4-dimethylphenyl glyoxylate

III Unclassified

Allantoin combined with aminobenzoic acid

5-(3,3-Dimethyl 2-nonrbornylidene)-3-penten-2-one

Dipropylene glycol salicylate

The absorption characteristic of a chemical is frequently expressed as the molar extinction coefficient. Sunscreen index based on optical density is shown in Table 28.2.

Table 28.2. Sunscreen index (SI) based on optical density of 1 per cent solution at 308 nm.

Compound	SI
Ethyl p-dimethylaminobenzoate	14.8
Ethyl p-aminobenzoate	9.6
Isobutyl p-aminobenzoate	9.2
Propyl p-aminobenzoate	9.0
n-Butyl p-aminobenzoate	8.0
ß-Methyl umbelliferone	7.7
p-Aminobenzoic acid	7.4
Dehydroacetic acid	7.0
3-Carbethoxycoumarin	6.6
Benzilidine camphor	6.6

(Cont'd)

Compound	SI
Heliotropine	6.5
Umbelliferone acetic acid	6.0
Salicylic acid	4.3
Sodium p-aminosalicylate	4.3
Menthyl salicylate	4.0
Salicylamide	3.9
Methylenedisalicylic acid	3.0
3-Carboxycoumarin	3.0
Thiosalicylic acid	2.7
Brightener W/450	2.7
p-Hydroxyanthranilic acid	2.6
Sodium salicylate	2.4
Digalloyltrioleate	2.3
α-Resorcylic acid	2.2
Salicylaldehyde	2.2
p-Aminosalicylic acid	1.9
Dipropyleneglycol salicylate	1.9
Pyribenzamine	1.8
Pyrianisamine maleate	1.7
Sodium gentisate	1.7
Fluorescent white	1.6
Ethanolamide of gentisic acid	1.5
Ethyl gallate	1.4
Sodium sulphadiasine	1.2
Ethyl vanillate	1.1
Sodium sulphapyridine	0.95
Lauryl gallate	0.85
Totaquine	0.80
Barbituric acid	0.19
Chlorophyll	0.15
Amberlite IR–4–B	0.15
Salicyl alcohol	0.05
Anisic acid	0.01
Uvitex RBS	0.01
Uvitex RS	0.005

Water resistance and perspiration resistance of sunscreens

The efficacy of sunscreen preparations also depends on their ability to form
substantive films on the skin, which are resistant to removal by water or

perspiration. Hence, the evaluation of sunscreen preparations is done by assessing their water and perspiration resistance.

An appropriate test for determination of resistance to perspiration is to obtain sweating within 30 min. under controlled environmental conditions. Such a claim will be allowed if the sunscreen preparation under test retains its original PCD after the test. The claim 'resists removal by sweating' is also appropriate if the product proves to be water resistant, since water immersion is a more severe test than sweating.

Tests for water resistance and waterproof is conducted. It is reproducible if carried out in an indoor pool. It is carried out after 40 minutes of moderate activity in water, and that to test the claim that a given prodcut is waterproof after 80 minutes of moderate activity.

Efficacy of sunscreens

The efficacy of a sunscreen preparation depends on the amount of harmful (erythemogenic) solar radiation, the absorption range of the active sunscreen, its absorption peak, concentration employed, its resistance to water and perspiration, the solvent employed and other constituents present.

The molar absorptivity of a sunscreen in the erythemogenic and photosensitisation ranges of the uv spectrum is another a criterian to know effectiveness of a sunscreen agent. The larger its value at a specified wavelength, the greater the ability of the sunscreen agent to absorb uv radiation at this wavelength. The effectiveness of sunscreens with a low molar absorptivity can be increased by raising the concentration of the active ingredient in the preparation. The factors which can influence the effectiveness of a sunscreen preparation are :

(1) pH;

(2) The solvent system employed;

(3) The thickness of the residual film on the skin; and

(4) The stability of the product during the time it is employed.

A change in the pH, will change the ratio of ionised and nonionised fractions of the sunscreen agent and result in a shift in the absorption range (and absorption peak) away from the erythemogenic range of 290–320 nm with a consequent reduction in efficacy.

In practically all aromatic uv absorbers, addition of solvents can also produce a shift of their absorption range. Thus the absorption peak of *p*-aminobenzoic acid in 100 per cent isopropanol was moved from 287.5 to 267.5 nm in 100 per cent water. Mineral oil can produce a shift of the absorption range of sunscreens to a lower wavelength. Change of solvent, adversely affect the protection given by the sunscreen agent. On the other hand, 2-ethylhexyl palmitate is very stable to uv radiation and has practically no effect on the absorption peak.

Formulation of Sunscreens

Consider the following points for formulation. It should be convenient to use effectively with sufficient quantity. It should be soluble in aqueous or non-aqueous solvent. The film of non-volatile material left on the skin must have desirable properties. The various composition of sunscreens is given in formulae 1–11.

Formula 1 (Lotion, primarily aqueous, with a small amount of thickener)

Constituents	per cent
Filtrosol	7.0
Methyl cellulose	0.5
Glycerine	2.0
Ethyl alcohol	10.0
Water	80.5
Perfume	q.s.

Formula 2 (Lotion, primarily alcohol, with oleyl alcohol as residual lipid)

Constituents	per cent
Giv-tan	1.5
Ethyl alcohol (denatured)	65.0
Oleyl alcohol	10.0
Water	23.5
Perfume	q.s.

Formula 3 (Clear sunscreen lotion)

Constituents	per cent
Isobutyl *p*-aminobenzoate	5.0
Tween	9.0
Alcohol SD-40	45.0
Water	41.0

Formula 4 (Alcoholic lotion including silicone as the residual vehicle)

Constituents	per cent
Sunscreen agent	1.0
L-43 silicone (Union Carbide)	5.0
Ethyl alcohol	94.0
Perfume	q.s.

Formula 5 (Gel with a hydrophilic residue)

Constituents	per cent
Sunscreen (water soluble)	5.0
1,2,6-Hexanetriol	15.0
Distilled water	80.0

in addition :

Carbopol	0.1
Sodium hydroxide 10%	0.42

Formula 6 (Non-greasy oil based on isopropyl palmitate and silicone)

Constituents	per cent
Sunscreen (oil-soluble)	1.0
Silicone fluid	10.0
Isopropyl palmitate	89.0

Formula 7 (Water-in-oil emulsion with an oil-soluble sunscreen)

Constituents	per cent
Sunscreen (oil-soluble)	3.0
Mineral oil	34.0
Atlas	2.0
Avitex	5.0
Beeswax	2.0
Isopropyl myristate	0.5
Petroleum jelly	7.5
Water	46.0

Formula 8 (Oil-in-water emulsion with a low oil content of now-greasy character)

Constituents	per cent
Giv-tan	1.5
Diethylene glycol monostearate	2.0
Stearic acid	3.5
Cetyl alcohol	0.4
Isopropyl myristate	4.0
Triethanolamine	1.0
Triethanolamine lauryl sulphate	0.75
Water	86.85
Perfume, preservative	q.s.

Formula 9 (Low oil content vanishing cream containing veegum)

Constituents	per cent
Filtrosol A	5.00
Stearic acid	6.00
Cetyl alcohol	0.50

Veegum	2.28
Water	85.42
Borax	0.50
Potassium hydroxide	0.30
Perfume, preservative	q.s.

Formula 10 (Oil-in-water cream with a moderate oil content of mixed character)

Constituents	per cent
Sunscreen, etc.	q.s.
Glyceryl monostearate	16.0
Cetyl alcohol	1.0
Mineral oil	10.0
Seasame oil	10.0
Glycerine	7.0
Water	to 100.0
Antioxidant, preservative, perfume	q.s.

Formula 11 (Water-in-oil cream with an oil : water ratio similar to formula 10)

Constituents	per cent
Ethyl p-diethylaminobenzoate	1.0
Mineral oil	9.0
Ozokerite	2.0
Micro-crystalline wax	1.0
Paraffin wax	5.0
Petroleum jelly	2.0
Lanolin	3.0
Isopropyl myristate	10.0
Arlacel	1.5
Glycerine	5.0
Preservative, perfume	q.s.
Water	to 100.0

Formulae 12 and 13 are aerosol sunscreen oils. Formula 12 is based on mineral oil, while formula 13 is alcohol-based, contains silicone and is non-greasy and aerosol quick breaking foam non-greases is shown in formula 14.

Formula 12

Constituents	per cent
Filtrosol	3.0
Mineral oil	75.0
Eutanol	8.0
Lantrol	2.0
Isopropyl myristate	12.0
Perfume	q.s.
Concentrate	40
Propellants	60

Formula 13
Constituents	*per cent*
Sunscreen agent	5.0
Isopropyl myristate	25.0
Silicone	5.0
Perfume oil	1.25
Lanolin oil	2.5
Menthol Racemic	0.25
Absolute alcohol	61.0
Concentrate	20.0
Propellants	80.0

Formula 14 (Aerosol quick-breaking foam, substantially non-greasy)
Constituents	*per cent*
Myristic acid	1.3
Stearic acid	5.0
Cetyl alcohol	0.5
Isopropyl myristate	1.3
Glycerine	5.0
Sunscreen agent	1.0
Silicone fluid	1.0
Triethanolamine	3.0
Distilled water	80.6
Benzyl alcohol	1.0
Perfume oil	0.3

The release of histamine due to exposure to sun increases sensitivity of the skin. It is well-known that perfume may have irritant effect. Therefore the perfume selected should be only below 0.2%.

Heavy sunscreens

Sunscreens are used to resist pigmentation completely. This class of products also includes preparations which are applied to the skin with the sole object of providing a mechanical barrier to sunlight, such as calamine lotion or heavy make-up. Many attempts have been made to develop cosmetic products which would inhibit the formation of freckles or any skin lesions resulting from exposure to sunlight or other uv light sources. Like other sunscreen preparations, these products are commercially available in cream, lotion or aerosol foam form. Heavy sunscreen creams constitutes :

(1) 2-hydroxy-4-methoxy-4'-methylbenzophenone;

(2) 3-benzoyl-4-hydroxy-6-methoxybenzophenone (1 %);

(3) p-aminobenzoic acid (3–15 %);

(4) aesculin (5–10 %);

(5) salol in yellow vaseline (10 %);

(6) methyl salicylate in an ointment base (10 %); and

Heavy sunscreen lotions constitutes :

(1) p-aminobenzoic acid in 70 per cent alcohol (10–50 %); and tannic acid in 25–30 per cent alcohol (5–10 %).

PALLIATIVE PREPARATIONS

Preparations for the relief of sunburn may be formulated on the basis of calamine lotion, or other zinc preparations.

Formula 15 (Calamine lotion type)

Constituents	per cent
Colloidal calamine	20.0
Glycerine	5.0
Water	75.0
Antiseptic	q.s.

The following improved calamine lotion (formula 16) is suggested :

Formula 16

Constituents	per cent
Zinc oxide	8.0 g
Prepared calamine	8.0 g
Polyethylene glycol	8.0 cm^3
Polyethylene glycol (monostearate)	3.0 g
Lime water	60.0 cm^3
Water	to 100 cm^3

Triethanolamine stearate milks are also soothing (Formula 17) :

Formula 17

Constituents	per cent
Triethanolamine stearate	4.80
Liquid paraffin	10.00
Water	85.20
Antiseptic	q.s.

Greasy preparations should not be used in the treatment of sunburn. They only retain the heat of the burn and prevent the use of an antiseptic capable of mixing with the secretions and preventing bacterial infection. These preparations should be either aqueous solutions or oil-in-water emulsions which are capable of exerting both a protective and cooling effect.

ARTIFICIAL SUNTAN PREPARATIONS

Stains

The enhanced colour of suntan may be regarded either as functional, to prevent skin damage by absorption of erythemal radiation, or as cosmetic, to indicate the health and well-being of the subject. In either case the effect may be obtained by the use of staining materials. From ancient times materials such as walnut juice have been used. Water-soluble dyes, however, will streak on exposure to rain or moisture, and oil-soluble materials have been recommended. A vegetable product may be obtained by extracting a mixture of cudbear and henna with ten parts of warm deodorised olive oil, the ratio of the vegetable dyestuffs being adjusted to give the desired shades. A more modern practice is to use suitable oil-soluble dyestuffs in less greasy lipid material. Heavily pigmented face powders or modifications of the 'cosmetic stocking' may also be used. In conjunction with suitable shades of make-up, excellent effects may be obtained.

Systemic Materials

Alkoxypsoralens, and methoxsalen (8-methoxypsoralen) photosensitiser has been investigated extensively both as a treatment for vitiligo and also as a possible cosmetic tanning preparation. Local application, probably the most pigment-stimulating method of use, causes severe oedema, erythema, blistering and pain when the skin is subjected to natural or artificial ultra-violet radiation. Increased pigmentation is said to occur after ingestion of 10-20 mg of methoxsalen when the skin is exposed to sunlight within 2-4 hours.

Chapter 29

Face Powders and Make-up

INTRODUCTION

The smooth finish to the skin can be obtained using face powder. The visible imperfections or shining due to moisture or grease either from perspiration or formulation used on the skin can be masked. The aim is to make the skin pleasant to touch. The face powder is blend of several constituents.

Covering Power

The covering power of face powders should be sufficient so as to conceal various defects of the facial skin including scars, blemishes, enlarged pores and excessive shine. It is increased by use of titanium dioxide, zinc oxide, kaolin and magnesium oxide.

Absorbency

The second important function of face powders is to eliminate shiny skin in certain facial areas by absorbing sebaceous secretions and perspiration. A high absorptive capacity is a essential property of material. The colloidal kaolin, starch, precipitated chalk and magnesium carbonate have this property hence used in face powders.

Hewitt method is used to determine water absorbent properties. In this method, filter, a known weight of powder shaken with excess water, through buchner funnel. Weigh the wet powder and find out increase in weight.

Besides absorbancy of water, powder must be grease absorbent. If a person's face is inclined to dryness a more greasy foundation is usually employed. A powder which is not grease-absorbent will show a shiny nose or face which will necessitate re-powdering. Constituents of higher opacity such as zinc and titanium oxides tend to mask greasiness, while starch, chalks and kaolin absorb only a certain amount of grease. Oil absorption and water absorption capacities of powder bases are shown in Tables 29.1 and 29.2.

Table 29.1. Oil absorption capacities of powder bases.

Substance	Oil take-up (ml per g of substance)	Saturation time (min)
Oracid (urea-formaldehyde foam)	11.11	15
Aerosil	6.00	15
Magnesium carbonate	5.40	15
Magnesium oxide	3.30	15
Kieselguhr	2.80	15
Kaolin	2.70	15
Talc	2.50	15
Rice starch	2.10	15
Zinc stearate	0.40	15

Table 29.2. Water absorption capacities of powder bases.

Substance	Water take-up (ml per g of substance)	Saturation time (min)
Oracid (urea-formaldehyde foam)	16.60	30
Aerosil	8.70	45
Magnesium carboante	4.03	28
Kieselguhr	3.20	12
Magnesium oxide	2.60	20
Titanium dioxide	2.30	30
Kaolin	1.50	5
Talc	1.40	10
Zinc oxide	1.10	18
Rice starch	0.75	15
Zinc stearate	0.05	120

Slip, Adhesion and Bloom

Slip

Slip is the quality of easy spreading and application of powder to produce a characteristic smooth feeling on the skin. Slip is mainly imparted by talc and also by metallic soaps such as zinc stearate and, to a lesser extent, starch.

Talc

Talc is a hydrated magnesium silicate. It is the major component of face powders, and in high class products it may be used in amounts of up to 70 per cent or more. Its main function is to impart slip and good adhesion. However, the covering power and the moisture absorbing capacity of talc are low. It must therefore be combined with other powders to modify these deficiencies. Talc is also used in talcum powders.

The most suitable physical form of talc for use in cosmetics is the foliated variety, the flat platelets of which slide readily over each other, thus accounting for the high slip characteristics of the product.

Adhesion

Adhesion is another important property of face powder constituents, determining how well the powder will cling to the face. The magnesium stearates are used in face powders in amounts ranging between 3 and 10 per cent. They also render the ultimate product soft and fluffy and furthermore impart to powders some water-repellent characteristics. The adhesion of powders to the skin can also be improved by the addition of 0.5 to 1.5 % of certain emollients such as cetyl or stearyl alcohols and glyceryl monostearate.

Powdered silica and silicates

Very finely divided pure silica has been introduced for cosmetic purposes. The use of this substance obviates the need for zinc and magnesium stearates. The addition of 10 per cent to an ordinary powder mix exerts a marked effect in increasing its fluffiness. Up to 20 per cent may be used in a face powder or 30 per cent or more in talcum and baby powders. The small amount of such an ultrafine silica is very efficient as an anticaking agent in body powders.

The number of ultrafine synthetic silicates having high oil and water absorption preparation can be added to face powders.

Bloom

The materials chiefly used to impart bloom, the requirement for which may vary according to fashion, are chalk, rice starch and prepared starch. These have been described above.

Powdered silk

The silk powder spreads easily and adheres tenaciously to the skin, producing a velvet matt finish. It will absorb as much as three times its volume of water and still retain the appearance of a powder. The addition of large amounts of such materials as kaolin or chalk is not advised, as it is claimed that they tend to make the final product too dense and compact.

Colour

The solubilisation by sweat and lipid secretions makes water soluble or oil soluble dye-stuffs unsuitable. Face powders are coloured due to inorganic, organic pigments and organic lakes. Inorganic pigments include natural and synthetic iron oxides which give yellows, reds, browns and black; ultramarines which give green and blue, and chrome hydrate and chrome oxide which give green.

The choice of colour is usually a matter of taste. It is advisable to keep colour formulations as simple as possible so that matching with fresh batches of raw material is made easier. The colour effect produced by a powder applied to the skin will depend *inter alia* upon the opacity of both tinted and white pigments, their particle size, the degree of dispersion, the thickness of the applied film and the colour of the skin.

Perfume

The importance of the perfume on the sales appeal of the product cannot be over-emphasised. Usually the powders are perfumed very lightly. The odour of the face powder must be fragrant and pleasant, and a preference is shown today for either a flowery fragrance or that of a synthetic bouquet.

The compatibility of perfume with other constituents of the product must be carefully checked. Talc, for example, usually contains a little free lime, magnesia or iron which may adversely affect the perfume depending on the amount of these substances present. The perfume may also be affected by precipitated chalk, by kaolin, magnesium carbonate or a metal stearate, if these contain impurities, or indeed by some of the pigments used in colouring the powders. Finally, it should be remembered that the perfume note in a powder will be different from that, for example, in an alcoholic solution, particularly in the case of floral bouquets, and any tests on perfume acceptance must be carried out on the final product.

Formulation

The powder given in formula 1 is very transparent, 'light' powder, and is favoured by persons who wish to impart some colour and bloom to the face without appearing to be 'made-up'. The starch may be replaced with precipitated chalk.

Formula 1

Constituents	*per cent*
Talc	80.0
Zinc oxide	5.0
Zinc stearate	5.0
Rice starch	10.0
Perfume, colour	q.s.

Formula 2 has high opacity and gives a very definite opaque matt finish which tends to hide minor skin defects. It is more popular with certain people who like to have a definite powdered appearance without being over-powdered.

Formula 2

Constituents	*per cent*
Talc	30.0
Zinc oxide	24.0
Zinc stearate	6.0
Precipitated chalk	40.0
Perfume, colour	q.s.

The formulae 3 and 4 are popular variations of powder.

Formula 3

Constituents	*per cent*
Talc	65.0
Precipitated chalk	10.0
Zinc oxide	20.0
Zinc stearate	5.0
Perfume, colour	q.s.

Formula 4

Constituents	*per cent*
Talc	60.0
Kaolin	20.0
Zinc oxide	15.0
Zinc stearate	5.0
Perfume, colour	q.s.

As previously stated, both starch and precipitated chalk tend to impart bloom; formula 5 is a powder of medium weight, that is medium opacity or coverage and bloom.

Formula 5

Constituents	per cent
Talc	50.0
Rice starch	15.0
Precipitated chalk	15.0
Zinc oxide	15.0
Zinc stearate	5.0

The following formula (formula 6) is prepared from precipitated chalk and a high proportion of zinc stearate is incorporated. To provide some grease-resistance, kaolin and titanium dioxide are employed.

Formula 6

Constituents	per cent
Water-proof chalk base	50.0
Zinc or magnesium stearate	10.0
Kaolin	20.0
Titanium dioxide	6.0
Talc	14.0
Perfume, colour	q.s.

The following formula (formula 7) gives a good medium powder, characterised by excellent slip, absorbency, adequate coverage, good velvety feel and adherence — and an indefinable improvement in mattness, etc., which is attributed to the rice starch :

Formula 7

Constituents	per cent
Zinc oxide	16.0
Talc	37.0
Zinc stearate	5.0
Precipitated chalk	18.0
Rice starch	8.0
Kaolin (best cosmetic grade)	16.0

A special kind of 'fatty powder' is much favoured by persons afflicted with a rough or dry skin (Formula 8) :

Formula 8

Constituents	parts
Vaseline	50
White beeswax	40

Petroleum jelly	40
Stearin	20
Glyceryl monostearate	75

The modern trend with face powder is to apply it over foundation to achieve special effects, for example, matt or shimmer. The following complete formula illustrates a light shimmering effect : (Formula 9)

Formula 9
Constituents

	per cent
Talc	77.00
Zinc stearate	5.00
Zinc oxide	2.00
Kaolin	5.00
Mica	10.00
Red iron oxide	0.36
Yellow iron oxide	0.36
Black iron oxide	0.03
Perfume	0.25

MANUFACTURE

In a horizontal mixer with a screw agitator mixing of ingredients in face power is completed. If lakes are used they may be mixed with a small quantity of one of the constituents, chalk, zinc oxide or talc, and the colour concentrate so formed mixed with the main bulk. Water-soluble or alcohol-soluble dyes are best sprayed on to the mix or alternatively on to one of the constituents which possesses good absorbency such as chalk, kaolin or magnesium carbonate, then dried and mixed into the main bulk.

Machines are available which mix, sift and spray the perfume automatically. The perfume can also be added by meter pump by feeding a long tube with a multitude of small holes fitted along the top of the blender (for example, on to magnesium carbonate or chalk in the ribbon blender) before mixing with the remainder of the powder.

The use of water-soluble or oil-soluble colours should be avoided as they lead to streaking, darkening and staining of the skin. In the pebble mill method of mixing which obviates the making of a colour base. All the ingredients, including the colour and perfume, are milled together in a pebble mill for six hours, discharged. Such a mill delivers an excellent product and is widely employed in preference to mixers. This method, however, is too slow to be recommended for large-scale production. Micro-pulverisers are being increasingly employed since the material obtained is finely ground and uniformly mixed and only a rough pre-mixing is required. Various other pulverisers such as disintegrators, hammer mills, attrition mills, etc., may also be used.

Pin-disc mills give good results, especially in respect of colour dispersion, provided that the material is first given a rough pre-mix. Such mills are often applied in conjunction with a turbine sifter. The *Air Spun* process is described briefly is as follows.

Pass purified and cold air, under great pressure (100 psi; 700 kPa) in a continuous stream into a closed drum-shaped chamber. This chamber or mill is called a microniser as it reduces particles to micron (μm) size. The speed of the air stream, the manner in which it is directed and the shape of the chamber itself, cause air to revolve about this chamber at a rate somewhat is excess of 1000 mph. In this super cyclone all the ingredients are hurled against each other until they reduce themselves to the desired size and fluffiness. When this is attained such particles are emitted through a central exit while the heavier particles are forced to remain. It is claimed that the particle size was selected after numerous experiments had shown it to be the best for the appearance effect and adhesiveness of a face powder.

There are various methods available for preparing powders of a desirable particle size. These depend upon elutriation by means of air or water. Control of such separated fractions may be carried out by microscopic examination of the fractions, but this method involves the visual or photo-micrographic inspection of a very large number of fields in the microscope and is tedious. A useful variation is the method of Andreason in which volumes of the suspension, after definite times of settling (calculated from Stokes' Law) are pipetted into a tared dish, evaporated to dryness and weighed. Various air permeability methods are also widely used.

The most modern method of particle size analysis applicable to cosmetic powders entails the use of a Coulter Counter. Packaging is usually carried out by automatic vacuum filling devices, which minimise dusting. In choosing powder boxes, care should be taken to see that these are prepared with an odourless glue, otherwise the fragrance of the powder will be ruined in storage.

COMPACT POWDER

Compact powders enjoy wide popularity. They are prepared by either a damp or a dry compression process. The moulding process used originally, which entailed the use of plaster of Paris, has fallen into disuse.

In the damp process, the powder, is milled to the required plasticity after mixing intimately mixed with a suitable binding agent. Then compressed into suitable containers, usually metal godets, and dried for the requisite period in a current of warm air. In the dry process the mass is subjected to compression without being wetted to any appreciable extent. This process, although difficult to achieve satisfactorily at first, is probably the best to use for manufacturing compacts on a large scale. Because it can be rigidly adhered to once the mix and suitable conditions have been determined. Hydraulic or reciprocating presses may be used for the manufacture of compact powders. They range from foot operated presses producing one cake at a time, to fully automatic presses which may produce up to 60 units per minute.

The requirement of good covering power, adhesion and uniformity in compositions mentioned in respect of conventional face powders also applied to compact powders. The adequate binding properties to the compressed powder mixture is attained in order to remove easily from cake for application using powder puff. Furthermore, the powders used for compacts should be free-flowing so that they do not adhere to punches or dies during compression; otherwise, air pockets will be formed which will result in an uneven compression and cause the cakes to break. From this it can be seen that one of the main aims during manufacture of compact powders is to ensure that the compressed cakes are of uniform density.

The actual pressing process can also affect the shade of the product so quality control can be a problem. In fact, if large volumes are anticipated it is always best to conduct an extended manufacturing trial rather than go straight from the laboratory bench into full-scale production.

The composition of compact powders is generally very similar to that of loose powders. The differences which exist arise from the need to meet the requirement of greater cohesion and are largely evident in terms of percentages of some of the components present. In compact powders, colloidal kaolin, zinc oxide and metallic stearates are usually present at a higher level than in loose face powders, and starch is sometimes incorporated to facilitate compression. A binding agent to improve their cohesion is added so that on compression a firm cake is produced. Water-soluble and water-insoluble binding agents may be used. The former are natural and synthetic gums which are used in amounts ranging from 0.1 per cent to about 3 per cent by weight of the product, and are usually of a 5–10 per cent aqueous solution is mixed. A very much favoured binding agent for this purpose is low-viscosity carboxymethylcellulose. A small amount of a humectant is usually added to the solutions. A water-insoluble binding agent should be use in the form of oil-in-water emulsion, for example glyceryl monostearate, cetyl or stearyl alcohols, isopropyl esters of fatty acids, lanolin and its derivatives or ozokerite, paraffin wax and microcrystalline waxes.

In the dry compacting process, it is usual to employ zinc or magnesium stearates at a level of 5–15 per cent by weight, as well as a lubricant such as mineral oil in similar proportions.

Considerable changes have taken place in formulation during the last fifty years or so; an ammonia–stearin–starch compound containing white petrolatum, ammonia, starch and stearin (Formula 10) for the manufacture of compact powder and rouge has been recommended.

Formula 10
Constituents

	per cent
Stearin	100 g
White petrolatum	20 g
Ammonium hydroxide solution, 0.97 s.g.	50 cm^3
Rice or maize starch	250 g

Formula 11 gives a formula to which the damp compression process is applicable.

Formula 11
Constituents

	per cent
Kaolin	20.0
Zinc oxide	15.0
Precipitated chalk	25.0
Talc	32.0
Compact binder	8.0
Perfume, colour	q.s.

The dry compression method is suitable for mass production, but necessitates the use of higher pressures than are employed in the damp compression process. The composition of a compressed face powder cake by the dry compression process is given in formula 12.

Formula 12
Constituents

	per cent
Talcum	61.25
Sodium lauryl sulphate	0.75
Titanium dioxide	7.50
Zinc stearate	11.25
Inorganic pigments	1.00

Mineral oil	4.50
Spermaceti	3.00
Cetyl alcohol	1.50
Lanolin	1.00
Glycerine	7.50
Hexachlorophene	0.25
Alkyldimethylbenzyl ammonium chloride	0.20
Methyl-p-hydroxybenzoate	0.09
Propyl-p-hydroxybenzoate	0.09
Perfume	0.12

Procedure : Talc, sodium lauryl sulphate, titanium dioxide, zinc stearate and inorganic pigments are first mixed together in a ribbon mixer for an hour. The premixed hexachlorophene, quaternary ammonium base and preservative are added to the mixer and the batch is mixed for about two hours, before passing through a micro-pulveriser, using a no. 0.013 screen. The temperature of the material should not be allowed to rise more than 37°C during this operation. After cooling, the powder blend is repassed through the micro-pulveriser. The mineral oil, spermaceti, lanolin and glycerine, mixed with heating until liquid, are then sprayed into the dry batch which is mixed in the ribbon mixer for a further half hour. The batch is passed through a comminuter and once again through a pulveriser, using a 3/16 in screen. When cool, the powder is passed through the micro-ulveriser using a no. 0.027 screen, and finally, having been cooled once more, is passed through a 40 mesh screen, ensuring that no heat is generated during this final screening. The resulting material is filled into containers and 40–50 psi pressure is applied to convert the powder into cake. It is usually advisable to keep the powders for several days in a suitably humid atmosphere before pressing, to facilitate the escape of any trapped air, and to ensure that the powder blend will not be too dry when compressed. During the dry compression process, it is usual to apply a small pressure initially to squeeze out the air, and thus avoid the formation of air pockets in the powder cakes. Pressure is then gradually increased up to 150 psi or even more 1000 + kPa before the die is removed from the surface of the cake.

CAKE MAKE-UP

The modern cake make-up originates from theatrical grease-paint and has acquired popularity because of its ease of application and stability and also because more product can be applied to the face and thus deeper shades and effects can be achieved. The cake make-up is also made from talc, kaolin, zinc oxide and precipitated chalk. In addition it has inorganic pigments such as titanium dioxide and iron oxides. Humectants such as sorbitol or propylene glycol may also be incorporated together with other additives such as sorbitan sesquioleate, lanolin or mineral oil and perfume. Humectants and other liquid constituents are usually combined and sprayed on to the powder constituents while these are mixed in the ribbon mixer. The resulting blend is granulated and finally compressed.

A dry cake-form make-up which can be applied with a moistened pad or sponge. The cake contains oily and waxy ingredients (0.8–24 per cent), a water-soluble dispersing agent (1–13 per cent) and fillers (35–80 per cent) and pigments (12–50 per cent) whose particles are coated with the oils and waxes to make them water-repellent. The make-up is prepared by adding the fillers (talc, chalk) and pigments (zinc, titanium and ferric oxides) to the oils and waxes dispersed in water. After drying mixture is pulverised and compressing it into cake-form.

In case a much higher ratio of powder to emulsion used drying is not needed. For example approximately 12 to 1, and mixing with a Duplex mixer. The mixing operation takes between 30 and 60 minutes and the mix is compacted after granulating through a fine screen. The following formulae can be made in this fashion (formula 13) :

Formula 13

Constituents	*parts*
Powder base (perfume)	100
Emulsion :	
Stearic acid	34.5
Mineral oil	21.5
Triethanolamine	10.5
Water	33.5

(Stearic acid, Mineral oil, Triethanolamine, Water) } 8.5

For deep shades with good coverage, zinc oxide is replaced by titanium dioxide (this also facilitates better pressing characteristics) formula 14 :

Formula 14

Constituents	*per cent*
Colour mix :	
Talc	89.75
Titanium dioxide	9.00
Yellow iron oxide	0.75
Red iron oxide	0.40
Black iron oxide	0.10
Base :	
Stearic acid	15.0

Acetylated lanolin	3.0
Stearyl alcohol	3.0
Glyceryl monostearate	2.2
Sulphonated castor oil	1.2
Triethanolamine	3.0
Mineral oil	35.0
Polyethylene glycol	10.0
Water	27.6
Completion ratio :	
Base	20.0
Colour mix	79.8
Perfume	0.2

Pressing of this product can be very rapid as for compact powders, although it is important to check that there is consistent dispersion of the base in the colour mix throughout long manufacturing runs.

Application of Cake Make-up

The application of cake make-up is recommended in the following manner: To apply cake make-up correctly, wet and squeeze out a sponge and rub it lightly over the face. Before drying smooth cake make-up on the face blend in the rouge. Carefully squeeze all water out of the sponge and rub it lightly over the face until the make-up is dry. It is recommended that people who do not like a heavy make-up should blot off the excess of water with a dry towel or facial tissue.

The cake make-up is satisfactory for the younger woman; and persons in the later thirties, and even younger persons who suffer from dry skin.

MAKE-UP CREAM

Foundation make-up preparations in cream form are essentially suspensions of pigments in an emulsified lotion. The suspension is usually homogenised and milled. The addition of the pigment is usually made at about 50–55°C as the emulsion is slowly cooled with agitation. The following formula for a cream make-up (formula 15) :

Formula 15	
Constituents	*parts*
Glycerol mono- and distearate (pure)	2.0
PEG monostearate	1.0
Stearic acid	11.5
Cetyl alcohol	0.5

Isopropyl myristate or palmitate or palmitate	2.0
Propylene glycol	12.0
Sorbitol syrup	2.5
Preservative	0.1
Titanium dioxide	2.2
Talc	8.0
Colour pigments	1.0
Water	57.4

LIQUID POWDER

Liquid powders have been used as a base for ordinary powder or to replace such powder for evening wear, dances or similar occasions. They have also included a theatrical product, 'wet white', which is used for whitening the neck and arms. Its basic components are zinc oxide and bismuth oxychloride, subnitrate and carbonate incorporated into a liquid consisting of a mixture of glycerine and water in varying proportions. Glycerine is employed in amounts of up to 30 per cent and the water is often replaced by triple rose water or other fragrant waters; starch is also used sometimes to suspend heavier constituents. To prepare 'wet white' the constituent powders are mixed and glycerine is added. Surfactants could also be included to assist in dispersing the powders. A 'wet white' formula on these lines, is given in formula 16 :

Formula 16

Constituents	*per cent*
Bismuth subnitrate	5.0
Starch	5.0
Zinc oxide	10.0
Glycerine	15.0
Rose water	65.0

Instead of bismuth salt and starch, pigments, fillers and colours are added in modern formulae (formulae 17 and 18) :

Formula 17

Constituents	*per cent*
Zinc oxide	3.0
Chalk	15.0
Kaolin	2.0
Colouring matter	q.s.
Glycerine	15.0
Water	65.0
Preservative, perfume	q.s.

Formula 18

Constituents	per cent
Zinc oxide	10.0
Titanium oxide	10.0
Talc	10.0
Colouring matter	q.s.
Gum tragacanth (0.5% solution)	25.0
Glycerine	15.0
Water	30.0
Preservative, perfume	q.s.

Liquid powder formulation is given in formula 19 :

Formula 19

Constituents	per cent
Sodium carragheenate	2.0
n-Propyl alcohol	2.0
Propylene glycol	5.0
Water	68.5
Veegum HV	0.5
Talc	10.0
Magnesium carbonate	4.0
Titanium dioxide	1.0
Colour pigments	7.0
Perfume	q.s.

LIQUID MAKE-UP

Liquid make-up consists essentially of pigments dispersed in a viscous base. The liquid make-up preparations were suspensions of pigments in an aqueous alcoholic solution. They require vigorous shaking prior to use to ensure uniform distribution of the product during application.

The main trouble in preparing this product is that to prevent the sedimentation of constituent pigment. It is done by dispersing them in a hydrocolloid base or in a liquid emulsion. The hydrocolloids used for thickening the preparations may be selected from cellulose derivatives, carragheenates, etc. The pigments used are talc, kaolin, zinc oxide, titanium dioxide, calcium or magnesium carbonates and others.

In emulsified products, raw materials used included propylene glycol monostearate, glyceryl monostearate, fatty alcohols such as cetyl or oleyl alcohols, isopropyl myristate, lanolin and its derivatives, polyethylene glycols, humectants and others. A liquid make-up composition is shown in formulae 20–21.

A liquid make-up formulae is given in formula 20. Viscosity may be adjusted by varying the amount of gum and bentonite.

Formula 20

Constituents	per cent
Propylene glycol	4.40
Polyethylene glycol	1.92
Preservative	0.32
Gum tragacanth	76.68
Bentonite	0.96
White oil	1.20
Oleyl alcohol	6.72
Stearic acid	4.20
Triethanolamine	1.92
Perfume	q.s.
Titanium dioxide plus powdered pigments	q.s.

Pigments and titanium dioxide are usually 5–10% as in the formula 21.

Formula 21

Constituents	per cent
Isopropyl lanolate	3.50
Isopropyl myristate	4.20
Squalane	1.40
Purcellin oil	2.10
Mineral oil	12.80
Sorbitan oleate	1.00
Veegum (5% solution)	30.00
Propylene glycol	8.00
CMC (1% solution)	20.60
Polyoxyethylene sorbitan mono-oleate	4.00
Water	6.04
Titanium dioxide	4.60
Yellow iron oxide	0.56
Red iron oxide	0.55
Black iron oxide	0.10
Perfume	0.25
Preservative	0.30

STICK MAKE-UP

This stick concept can be extended further to give the cover-up product, used to disguise birthmarks and blemishes. The product is heavily pigmented and is applied like a lipstick.

Formula 22

Constituents	per cent
Lanolin alcohols	2.8
Ozokerite wax	8.0
Paraffin wax	6.0
Mineral oil	20.2
Isopropyl myristate	10.0
Lanolin	2.8
Titanium dioxide	36.8
Kaolin	8.0
Yellow iron oxide	2.5
Black iron oxide	0.6
Red iron oxide	2.0
Perfume	0.3

FACE PACKS AND MASKS

The use of face packs by women dates back to early antiquity when some of the earths used in them were credited with almost miraculous healing powers. These preparations are applied to the face in the form of liquids or pastes. They are then allowed to dry or to set with the object of improving the appearance of the skin, by producing a transient tightening effect as well as by cleansing the skin. Such a preparation should possess the following properties :

(1) It should be a smooth paste without gritty particles and without an 'earthy' or other objectionable odour.

(2) When applied to the face by immediate drying must form adherent coating on the skin which can be peeled off the face.

(3) It should produce a definite sensation of tightening of the skin after application and cleansing of the skin.

(4) It must be dermatologically innocuous and non-toxic.

The five basic systems based on wax, rubber, vinyl resins, hydrocolloids and earth respectively gives satisfactory properties.

Wax-based Systems

Wax-based masks generally consist merely of paraffin wax of a suitable melting point. It may be mixtures of wax with the addition of a little petroleum jelly and cetyl and stearly alcohols. The use of microcrystalline waxes may assist the continuity of a suitable 'wax mask'. The following formula 23 illustrates the use of this material :

Formula 23

Constituents	per cent
Micro-crystalline wax	13.0
Paraffin wax	60.0
Cetyl alcohol	5.0
Mineral oil	20.0
Bentone	1.4
Isopropyl alcohol	0.6

Rubber-based Systems

Rubber masks are removed from the skin very easily by mere pulling. Slight plumping of the skin becomes noticeable which disappears after skin respiration has been restored to normal. Formula 24 illustrates the composition of face packs based on rubber latex.

Formula 24

Constituents	per cent
Latex emulsion	25
Sorbitol	5
Methylcellulose	10
Kaolin	3
Borax	1
Water	56
Preservative	q.s.

Vinyl-based Systems

Vinyl-based systems are generally based on polyvinyl alcohol or vinyl acetate resin as film-former. The use of polyvinyl alcohol in face packs is illustrated by formula 25.

Formula 25

Constituents	per cent
Veegum	0.5
Kaolin	0.5
Titanium dioxide	0.3
PVA	12.0
Propylene glycol	8.0
Ethanol	20.0
Water	58.7

Hydrocolloid-based Systems

Hydrocolloid-based systems are high-viscosity sols which after application lose water and form a flexible gel film. They are also solid gels which are melted before application to the face. The sensation of tightness is produced by the eventual shrinking of the gel on loss of further moisture. The formula 26 for a 'gelanthum' mask containing gelatin and tragacanth is given below :

Formula 26

Constituents	*per cent*
Gum tragacanth	2.2
Glycerine	2.5
Gelatin (white)	2.3
Water	90.5
Zinc oxide	2.5

For a 'gelanthum mask with honey', is replacement of about 4.5 parts of the water in the above formula by honey is recommended. The formula 27 is given below :

Formula 27

Constituents	*parts*
Casein	20.0
Borax	0.5
Glycerine	5.0
Water	100.0
p-Hydroxybenzoic acid propyl ester	0.1

The following composition for a face pack is given in formula 28.

Formula 28

Constituents	*parts*
Gelatin	10.00
Water	50.00
Camphor	0.05
Zinc oxide	3.00
Kaolin	5.00
Titanium dioxide	2.00

The preparations given in formulae 27 and 28 are heated before use and applied in a warm state. The use of alcohol as part solvent is shown in formula 29.

Formula 29

Constituents	*per cent*
Carbopol	2.0
Water	42.0
Alcohol	50.0
Glycerol	2.0
Di-isopropanolamine	4.0

Earth-based Systems (Argillaceous Masks)

Earth-based systems are referred as paste masks. They include clay facial packs and the so-called mud packs and a high percentage of solids. These products can either be presented in bulk, or packed in sachets, for mixing with water when required, or they can be presented ready mixed for use. The ready mix should be sterilised using heat or ethylene oxide and should contain preservative. The presence of bentonite produces a genuine cleaning effect, particularly on very greasy skins. China clay, colloidal kaolin, fuller's earth, bentonite, etc., may be used as the 'argillaceous' material. The colour of fuller's earth or bentonite is modified by blending with kaolin, and/or the addition of zinc oxide or titanium dioxide.

Bentonite has strong affinity for water and its thixotropic properties as it is derived from volcanic clay. Certain bentonites will absorb up to fifteen times their volume of water. The addition of a small quantity of magnesium oxide, or some other substance possessing a similar pH increases absorption capacity.

The consistency of bentonite gels is influenced by concentration and pH of the gels. A gel containing 6 per cent bentonite has the consistency of glycerin, while a 20 per cent gel has the consistency of lanolin. The formulae 30–31 are examples of earth based systems.

Formula 30 (All-purpose masks)

Constituents	*per cent*
Kaolin	35.0
Bentonite	5.0
Cetyl alcohol	2.0
Sodium lauryl sulphate	0.1
Glycerine	10.0
Nipagin M	0.1
Perfume	q.s.
Water	to 100.0

Formula 31

Constituents	per cent
Glyceryl monostearate	3.0
Lanolin oil	2.0
Sodium lauryl sulphate	2.0
Veegum	8.0
Kaolin	10.0
Propylene glycol	7.0
Titanium dioxide	4.0
Ethanol	6.0
Isopropyl myristate	2.0
Water	56.0

The following two formulations (formulae 32 and 33) for face packs for dry and greasy are given below :

Formula 32 (Face pack for dry skin)

Constituents	parts
Kaolin	80.0
Starch	10.0
Cold cream	20.0
Cetyl alcohol	2.0
Hydrophilic oil	5.0
Water, boric water or infusions	q.s.

Formula 33 (Face pack for greasy skin)

Constituents	parts
Kaolin	80.0
Magnesium carbonate	15.0
Starch	5.0
Tragacanth gum	1.0
Water	q.s.

For oily, sallow skin or skin blemishes 'oxygenated' mask is recommended, after conducting trial on a small area of the skin. An 'oxygenated' face mask, based on kaolin. Bergwein has the following composition (formula 34) :

Formula 34

Constituents	*parts*
Colloidal kaolin	800
Salicylic acid	20
Aluminium lactate	5
Magnesium peroxide	200

Anti-wrinkle Preparations

Face packs are usually left on the face for about 10–25 minutes, to allow most of the water to evaporate and the resulting film to contract and harden, after which they are removed. The anti-wrinkle preparations are based on bovine serum albumin, and form an 'invisible' mask over the skin. They are allowed to remain in contact with the facial skin for about 6–8 hours, that is, for the period of time during which they remain effective, and are then removed by washing.

Sterile solution containing a suitable preservative and ready for immediate use without dilution; as a 30 per cent solution to be diluted with an equal volume of water prior to use; and also in the form of a freeze-dried powder which before use is reconstituted with water. The smoothing effect would seem to be a purely physical one entailing the 'filling-in' of facial wrinkles and the formation of a tight occlusive film which stretches the skin. How long this effect lasts will largely depend on how much the facial muscles are used.

A wrinkle-smoothing composition disclosed in an American patent contains proteins obtained from cow's milk whey, namely α-lactalbumin and some ß-lactoglobulin. The film produced with this composition is claimed to be effective for about four hours and can be re-activated by moistening it with a little water. Both the component proteins are undenatured and water-soluble, are present in the aqueous composition in a ratio of 4:1. It may constitute 10–30 per cent by weight of the total composition. Water may be present within a range of 70–90 per cent by weight, together with 2–4 per cent by weight (of the composition) glycerin or propylene glycol, to increase the flexibility of the dried films. Bacteriostatic and Fungistatic preservative must not precipitate the lactalbumin or the lactoglobulin. The pH of the final preparation is maintained within the 5–7 range.

Apart from the ability to re-activate films produced, other advantages claimed for this preparation are that when it is dry it is not shining in appearance, that it does not flake away after drying and that no make-up is required to conceal the film.

Chapter 30

Coloured Make-up Preparations

LIPSTICK

Lipsticks, the lip cosmetics moulded into sticks They are essentially dispersions of colouring matter in a base consisting of a suitable blend of oils, fats and waxes.

Lipstick is used to impart an attractive colour and appearance to the lips, accentuating their good points and disguising any bad ones. Narrow bad-tempered lips may be widened, and broad sensual lips made to appear narrower by its use. In fact, if applied intelligently it is capable of entirely altering the apparent facial characteristics.

Since lips are considered to be more alluring when they possess a slightly moist appearance, this is always achieved by the use of a greasy base which also exerts an emollient action.

Characteristics Required in a Lipstick

A good lipstick should have the following characteristics :

(1) It should have retainable smooth surface of uniform colour, free from defects such as pinholes or grittiness due to colour or crystal aggregates. It should not exude oil, develop a bloom, flake, cake, harden, soften, crumble nor become brittle over the range of temperatures likely to be experienced.

(2) It should be innocuous, both dermatologically and if ingested.

(3) It should be easy to apply. A film on the lips that is neither excessively greasy nor too dry, that is reasonably permanent. But capable of deliberate removal, and which has a stable colour.

Ingredients of Lipsticks

Colour

The colour of a lipstick is one of the major selling points. It can only be dealt with in general terms, since the precise shades are dictated by ephemeral fashion. It is usual for the colour to contain red and this allows shades ranging between orange-yellow and purple-blue, although even greens are not unknown. The colour is imparted to the lips in two ways :

(1) By staining the skin, which requires a dyestuff in solution, capable of penetrating the outer surface of the lips.

(2) By covering the lips with a coloured layer which serves to hide any surface roughness and give a smooth appearance. It is met by insoluble dyes and pigments which make the film more or less opaque.

Typical proportions for the colours in a lipstick are as under :

Constituents	*per cent*
Staining dyes (bromo acids)	1/2–3
Oil-soluble pigment	2
Insoluble pigment	8–10
Titanium dioxide	1–4

Staining dyes

The most widely used staining dyes are water-soluble eosin and other halogenated derivatives of fluorescein called as 'bromo acids', a term originally applied to acid eosin, tetrabromofluorescein.

Eosin

Eosin is an insoluble orange compound which changes to an intense red salt at pH above 4. It is also known as D & C Red No 21. When applied to the lips in the acid form, it produces a relatively indelible purple red stain on neutralisation by the lip tissue.

Eosin and some of its derivatives can give rise to sensitisation or photo-sensitisation, leading to cheilitis (inflammation of the red portion of the lips) or more general allergic reactions.

Difficulties may be encountered when using bromo acid dyestuffs in obtaining entirely homogeneous dispersions of these dyes in the lipstick mass. This could result in shade variations in lipsticks. Novel compositions have been disclosed which aim to overcome these difficulties and to ensure that the colours of the lipsticks sold are identical with those conferred by them on the skin. These compositions comprise amine salts of bromo acid dyestuffs and fatty substances which are at least partial solvents for such components. The amines employed in the preparation of these compositions are selected from non-aromatic primary, secondary or tertiary monoamines, in which the groups attached to the nitrogen atom contain at the most 6 carbon atoms. Some of them may contain hydroxy radicals, others may be heterocyclic compounds. Specifically mentioned in patent claims have been triethanolamine, diethanolamine, 2-amino-2-methyl-propanediol-1,3, monoisopropanolamine, trihydroxymethylaminomethane, diglycolamine and morpholine.

To prepare these compositions the bromoacids are dispersed in a fatty material, preferably soya lecithin, and the amine is added with mixing. The resulting mixture may be optionally heated and allowed to cool. The amine salts of the bromoacid dyestuffs are then dissolved in a mixture of waxes (for example carnauba wax, beeswax and ozokerite) and oils (for example various lanolin derivatives and vaseline).

Non-eosin staining dyes

Non-eosin staining dyes are an interesting development in view of the restrictions applying to some of the fluorescein derivatives. It is claimed that water-soluble FD&C and D&C dyes. They are useless in lipsticks, when converted to the free sulphoacid form become water-insoluble, lipophilic, and suitable for use as staining dyes in lipsticks, covering a wide colour range.

Pigments

Inorganic and organic pigments are metallic lakes used to give intensity and variation of colour. When selecting lakes the possibility of reaction with the base, for example soap formation with free fatty acid, must be borne in mind.

Titanium dioxide: It is the most effective white pigment for obtaining pink shades and giving opacity to the film on the lips. Its concentration used is upto 4 per cent. However, the use of titanium dioxide requires great care in the grade of material selected (anatase or rutile) and the surface treatment it has received to make it lipophilic, and also in the method of incorporation, if unexpected troubles such as oily exudation, streaking, dullness and coarse texture are to be avoided.

Although two D&C colours, D&C Red No. 36 and D&C Orange no. 17, are metallic lakes, insoluble in both water and oil. They may be considered as pigments. Similarly, the amount of staining dye used often exceeds the solubility in the base and the insoluble portion will act as a pigment.

Lakes: Of many of the D&C dyes with metals such as aluminium, barium, calcium and strontium are potential pigments for lipsticks. However, some strontium and zirconium lakes are banned. Aluminium lakes are not usually favoured because of their lack of opacity, but this very property would seem to suggest their use in transparent lipsticks.

The following lakes are considered to be the most useful lipstick colourants :

(1) Calcium lakes of D&C Red Nos 7, 31 and 34.

(2) Barium lakes of D&C Red No. 9 and Orange No. 17.

(3) Aluminium lakes of D&C Red Nos 2, 3 and 19 and FD&C Yellows Nos. 5 and 6.

When the parent D&C colour is subject to restriction the lakes are also restricted in the same way.

As noted above, pigments and lakes are used at levels between 8 and 10 per cent or perhaps over a wider range, say 5–15 per cent. Iridescent lipsticks utilise either mica coated with titanium dioxide or bismuth oxychloride at levels of up to 20 per cent, depending upon the effect desired.

Base

The quality of the lipstick during manufacture, storage and use will be determined for the most part by the composition of the fatty base and colour. The quality is also concerned with the rheology of the mixture at various temperatures. For instance, during manufacture (usually while warm) it must be possible to mill and grind the mass, and to pour and mould it while holding the insoluble colours evenly dispersed without settling. In the moulds it must set quickly with a good surface and good release properties. During shelf life and usage life the stick must remain rigid and stable, and generally in good condition. In use the stick must soften sufficiently in contact with the lips, and be sufficiently thixotropic to spread on the lips to form an adherent film which will not smear nor, ideally, transfer to cups or glasses.

Dyestuff solvents

Some part of the base must act as the necessary solvent for the staining dyes. Many of the normal fatty materials are too non-polar to dissolve the dyestuffs. It is convenient to consider first those ingredients which do have solvent properties for eosin, and which must form some part of the base.

Caster oil

It is a traditional material for dissolving bromoacid. This is due to its high content of ricinoleic (hydroxyoleic) acid, which is unique among natural oils. Its other properties include a high viscosity, even when warm, which delays pigment settling. The oiliness which helps with gloss and emollience, although too high a quantity causes drag during application and an unpleasant greasy film. The level of castor oil is 25–50%. The disadvantages of castor oil include an unpleasant taste and potential rancidity.

Fatty alcohols

The hexadecyl alcohol is a good solvent for bromoacid dyes. Lipsticks containing it is applied with very little drag do not bleed or smear. Its presence lessens any tendency to develop an unpleasant taste during storage.

Esters

They are lipophilic liquids of low oiliness giving lubricating and emollient effects with some dye solvent properties.

Glycols

They are more polar than the fatty alcohols and might be expected to be better dye solvents with two hydroxy groups.

Polyethylene glycols (Carbowaxes)

They have good dyestuff solvent power, which correlates to some extent with water solubility and this is a detraction. These materials would seem to have considerable possibilities in lipstick formulation, if correctly selected.

Monoalkanolamides

They have good dye-dissolving properties. They have also no action on the plastic material of the lipstick cases. The pre-dissolved in Loramine Wax is kept as a concentrate for more convenient dispensing when required.

Other base ingredients. The ingredients quoted so far have had the high melting points or hardness which give satisfactory moulding properties, i.e. quick setting and good release with a glossy surface and a rigid stick. This function is generally fulfilled by the inclusion of waxes or wax-like materials.

Carnauba wax

It is a very hard vegetable wax used for raising the melting point, imparting rigidity, hardness and contraction properties in the moulding process.

Candelilla

It is another hard vegetable wax serving the same functions as carnauba wax but has a lower melting point and is less brittle.

Amorphous hydrocarbon waxes

In mineral oil, for example ozokerite wax, gives a short-fibred texture to the product.

Petroleum-based waxes

Microcrystalline wax, are also used to modify the rheology of the product.

Beeswax

It is the traditional stiffening agent for castor oil, but it can give a grainy and dull effect if used in large quantities.

Hydrogenated vegetable oils having wax like properties are more solid and less prone to rancidity than the unhardened oils. Of particular interest is hydrogenated castor oil, still retains eosin solvent properties. Some other softer materials will be found to be of use in lipstick manufacture.

Lanolin and lanolin absorption bases

They have eosin solvent properties and also act as binding agents for the other ingredients, tending to minimise sweating and cracking of the stick and acting as plasticisers. Absorption bases are recommended to enhance the gloss on the lips. They are useful upto 10 %.

Petroleum jelly

Paraffin oils they act as lubricants and improve spreading properties. Large amounts tend to impair the adhesion properties and can be difficult to blend if much polar material such as castor oil is present.

Lecithin

It acts as a dispersing agent for pigments, in addition to facilitating the application of the lipstick and improving the adhesion to the lips.

Silicone waxes

They are largely insoluble in water, ethanol and organic fats and oils. The compositions containing them have improved viscosity, stability and a sharper melting point. They are said to be superior to compositions with no silicone wax, retaining their form plus all the properties of the silicone wax over a wider temperature range. They are also said to have good shelf stability.

Perfumes

Perfumes selected should mask the fatty odour note of the base and should be non-irritant to the lips. The flavour as well as odour must be considered as consumer is likely to apprehend it in the mouth and nose. Perfumes should be stable and compatible with the other constituents of the lipstick base. The amount of perfume used is 2–4%.

The preferred perfumes are of the light floral or light sophisticated type with no single essential oil predominating. Rose alcohols and esters are often used as well as other essential oils, preferably terpeneless, such as aniseed, cinnamon, clove, lemon, orange and tangerine. The following formulae illustrate some lipstick formulations in existence. It is prepared by: first melt the ingredient by heating. Then add 7 parts of pigment to 92 parts and run the mixture through heated mill. Run the final blend into the mould with 1 part of perfume.

Formula 1

Constituents	*per cent*
Loramine	20.0
Lanolin	10.0
Cocoa butter	5.5
Refined beeswax	4.0
Ozokerite	18.0
Carnauba wax	4.2
Oleyl alcohol	7.0
Mineral oil (high viscosity)	29.3
Perfume	2.0 or *q.s.*
Insoluble lake	10.0 or *q.s.*
Bromo acid (or eosin salt)	2.0

Formula 2

Constituents	*per cent*
Castor oil or bromoacid solvents	30.0
Mineral oil	15.0
Beeswax	15.0

Paraffin	10.0
Carnauba wax	10.0
Ceresin wax	10.0
Silicone Fluid	10.0
Perfume-flavour compound	q.s.

Formula 3

Constituents	per cent
Hexadecyl alcohol	26.0
Castor oil	20.0
Propylene glycol monolaurate	15.0
Anhydrous lanolin	2.0
Ceraphyl	5.0
Candelilla wax	32.0

Formula 4

Constituents	per cent
Hexadecyl alcohol	44.0
Butyl stearate	2.0
Isopropyl palmitate	3.5
Anhydrous lanolin	7.0
Petrolatum	12.0
Candelilla wax	11.0
Carnauba wax	11.0
Stearic acid, triple pressed	8.0
Cetyl alcohol	1.5
Nordihydroguaiaretic acid (antioxidant)	0.02
Citric acid	0.006

Formulae	5 *Deep stain lipstick*	6 *Creamy lipstick*
Constituents	*per cent*	*per cent*
Castor oil	60.0	65.0
Lanolin	5.0	10.0
Isopropyl myristate	—	5.0
Propylene glycol monoricinoleate	10.0	—
Polyethylene glycol 400	5.0	—
Beeswax	7.0	7.0
Candelilla wax	7.0	7.0
Carnauba wax	3.0	3.0
Ozokerite	3.0	3.0
In addition:		
p-Hydroxybenzoic acid propyl ester	0.2	0.2

Bromo acids	3.0	3.0
Colour lakes and pigments	12.0	12.0

Manufacture of Lipsticks

Lipstick manufacture employed will depend to some extent on the formulation and the plant available. In general, the manufacture of lipsticks consists of three stages :

(1) The preparation of component blends, that is the oil blend, colour dispersion and wax blend.

(2) The blending of these intermediates to form the lipstick mass.

(3) The moulding of the lipstick mass into sticks.

Preparation and blending

The colours, that is the bromoacids, pigments and lakes, are blended separately and dispersed in the complete lipstick base. The object of the operation is to produce a completely homogeneous dispersion of colourants conducive to the preparation of a smooth stick. A finely ground colours are used to wet fatty mixture and taken to milling to grinding process. Ball mills, sand mills, roller mills, colloid mills, edge or end runner mills has been used. It is desirable that the portion of the base used should be a plastic semi-solid during the milling and it may sometimes be necessary to mill under warm conditions.

The soluble dyes are first dissolved in dye-solvent using heat. If solution is complete, this portion can be left on one side while the pigments are dispersed, but if not, the dye blend must at some point be ground with the pigment dispersion.

In preparing the pigment dispersion the colours are first milled with wetting chemicals like lanolin, polyglycol compounds and then add the completely non-polar materials.

The remainder of the base materials, for example the high melting waxes which may not have been used in the dye or pigment blends, are melted and blended together and then blended with the others. In order to secure an intimate mix a final milling operation is desirable.

Vacuum processing of the lipstick melt facilitates dispersion of pigments by removing the film of adsorbed gas on the pigment particles, which is a barrier and causes incomplete wetting. It has been established that petroleum jelly will dissolve 10 per cent air by volume at 90°C but only 5 per cent at 20°C. Therefore the mix has been held at high temperature and allowed to absorb its full quota of air and then cooled. The air is not released until the viscosity of the mass is too great to allow to escape completely. It may collect round the pigment particles, displacing the oil and giving the effect of incomplete wetting.

The manufacture of lipsticks conducted at the lowest temperature. It is desirable that when mixing is complete the mixture should be transferred to a jacketed closed vessel. The fluid mixture is stirred gently while sufficient vacuum is applied to remove all occluded air. When the air has been removed, vacuum is shut off, the mass is stirred and the perfume is added and mixed through. The mass is then ready for moulding immediately. Else can be set into blocks which are stored and moulded as required.

In one of the two methods described, the wax blend is prepared and stored in a molten state while all constituent oils are metered into a storage tank fitted with a suitable mixer. A part of the oil blend is then mixed with the dry pigments in a pre-mix pan attached to a sand mill and the resultant slurry passed through the sand mill into the steam pan fitted with a combined mixer and pump. The remainder of the oil blend is then passed through the pre-mix pan and sand mill to flush any residual colour.

The wax blend is metered into the steam pan simultaneously and thoroughly mixed with the other constituents. The resulting base is tested for shade, perfume and flavour are added, and the lipstick mass is pumped into storage trays, using the mixer as a pump.

Moulding

For moulding first lipstick mass is remelted in a jacked pan. The entrapped air is removed by slow agitation for 30 minutes. The molten mass is then run into moulds for casting. Moulds are usually filled to excess to prevent the formation of a depression in the centre of the stick. This excess is scraped off and the mould is then carefully cooled to allow the mass to set. The overcooling will affect the removal of stick from the moulds.

The moulds are made of brass and aluminium. They may be of the vertical split type (with a split down the centre to allow for the easy removal of the sticks) or the automatic ejection type. A cooling cabinet is required with the split moulds. A preheating device in moulds can raised a temperature of about 40°C before filling with the lipstick mass. This prevents 'flow marks' which would otherwise be visible on the moulded stick. With the water-jacketed automatic ejection mould a cooling cabinet is unnecessary. The jacket is filled with warm water prior to pouring the lipstick mass. Then the cold water is introduced and kept running to chill the lipstick mass. When cooled, the moulds are opened and the sticks are pushed out automatically or with a rubber-covered finger.

Automatic ejection mould (whether air-cooled or water-cooled) will not produce a good bullet–shaped stick, because a small ring or ridge is left near the tip of the stick caused by the thickness of the metal edge of the ejection plunger. It is considered that such moulds are best for the wedge-shaped stick. After moulding, the sticks should be stored for about a week before being filled into the lipstick holders. Then lipstick is passed rapidly through a small gas flame to melt the surface layer, in order to remove any surface spots and to produce a bright, smooth and glossy surface. This process is called as flaming.

Transparent Lipsticks

A lipstick can be described which uses soluble or solubilised dyes. This allows light to shine through it, giving it a sparkle. The suitable solvents such as a monoalkyloamide of mixed fatty acids or dipropylene glycol methyl ether enhances staining action of dyes. Standard lipstick case is shown in Fig. 30.1.

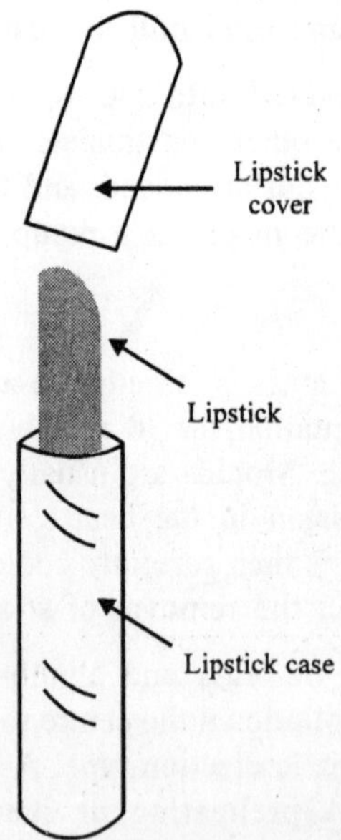

Fig. 30.1. Lipstick case.

Water-soluble dyes may be used for additional colour effect. The compatibility of such dyes is improved by using anhydrous lower alcohols such as ethanol or isopropanol as co-solvents. These alcohols control syneresis and improves stability at 2 to 10 per cent.

Polyamide resins or blend of solid polyamide resin with liquid polyamide resin of molecular mass 2000 – 10,000 and 600 – 800 are used for moulding lipstick to give solid stick.

Ensure that the resulting sticks are not brittle. These materials can be used with softening agents like a lower aliphatic alcohols in conjunction with other polyamide solvents like fatty acid esters, to prevent brittleness the glycol ester of higher fatty acids, particularly those used are between C_{12} and C_{18}. The formulations are given to illustrate transparent lipsticks in formula 7 :

Formula 7

Constituents	*per cent*
Polyamide resin (average MW 8000)	20.0
Polyamide resin (average MW 600-800)	5.0
Propylene glycol monolaurate	28.0
Castor oil	12.6
D&C Red 21	0.3
Lanolin alcohols	8.0
Dipropylene glycol methyl ether	10.0
Ethoxylated lanolin alcohols (5 mol Et_2O)	10.1
Anhydrous isopropanol	5.0
Perfume	1.0

Lip Salves

The lip salve is used, not for decoration, but for protection against exposure to cold, in winter or sub-arctic conditions. The requirement is simply for a fairly substantial, flexible, adherent, moisture-resistant film on the lips and there is no need for staining dyes. The base can be made largely from mineral oils, jelly and waxes. It is necessary to include a proportion of a more hydrophilic material to promote adherence, perfume blending and general properties. In some cases a small amount of antiseptic can be added. Some users prefer a coloured salve in which case the colour is provided by a small amount of oil-soluble dye or dispersed lake not necessarily of a staining type. Suitable base formulation are given in formulae 8 and 9 :

Formula 8

Constituents	*per cent*
Paraffin wax	30.0
White petroleum jelly	35.0
Technical white oil	20.0
Beeswax	15.0

Formula 9

Constituents	per cent
Cetyl alcohol	5.0
Eutanol	30.0
Beeswax, white	25.0
Paraffin wax, 52°C	15.0
Liquid paraffin	25.0

A lip-protective pomade is made up as follows formula 10 :

Formula 10

	Constituents	per cent
A	Mineral oil	25.0
	Petroleum	7.0
	Ceresin wax	24.0
	Beeswax	7.0
	Atlas G-2859	3.5
	Tween 60	0.5
	Union Carbide L-45 Silicone Fluid (1000 cS)	5.0
B	Water	28.0
	Preservative	*q.s.*
C	Perfume	*q.s.*

A lip salve which is used as a decorative item has become known in recent years as a lip gloss. This popular item can be applied to the lips without other make-up or over the normal lipstick. Lip gloss preparations take the form of a stick, or a salve for finger application, or a liquid for roll-on or brush application. These compositions may also contain a pearl pigment to confer a transparent sheen in addition to the normal gloss. Formulation of finger applied lip gloss & liquid lip gloss are given below in formulae 11 and 12 :

Formula 11 (Finger-applied lip gloss)

Constituents	per cent
Syncrowax HRC	5.0
Syncrowax HGLC	5.0
Timica silk white	8.0
Fluilan	8.0
Liquid paraffin	74.0
Perfume, antioxidant	*q.s.*

Formula 12 (Liquid lip gloss)

Constituents	per cent
Polybutene	30.0
Lanolin oil	20.0

Mineral oil	24.8
Oleyl alcohol	25.0
Saccharin	0.2
Perfume, antioxidant	*q.s.*

Liquid Lipsticks

Liquid lipsticks have been developed with the object of providing more permanent films than can be obtained with conventional lipsticks. These liquid lipsticks consist of ethanolic solutions of alcohol-soluble dyes, film-forming resins like ethyl cellulose, polyvinyl alcohols and polyvinyl acetate and plasticisers like triethyl citrate, dioctyl acetate, methyl abietate or polyethylene glycols. The dyestuffs employed are alcohol-soluble halogenated fluoresceins and also other alcohol-soluble dyes.

The dyes and perfume are dissolved in the alcohol. Then resulting solution is filtered. The disadvantage associated with the use of these alcoholic preparations is 'smarting' produced on the lips on application. Excessive drying of the lips may also result. Persons affected in this way should discontinue the use of these preparations. Apply occasionally a 40 per cent solution of aqueous glycerine and use an emollient lip salve. The manufacture of liquid lipsticks based on ethyl cellulose, the formulations are given in the following illustrative formulae 13 and 14 :

Formula 13

Constituents	*per cent*
Ethyl cellulose	3.1
Ethyl alcohol	68.4
Petroleum ether	20.0
Hydrogenated methyl abietate	7.5
Rhodamine	1.0

Formula 14

Constituents	*per cent*
Ethyl cellulose	3.06
Bleached wax-free shellac	4.94
Ethyl alcohol	65.00
Petroleum ether	14.66
Hydrogenated methyl abietate	12.00
Fuchsine	0.30
Saccharin	0.04

The accessory agents like perfumes, preservatives, antioxidants, etc., may be added. The preparations are preferably clear and transparent and are of a viscosity ranging between 3 and 500 cP, while best results are obtained in a range between 20 and 50 cP. Another lipstick formulation is given below in formula 15 :

Formula 15
Constituents   *per cent*

Constituent	per cent
Ethyl cellulose	1.0
PVA (high viscosity)	3.0
Triethyl citrate	1.0
Alcohol-soluble lanolin	0.5
PEG	1.0
Isopropyl alcohol, pure	40.0
Bromoacid dyes	3.0
Flavour	0.5
Ethanol	50.0

ROUGE

The oldest make-up preparations used to apply colour to the cheeks is rouge. The preparations are available in liquid, cream and solid forms, of which the pressed powder or dry rouge is the most popular.

Dry Rouge (Compact Rouge)

Compact rouge differs from an ordinary compact powder in being more highly tinted. The properties of compact rouges are therefore virtually identical with those of ordinary compact powders. The finished product must be smooth, free from grittiness and should be easy to apply. It should have good adhesion to the skin and a good covering power. Because these products are highly tinted and may be conspicuous if not adequately blended. It is essential that component colours are distributed evenly throughout the product. A component liable to impart undesirable opacity must be used in moderate amount which has no deleterious effect on the appearance of product. e.g. at lower concentration zinc oxide is used. The hygroscopic nature of kaolin are liable to produce colour streaking in damp weather.

Grinding and shade-matching stages are included in the manufacture of compact rouges. The component raw materials must be very finely divided to facilitate their intimate blending as a prerequisite to a uniform distribution of component colours. The constituents are, therefore, intimately mixed and simultaneously ground in hammer mills or attrition mills rather than just blended in ribbon mixers. This operation is carried out in the absence of large quantities of liquids. The presence of a little water or oil will ensure better colour distribution. The binder is either sprayed into the powders while they are being mixed, or incorporated in dry form.

The raw materials used in the manufacture of compact rouges are talc, kaolin, precipitated chalk, magnesium carbonate, titanium dioxide, zinc stearate, inorganic oxides, certified colourants and perfumes. Zinc oxide between 5 to 10 % is used to increase the adhesion of the rouge, confers opacity to the product. Titanium dioxide has an appreciably greater covering powder than zinc oxide, and it also gives brighter and more stable colour shades.

Metallic stearates between 4–10 % are used extensively as dry binders for compact rouges as they improve the adhesion with skin.

Compact rouge formulation

Typical compact rouge formulations are given in formulae 16 and 17 :

Formula 16

Constituents	*per cent*
Talc	48
Kaolin	16
Zinc stearate	6
Zinc oxide	5
Magnesium carbonate	5
Rich starch	10
Titanium dioxide	4
Colours	6

Formula 17

Constituents	*per cent*
Talc	67.5
Zinc stearate	5.0
Titanium dioxide	4.0
Red ion oxide	11.5
Black ion oxide	0.3
D&C Red 9 — barium lake	0.2
Mica coated with titanium dioxide	6.8
Perfume	0.2
Base	4.5

Base:

Beeswax	12.00
Lanolin	2.00
Mineral oil	86.00

The colour lakes used for lighter and deeper shades are 1.5 and 6 %. Colour lakes provide wide range of shades. Water-soluble and oil-soluble dyes are not suitable as colourants in compact rouges.

Wax-based Rouge

Wax-based rouges are in many respects similar to lipsticks. Formulae for bases resemble the drier type of lipstick. The shape is sometime larger in diameter ($\doteq$ 2 cm), colour is red or pink shade, or white. It is known as 'glemers' as it is used to disguise the colours of the face. The bromoacid type staining dye is not not due to possiblity of dermatitis. The formulae 18 and 19 are examples for wax-based rouges :

Formula 18

Constituents	per cent
Candelila wax	8.6
Carnauba wax	5.4
Beeswax	4.0
Mineral oil	17.0
Lanolin	2.0
Isopropyl myristate	33.0
Inorganic pigments and colour lakes	30.0

Formula 19

Constituents	per cent
Carnauba wax	10.0
Paraffin wax	2.5
Ozokerite wax	5.0
Petrolatum	4.0
Mineral oil	68.5
Titanium dioxide	1.9
Red iron oxide	1.0
Bismuth oxychloride	7.0
Perfume	0.1

Cream Rouge

Two types of cream rouge are available: anhydrous products and emulsified products.

Non-aqueous cream rouge

Cream rouges are more difficult to apply than compact rouge. It is claimed that their use gives better results, in that there is no obvious line of demarcation when they are correctly applied. This gives a more natural effect. An effective anhydrous cream rouge has following formula 20. It is prepared by first grinding colours with the powder constituent, and then mill this into the warmed fatty base. Alternatively, mix all the ingredients and mill in a warmed ointment mill.

Formula 20

Constituents	per cent
Petroleum jelly	70.0
Kaolin	30.0
Colour, perfume	q.s.

Hexadecyl alcohol is claimed to impart a smooth and soft feeling to the skin during application and use. It also facilitates the spreading of a cream rouge. Such an anhydrous rouge may be prepared with the composition given in formula 21 :

Formula 21

Constituents	per cent
Hexadecyl alcohol	27.0
Light mineral oil	23.5
Talc	10.0
Ozokerite	10.0
Carnauba wax	6.0
Titanium dioxide	20.0
D&C Red No. 9 lake	3.0
Perfume	0.5

Emulsions

Cold cream or vanishing cream are emulsion products. The composition of emulsified cold cream is given in formula 22. The process involves dispersion of finely powdered colour base with melted fat. Add borax solution at 75°C, mix and finally mill.

Formula 22

Constituents	per cent
Lanolin	5.0
Cocoa butter	5.0
Beeswax	14.0
Liquid paraffin	30.0
Cetyl alcohol	1.0
Water	44.2
Borax	0.8
Colour	q.s.

The composition shown in formula 23 of cream rouge provides a product which is soft, creamy, and possesses excellent spreading properties. Because it is of the water-in-oil type it has less tendency to dry out than emulsified creams of the oil-in-water type.

Formula 23 (Oil phase)

Constituents	per cent
Arlacel C	2.0
Lanolin (anhydrous)	2.0
Mineral oil 65/75	16.0
Petrolatum	30.0
Preservative	q.s.

Cream rouge of the vanishing cream type has following composition is shown in formula 24 :

Formula 24

Constituents	per cent
Stearic acid	15.00
Water	76.32
Potassium hydroxide	0.50
Sodium hydroxide	0.18
Glycerine	8.00
Colour, preservative	q.s.

The solution of water soluble-dye in water or glycerin is used in emulsion products. After saponification add to main bulk of cream. In formula 25, diglycol stearate is employed as emulsifying agent.

Formula 25

Constituents	per cent
Diglycol stearate	20.0
Water	70.0
Glycerine	10.0
Water-soluble colour	q.s. (from 0.6 to 1.0)
Preservative	q.s.

The concentration of the colour darkens the cream surface as water soluble dyes are difficult to evaporate. Therefore it is necessary to incorporate glycerine, or *d*-sorbitol. Since it possesses a much narrower humectant range than glycerine, that is, it absorbs less moisture in a moist atmosphere, but holds it better in a dry one. A further improvement might be the packing of such products in a tightly capped tube.

Most of the colours employed in cream rouge are insoluble lakes, although sometimes oil colours are used. In the vanishing cream type, however, these lakes may be supplemented by fluorescein dyestuffs it is necessary to incorporate a non-volatile solvent.

Example of a vanishing cream type of rouge, has the following composition in formula 26 :

Formula 26

Constituents	*per cent*
Cetyl alcohol	2.00
Stearin	18.00
Propylene glycol	4.00
Isopropyl myristate	8.00
Carbowax	4.00
Potassium hydroxide	0.70
Water	54.65
Perfume	0.50
Preservative	0.15
Lakes	8.00

Liquid Rouge

Liquid rouge yield extraordinarily good results when skilfully applied and maintain a degree of favour. The essential requirements of such a preparation are that it has sufficient viscosity to permit easy and even application and that it dries reasonably rapidly.

The requisite amount of water soluble dye is dissolved by adding a gum or synthetic thickener for viscosity. A small portion of wetting agent helps to promote spreading. Formulae 27 and 28 shows composition of liquid rouge :

Formula 27

Constituents	*per cent*
Methylcellullose	2.0
Wetting agent	0.1–0.2
Water	to 100.0
Water-soluble red, preservative	*q.s.*

Formula 28

Constituents	*per cent*
Sodium alginate	0.45
Calcium citrate	0.13
Wetting agent (dermatologically acceptable)	0.1–0.2
Water	to 100.0
Water-soluble red, preservative	*q.s.*

In order to shorten the drying time, 10–20 per cent alcohol or more may be added to gum mucilages, but high concentrations must not be used with alginate products otherwise precipitation may occur.

The preparations of liquid lipstick with a plasticiser to obtain the requisite flexible film, can also serve as liquid rouge. No wetting agent is necessary. By the use of suitable dyes more permanent preparations are obtainable than with aqueous mixtures. Since evaporation of alcohol will lead to an increase in the viscosity of the product. It is particularly important that containers should be provided with a well-fitted wick built into, or screwed into, the neck of the container, the whole being enclosed in a screw dust-cap. This enable ready and even application without sticking users hand with varnish.

The resultant rouge film is readily removed with a little alcohol with little high boiling solvent to prevent too rapid evaporation of the rouge remover. The following formula represents composition.

Formula 29

Constituents	*per cent*
Tincture of styrax, 5%	38.0
Alcohol, 95%	60.0
Castor oil	2.0
Spirit-soluble red	*q.s.*

The above solutions are prepared by simple mixture. Proportions of film-forming material, plasticiser and dye may be altered to suit particular requirements.

EYE MAKE-UP

Since many years eye-make-up is us used. Now-a-days Mascara, eyeshadow and eyebrow pencils are effective if applied properly. Mascara pencil is shown in Fig. 30.2.

Mascara (Eyelash Cosmetic)

Egyptian women soon discovered that smearing the eyelids with unguent and stibium (that is, powdered antimony) not only eased the pain in the eyes and rested them but also added to the natural beauty of their faces. The whiteness of the whites of the eyes was emphasised by the darkened eyelids and the large pupils appeared like black pools in their midst, the whole effect being very striking.

Mascara is a black pigmented preparation for application to the eyelashes or eyebrows to beautify the eyes. It has been claimed that by correctly applying mascara to darken the eyelashes and to increase their apparent length, the brightness and expressiveness of the eyes is enhanced. Mascara bottle is shown in Fig. 30.3.

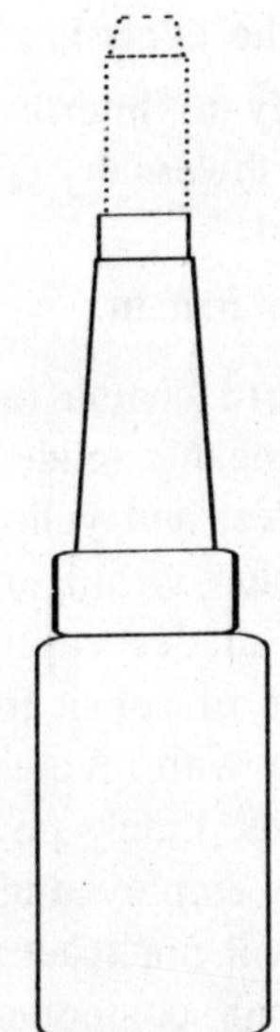

Fig. 30.2. Mascara Pencil.

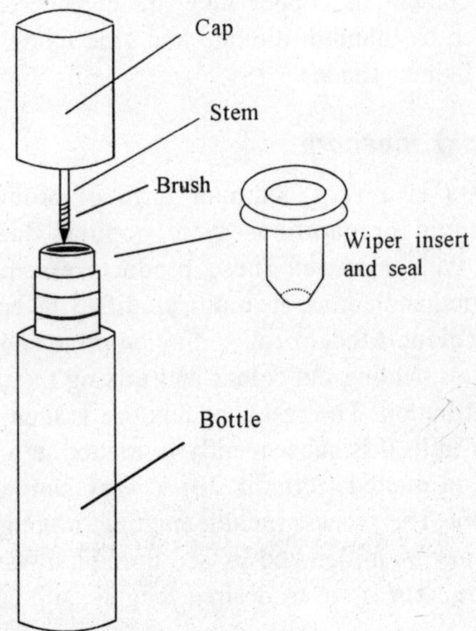

Fig. 30.3. Mascara bottle.

Characteristics of Mascara :

(1) It must be capable of easy and even application;

(2) It must have no tendency to run and thereby cause smudging;

(3) It must not cake, causing the eyelashes to stick together;

(4) It must not dry too rapidly to interfere with the evenness of the application, but should nevertheless dry fairly rapidly and be reasonably permanent once applied; and

(5) It must be neither toxic nor irritant.

Umbers (brown ochres), burnt sienna (a mixture of hydrated ferric oxide $Fe_2O_3 \cdot H_2O$ with some manganic oxide) and synthetic brown oxides have been used for brown shades, and yellow ochres for yellow shades. For bluish shades soluble oil blue is employed, while for green shades chromium oxides and for red shades carmine, the aluminium lake of cochineal, have been used. Salts of cobalt may be used only if they are insoluble salts and will not react with other ingredients of the preparation to form soluble cobalt compounds. Finely powdered metals, such as silver and aluminium, have been widely employed in eyeshadow and may be used if they are insoluble salts and will not react with other ingredients of the preparation to form soluble cobalt compounds. Finely powdered metals, such as silver and aluminium, have been widely employed in eyeshadow and may be used if they consist entirely of the pure metal. If they contain an appreciable content of copper they are considered to be harmful. White pigments such as titanium dioxide and zinc oxide may also be sometimes included to lighten shades.

Cake (block) mascara

Cake mascara is a very common form of product. Since 1960s softer cream or liquid preparations were popular due to special container-dispensers with a brush. These products essentially consist of a soap (generally triethanolamine stearate) modified in consistency with oils and waxes, and colour. Modern cakes may be produced by melting together the waxy materials, adding the colour and mixing them in thoroughly to obtain an even distribution. The resultant mixture is then cooled and worked in a heated roller-mill. It is subsequently re-melted in a pan and cast into cakes. A composition quoted (formula 30) is very similar to a large majority of block mascara. The process include melting, mixing and extruding in proper form. Or it may be milled and passed through a warm plodder after mixing. Then mascara strip is cut to desired lengths, and cake mascara in formulae 31–33 :

Formula 30

Constituents	per cent
Staric acid	27.0
Triethanolamine	12.0

Beeswax	30.0
Carnauba wax	50.0
Colour (bone black)	25.0

Formula 31

Constituents	per cent
Triethanolamine stearate	40.0
Paraffin wax	30.0
Beeswax	12.0
Anhydrous lanolin	5.0
Lampblack	13.0

Formula 32

Constituents	per cent
Glyceryl monostearate	60.0
Paraffin	15.0
Carnauba wax	7.0
Lanolin	8.0
Lampblack	10.0

Formula 33

Constituents	per cent
Triethanolamine stearate	50.0
Carnauba wax	24.0
Paraffin wax (MP 45°C)	12.5
Anhydrous lanolin	4.5
Silicone fluid	5.0
Carbon black	3.8
Propyl-p-hydroxybenzoate	0.2

Cream Mascara

Cream mascara may be prepared by milling the pigment into a vanishing cream base, or by the use of a suitable oil-soluble dye. In such cases a suitable wetting agent lowers surface tension and prevents adherance of colour to brush. The coloured pigment may be incorporated into the base immediately after emulsification has been completed. When the product is cooling continue the agitation which allows air to escape and product is filled into tubes. Alternatively, a previously prepared base is melted in order to incorporate the pigment.

The composition of reliable cream mascara which does not dry up in the tube and which adheres well to the eyelashes is given in formula 34. A modern equivalent of this formula would employ a synthetic hydrocolloid such as hydroxyethylcellulose to replace the quince seeds. A process include grinding the gum and the ivory black (or umber, etc.) with the syrup (3 per cent sugar to 2 per cent water). Then adding the mucilage containing a preservative, such as methyl-*p*-hydroxybenzoate.

Formula 34

Constituents	per cent
Mucilage of quince seeds	35.0
Sugar syrup	35.0
Gum arabic	7.5
Ivory black (or umber)	22.5

Recent cream mascara are shown in formulae 35, 36 and 37. The application on eyelashes is free from the tendency to run or cake have good performance properties.

Formula 35

Constituents	per cent
Polyethylene glycol 400 stearate	10.0
Diglycol stearate	8.0
Lanolin	3.0
Stearyl alcohol	13.0
Isopropyl palmitate	2.0
Triethanolamine lauryl sulphate	1.5
Water	54.5
Colour	8.0

Formula 36

Constituents	per cent
Triethanolamine stearate	45.0
Carnauba wax	15.0
Glyceryl monostearate	5.0
Anhydrous lanolin	10.0
Unbleached beeswax	5.0
Lampblack	20.0

Formula 37

Constituents	per cent
Hexadecyl alcohol	7.3
Propylene glycol	9.1
Stearic acid	11.2
Glyceryl monostearate	4.5
Triethanolamine	3.6
Ultramarine blue	9.1

Methyl-*p*-hydroxybenzoate	0.2
Water	55.0

The incorporation of 10 % pigment or dyestuff into greasy base renders easy removal formulae 38 and 39. It is warmed for application with a brush.

Formula 38

Constituents	per cent
Beeswax	4.0
Spermaceti	4.0
Cetyl alcohol	2.0
Cocoa butter	6.0
Petroleum jelly	64.0
Oil-soluble blue	20.0
Preservative	*q.s.*

Formula 39

Constituents	per cent
Cocoa butter (odourless)	5.0
Petroleum jelly	50.0
Paraffin wax (uncrystallisable)	5.0
Lampblack	40.0

Procedure: Melt and mix.

The new process include, suspension of carbon black in alcoholic solution of resins is formed. A little castor oil is often included which dry rapidly, producing a semi-permanent waterproof colour. The composition is shown in following formula (formula 40):

Formula 40

Constituents	per cent
Rosin (10% alcohol solution)	3.0
Castor oil	3.0
Ethyl alcohol	84.0
Lampblack	10.0

Eyeshadow

An attractive 'moist'-looking background for the eyes is produced using eyeshadows. The use of gold leaf, bronze powder and powdered aluminium produce silver or gold effect. Various pearlescent pigments, for example those based on bismuth oxychloride or mica coated with titanium dioxide may also be used. Eyeshadow preparations available, are for example ·

Liquefying or emulsifying type; emulsions; stick eyeshadow; liquid suspensions or dispersions; and as pressed powders.

Cream eyeshadow

Cream eyeshadow may be made by mixing first of all the selected colours and blending these pigments with petroleum in a roller mill. This mass is then stirred into the blend of fatty and waxy constituents which have been previously blended by melting them together in a pan. After thorough agitation, the product is then poured into suitable containers. Alternatively, the pigments in powder form may be stirred directly into the molten mass of fats and waxes. Then passed through a roller mill to ensure thorough distribution of colours and remove any lumps from the formulation. The second mixing operation may follow before the product is poured into the containers. As in the case of mascara, the eyeshadow base should be capable of easy and smooth application. It should also be waterproof to avoid streaking. Sometimes, in order to facilitate liquid filling of the product into containers, the product requires to be kept warm in a jacketed steam-heated pan and, gently stirred to prevent the sedimentation of any heavy pigments. The composition of eyeshadow creams of the anhydrous type include cocoa butter. Mix 85 parts of this base with (say) 12 parts of titanium dioxide and 3 parts of a mineral pigment or earth colour as given in formulae 41–43 :

Formula 41

Constituents	*per cent*
Cocoa butter (odourless)	2.0
Beeswax	3.0
Spermaceti	5.0
Lanolin	5.0
Petroleum jelly	55.0
Zinc oxide	30.0
Cosmetic lake, preservative	*q.s.*

Formula 42

Constituents	*per cent*
Petroleum jelly	75.0
Cocoa butter (odourless)	8.0
Lanolin	7.0
Cetyl alcohol	3.0
Paraffin wax (uncrystallisable)	7.0
Cosmetic lake, preservative	*q.s.*

Formula 43
Constituents

	per cent
Bleached beeswax	4.5
Spermaceti	9.5
Lanolin absorption base	13.0
Paraffin of low melting point	73.0

To obtain a grey shade, suggested a mixture of 1 part of ultramarine blue, 1 part of lampblack and 2 parts of titanium dioxide for a brown shade, a mixture of 3 parts of an umber and 1 part of titanium dioxide. And then heating the colours with the mixture of fats and waxes (Formula 43) and passing them through a roller mill. Another eyeshadow cream formula is also suggested.

Formula 44
Constituents

	per cent
White petroleum jelly	59.0
Glycerol monostearate	17.0
Lanolin	4.0
Beeswax	8.0
Candelilla wax	4.0
Pigment	8.0

Emulsion-type eyeshadow creams are produced by mixing suitable pigments into an emulsion and distributing them evenly throughout the base. The products should be filled into containers when cold. Emulsions based on triethanolamine stearate can be used as bases for such preparations (Formula 45).

Formula 45
Constituents

	per cent
Stearic acid	1.5
Glyceryl monostearate	1.5
Lanolin	4.0
Isopropyl myristate	5.0
Veegum (5% solution)	30.0
Triethanolamine	0.75
Water	38.25
Propylene glycol	4.0
Ultramarine blue	4.5
Black iron oxide	1.2
Chrome hydrate	0.8
Mica coated with titanium dioxide	8.5

Stick eyeshadow

The sticks contain a fairly high proportion of waxes such as ceresin, ozokerite or carnauba. The formulation is given in formula 46.

Formula 46 Constituents	*per cent*
Castor oil	43
Mineral oil 75/85	6
Hydrogenated cottonseed oil	5
White ceresin MP 76°C	26
Carnauba wax	4
Titanium dioxide	8
Iron oxide ochre shade	4
Iron oxide sienna shade	4

Pressed powder eyeshadow

Pressed powder eyeshadows can be regarded as compact rouge with a much higher colour levels than the rouge products. For example, with products containing high levels of an iridescent material such as mica coated with titanium dioxide, care must be taken that, in dispersing the product, the platelets are not shattered by the hammer mills. The product is usually applied by means of a sponge-tipped plastic rod, although finger-tip application is popular. Typical formulae illustrating a matt shade (formula 47) and a high iridescent shade (formula 48) are given below :

Formulae Constituents	*47* per cent	*48* per cent
Talc	82.5	41.7
Zinc stearate	6.0	7.0
Ultramarine blue	5.4	—
Black iron oxide	0.1	—
Chrome hydrate	2.0	—
Yellow iron oxide	—	2.0
Mica coated with titanium dioxide	—	40.0
Base (see formula 17 above)	4.0	9.3

Liquid eyeshadows

Liquid eyeshadows take the form of either a liquid suspension or a liquid dispersion. In the former case the pigments are suspended in a mixture of oils, but they usually settle and have to be shaken before use. The liquid dispersions are prepared from dilute solutions of alcohol in water thickened with a synthetic gum such as methyl cellulose. It is suitably preserved and containing a wetting agent in which the pigments have been dispersed. However, liquid eyeshadows are not very popular.

Eyeliner

Eyeliners used on eyelids, particularly the upper eyelids, close to the eyelashes to enhance expressiveness. They are available in liquid, cake and pencil form. The compressed powder form is similar in composition to cake mascara. The following composition is suggested (formula 49) for dry eyeliner is given as under :

Formula 49

Constituents	*per cent*
Mineral oil	5
Colour	30
Titanium dioxide	5
Talc	60

Liquid eyeliners may be used to apply colour to the tissue around the eyes. Usually a brown colour is considered a good colour for daytime wear. A liquid eyeliner composition is given in formula 50 :

Formula 50

Constituents	*per cent*
Water	40.00
Methocel HG 60–50V	1.00
Veegum	1.00
Shellac	1.08
Oleic acid	0.50
Triethanolamine	0.40
Water	3.02
Pigment	18.20
Alcohol	5.00
Water, preservative	to 100.00

Eyebrow Pencils

Eyebrow pencils may be either of a wax crayon type pigmented black or brown, in which the pigments are present at a higher level than in cream eyeshadow preparations. They may be in the form of an extruded eyebrow pencil similar to an ordinary lead pencil, in which a crayon-type formulation has been enclosed in a wooden casing. In the latter form they are even more readily applied than the stick type. Eyebrow pencils are used either to accentuate the line of the eyebrows, or to modify their outline after plucking. The proportion of beeswax or other high-melting constituent may be optionally increased to produce a somewhat firmer pencil in the lipstick type. However, such sticks must be capable of ready and uniform application as given in formula 51 below :

Formula 51

Constituents	per cent
Hydrogenated castor oil	46.0
Hydrogenated cottonseed oil	12.0
Cocoa butter (odourless)	8.0
Castor oil	8.0
Lanolin absorption base	17.0
Black iron oxide	9.0

If a brown stick is required the black iron oxide may be partially replaced by the red and yellow versions. An eyebrow pencil composition prepared by stirring the colours into the molten fat phase and mill until uniform. Fill the warm batch into the moulds.

Formula 52

Constituents	per cent
Ozokerite (MP 70°–75°C)	45.0
Unbleached beeswax	24.0
Cocoa butter	22.5
White petrolatum	6.0
Absorption base	1.5
Anhydrous lanolin	1.0
In addition:	
Pigment or lake colours	10.0
Oil-soluble colour	1.0

Formula 53

Constituents	per cent
Beeswax	21.0
Carnauba wax	5.0
Paraffin	29.0
Cetyl alcohol	8.0
Vaseline	18.0
Lanolin	9.0
Pigments	10.0

Chapter 31

Nail and Manicure Products

INTRODUCTION

The top surfaces of the end joint of each finger and toe are protected by translucent plates, composed of hard keratin are called as nails. Each lies upon a *nail bed*, to which it appears to be fused, except at the far or *distal* end where there is a free margin. At the near or *proximal* end a paler *lumina* is visible, and the posterior margin is slightly overlapped by a narrow extension of the horny layer of the epidermis which is known as the *cuticle*.

Many variety of cosmetic preparations related with the cleansing and decoration is complete manicure treatment. The discussion is followed on those treatments.

CUTICLE REMOVER

When the free edge of the nail has been shaped, by mechanical means such as cutting or filling, the next problem is to shape the base of the nail. Where the skin adjoins the nail it becomes cornified and the dead cells, together with sebum, form an irregular appendage which grows thick and ragged and partially obscures the 'halfmoon' or lunule. Some improvement can be effected by mechanically loosening and pushing back the cuticle, but removal of excess cuticle by cutting is unsatisfactory and cuticle removers are used.

Cuticle removers are basically alkaline materials in liquid or cream form. To 2-5 % of aqueous or alcoholic solution of potassium hydroxide add 10-20% of glycerine or propylene glycol. The humectant added serve to counteract the irritation potential of alkali hydroxides and also to retard evaporation and increase viscosity. The latter effects can also be achieved by the use of suitable water-soluble gums or hydrocolloids.

Formulation is given in formula 1 :

Formula 1

Constituents	*per cent*
Trisodium phosphate	8.0
Glycerine	12.0
Water	80.0
Perfume	*q.s.*

The milder products incorporate 8–10 per cent of monoethanolamine or isopropanolamine or 10–12 per cent of triethanolamine. Formula 2 gives a basic formula suggested for such a preparation :

Formula 2

Constituents	*per cent*
Sorbitan monopalmitate	2.0
Polyoxyalkylene sorbitan monopalmitate	2.0
Mineral oil	15.0
Water	81.0
Cuticle remover	*q.s.*

The process to prepare cuticle remover is heat first three items at 65°C. The emulsify using water without using cuticle remover. Now mix them when cold. The special emulsions prepared has following composition :

Formula 3

Constituents	*per cent*
Cetyl alcohol	2.5
Myristyl alcohol	3.5
Polyoxyethylene (5) lauryl ether	1.0
Glycerine	4.0
Potassium hydroxide	1.6
Water	87.4

NAIL BLEACH

Nail bleaches are solutions or creams used for the removal of ink or tobacco stains, vegetable stains, etc., from the nails due to oxidation or reduction. Oxidation may be achieved by the use of hydrogen peroxide, either at 20 vols or diluted to about 1 : 4, by chlorine compounds such as the hypochlorites and the cyanurates and by sodium perborate and zinc peroxide (the latter two substances decompose in aqueous solution). Use of sulphites with dilute acid is probably the easiest method for achieving reduction in such preparations. A good quality nail bleaching lotion has the following composition :

Formula 4
Constituents	*per cent*
Hydrogen peroxide (10 vols — 3% w/w)	73.5
Ammonia (SG 0.93)	0.5
Rose water	25.0
Preservative	1.0

NAIL CREAM

At the beginning the relief of brittle nails was obtained by massaging them with olive oil after having bathed in warm water. Emollient cream is practical solution which contains a suitable humectant. The base of emollient creams is lanolin or beeswax-borax emulsion. These creams should be applied after soaking the hands in warm soapy water and thoroughly drying them, just before retiring for the night. This treatment should be carried out at least once and preferably 2–3 times per week (that is, on alternate nights) depending upon the condition of the nails. Wearing fingerstalls or gloves may assist the treatment.

It is found that cholesterol, appear to assist in maintaining the natural elasticity of the nail. The requisite degree of moisture in the keratin is maintained due to water in oil emulsion of cholesterol. It would appear that the use of a nail cream containing oils in an emulsified form with water (and possibly incorporating a mild antiseptic) would be of immense value. The inclusion of a little lanolin or emollient or sufficient humectant to cold, vanishing, foundation or all purpose cream will result into nail cream.

NAIL STRENGTHENER

The appearance of nails is affected due to splitting and breaking which makes manicuring difficult and painful. This had led to the development of various products. For strengthening brittle nails and for eliminating brittleness and dryness the following composition is suggested (Formula 5):

Formula 5
Constituents	*per cent*
Potash alum	3.00
Glycerine	10.00
Formaldehyde	0.01
Menthol	0.001
Water	to 100.000

NAIL WHITE

Nail whites are preparations used to produce an even white edge to the nails. They are based on an inert white pigment such as zinc oxide, titanium dioxide, kaolin, talc or colloidal silica, of which the first two are the best. The pigment forms 20–30 per cent by weight of the preparations, which is generally presented as a stiff paste. For general purposes a fatty base is preferable. The 38% of titanium dioxide along with 62% of petroleum jelly is best example.

It is readily produced by milling the titanium dioxide into the melted petroleum jelly. Various mixtures of suitable fats and waxes may replace the petroleum jelly as desired.

Nail white pencils have largely replaced nail white creams. They are easier to apply, are less messy in use, and simply require moistening with water before application to the under-surface of the nail free edge. The pencil then being recapped. A wax may be used as a base formula 6 :

Formula 6	
Constituents	*per cent*
Beeswax	55.0
Hydrogenated cottonseed oil	9.0
Cocoa butter	9.0
Caster oil	8.0
Lanolin base	9.0
Titanium dioxide	10.0

Such a base is ready for use and is not wetted before application. The usual type, which requires preliminary damping, is prepared by massing casein or gum arabic with the inert pigment, or by suspending it in sodium stearate base.

NAIL POLISH

A distinction is drawn between those polishes which by abrasive action bestow a gloss on the nail surface, and a nail varnish. The former, by reason of the friction set up in the buffing process, draw the blood to the numerous capillaries of the nail bed and, by increasing the blood supply, may exert some slight stimulating effect upon the growth of the nail. The latter, which depend upon the deposition of a thin film of highly lustrous cellulose lacquer upon the nail plate, have increased enormously in popularity and have largely replaced the abrasive type, although abrasive polish may also be used between two successive coats of varnish to enhance the lustre, in the manner of the French polisher.

Abrassive polishes consist of a suitably fine abrasive powder which is applied to the nail and then polished with a shaped chamois leather pad. The principal constituents of such powders are stannic oxide, talc, silica, kaolin, precipitated chalk, etc., the first being an excellent abrasive but rather more expensive than the others listed. The composition of a abrasive is given in formula 7 :

Formula 7

Constituents	per cent
Stannic oxide	90.0
Powdered silica	8.0
Butyl stearate	2.0
Pigment, perfume	q.s.

Butyl stearate render the product less gritty. It may be replaced, by oleic acid. The method of preparation is by simple trituration in a mortar or revolving mill. The liquid or paste or pencil form may be prepared by addition of methyl cellulose or tragacanth with glycerine or glycol ethers.

Wax polishing pencils embodying a large proportion of abrasive powder replaced the older powder and block type. They were effective and could be carried and used without the danger of split contents. The formation is shown in formula 8 :

Formula 8

Constituents	per cent
Hardened palm kernel oil	22.0
Synthetic wax (high melting point)	7.0
Stannic oxide	71.0

A softer stick may be obtained by omitting the synthetic wax and making a simple mixture of abrasive powder with the hardened oil. Alternatively, a base of rosin, beeswax, ceresin and petroleum jelly, or other suitable combination of waxes, may be employed. Nail polish should be easy to apply, giving a good cover of the nail with two coats, and should not steak or apply unevenly. The polish should dry quickly to the touch and leave a high-shine finish film that will not stain the nail when removed. The polish is mostly dispensed from a bottle with a brush applicator Fig. 31.1.

NAIL LACQUER (NAIL VARNISH)

The largest and most important group of manicure preparations is that of nail varnish or lacquers. A nail lacquer should fulfil the following properties :

(1) It should be innocuous to the skin, nails and stable on storage.

(2) It should be easy and convenient to apply and give a film with satisfactory characteristics.

The characteristics of a satisfactory film are :

(i) Even thickness; (ii) Uniform colour; (iii) Good gloss; (iv) Good adhesion; (v) Sufficient flexibility; (vi) Hard, non-tacky surface; (vii) Satisfactory drying properties; and (viii) The film should maintain these properties for approximately 1 week.

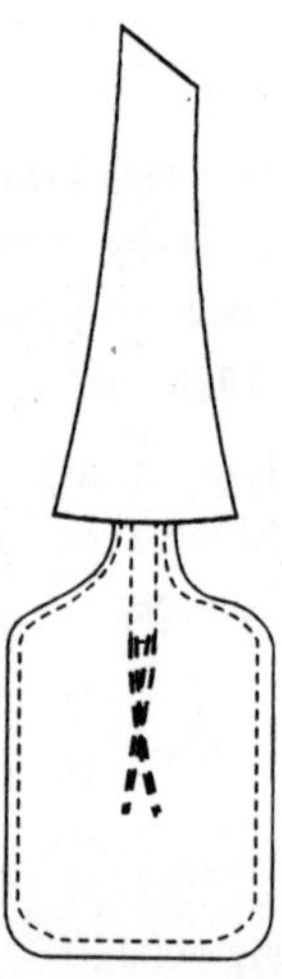

Fig. 31.1. Nail polish bottle.

Ingredients of Nail Lacquer

The main essentials for the manufacture of nail lacquer are a lacquer base with suspending properties and a colouring system. These two components can be broken down into necessary constituents, and these are described before manufacture of the product is considered.

Film-formers

The basic film-forming material in nail lacquers is nitrocellulose. A cellulose nitrate obtained by the reaction of mixtures of nitric acid and sulphuric acid with cotton. Film formed are waterproof, hard, tough and resist abrasion.

Resins

The resins are used to impart adhesion and to improve gloss. With a dry nitrocellulose resin ratio of 2:1, the moisture permeability of a film has been reduced to about half of that of an unmodified nitrocellulose film of the same thickness. These resins are moderately stable to light and soluble in the majority of lacquer solvents and diluents.

Plasticisers

Plasticisers along with nitrocellulose are included in nail lacquer formulations. This ensures adhesion of a film on the nails after evaporation. The film is flexible and does not flake off. The film becomes pliable due to high boiling point of plasticisers. Plasticisers, even at low concentration, will furthermore enhance the gloss of resultant films and will also improve the flow properties of lacquers.

Solvents

It will be seen from the list of desirable film properties that the properties of the solvent, particularly the evaporation characteristics, are of the utmost importance. It is common practice to rank solvents by their boiling points, which also appear to correlate with the viscosities of the resultant nitrocellulose solutions and hence with the spreading characteristics.

Diluents

Diluents are organic solvents miscible with nitrocellulose. They are used as solvents for modifying resins used in lacquers, since true nitrocellulose solvents are expensive. They also help to stabilise the viscosity of lacquers. Their use reduces the overall cost of formulations.

Colours

Insoluble lakes with titanium dioxide are incorporated in the majority of nail lacquers to produce a suitable shade, opacity and creaminess and to produce pastel shades.

Soluble dyestuffs such as erythrosine, carmoisine and rhodamine, which were used at one time in nail lacquers, were found to stain the surrounding skin and the use of such dyestuffs has been virtually abandoned.

Pearlescent (nacreous) pigments

In addition to the conventional pigments, nail lacquers can incorporate iridescent materials Guanine (2-amino-6-hydroxy purine), a crystalline substance from the scales and body of various fish, has been one of the most frequently suggested. In its pure form it is non-toxic, but there have been references to this material being the cause of dermatitis arising from nail lacquers.

Suspension agents

The modern trend towards highly pigmented and pearlescent nail lacquers has led to a critical assessment of the ability of traditional nail lacquer formulae to suspend materials at high concentration. The suspension properties are obtained by creating a thixotropic system with the use of pretreated colloidal clays such as benzyl dimethyl hydrogenated tallow ammonium montmorillonite, dimethyl dioctadecyl ammonium bentonite, or dimethyl dioctadecyl ammonium hectorite. The clays increase the viscosity of the system to such an extent that the heavy oxide pigments remain in suspension.

Formulation

The flow characteristics of formulations were ascertained by finding out rate of drying of lacquer films, their appearance, hardness, adhesion and resistance to soap, detergents. Tests should also be carried out on the flow and evenness of application of lacquer films, their hardness and gloss. Any drag observed on the applicator brush should also be recorded. Formulation of base nail lacquers can be illustrated by the following formulae 9 and 10 :

Formula 9 (Typical nail lacquer base)

Constituents	*per cent*
Nitrocellulose (about 1/2-second viscosity)	10.0
Resin	10.0
Plasticiser	5.0
Alcohol	5.0
Ethyl acetate	20.0
Butyl acetate	15.0
Toluene	35.0

Formula 10

Constituents	*per cent*
Nitrocellulose (1/2-second)	18.28
Santolite (resin)	6.43
Dibutyl phthalate (plasticiser)	2.15
Camphor (plasticiser)	1.88
Ethyl acetate (solvent)	6.45
Butyl acetate (solvent)	25.42
Toluene (diluent)	29.56
Isopropanol (diluent)	9.83

Resins consisting of polymerisation products of acrylic acid and its derivatives may be used in nitrocellulose compositions. It overcomes inherent disadvantages of nail varnishes like shrinkage during drying, peeling at the edge of the nail, discoloration, thickening of the enamel by evaporation of volatile ingredients when stored in poorly stoppered containers, and failure to achieve proper gloss and durability. A typical formula cited is the following formula 11 :

Formula 11

Constituents	*parts*
Cellulose nitrate	3.2
Denatured alcohol	1.4
Butyl acetate	19.3
Ethyl alcohol	19.3
Toluene	19.3
Propyl methacrylate resin	15.0
Plasticiser	3.2

Both high-boiling and low-boiling solvents are said to be eliminated in the above formula and the enamel produced is said to afford excellent adhesion and to be remarkably resistant to soapy water, alkali, marking and scratching. An interesting formula containing a chlorinated biphenol plasticiser is given in formula 12 :

Formula 12

Constituents	*parts*
Polyvinylchloride-acetate copolymer	10.7
Polymethylmethacrylate	7.0
Methyl ethyl ketone	21.5
Sorbitol	20.0
Arochlor	15.0
Tricresyl phosphate	8.0
Ethyl acetate	22.3
Colours	0.5

Manufacture of Nail Lacquer

In practice it is more usual to buy a base lacquer which incorporates the particular requirements for plasticiser, resin, etc., and mix with it pearlescent materials and pigments, etc. The specialist approach is needed to handle and process nitrocellulose and bentone. Hence, most cosmetic manufacturers handle nail lacquer production along these lines. Many just fill the finished product as supplied to them from a specialist contract manufacturer, although they themselves may well have developed the shades, etc. An outline of all the manufacturing procedures involved in nail lacquer production is given in Fig. 31.2.

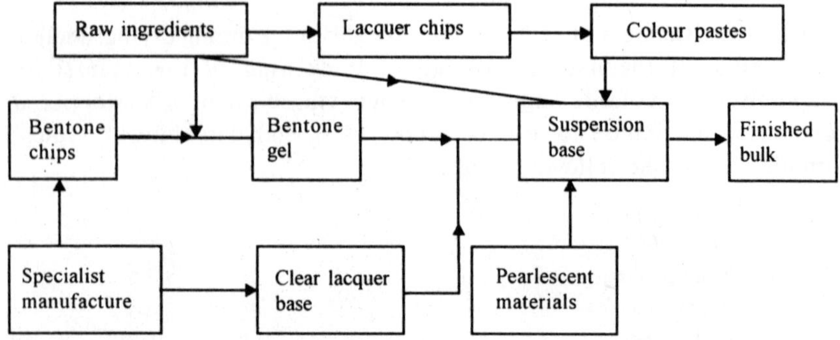

Fig. 31.2. Manufacturing routes for nail lacquer production.

Base Coats and Top Coats

The chipping of the nitrocellulose film is prevented by base coat. Here the film is formed on the nail is to which nitrocellulose film adheres closely. Phenol-formaldehyde resins, sulphonamide-formaldehyde resins, alkyd resins and methacrylate are used for this purpose.

A larger amount of resin is used in base coat for increased adhesion. They dry more rapidly and give a harder film that the normal lacquer film. Films built on a base coat are far less liable to crack than ordinary nail enamels. Clear and pigmented base coat formulations are available with a lower solid content and lower viscosity than normal lacquer formulations.

Top coats are usually clear lacquer films. They are applied over the pigmented lacquer film in order to increase film thickness and thus resistance to abrasion and wear, and also to improve gloss. Top coats have a higher nitrocellulose and plasticiser content and a lower resin content than normal nail enamel. They also have a relatively high diluent content and tend to dry rapidly. Three formulae for a base coat, a clear lacquer and a top coat are given in formulae 13–15 :

Formulae	13	14	15
	Base coat	*Clear lacquer*	*Top coat*
Constituents	*per cent*	*per cent*	*per cent*
Nitrocellulose	10	15	16
Santolite resin	10	7.5	4
Dibutyl phthalate	2	3.75	5
Butyl acetate	—	29.35	10
Ethyl acetate	34	—	10
Ethyl alcohol	5	6.4	10
Butyl alcohol	—	1.1	—
Toluene	39	36.9	45

Enamel Remover

The enamel removers consist of simple mixtures of solvents, such as acetone, amyl acetate or ethyl acetate, which may contain small amounts of fatty material to counteract any excessive drying action of the solvents on the nails. Originally castor oil was used, but now-a-days esters such as butyl stearate or dibutyl phthalate, fatty alcohols or soaps may be used for the same purpose. An example of an enamel remover is provided by the following formula :

Formula 16

Constituents	*per cent*
Butyl stearate	5.0
Diethylene glycol monoethyl ether	10.0
Acetone	85.0

Cream nail enamel removers

A number of creamy nail enamel removers have been introduced. These consist of a fairly high-boiling solvent for the lacquer, introduced into a cream having lanolin or other emollient material. Alternatively they may be composed of a lacquer solvent solidified by means of a suitable wax. The different composition of enamel remover is shown in formulae 17, 18 and 19.

Formula 17

Constituents	*per cent*
Paraffin or beeswax	11.5
Lanolin	4.0
Sodium or potassium linoleate	2.6
Methyl ethyl ketone	to 100.0

Formula 18

Constituents	*per cent*
Sulphonated olive oil	10.0
Sodium hydroxide (10%)	1.0
Benzyl acetate	15.0
Acetone	15.0
Titanium dioxide	0.5
Bentonite	2.0
Water	56.5

Formula 19

Constituents	*per cent*
Stearic acid	9.5
Triethanolamine	3.5

Mineral oil	10.0
Butyl acetate	50.0
Butyl stearate	5.0
Carbitol	5.0
Water	17.0

Nail Drier

Nail driers are aerosol formulations. It makes use of the rapid evaporation of a propellant to speed up the drying of freshly enamelled nails, by drawing off the solvent present in the nail varnish. The drying is also combined with the deposition of a transparent film of oil over the freshly applied enamel. This reduces tackiness and prevents it from smearing if touched.

PLASTIC FINGERNAILS AND ELONGATORS

For improving the cosmetic appearance of damaged or short stubby nails plastic fingernails and elongators are used. They are produced by the polymerisation or copolymerisation promoter. A plasticiser, an opacifier, a pigment and a filler may also be included.

The most frequently employed monomer has been methyl methacrylate while poly(methyl methacrylate), polystyrene, polyisobutylene and cellulose acetate have been used as the polymer. If methyl methacrylate is copolymerised with a crosslinking agent such as ethylene dimethacrylate, more brittle polymers with improved solvent resistance are produced. The crosslinking agent polymer is usually employed in amounts ranging between 2 and 10 per cent by weight of the monomer. The actual level at which it is used largely determines the hardness and brittleness of the product. Compounds of improved flexibility may be produced by forming a copolymer containing 5–20 per cent by weight of a higher-molecular-weight ester of acrylic or methacrylic acid such as lauryl methacrylate. Triphenyl or tricresyl phosphate, dimethyl, diethyl or dibutyl phthalates (1–10 %) may be used as plasticiser.

Polymerisation of methyl methacrylate is a free radical-catalysed reaction entailing the dissociation of the catalyst to free radicals which initiate polymerisation. The catalysts which are preferred in the production of nail elongators are benzoyl peroxide or lauroyl peroxide, and they may be used in amounts of up to 3 per cent by weight of the solid constituent of the preparation; usually, however, 1 per cent of the catalyst on the weight of monomer will be adequate. A polymerisation promoter is also usually present to induce free radical release and to allow polymerisation of component monomers to take place at room temperatures. that is at temperatures appreciably lower than those at which polymerisation usually takes place. Among fillers, silica and aluminium silicate, mica and metallic oxides (25%) have been used. Pigments and opacifiers (0.1 %) are generally used.

Prepare the surface of the nail by removing any excess cuticle and traces of old lacquer. This makes nail ready for applying nail elongator. It also ensure good adhesion between the nail surface and the plastic material applied. The skin surrounding the nail is protected from polymerisation taking place on its surface by applying to the skin a material with little adhesion for the plastic, usually a water-soluble resin easily removable with water.

The procedure used to form a nail elongator *in situ* entails successive applications, first of the liquid monomer and then of the powder containing the catalyst. This is repeated until the nail has been adequatedly lengthened. Once the tip of the nail has been strengthened, the nail is then built up from the centre of the nail to the outer edge. The polymerisation of methyl methacrylate and vinyl monomers is inhibited by addition of hydroquinone. The amount of catalyst and promotor is also kept under control. The composition of solid and liquid components, the following formulation given in formula 20 :

Formula 20

Constituents	*per cent*
Solid component	
Poly(methyl methacrylate) (granules)	75
Aluminium silicate	23
Benzoyl peroxide	2
Liquid component	
Methyl methacrylate (monomer)	83
Lauryl methacrylate (crosslinking agent)	15
Diethyl aniline (promoter)	2

The artificial nails are produced *in situ* by rapid polymerisation or copolymerisation of a constituent monomer at room temperature in the presence of a polymer and a redox catalyst system.

They are formed from two components, one of which is a primer. It contains a mixture of poly(methyl methacrylate), a copolymer of vinyl chloride vinyl acetate, and the oxidising component of a redox catalyst system. This component is used to form coatings on the exposed nail bed and the surrounding skin and protects them from irritation by the monomer contained in the second composition. The film-forming solids content of the primer composition should not exceed 25 per cent.

The second composition contains most of the resins for the artificial nails. It consists of a solution of a poly(methyl methacrylate). It also contains the reducing component of the redox catalyst. Both compositions are stable as long as they are stored separately, but when they are combined the monomer begins to polymerise.

NAIL MENDING COMPOSITIONS

Mending compositions are basically mixtures of an adhesive, a fibre reinforcing material and a solvent. The first two components produce a film which forms a strong bond between the broken parts of the damaged fingernail, improving its appearance and preventing further damage. Whereas the solvent will quickly evaporate and allow the reinforced adhesive film to set and to mend the damaged fingernail. These preparations are preferably applied in four coats. Each coat is applied before drying the next coat in a direction perpendicular to the previous one. The nail polish is applied about one hour after the application of the final coat of the mending preparation, to allow all the coats to dry thoroughly.

The nitrocellulose is preferred organic film-forming material of the mending composition. An aryl sulphonamide formaldehyde is also included to make the film produced by the composition more adherent, flexible and tough. A plasticiser may be added to make the film flexible. Short rayon fibres (1.5 mm long) and of small diameter (1.5–5 denier) are used as reinforcing agent. Other fibres may also be used if they are not soluble in the mending composition. A gelling agent (preferably silica) is also present, to ensure that the fibres used remain suspended within the adhesive. Solvents are ethyl acetate, butyl acetate and toluene used in a combination of 30–50 per cent, 5–20 per cent and 20–30 per cent respectively. Pigments or dyes may also be included to confer the desired colour. The compositions just described are illustrated by formula 21. It has been claimed that nail polish can be applied to and removed from such mended nails without difficulty and without disturbing the 'repair'.

Formula 21
Constituents   *per cent*
Nitrocellulose   10.3
Aryl sulphonamide formaldehyde resin   4.1

Silica	2.0
Dibutyl phthalate	0.5
Rayon fibres	0.5
Ethyl acetate	46.2
Butyl acetate	5.1
Toluene	31.3

Chapter 32

Hair

INTRODUCTION

Millions of fine and downy hairs cover the human body. The cosmetics generally are prepared for applications of darker hairs on the scalp. The growth of the scalp hair is double than average hair growth. The follicles of the skin produce hair about 30 centimetres in three years. After resting for three weeks falls down. Every day about 100 scalp follicles go into resting phase. Protein gives hair unusual strength, flexibility both in length and width. There fore it is building block of hair.

PARTS OF HAIR

Hair has two main parts : the cuticle and the cortex. The hair of some individuals also has a third part, the medulla. Figure 32.1. Shows 2500 times magnified image of hair. On the outside of the hair lies the cuticle, a protective armor formed of colourless, translucent layers of scales. Normally, about six layers of cuticle protect a hair, overlapping much like shingles on a roof or scales on a fish. When the hair bends, these layers flex smoothly. The cuticle is lubricated and protected by a natural lipid (oily) substance called sebum, which is secreted into the hair follicle and onto the base of the hair by the neighbouring sebaceous gland. Sebum coats the hair, providing smoothness, flexibility, and sheen. However, too much sebum causes oiliness and greasiness. Brushing hair improves its appearance because it moves the sebum away from the scalp and distributes it throughout the hair, giving it luster.

The next part of the hair is cortex, which is held together by the surrounding protective cuticle. The cortex contains keratin, which impart the particular colour and shape (curliness or straightness) to an individual's hair. Melanin molecules present in it and determine hair shade. While those who have melanin in large granules will have brownish hair. The structure of keratin gives hair its strength, shape, and elasticity.

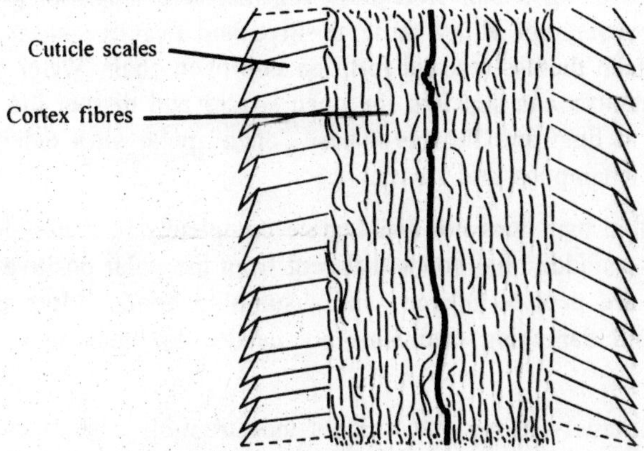

Fig. 32.1. A cross-section of a normal hair. The dark centre line is the medulla.

As a person ages, certain changes normally occur in the hair. First, the production of sebum decreases, leaving the hair dry and brittle. Without lubrication, the scales of the cuticle are prone to breakage, giving the hair a drab appearance. Production of melanin usually decreases, too, which causes the hair to lose its usual colour.

The natural changes which occur with aging are not the only ones which can alter the quality of hair. The cuticle of the hair can be damaged both mechanically and chemically. Researchers have discovered that even combing the brushing can cause the cuticle to break off, exposing the cortex to chemical or light damage. The attempts to alter hairs appearance cosmetically affects it to be fair aging process commerce. Few people straighten hair, while others wave it; some bleach colour out, while others try to put colour in; some literally rip hair out of the skin, while others shave it off or remove it chemically. Permanent waving, bleaching, and dyeing will all cause progressive deterioration of the hair fibre.

Properties of Hair

The physical properties of hair (lustre, colour, straightness) are changed due to hair preparations. There are two main types of linkages, or "bonds," which are critical for cosmetics. The one high temperature or powerful chemicals. Some of the most important bonds, providing about 70 per cent of hair's strength, are disrupted by water. Wet hair can be easily curled, straightened, and stretched; people take advantage of water's bond-breaking quality when they moisten hair to reshape it. Water has this powerful effect because it has positive and negative poles, like a magnet. So does hair. The opposite poles along a hair fibre hold together like magnets, giving hair strength. However, not all of these positive and negative sites are close enough to attach themselves and form an unbroken chair. Water and other charged substances can slip into the open spaces and disrupt the bonds in hair, making it limp and flexible. These "open" polar sites determine the behaviour of shampoos and rinses.

The fact that hair does not disintegrate completely in water shows that it also possesses additional bonds different from the polar positive-negative ones. These are peptide-linkages and disulphide bonds. More powerful, dangerous, and damaging chemicals are needed to break these stronger bonds.

Hair is extremely strong, but it is not indestructible. Use of excessively harsh cleaning agents or of powerful chemically-active preparations may cause the scales of the cuticle to break off, resulting in split ends, and a brittle, strawlike texture. Once the damage is done, it is permanent until the hair is replaced. Some hair products temporarily mask this damage, but they cannot cure it.

Chapter 33

Shampoos

INTRODUCTION

Almost all the people use shampoos which are required for washing of hair. They constitute suitable detergents. They are packed in special package convenient to use. The primary function remains that of cleansing the hair of accumulated sebum, scalp debris and residues of hair-grooming preparations. The cleansing should be selective and should preserve a quantity of the natural oil that coats the hair, and above all, the scalp.

The process of shampooing is a cleansing, purifying activity, designed to free them from daily accumulation and maturation of grease, dirt, perspiration, cooking smells, dandruff, environmental pollution and so on. In fact, it seems sensible to retain the definition of a shampoo as a suitable detergent for washing the hair with the corollary that it should also leave the hair easy to manage and confer on it a healthy look.

DETERGENCY

The ambiguous objective, variability of substrate and the process made the development of detergent system adopted to hair complex. The substrate to be cleaned is made of the relatively hard but porous keratin of the hair. The soft keratin of the scalp is sensitive to drying and defatting. While there exists very large individual variation in fibre number and diameter, the average surface area of a female head of hair is calculated to be between 4 and 8 m^2—that is, 50 to 100 times the average scalp area.

The kind of soil to be removed, either natural or captured, varies greatly according to the weather, the life style, the type of work, the physiological functions, the haircare practice and so on.

539

The problem of cleaning hair is particularly one of grease removal although shampoo removes more soil than oil. Hair pick-up dirt without the help of greasy layer as it represents hard surface. As long as the grease can be removed, it is quite easy to remove the dirt. To remove grease from the hair, one must find some agent with a greater affinity for the grease found suitable. This function can be accomplished by absorbent solids such as fuller's earth or flour, but solutions of surface-active materials (Surface-active Agents) are much more convenient and more commonly employed.

The mechanism of detergent action involves a number of complex physical phenomena—wetting, foaming, emulsifying and peptisation—several of which are imperfectly understood. The removal of dirt, involves the following processes :

(1) The detergent solution must be able to wet both the dirt and the substrate, hence it must lower the surface tension.

(2) The dirt or oil particles are displaced by the detergent solution due to reduction of interfacial tension.

(3) The dirt particles to be kept dispersed in order that they may readily be washed away.

The most important property in hair washing is the ability of polar group of the detergent to displace oil from a surface.

EVALUATION OF DETERGENTS AS SHAMPOO BASES

The screening test for evaluation of detergents and as shampoo bases is measurement of surface tension and interfacial tension. The trial on the head must be conducted for detergent selection because the after-effects of the shampoo may often be the deciding factor. With the wide range of materials available it is a relatively easy task to find materials that will adequately clean hair and allow for adequate lather. The final criterion which then applies in the selection of the detergent is its effect on the hair. These effects are best observed in comparative tests on the same head, as hair diameter, quantity, greasiness and previous treatments may all affect the results.

RAW MATERIALS FOR SHAMPOOS

Originally the shampoos were made from soap or soap mixtures. But today, the shampoos are made from synthetic detergents. Ingredients of present shampoos can be classified as under :

Surfactants, Foam boosters and stabilisers, Conditioning agents, Special additives, Preservatives, Sequestering agents, Viscosity modifiers, Opacifying or clarifying agents, Fragrances, and Colour.

Non-ionic detergents does not have sufficient foaming properties. hence it can not be used in shampoos. While cationic detergents are ideal due to foaming, cleaning power. Anionic and ampholytics are suitable due to low cost, foaming and conditioning properties.

From the list of anionic surfactants alfa olefin sulphonates. alkyl sulphates, alkyl polyethylene glycol sulphate. Sulphoccinates and acyl lactylates are used in shampoos. Among the non-ionic surfactants fatty acid alkanolamides, polyalkoxylated derivatives and amine oxides are used. N-alkylamino acids, betains are amphoteric surfactants used in shampoos.

Additives

Conditioning agents

Conditioning agents are placed to improve manageability, feel and lustre of the hair. The materials which have been claimed to have conditioning effect on hair include fatty materials like lanolin, mineral oil; natural products like polypeptides, egg derivatives, herbal extracts; and synthetic products like specifically designed surfactants and resins.

Viscosity modifiers

Thickening of shampoo can be achieved by a number of compounds. Compounds which can be used to modify viscosity of shampoos include: Electrolytes, Natural gums, cellulose derivative and carboxyvinyl polymers and others. The list includes, ammonium chloride, sodium chloride, gum karaya, gum tragacanth, alginates, hydroxycellulose, hydroxypropyl cellulose, carboxy methyl cellulose, carbopol 934, phosphate esters, TEA soap, poly vinyl alcohols, and alkanolamides.

Opacifying and clarifying agents

A number of materials can be used for opacifying or giving pearlescent effect. Non-ionic solubilisers, alcohols, phosphates improve transparency. The opacifying agents are : Alkanolamides of higher fatty acids, glycol mono- and di- stearates, propylene glycol - and glycerol-monostearates and palmitates, fatty alcohols (cetyl, stearyl), Milky emulsions of vinyl polymers and latexes, magnesium calcium or zinc salt of stearic acid, finely dispersed zinc oxide or titanium dioxide, and magnesium aluminium silicate.

The clarifying Agents are: Ethanol, isopropanol; propylene glycol, hexylene; glycol, dimethyl-octyne diol phosphate; polyethoxylated alcohols and esters.

Sequestering agents

These materials are used to prevent deposition of calcium and magnesium salts of soaps on to the hair. EDTA salts (Ethylene diamine tetracetic salts) and polyphosphates are the most commonly used sequestering agents in shampoos.

Preservatives

Bacterial growth in shampoos can lead to breakdown of detergents in shampoos resulting in discolouration of the product. The surfactant in shampoos interfere with bactericidal activity of anti-bacterials. This fact should be considered while selecting preservatives. Therefore, higher concentration of preservatives are necessary in shampoos. The most effective antimicrobial agent is formaldehyde. Formaldehyde remains unaffected by surfactants and can be used in concentration of 0.1 - 0.15%. But it is not compatible with certain ingredients of shampoos, e.g., protein hydrolysates.

Esters of parahydroxy benzoic acid are quite effective against fungi, but these are inactivated by non-ionics. These are also not effective against Pseudomonas.

Perfumes

The fragrance selected must be soluble and compatible with shampoo. It means that, viscosity and stability should remain unaffected. Herbal, fruity or floral fragrances can be used in shampoos.

FORMULATION

The shampoos can be classified into the following main groups : (i) Clear liquid shampoos; (ii) Liquid cream/lotion shampoos; (iii) Cream shampoos; (iv) Jelly shampoos; (v) Powder shampoos; (vi) Aerosol foam shampoos.

Clear Liquid Shampoos

Clear liquid shampoos can be prepared with detergents having lower cloud point. These can also be made with detergents and alkanolamides. The different compositions of clear liquid shampoos is given in formulae 1-3.

Formulae	*1*	*2*	*3*
Constituents	*per cent*	*per cent*	*per cent*
Sodium lauryl ether sulphate	40.0	–	–
Sodium chloride (To desired viscosity)	2.0–4.0		
Triethanolamine lauryl sulphate		50.0	45.0
Lauric isopropanolamide		2.0	–

Coconut monoethanolamide		–	–	2.0
Water	to make	100	100	100
Colour, Perfume and Preservatives		q.s.	q.s.	q.s.

Concentrated clear liquid shampoo can be prepared with more soluble monoethanolamine lauryl sulphate. These shampoos may be used by hairdressers by diluting at the time of use. The composition of such shampoo is monoethanolamine lauryl sulphate (70.0%), coconut diethanol amide (2.0%), colour, perfume, preservative in small quanitty made to 100 % with water.

Liquid Cream or Lotion Shampoos

These can be prepared by adding opacifier like ethylene glycol monostearate or ethylene glycol distearate to clear liquid shampoos. Detergent manufacturers also make preformulated concentrated materials which can be diluted, coloured and perfumed. Ethylene glycol monostearate, ethylene glycol distearate can be used as opacifiers. These however, have tendency to redissolve in shampoo in hot weather. Insoluble magnesium stearate can be added to overcome this problem. Formulae 4 and 5 are examples of liquid cream shampoo with pearly sheen.

Formula 4
Constituents	*per cent*
Monoethanolamine lauryl sulphate	45.0
Ethylene glycol monostearate	5.0
Water	to make 100.0
Perfume, Colour & Preservatives	q.s.

Formula 5
Constituents	*per cent*
Triethanolamine lauryl sulphate	35.0
Glyceryl monostearate	2.0
Magnesium stearate	1.0
Water	to make 100.0
Colour, Perfume & Preservatives	q.s.

An alkanolamide can be added to such formulation as foam booster as exemplified in formula 6.

Formula 6
Constituents	*per cent*
Monoethanolamine lauryl sulphate	38.0
Ethylene glycol monostearate	5.0
Lauric isopropanolamide	2.0

Water	to make 100.0
Colour, Perfume and Preservatives	q.s.

Cream Shampoos

Professional hair dressers demand cream shampoos. These products can be packed in collapsible tubes. These products are also known as shampoo pastes. Generally, these constitute sodium alkyl sulphates (38%) which give products of firm consistency. Cetyl alcohol (7.0%) can be added as builder. An alkanolamide such as coconut diethanolamide, lauric isopropanolamide can be added in concentration of 1–2% as foam stabiliser.

The disadvantage is that mass of crystals present in shampoo may dissolve in hot weather making the product runny and translucent. On cooling, recrystallisation takes place and crystals formed may be large appearing as lumps or fibrous mass.

Jelly Shampoos

Jelly shampoos are formulated with detergent alone or in combination with soap. By varying the proportions of triethanolamine coconut soap, triethanolamine lauryl sulphate and sodium lauryl sulphate different textures can be prepared.

Jellies can also be prepared by thickening the conventional clear liquid shampoos with hydroxy-alkyl or methyl cellulose ethers or alkanolamides. Appropriate combination of anionic and amphoteric surfactants can also produce gels. Formula 7 is example where methyl cellulose ether has been used as thickening agent.

Formula 7

Constituents	*per cent*
Alkyl dimethyl benzalammonium chloride	15.0
Triethanolamine lauryl sulphate (40%)	28.0
Coconut diethanolamide	7.0
Hydroxy propyl methyl cellulose	1.0
Water	to make 100.0
Perfumes, Colour & Preservatives	q.s.

Powder Shampoos

The powder shampoos were prepared with soap powder and diluents like sodium carbonate, sodium bicarbonate, disodium phosphate or borax or their combination. The composition is given below in formula 8 :

Formula 8
Constituents	per cent
Soap Powder	35.0
Sodium Bi-carbonate	45.0
Di-sodium Phosphate	20.0
Perfume	q.s.

Such formulations were also available with herbal ingredients which were supposed to give slight colouring effect to the hair. Formulae 9 and 10 are examples of these.

Formula 9
Constituents	per cent
Henna Powder	5.0
Soap Powder	50.0
Sodium Carbonate	22.5
Potassium Carbonate	7.5
Borax	15.0
Perfume	q.s.

Formula 10
Constituents	per cent
Camomile Flowers Powder	5.0
Soap Powder	55.0
Sodium Carbonate	15.0
Borax	25.0
Perfume	q.s.

The soap powder was replaced with detergents. Formula 11 is an example.

Formula 11
Constituents	per cent
Sodium lauryl sulphate	20.0
Sarcoside	5.0
Sodium bicarbonate	10.0
Sodium sulphate	65.0
Perfume	q.s.

The disadvantage of powder shampoo is difficult to use and these leave the hair in poor condition.

Aerosol Shampoos

Shampoos available in aerosol packaging are called aerosol shampoos. Shampoos for aerosol packing are required to be specially formulated.

When sprayed, the foam should be manageable and should have enough staying power. The aerosol shampoos are applied by spraying on the palm of the hand. Then it is thoroughly rubbed into the hair which have been wetted in advance. Thereafter, these may be removed as is done in the case of conventional shampoos. They constitute shampoo base 90% and propellent 10 %.

Shampoo base may be prepared by mixing : ammonium/triethanolamine lauryl sulphate 60.0 %; coconut diethanolamide 2.0 %; and water, perfume, colour and preservatives.

Some Special Shampoos

Shampoos formulated for a specific purpose are dandruff and shampoos for conditioning effect. These two are, by far the most important groups of shampoos which are marketed for their special effects.

Anti-dandruff shampoos

As a result of cell division in deeper layers of epidermis, some cells are being always pushed up towards the surface of the skin. The cells farthest from the dermis are keratinised and subsequently are shed off. This is true of scalp also. A process of keratinisation and shedding off increases to an abnormal rate due to any reason in case normal function do not synchronise. Such a condition causes dandruff. In another condition there is abnormal secretion of sebum from sebaceous glands. The former condition is known as *seborrhoea sicca* and the latter as *seborrhoea oleosa*. These abnormal conditions of scalp are generally associated with an increase in the growth of bacteria and fungi. This may lead to skin diseases like acne and psoriasis. The above-mentioned conditions should, therefore, be treated. One of the approach is, preparation of medicated shampoos for dandruff. A medicated shampoo is required to have the following functions :

(1) It should be clean both the hair and the scalp and should leave the hair manageable.

(2) It should not irritate sebaceous glands, and contain and anti-microbial to prevent growth of increased incidence of microbes.

(3) The active material used should not sensitise the scalp.

(4) It should reduce the degree of itching and scaling.

The component used for treatment of dandruff are: Thymol, 1,2 Bithionol, Pyrithione Zinc, Resorcinol, Selenium Sulphide, Prepared Coal tar,

Quaternary Ammonium Compounds when used as shampoo base. Use of thymol, camphor and menthol has been examplified in formulae 12 and 13.

Formula 12
Constituents

	per cent
Thymol	0.1
Camphor	0.1
Triethanolamine lauryl sulphate	60.0
Water	to make 100.0
Colour, Perfume and Preservatives	q.s.

Formula 13
Constituents

	per cent
Thymol	0.05
Camphor	0.1
Menthol	0.1
Triethanolamine lauryl sulphate	55.0
Water	to make 100.0
Colour, Perfume and Preservatives	q.s.

Thymol, menthol and camphor are dissolved in alcohol and incorporated in the formulation. Cationic surfactants with high molecular weight possess lower degree of bactericidal action. While selecting these compounds ensure that defatting of scalp and irritation do not occur after use for long time. Of these, quaternary ammonium compounds are of interest. Cetrimide, a mixture of dodecyl, tetradecyl and hexadecyl-trimethyl ammonium bromides is non-irritant and non-toxic in dilutions which are generally used. Other quaternary ammonium compounds are benzalkonium chloride and stearyl dimethyl benzyl ammonium chloride. Though stearyl dimethyl benzyl-ammonium chloride is less bacterio-static than centrimide or benzalkomium chloride, but its use improves appearance of hair. Centrimide is soluble in two parts of water but alcohol is used so that clarity is maintained during storage. In shampoos based on centrimide-diethylphthalate is also added to increase the viscosity. Formula 14 is an example of such a formulation. A cream shampoo can be prepared by adding cetyl alcohol as given in formula 15.

Formula 14
Constituents

	per cent
Centrimide	15.0
Alcohol	10.0
Diethylphthalate	2.0
Water	to make 100.0

Formula 15

Constituents	*per cent*
Centrimide	15.0
Cetyl alcohol	12.0
Water	to make 100.0

Conditioning shampoos

The two terms "conditioning shampoos" and "shampoo conditioners" appear to mean the same thing. Their applications are same. The prime function of conditioning shampoo is cleaning. Also it improves manageability, feel and appearance of air. Whereas conditioner untangle wet hair and to improve manageability of both wet and dry hair.

Quaternary ammonium compounds have conditioning and bactericidal properties. Conditioning properties in quaternary ammonium compounds are imparted by the relatively large complex cation which is substantive to protein. The static charge is removed due to absorption of quaternary ammonium compounds on to the hair fibre. This reduces friction between individual hair strands. Thus hair become more manageable. Satisfactory products can be prepared using stearyl dimethyl benzyl ammonium chloride. Formulae 16 and 17 illustrate use of stearyl dimethyl benzyl ammonium chloride.

Formula 16

Constituents	*per cent*
Stearyl dimethyl benzyl ammonium chloride	5.5
Ethylene glycol monostearate	2.0
Cetyl alcohol	2.5
Water	to make 100.0
Perfume and Preservatives	q.s.

Formula 17

Constituents	*per cent*
Stearyl dimethyl benzyl ammonium chloride	4.0
Cetyl alcohol	3.0
Mineral oil	0.5
Water	to make 100.0
Perfume and Preservatives	q.s.

MANUFACTURE OF SHAMPOOS

Manufacture of Liquid/Cream Shampoos

Clear liquid shampoos can be made from preformulated concentrated solutions which are many a time available from detergent manufacturer. These are to be diluted, coloured and perfumed. The dilution can be done in a stainless steel tank with a low speed stirrer.

There is a general method, if an alkanolamide is being used. Alkanolamide should be dissolved in about half the total amount of the detergent with gentle heating. Then the remaining detergent solution is added slowly. Perfume is dissolved in cold concentrated detergent. Colour and preservative(s) are dissolved separately in water and are added to the mix. The remaining water is added with slow stirring finally to make up the volume. Rapid stirring will cause froth and will incorporate air bubbles. The mixing operations are usually carried out in stainless steel tanks with a stirrer having speed regulator.

Viscosity of solutions of lauryl alcohol ether sulphates is usually controlled by adding a solution of sodium chloride (1 part sodium chloride and 3 parts water) to the shampoo in small proportions. The shampoo is stirred well after each addition. With the addition of salt solution viscosity increases and reaches a maximum.

In case of cream shampoos where glycol stearates are used as opecifiers, glycol stearates are heated gently with a part of detergent to dissolve them. To this solution are then added detergents and other additives. In a cream shampoo where waxes are to be incorporated, the following procedure can be used.

First in a steam or water jacked vessel detergent is heated to 80°C. Then melted wax is added with stirring. Cool the mixture to 40–45°C and add perfume, colour, preservative with stirring. The product is filled into containers while it is still warm.

Chapter 34

Hair Setting Lotions and Conditioners

INTRODUCTION

The hair setting is required for men and women, to improve the control and manageability of hair, to impart some lustre, and to maintain a hairstyle and despite the various environment conditions to which hair is submitted (wind, humidity, dryness, cold, heat, sun, etc.).

The relative importance of these factors varies from one country to another, from age to age, and according to the state of the hair. It also differs notably between male and female users: men generally consider adequate control as the prime requisite for a hair dressing, with gloss as secondary, and as such men's products have been classically based on the use of oleaginous materials. Women first look for products that give a pleasing appearance to the hair, but at the same time they require good hold; they do not want product which render hair heavy and which tend to make it lank or greasy.

The trend in women's fashions for softer and freer styles. The increasing use of the brushing technique have led to an increase in the demand for products. These products enhance disentangling and a greater need for products that protect, strengthen and/or improve the condition of the hair. This has occurred at the same time as the extraordinary development of conditioners.

HAIR SETTING

All the methods attempt to form a film or coating on the hair for maintenance. The methods that could be used are :

(1) A film on the hair is obtained on evaporation of water from water soluble materials like gums, protein derivatives, water soluble synthetic polymers.

(2) A film/coating of anhydrous materials can be achieved by purely physical means. This method is applicable to fats of vegetable, mineral or animal origin. However, the film/coating is quite greasy. This problem has been overcome by availability of some fatty esters.

(3) Emulsions can form film or coating with much more ease. Dilution of any phase can be accomplished in a w/o or o/w type emulsion.

(4) Another method that can be utilised is to make a solution of water insoluble materials in volatile solvents like alcohol and apply to the hair. Volatile solvent(s) evaporate(s) leaving behind a film on the hair.

(5) Using adsorption process film can be deposited on hair. Properly formed films made by this method resist rinsing, handling and sometimes even washing. Such a phenomenon is exhibited by cationic substances.

It may also be noted that sometimes the effect can be achieved with lower concentrations and increase in the concentrations dissipates the effect. The points of concern in such preparations are firmness of dispersion of material(s), freedom from irritation or sensitisation.

Formulation

Wave-sets

The former formulations of wave-sets were based on natural gums like gum karaya, acacia, tragacanth and quince seed. These natural products have varying physical properties according to climate, soil, method of collection. In addition to this, they yield dull and brittle films which may crumble and may become sticky in humid weather. Formula 1 is given as an example of such products.

Formula 1

Constituents	*per cent*
Gum karaya	1.5
Sodium alginate	0.5
Glycerine	5.0
Ethanol	5.0
Water	to make 100.0
Perfume, Colour and Preservatives	q.s.

Synthetic polymers soluble in water-alcohol mixture have replaced natural gum. These products can be formulated with a film forming material, plasticiser, solvent (water-alcohol), perfume and colour. Formulae 2 and 3 are examples of such products.

Formulae	*2*	*3*
Constituents	*per cent*	*per cent*
Polyvinyl pyrrolidone (PVP)	3.5	1.0
Propylene glycol	2.5	2.0
Triethanolamine	–	1.0
Ethanol	40.0	–
Carboxyvinyl polymer	-	1.5
Water	to make 100.0	to make 100.0
Perfume and Colour	q.s.	q.s

Later on hair setting or styling gels became popular. Hair setting lotions due to low solid content cannot form gels. Thickening agents like carboxy polymethylene resins (Carbopols-Goodrich) have been extensively used for making gels. Propylene glycol and phthalates (e.g., di-methyl phthalate) are used as plasticisers. See formulae 4 and 5 :

Formula 4

Constituents	*per cent*
Carbopol	0.25
PVP	3.00
Di-isopropanolamine	0.25
Di-methyl phthalate	0.50
Ethyl alcohol	10.00
Water	to make 100.0
Preservatives, Perfume and Colour	q.s.

Formula 5

Constituents	*per cent*
PVP/VA (50 : 50)	5.0
Carbopol 940	0.3
Propylene glycol	3.5
Di-isopropanolamine	0.5
Water	to make 100.0
Preservatives, Perfume and Colour	q.s.

Hair rinses

Since centuries vinegar and lemon juice have been used to remove lime soap after shampooing. Subsequently, solutions of polybasic acids like citric acid, tartaric acid were used. These were marketed in solid formulations with instructions to make solutions for rinse. Single application package were also available. These also contained colour and were called temporary colour rinses.

The products made with quaternary ammonium compounds are available as emulsions or creams. The earliest quaternary ammonium compounds that were used are alkyldimethyl benzyl ammonium chloride. A simple cream rinse can be prepared with alkyl dimethyl benzyl ammonium chloride and glyceryl monostearate. This cream rinse should be diluted to about 15 times before use. While selecting quaternary ammonium compounds their skin irritancy should be taken into consideration. Usually concentrations of 0.1% lose their irritancy and are still effective in their action on hair shaft. Formula 6 is an example of cream rinse.

Formula 6

Constituents	*per cent*
Alkyl dimethyl benzyl ammonium chloride	2.5
Glyceryl monostearate	2.5
Water	to make 100.0
Colour, Perfume	q.s.

HAIR-GROOMING PREPARATIONS

The groups of hair dressing are : brilliantines; alcoholic solutions; two layer systems; hair tonics; gum-base hair dressings; oil-in-water emulsions; and water-in-oil emulsions.

Formulation

Brilliantines

Brilliantines overcome dull appearance of hair. The main function of a brilliantine is to impart a measure of grooming and sheen to the hair. There are two types of brilliantines, liquids and solids. The mineral oil with small quantity of colour and perfume is liquid brilliantines. Another liquid constitute mineral oil (75 %) and deodorised kerosene (25 %).

The coupling agents may be added to keep the product clear on storage. Materials like vegetable oils, fatty alcohols/esters, non-ionic surfactants can solubilise perfume oils.

Vegetable oils are liable to become rancid, therefore, use of anti-oxidants in hair oils containing vegetable oils is necessary. Formulae 7 and 8 are examples of hair oils containing vegetable oils.

Formulae Constituents	7 per cent	8 per cent
Castor oil	20.0	60.0
Deodorised kerosene	to make 100.0	–
Colour, Perfume and Preservatives	q.s.	q.s.
Almond oil	–	40.0

Formula 9 is related to hair oils based on mixtures of mineral oil and vegetable oil(s) have also been formulated as :

Formula 9 Constituents	per cent
Mineral oil	70.0
Peanut oil	30.0
Colour, Perfume and Preservatives	q.s.

Solid brilliantines are used for unruly or kinky hair where greater grooming is required. These products are invariably opaque. Opacity of these products increases with increasing proportion of waxes. Formula 10 is an example of typical solid brilliantine.

Formula 10 Constituents	per cent
Mineral oil	75.0
Stearic acid	25.0
Colour, Perfume	q.s.

Formulae 11 and 12 are examples of solid brilliantines having vegetable oils in them.

Formulae Constituents	11 per cent	12 per cent
Castor oil	40.0	77.0
Cocoa butter	10.0	–
Petroleum jelly	50.0	–
Ceresin	–	23.0
Colour, Perfume and Preservatives	q.s.	q.s.

Alcoholic Solutions

Alcoholic lotions as hair dressing can be prepared diluting viscous oils with alcohol. It is possible to get wetting action first and deposition of thin layer of oil next after evaporation of alcohol. Castor oil is soluble in alcohol in all proportions and therefore, enjoys the unique place amongst fixed oils. Formulae 13 and 14 are examples of alcoholic lotions.

Formulae Constituents	13 per cent	14 per cent
Castor oil	30.0	20.0
Glycerine	10.0	–
Ethanol	60.0	78.0
Tincture benzoin	–	2.0
Colour, Perfume	q.s.	q.s.

Oil-in-water emulsions

Good pouring, spreading and attractive cream like properties made oil-in-water (o/w) emulsion immediately acceptable. It does not give greasy feeling and can be easily removed by shampooing. Since emulsions are vulnerable to microbial growth, preservatives soluble in oil phase and soluble in water phase should be added. Desired grooming can be achieved using mineral oil of low or medium viscosity is generally used. Other materials like petroleum jelly, waxes are added to increase the body of the emulsion. Gums have also been used to increase fixative properties, viscosity and stability of emulsions. A simple formulation (Formula 15) of this type comprises of oil and water with a suitable emulsifier.

Formula 15 Constituents	per cent
Mineral oil	45.0
Stearic acid	5.0
Triethanolamine	1.0
Water	to make 100.0
Perfume, Colour and Preservatives	q.s.

Method of Preparation

Separately heat mineral oil, stearic acid and preservative (oil and water soluble), water ethanolamine to about 70°C. Add aqueous solution to oil with slow stirring. Continue stirring and allow to cool to about 45°C. Add perfume and colour and blend thoroughly. The inclusion of lanolin, lanolin esters, fatty acids/esters for emollience, glycerine, propylene glycol gives humectancy, gums to increase stability and fixative properties.

Water-in-oil emulsions

Inspite of difficulty to prepare water-in-oil emulsion hair dressing of this type are popular among men and women. These products are smooth lustrous creams which give excellent grooming and high sheen to the hair. The film deposited on hair is water resistant and this makes this type of hairdressing useful to swimmers. W/O emulsions have greasier feel but not as pronounced as in the case of brilliantines.

Several types of emulsifying agents have been used to prepare w/o hair dressings, polyvalent soaps being the oldest of them. Calcium stearate and oleates were used earlier but subsequently it was found out that magnesium is that more stable. A combination of emulsifiers has been found more useful than any single emulsifier. Zinc and aluminium stearates have been used in combination with other emulsifiers. Vegetable oils, waxes, lanolin can be added for desired attributes in the product. Formula 16 can be considered as an example.

Formula 16

Constituents	*per cent*
Mineral oil	38.0
Petroleum jelly	20.0
Beeswax	18.0
Oleic acid	0.6
Lanolin absorption base	1.0
Sodium hydroxide	0.4
Magnesium sulphate	1.0
Borax	0.5
Water	to make 100.0
Perfume and Preservatives	q.s.

Manufacture of Hair Grooming Preparations

Brilliantines

Liquid brilliantines or hair oils are manufactured by mixing of miscible ingredients in suitable mixing tank. However, when liquid brilliantine is based entirely on low viscosity mineral oil, solubility of perfume may be a problem. Therefore, solubility of the perfume should be pre-determined. Perfume can be made oil soluble by using deodorised kerosene or isopropyl myristate. Before adding perfume, it is first dissolved in deodorised kerosene or isopropyl myrisate. Then added to the bulk. Similarly colour (oil soluble colour) is first dissolved in a part of oil and added to the bulk. Perfume and colour are thoroughly blended. For bright and crystal-clear appearance the bulk finished product may be filtered. These products should be processed either in stainless steel or high density polymer vessel. Anti-oxidants (oil soluble) should be mixed in a part of oil and should be added to the mix with stirring.

Solid brilliantines and pomades are prepared by melting and mixing all the ingredients except perfume, colour and preservatives (anti-oxidants) mixed with a stirrer. The heating is carried out in a water jacketted vessel. Colour and anti-oxidants are mixed in warm oil which will not cause solidification of the bulk. This mixture and perfume is added, as the cooling occurs but well before solidification in mixing vessel. Materials should not be worked when partially solidified, otherwise, air bubbles will be permanently incorporated. The product, while hot is run into warmed jars and these filled containers are allowed to cool slowly in a warm room.

Alcoholic solutions

Alcoholic solutions of fatty materials, are manufactured by process of mixing. However the equipment should be such that would not permit evaporation of significant amount of alcohol.

PERMANENT HAIR WAVING PREPARATIONS AND HAIR STRAIGHTENERS

People having curly hair desire to have straight hair and vice-versa. This has resulted into generation of hair straighteners.

Heat Waving Preparations

General

As the name suggests these preparations are used when waving effect is achieved by heat. The hair are wound, moistened with heat waving preparations and are heated to about 93 to 104°C electrically or by means of heating pads. Scalp is protected from heat by so called "spacers" which are generally made of felt. The residual alkali is removed after heat treatment from hair by rinsing with water, or dilute acid or oxidising agent which are called as neutraliser.

Formulation

To protect the hair from drying and damage, non-volatile alkalies were replaced with volatile alkalies like ammonia, ammonium carbonate, morpholine etc. These alkalies initially provide pH 9–11 but being volatile during heating is dropped. Thus the chances of overwaving or partial hydrolysis of hair protein are minimised. These have disadvantage of ammonia odour and their premature loss from solution. The addition alkanolamines control premature loss of alkali. The use of sulphites in the heat wave preparations occurred about the same time as the use of volatile bases. Addition of a surface active agent helps in wetting of the hair. Formula 17 is an example of such a formulation. A similar formulation can be incorporated in absorption base to make a creamy preparation.

Formula 17

Constituents	*per cent*
Ammonium Carbonate	3.5
Potassium Carbonate	2.0
Borax	0.5
Potassium sulphate	2.0
Monoethanolamine	5.0
Sulphonated Castor oil	1.0
Water	to make 100.0

Cold Waving Preparations

General

In the cold waving process, external heat is not applied to the hair. The process is performed at room temperature. First, the hair are shampooed. Then moistened with waving lotion and wound on to curlers. After winding, the hair is left for 10–40 minutes. The wound curls are rinsed with oxidising solution (neutralisers) afterwards. The curls are unwound and usually a further rinse with neutraliser. The hair is finally rinsed with water and is set into the desired style.

Formulation

Most of the cold wave preparations are based on thioglycollic acid. Lotions are prepared with different concentrations of thioglycollic acid. The product is meant for professional use has 10 % thioglycolic acid by the hair dresser. Formula 18 is an example of liquid cold waving preparations.

Formula 18

Constituents	*per cent*
Calcium thioglycollate	80
Ammonium chloride and/or	47–65
Ammonium formate	55–77

Creams containing thioglycollates can also be prepared. They have been used for hair straightening rather than hair waving. A typical product is shown in formula 19 :

Formula 19

Constituents	*per cent*
Modified & stabilised polyglycerol monostearate emulsifier	21.0
Polyglycerol monostearate	18.0
Stearyl amide	2.1

Thioglycollic acid (38%)	13.0
Ammonia (26%)	1.0–2.0
Ammonium sulphite	1.0
Sodium lauryl sulphate	1.0
Water	to make 100.0

HAIR TONICS AND CONDITIONERS

Conditioners are the result of women's need for attractive and healthy looking hair. Conditioners should give to the hair many characteristics, life, spring, softness, volume, body, sheen, a silky touch, fly-away control and ease of styling. The hair look dull after various treatments like bleaching, permanent waving, frequent shampooing, handling abuse and weathering. The hair may also become porous and brittle. Conditioners take care of such hair. The base of conditioning is the concept of substaintivity. i.e., sorption of adequate ingredients to modify the surface properties of the hair. Keratin is anionic in nature and therefore, shows preference for cationic substances. Some treatments like bleaching enhance anionic nature by creating strongly anionic sulphonic acid sites. Though to less extent, long exposure to sun and environmental atmosphere also enhance anionic feature of keratin.

Formulation

Medicated products

Excessive scaling and proliferation of micro-organisms are associated with dandruff, an anti-microbial agent and a keratolytic agent are included. Some anti-microbial agents which have been used are : quaternary ammonium compounds; chlorophenols; PVP - iodine complexes; quinolines.

The keratolytic agents are: colloidal sulphur; resorcinol; salicylic acid; selenium disulphide.

Derivatives of 2-pyridine-thiol-N-Oxide have strong anti-microbial activity. According to Kligman, these derivatives besides anti-microbial activity exert cytostatic activity also. Cytostatic activity results in reduction in turnover rate of epidermal cells and induction of more complete keratinisation.

Anti-dandruff formulations can be prepared in aqueous alcoholic-base or in a rinse or setting lotion (formulae 20 and 21).

Formula 20
Constituents

	per cent
Bis (2-pyridyl-1-oxide) disulphide soluble complex	0.15
Camphor	0.75
Menthol	0.05
Isopropyl alcohol	51.0
Water	to make 100.0

Formula 21
Constituents

	per cent
Stearyl dimethyl benzyl ammonium chloride	5.0
Cetyl trimethyl ammonium chloride	0.2
Cocoamido propylamine oxide	2.0
Propylene glycol	5.0
Hydroxypropyl methyl cellulose	0.5
Water	to make 100.0

Conditioners

These were based on cationic surfactant, cetyl-trimethyl ammonium chloride. Subsequently, other cationics, alkyl dimethyl benzyl ammonium halides, alkyl isoquinolinium and alkyl pyridinium halides were used. Longer chain alkyl compounds of exhibited best conditioning properties. Important cationic surfactants introduced in the conditioners are : ethoxylated quaternary ammonium phosphates (SP); quaternised fatty acid amino-amides; N-acyl colaminoformyl-methyl pyridinium derivatives.

The high substantivity of cationic polymers resulted into excessive build-up. Another group of compounds that has been used is partially or totally hydrolysed proteins. Use of some of the compounds have been illustrated in formulae 22 to 24.

Formula 22
Constituents

	per cent
Stearyl di-methyl benzyl ammonium chloride	5.0
Cetyl trimethyl ammonium chloride	0.2
Cocoamidopropylamine oxide	2.0
Propylene glycol	5.0
Hydroxy propyl methyl cellulose	0.5
Water	to make 100.0

Formula 23
Constituents

	per cent
PVP - VA Copolymer	1.0
Cetyl dimethyl (hydroxyethyl) ammonium chloride	0.5

Di-methyicone	0.2
Ethyl alcohol	23.3
Water	to make 100.0

Formula 24

Constituents	*per cent*
Stearyl amidopropyl-dimethylamine	1.5
Cetyl alcohol	2.5
Stearyl stearate	0.5
Hydrolysed protein	2.0
Lactic acid	0.5
Sodium chloride	0.5
Water	to make 100.0
Preservatives, Perfume and Colour	q.s.

Anti-seborrhic Preparations

After shampooing hair become greasy quickly due to excess secretion of sebum from sebaceous glands. Bunches are formed and dust deposites on them. Greasy uptake by the hair can be reduced by applying an oil repellent treatment. Perfluorinated compounds such as $CF_3 (CF_2) X (CH_2) X (CH_2) y Z$ (where Z is water-or oil-solubilising group) have been used. These compounds are used in concentration of about 0.05% to 0.2%. Formula 25 substantially retards the oil uptake by the hair.

Formula 25

Constituents	*per cent*
Perfluorononanoic acid	0.1
Polyoxyethylene polymethyl siloxane	0.2
Alcohol	30.0
Water	to make 100.0
Perfume	q.s.

Oil adsorbing materials have also been used. A number of compounds have been recommended for seborrhea and hair growth. But classes of compounds which seemingly have more selective activity include thio-ether or sulphoxide derivatives of cysteine, cysteamine, glutathione, pyridoxine and hydroxy aminothiols; thiolanediol derivatives; poly-unsaturated acids.

Manufacture of Hair Tonics & Conditioners

Most of the hair tonics are hydroalcoholic solutions. Hair conditioners are either dispersions or emulsious. The following manufacturing conditions need specific mention as these will greatly affect the conditioners. These are :

(1) In case of hair products, if oils or fats are heated to a too high temperature, they tend to form poor film and also lose their sparkle and become dull. Therefore, temperature should be only as high as is consistent with achieving a proper emulsification.

(2) Better results are obtained if a cream conditioner is to be manufactured with fats like petroleum jelly, lanolin and emulsifying agents like Span 60, Tween 60, by adding water phase to oil phase slowly.

(3) Viscosity, stability and shelf-life are greatly affected by rate and manner of agitation used. Reader may refer to the reference given in the foot note.

(4) Where dispersion is of importance, colloid mill should be used because smaller the particle size, better the dispersion. Superior covering film can be achieved with smaller amount of product.

(5) In case of creams, rate of cooling and length of cooling time are important factors.

Chapter 35

Hair Colourants

INTRODUCTION

Since origin of man one of the way to beautify hair is colouring. The quest for a change of external appearance has found expression in various ways in all civilisations. The main reasons for colouring the hair are the following: to change the natural colour, to colour the white hairs which begin to appear with age, or to change the colour of the hair temporarily on a particular occasion. It is necessary to use a range of colourants with varying composition and behaviour for colouring. The various hair colouring systems, the characteristics of an ideal hair colourant and the various practical hair colouring processes are described in this chapter.

HAIR COLOURING SYSTEMS

The hair colouring is divided into three classes on the basis of duration.

(1) The fugitive colours applied on hair have short duration hence, easily removed by shampooing. In this category the commercial products commonly designated 'colour rinses'. These products utilise colours of a high molecular weight which are in effect deposited on the surface of the hair without being able to penetrate into the cortex.

(2) *Semi-permanent Colouring:* These colours are removed after several shampooings (three to six), but whose fastness is poorer than that of permanent colours. The colours used are direct dyes of low molecular weight, having a good affinity for hair keratin. Hence, they are capable of penetrating the cortex.

(3) *Permanent Colouring:* This category provides effectively permanent colouration, resistant to shampooing and other external factors such as

brushing, friction, light, etc. This is the process most widely used and represents at least 80 per cent of the total colourings effected.

In this system, uncoloured intermediates undergo series of chemical reactions produce the desired colour *in situ* in hair.. The process is one of oxidation (almost always effected by hydrogen peroxide), followed by coupling and further oxidations.

Characteristics of an Ideal Hair Colourant

The ideal hair colourant should possess the following properties :

(1) *Harmlessness:* It should colour the hair without impairing the natural texture and gloss and be non-injurious to the hair shaft. It should produce no toxic effect non-irritant with the skin and should not be a non-determatitic agent.

(2) *Adequate Physical and Chemical Stability on the Hair:* The stability of dye used for hair should be good to air, sunlight, friction and sweat.

(3) *Compatibility with other Hair Treatments:* It should not change colour, nor bleach out, on the application brilliantines, setting lotions, hair lacquers, hair waving preparations, soaps or shampoos.

(4) *Stability in Solution:* The stability of colour and formulated products in aqueous solution should be good.

(5) *Selectivity:* Because it is always necessary to use a mixture of dyestuffs. The problem of selectivity plays a most important part. The dyestuff belonging to same chemical class has similar physico-chemical behaviour, hence they should be avoided.

(6) *Affinity for Hair Keratin:* The physico-chemical characteristics of affinity, in conjunction with the penetration of the dyestuff into the hair shaft, can be seen to be very important when account is taken of the technical limitations controlling the dyeing of hair, such as: temperature not greater than 40°C; short time of contact of hair with the dye, up to say 40 minutes; very weak dye solutions, etc.

THE PROCESS OF HAIR COLOURING

The range of commercial hair colourants gives various shades and tints to hair. It ranges very light blonde to black, passing through a range of tones: golden, ash, reddish, mahogany, violets, etc. There are more than sixty shades. The factors influencing process colouring hair are :

(1) *Commercial Products are Mixtures:* Each particular colour is the overall result of the superposition of individual colourations (red, yellow, violet, blue, etc.) supplied by each of the dyestuffs in the commercial mixture.

(2) *Concentrations of Dyestuffs:* The total quantity of all the dyestuffs used to obtain a shade is between 0.01 and 5 per cent by weight of the tinting medium applied to the head. The concentration is a function both of a dyestuffs used and of the procedure involved.

(3) *Duration of Colouring Process:* The time of contact of the dyeing solution with the scalp and the hair is of the order of 5 to 40 minutes.

(4) *Quantity of Solution Applied:* The amount of dyeing solution applied to a female head varies between 15 and 100 ml.

(5) *Frequency of Application:* For temporary colourants, this is of the order of once a week. On the other hand, for more permanent colourants, it is about once a month. In fact this is controlled by the regrowth rate of hair, which is about 1 cm per month.

(6) *Treatment after Colouration:* Colourants must be conceived and formulated so as to avoid to the maximum extent the staining of the scalp. It is necessary to rinse with abundant of water and shampooing once or twice. This will remove the dye which is not adsorbent on the hair.

(7) *Uptake of Dyestuff by the Scalp:* In order to achieve the desired colour, it is sufficient to place on the head a maximum of about 60 g of dye solution. This quantity is spread over a total surface of 700 to 1000 cm^2 of scalp and about 50,000 cm^2 representing the surface of the hair.

Permanent Colours (Oxidation)

It concerns the first stage of about 10 to 15 minutes, during which half the quantity of product is applied to the roots of the hair, the first 2 cm, close to the scalp, before applying the remainder to the rest of the hair and leaving in contact for a further 15 minutes.

TEMPORARY HAIR DYES

Dyestuffs

The dyestuffs used are basic dyes, acid dyes, disperse dyes, pigments or metallised dyes. The chemical class is azo, anthraquinone, triphenylmethane, phenazinic, xanthenic or benzoquinoneimine dyes. Table 35.1 lists dyestuffs with colour index number according to category.

Table 35.1. Dyestuffs used in temporary hair dyes.

Chemical class	Colour	Colour Index Number
Acid Dyes		
Azo	Yellow	13 065
	Yellow	19 140
	Red	14 720
	Red	15 620
	Red	16 185
	Red·	16 250
	Red	17 200
	Orange	15 575
	Orange	16 230
	Brown	14 805
Triphenylmethane	Green 22	42 170
	Violet acid 49	42 640
	Acid blue	42 735
Xanthene	Acid violet	45 190
Azine	Nigrosines (violet)	50 420
Anthraquinone	Acid violet 43	60 730
	Blue 62	62 045
Basic Dyes		
Azo	Yellow 57	12 719
	Orange	11 270
	Orange	11 320
	Red 76	12 245
	Brown 16	12 250
Triphenylmethane	Basic blue 5	42 140
	Violet 14	42 510
	Violet 3	42 555
Azine	Red	50 240
Indoaniline	L'Oreal 14-22	
Indophenol	L'Oreal 23-25	
Indamine	L'Oreal 26-32	
Disperse Dyes		
Azo	Yellow	11 885
	Orange solvent 45	11 700
	Orange solvent 9	11 005
	Disperse red 17	11 210
Anthraquinone	Orange 11	60 700
	Red 15	60 710
	Violet	60 725
	Blue 3	61 505
	Blue	60 500

(Cont'd)

Chemical class	Colour	Colour Index Number
	Black acetoquinone	—
	Black celliton	—
Metallised Dyes		
Azo	Cibalane blue F.B.N.	—
	Solvent yellow 90	—
	Solvent brown 43 etc.	—

Types of Commercial Temporary Product and their Formulation

Rinses and Coloured setting lotions are two main types of temporary hair colouring product. In rinses the dyestuffs are used in the form of simple aqueous or aqueous-alcoholic solutions. In order to increase the substantivity to hair, organic acids, special solvents are added or the hair may be pretreated with cationic compounds to increase the substantivity to hair. It should be noted that such tinting solutions may be sold as ready for use or alternatively be prepared by the user from concentrated products by simple dilution in water.

Crayons have been used from time to time for temporary colouring, and are employed like mascara—either rubbed direct on to the wet hair or transferred to the hair with a brush. They can be formulated with soaps and waxes, so as to give a product rather like a lipstick in formula 1 :

Formula 1

Constituents	per cent
Stearic acid	15.0
Triethanolamine	7.5
Glyceryl monostearate	4.0
Beeswax	46.0
Paraffin wax	10.0
Microcrystalline wax	10.0
Coconut diethanolamide	7.5
Colour	*q.s.*

SEMI-PERMANENT COLOURANTS

Dyestuffs

These dyes occupy an important place in formulation practice. It is because their presence is often indispensable in the formulation of the permanent colours, the oxidation dyes. The three main classes are :

(i) Nitrophenylenediamines; (ii) Nitroaminophenols; (iii) Aminoanthraquinones.

Various dyes

Reactive dyes

The use of reactive dyes has been a relatively new approach in the field of textile dyeing. These include materials such as dichlorotriazines (for example Procion dyes *ex* ICI), monochlorotriazines (for example Procion dyes *ex* ICI and Cibacron dyes *ex* Ciba) and trichloropyrimidines (Reactone dyes *ex* Geigy, Drimaren dyes *ex* Sandoz). These dyes work by actually reacting with the fibre and thus the dye part of the molecule is firmly held by the fibre.

It is now believed that conventional oxidation dyes such as *p*-phenylenediamine develop their final colour by interacting with reactive sites in the hair. Oxidation dyestuffs may thus be regarded as being a particular type of reactive dyestuff.

Metallised dyes

The preparation of such dyes may be based on the formation of complexes *in situ* in the hair by means of nickel or cobalt ions and various complexing agents.

Azo dyes obtained by coupling on the hair

Diazonium salts, together with various coupling agents, may be applied to the hair.

The various dyes used as auxiliaries to modify the shade are : *heterocyclid azo-derivatives*, also derivatives of *diazamerocyanines* and *quarternary derivatives of aminophenoxazinium.*

Commercial semi-permanent products and their formulation

Semi-permanent colourants can be presented either in the form of foaming lotions (anionic, cationic or non-ionic) or as anionic or cationic shampoos. According to the type of medium chosen and the product performance sought, various approaches to formulation have been developed, all having the goal of promoting the penetration of the dyes into the hair.

PERMANENT HAIR DYES

Permanent hair dyes are based almost exclusively on the use of reactive dyes, the so-called *para*-dyes. These are substances that are colourless at the time of their application to the head (the precursors) and are transformed into a coloured material *in situ* on the hair as a consequence of chemical reactions.

The precursors can be classified into two categories: the compounds called oxidation bases or primary intermediates and those called couplers or modifiers.

The chemical reactions in the formation of dyes are oxidation reactions and couplings or condensations. These are effected at alkaline pH (essentially due to the presence of ammonia) by the action of hydrogen peroxide or one of its solid derivatives—urea peroxide or melamine peroxide. The hydrogen peroxide is selected because its ability to promote the simultaneous decolouration of the hair to be tinted.

The bleaching which occurs simultaneously with but independent of the dyeing, results in the hair being rendered lighter. It permits to give new shades with the help of the new pigments. Thus it can thus be seen this system of colouration is unique and irreplaceable; this explains its widespread use and its name of colour lightening ('teinture eclaircissante').

Bases

The bases are benzene derivatives, substituted by at least two electron-donor groups, such as NH_2 and/or OH. This confers the property of easy oxidation due to *para* or *ortho* position. The most important compounds of this class are thus *p*-phenylenediamine and *p*-aminophenol, and *o*-aminophenol, to which one could add the *p*- and *o*-dihydroxybenzenes.

Couplers or Modifiers

Couplers or modifiers are benzene derivatives, substituted by the groups (—NH_2 and —OH) in *meta* position to each other. In this position it should be noted that the couplers do not have the property of easy oxidation by H_2O_2. The most usual couplers are among the following: (i) *m*-phenylenediamine; (ii) 2, 4-diaminoanisole; (iii) resorcinol; (iv) *m*-chlororesorcinol; (v) *m*-aminophenol; (vi) 1,5-dihydroxynaphthalene; (vii) 6-methyl-3-aminophenol; and (viii) 2-methylresorcinol.

Formation of Colours in the Hair

A very large amount of work has been devoted to elucidating the merchanism of oxidation dyeing and to the structure of the dyes. But all is not yet clear.

In fact, the number of parameters affecting the overall process is very great. For example, consider the influence of pH on the speed of reaction, the presence of the hair keratin itself which can affect the orientation of reactions, the complexity of the reaction mixtures (it is not unusual to use up to 10 couplers and bases), the possible hydrolyses of intermediate products, etc. All these variables make it practically impossible to specify exactly all the compounds which could potentially be formed. Thus empiricism in formulation still plays a very important part in the technology of oxidation dyeing, and no mean effort is required to arrive in practice at a formula that gives commercially reproducible performance.

FORMULATION OF PERMANENT HAIR DYES

For developing suitable formulation or note the following factors must be considered.

(1) The formulation base–solution, emulsion, gel, shampoo or powder.

(2) The selection of the dye components – the oxidation base, the coupling agent or the addition of a direct colourant.

(3) The selection of an alkali – ammonia is usually used.

(4) Antioxidants – usually a sulphite or ammonium thioglycollate is used to prevent oxidation of the dye before the product is to be used.

(5) The pack – required to be attractive and convenient in use.

The most convenient medium for colouring the hair is a shampoo. The system consists of a solution of the dye precursors together with an ammonium oleate soap or other surfactant, to which is added at the time of use another solution of stabilised hydrogen peroxide contained in a separate vessel. This mixture is applied direct to the hair and left in contact for 20 to 40 minutes. Afterwards the hair is rinsed with water and then washed again.

By use of fatty alcohol sulphates, fatty acid dialkanolamides, non-ionic or amphoteric surfactants, fatty alcohols, amine oxides, fatty amines and cationic surfactants, a whole range of emulsions can be formulated in order to produce the dyes in the form of creams, gels, shampoos, etc.

Sodium sulphite or thioglycollic acid with or without hydroquinone can be used as antioxidant. The use of ascorbic acid has also been recommended as has that of pyrazolones. The hydrogen peroxide, can be used in various procedures.

The modern version of this process is the combined bleach and dye, in which oxidation dyes are used in the presence of enough hydrogen peroxide to bleach the hair while the dyes are penetrating. In this way, dark mousy hair can be dyed to a warm blonde colour several shades lighter than the original colour, yet not apparently bleached. The difficulties found with early formulae have been overcome and suitable products are now available for both home and professional use.

The formulae which may be used are similar to those used for conventional oxidation dyeing, except that more alkali and hydrogen peroxide are used. The degree of lightening depends on the concentration of hydrogen peroxide and ammonia. Thus, for example, if 15 per cent ammonia and 20 per cent hydrogen peroxide are used a lightening of one shade may be achieved (say from black to brown), while if 15 per cent ammonia and 30 per cent hydrogen peroxide (20 vol) are used, then much greater effects can be produced (for example, dark auburn to blonde). The two end points on formulation are :

(1) The improvement of essential dyeing properties can be obtained by better storage stability, ease of application, shortening of time of application, increased covering power and penetration, fixation in the hair, reduction in potential for skin damage and protection against eventual hair damage.

(2) The conferring of new properties on the hair colourant are protection of hair structure, improvement of aesthetic quality of the hair, brightness, bulk, combability, general appearance, addition of antiseptics, antidandruff agents, antiseborrhea agents for the scalp, deodorants and addition of substances specific to operations other than dyeing, for example, film-formers, etc.

OTHER DYES FOR HAIR

Vegetable Hair Dyes

Henna

Of the vegetable hair dyes, only henna is of any real importance today. It consists of the dried powdered leaves of *Lawsonia alba*, *Lawsonia spinosa* and *Lawsonia inemis*, which are removed from the plants prior to flowering.

Henna owes its hair-dyeing properties to the presence of 2-hydroxy-1,4-naphthaquinone, often termed *lawsone*, which is soluble in hot water and is, in acid solution, a substantive dye for keratin. In dyeing hair with henna, a paste of the powdered henna and hot water, slightly acidified with citric, adipic or other suitable acids to an optimum pH of about 5.5, is applied to the washed hair.

This 'henna pack', which is kept in place by means of a towel, is allowed to remain on the head for the required time, which may vary from five to sixty minutes. Then hair is thoroughly shampooed, rinsed and dried.

Henna reng

Henna rengs produces blue-black shades. The addition of other substances to henna other than auburn i.e. mixture of indigo leaves and henna.

Chamomile

Of the various species of chamomile, only *Anthemis nobilis* and *Matricaria chamomillae* have a cosmetic use; they appear to be equally useful in tinting hair. The active ingredient in these flowers is 1,3,4-trihydroxyflavone, known also as apigenin. Either an aqueous extract or a paste of the ground flower heads may be used. To lighten the hair, a paste consisting of 2 parts chamomile and 1–2 parts kaolin mixed to a thin cream with hot water is applied to the head for a period varying from 15 minutes to 60 minutes depending on the shade required. Chamomile is also used as a constituent of hair-brightening rinses and shampoos.

Metallic Hair Dyes

In metallic dyes, compounds of lead are the most frequently used; compounds of silver, copper, iron, nickel and cobalt are sometimes employed, and less frequently salts of bismuth.

The colours produced by such compounds are due to sulphides formed by a reaction between the sulphur in the keratin and the metallic salts, or to metallic oxides formed by the keratin reducing the metal salts is not clear. It is possible that both reactions occur to some extent. Whatever the mode of action, the result obtained is the deposition of a coloured film along the hair shaft. This eventually, gives the hair a characteristic dull metallic appearance, renders the hair brittle and diminishes the efficiency of a subsequent permanent wave.

Lead dyes

Lead acetate precipitated sulphur, glycerine and water are active ingredients. Example 2 is a typical formula. Sodium thiosulphate may be incorporated in place of precipitated sulphur. Such preparations have poor stability.

Formula 2

Constituents	*per cent*
Precipitated sulphur	1.3
Lead acetate	1.6
Glycerine	9.6
Rose-water	87.5

The action of lead dyes is slow and progressive. The shades produced in grey hair usually pass from yellow through brown to black. The shades achieved depend upon the concentration of lead salts in the preparation, the number of applications, the original colour of the hair and the time during which the colour has been allowed to develop.

Lead dyes interact with skin proteins. Hence, lead solutions are relatively non-toxic under normal conditions of use. However, it must be remembered that these preparations are usually sold for home use and systemic effects may follow if lead remains on the hands and contaminates food. The ingestion by children might be fatal and preparations containing lead should carry adequate precautionary notices.

Other metallic dyes

Other metallic hair dyes, such as bismuth, silver, copper, nickel and cobalt salts, have been proposed, however, due to toxicity it requires precautions.

HAIR DYE REMOVERS

Sometimes user may have a lighter shade, hence removal of hair dye is necessary. In the case of metallic dyes it is usually dangerous to remove the dye by chemical means since the metals catalyse many reactions and may cause a violent production of heat which will probably damage both hair and scalp. The only remedy for an unwanted metallic dye is to let the hair grow.

Oxidation dyes can be removed by treating the hair with reducing agents such as sodium hydrosulphite or sodium formaldehyde sulphoxylate, usually at a concentration of 5 per cent. The use of formamidine sulphinic acid has also been described. A formula for hair remover is shown below :

Formula 3

Constituents	per cent
Formamidine sulphinic acid	1.5
Polyvinylpyrrolidone	5.0
Ethylene glycol monobutyl ether	5.0
Ammonium carbonate	1.0
Ammonia (25%)	0.5
Carboxymethylcellulose	2.5
Water	to 100.0

Semi-permanent dyes can often be removed by vigorous washing with shampoos containing little ammonia. Some dyestuffs, however, prove very resistant and a mixture of shampoo, reducing agent and bleach in the proportions 1 : 1 : 2 has been recommended for such cases.

BLEACHING AND LIGHTENING

Blonde shades are obtained by bleaching the hair in the usual manner to the palest possible shade of blonde with ammonia and hydrogen peroxide. The hair is afterwards given a rinse with a blue rinse containing about 1 : 100 000 of methylene blue (Ext. D&C Blue No. 1), or other suitable blue certified colour. The addition of the blue colour is necessary since the human eye considers a substrate which is very slightly blue in colour as whiter than white.

Bleaching of the hair must be thorough or the combination of a deep yellow hair shaft and a blue rinse may give the hair a distinctly greenish appearance.

Powder products obtain better control in the application of peroxide to the hair and extend the bleaching time. These powders kaolin or magnesium carbonate, used with peroxide and ammonia. There are powders which themselves provide ammonia and some form of active oxygen, when wetted with water or hydrogen peroxide. Typical early formulae contained up to one-third of sodium perborate or percarbonate, the balance being kaolin and/ or magnesium carbonate. More complex formulations are given in formulae 4 and 5.

Formulae	*4*	*5*
Constituents	*per cent*	*per cent*
Ammonium persulphate	3.0	—
Potassium persulphate	—	8.0
Potassium hydrogen tartrate	3.0	—
Potassium hydrogen oxalate	—	8.0
Sodium carbonate	3.0	13.0
Surfactant	1.0	1.0
Thickener	5.0	—
Magnesium hydroxide and/or aluminium hydroxide	to 100.0	to 100.0

Professional hair dressers use above preparation hence the degree of skill and care is required to get good colour.

For home use, products are formulated on much more simple lines. A suitable two-bottle pack would be :

Formula 6	
Constituents	*per cent*
A Hydrogen peroxide 20 vol	98.6
Tartaric acid	0.8
Sodium stannate	0.6
B Ammonia	4.5
Surfactant, e.g., ammonium soap	3.0
Water	92.5

Chapter 36

Permanent Waving, Hair Strengtheners and Straighteners

INTRODUCTION

In permanent waving the normal anatomy of the hair is seriously disrupted. The keratin fibre of the hair are attacked by chemicals present in waving lotions. Here permanent waves break fewer bonds and then re-form them. Since not all of the broken bonds are restored by the setting agents, repeated use of permanents makes the hair weaker and much more brittle.

There are four steps involved in permanent waving. First, detergents strip away the chemically resistant sebum and expose the cuticle (outer layer of cells). Secondly, swelling agents then enlarge and open up the hair shaft, thus enabling chemicals to reach the keratin in the cortex (inner core of fibres), the backbone of the hair. Thirdly, the bond-breaking agents (usually the same chemical found in depilatories) enter the shaft and break the hair's "cross-linking" bonds. In the final step, a bond-forming agent (called a "neutraliser" or "oxidiser") enters to re-form the bonds in a new configuration, which gives the hair a permanent curl. These new bonds hold the hair in the curled position until it grows out, is cut off, or until extremely harsh chemicals are applied again.

Various types of hair respond differently to the waving process. A thin hair will not have as many fibres or as great a distance between cross-linked strands of keratin. Thus, when it is bent, there is not as great a stretching difference between the inside and outside strands of keratin as in a thick hair. The resulting curl will not be as small or tight. To offset this, stronger solutions or longer times are used to break as many bonds as possible. This increases the danger to the user. Special caution must be used when giving children permanents; their hair is usually fine, and the waving chemicals must be very strong to do the job. The original condition of hair can not be obtained although reforming of bonds may occur. As the cuticle is permanently eroded by waving temporarily some hair rinse and sprays patch the hair.

The most extensive damage can be found near the ends of permanent waved hair since that part of the hair is the oldest and has been waved the most. The cracking and eroding of the cuticle will split the ends and make the hair look drab, dull, and frizzy. For permanent wave or reshape hair there are three methods :

(1) In cold waving powerful chemicals break the cross-linking bonds. The bonds are re-formed by "oxidising" or "neutralising" chemicals.

(2) Heat waving, which is based on Nessler's original method and uses both heat and water in the presence of bond-breaking chemicals to curl hair.

(3) In tepid waving, which combines hot and cold processes to allow lower waving temperatures than heat waving and lower concentrations of dangerous chemicals than cold waving.

HOW COLD-WAVING PREPARATIONS WORK

Cold waving requires two separate applications of chemicals. First, a waving lotion strips, softens, swells, and breaks the bonds in the hair. Second, a neutralising lotion reverses the effects of the powerful waving lotion by re-forming the bonds in a curled position. Normally, hair is first washed to remove the sebum, placed in curlers, and the waving lotion is applied to each curler. After about ten minutes, a control curl is taken to determine if enough bonds have been broken to leave the hair contoured around the curler. If the curl tends to return to its original state, the waving lotion is left on the hair for several minutes more to break additional bonds. Once enough bonds are broken, the waving lotion is rinsed out and neutralising lotion is applied. The longer the chemicals left tight curl results whereas shorter length yield soft curl. Principal ingredients of cold-waving lotions and their functions are shown in Table 36.1.

Table 36.1. Principle ingredients of cold-waving lotions and their functions.

Class of Ingredients (concentration)	Function	Ingredient Names (as they appear on product labels)
bond-breaking agents (4.6%–8.3%) antioxidants	disrupt the structures that give hair its shape prevent waxing lotion from reacting with air before it reaches the hair	calcium thioglycolate ammonium thioglycolate sodium hydrosulphite

(Cont'd)

Class of Ingredients (concentration)	Function	Ingredient Names (as they appear on product labels)
sequestering agents	prevent trace metals dissolved in tap water from reacting with the thioglycolates	tetrasodium EDTA tetrasodium pyrophosphate
pH adjusters (10%)	hold the pH around 9.3 so the thioglycolates will work properly	calcium hydroxide ammonium carbonate triethanolamine (TEA) monoethanolamine (MEA) ammonium hydroxide
emollients	replace some of the oils stripped off the hair by the strong surfactants and caustic chemicals in the waving lotion	lanolin, cetyl alcohol, synthetic spermaceti, mineral oil, propylene glycol
emulsifiers	keep the emollients dissolved in water	sodium lauryl sulphate sorbitan palmitate polysorbates
surfactants	remove sebum from hair	TEA & sodium lauryl sulphates
conditioners	cover some of the damage done to the hair by harsh chemicals and surfactants in the waving lotion	hydrolysed animal protein quaternium-40 potassium hydrolysed animal protein quaternium-19 acrylic copolymer

Waving Lotions

It is tricky to formulate safe and pleasant waving lotions. The chemicals present in depilators are similar to waving lotions, the little change in composition will result into destroying rather than waving hair. The bond-breaking agent in waving lotion is usually calcium thioglycolate or ammonium thioglycolate; both are highly alkaline, caustic chemicals that can burn or corrode the skin. In order for the bond-breaking agent to work properly in a cold-wave lotion, the pH must be around 9.3. This is ideal pH, as above 9.3 hairs are removed and below 9.3 bonds are not broken. Calcium hydroxide, ammonium carbonate, ammonia, or triethanolamine are commonly used as pH adjusters. These chemicals can harm the user.

Other less serious problems are involved in formulating a waving lotion. The main ingredient, calcium thioglycolate, smells like vinegar and develops a rotten-egg odour if it is left standing. This odour also arises when the lotion reacts with the hair. Moreover, wave preparations normally contain ammonia or other compounds, all of which have distinctive unattractive odours. Masking these odours is difficult. Usually a fresh cologne smell of higher value must be used. Lotion cosmetics form similar, but longer-lasting emulsions because emulsifiers hold the drops of oil in the water. Otherwise, the oil and water would quickly separate into distinct layers. Cloudy lotion-type waving liquid does not guarantee is gentle, soft, or more exotic formulation than a clear one. It means that it is costlier and cloudier.

Waving lotions also include surfactants (detergents) which remove sebum from the hair. The harsher surfactants are commonly found in permanent-wave preparations. Most waving lotions also contain emollients, which replace some of the oil stripped off the hair by the strong surfactants and caustic chemicals. Hair conditioners are now incorporated into some waving lotions. A combination of emollients and conditioners can patch damaged hair due to permanent waving.

Neutralisers

Once the hair's cross-linking bonds have been broken, they must be re-formed. The derivatives of citric acid, tartaric acid and lactic acid neutralise the waving lotions, hence neutraliser.

They also contain a bond forming agent to speed up the work of the oxygen generated by H_2O_2 in air, which forms bonds naturally, but too slowly. Peroxide can bleach hair, but not very easily in acid solutions. Powder preparations must be mixed with water at home, may contain powerful, flammable toxic chemicals such as sodium or calcium bromate. If these bromates are used, flame retarders such as urea are added. Sodium perborate or ammonium persulphate are less flammable can be used. Principal ingredients are shown in Table 36.2.

Table 36.2. Principal ingredients of permanent-wave neutralisers and their functions.

Ingredients Name	Function
Citric acid	neutralise the alkalanity of the waving lotion,
Acetic acid	usually around 0.1%–0.2%
Tartaric acid	
Lactic acid	
Phosphoric acid	
Hydrogen peroxide	re-form broken hair bonds 3%–6%
Sodium bromate	
Calcium bromate	
Sodium perborate	
Ammonium persulphate	
Urea	flame retardant
Methylparaben	preservative
Ammonium lauryl sulphate	detergents (remove excess oil and help hold
Polysorbate 20	waving lotion on the hair)
Polysorbate 40	

Some home permanent kits rely on self-neutralisation. The oxygen in the air reacts with the calcium with the calcium thioglycolate once it is placed on the hair. This method is too slow to be used in beauty parlours. However, it can be used at home with some advantages Self-neutralisation controls the overwaving of the hair that may occur if the user gets distracted and does not apply a neutralising liquid at the proper time. Unfortunately, self-neutralising products can make the hair frizzy.

HOW HEAT- AND TEPID-WAVING PREPARATIONS WORK

The effect of the heat-waving process is also breaking and reforming of hair. The most important difference between steam heat and cold waving is that, heat is used instead of thioglycolate for bond breaking and reformed by cooling rather than neutraliser. Since steam, like water, can break only about 70 per cent of the cross-linking bonds, some chemicals must still be used to achieve a permanent heat-wave. Accelerators-alkaline compounds such as ammonia and monoethanolamine-increase the used speed with which water breaks the cross-linking bonds.

Sodium bisulphite, initiator, increases speed of the bond breaking. These chemicals can create consumer problems, however, as ammonia and monoethenolamine are both toxic, and ammonia has the added drawback of giving the hair a reddish tint.

But the greatest danger to the user is the risk of hair and scalp burns. Because of the high temperatures (200° F.) involved. Tepid waves, in which heat is applied at a lower temperature along with the same chemicals used in cold waves. The advantage of tepid waving is that heat speeds chemical reactions. Thus lower concentrations of toxic chemicals can be used without increasing the total time needed to wave the hair. The ingredients for a tepid wave product will be similar to the chemicals in a cold wave but will occur in lower concentrations.

Permanent-Wave Pretreatments

People who have permanently waved, bleached, or dyed their hair or have exposed it to excessive sunlight or to extremely strong shampoos probably have damaged hair. Such weakened hair absorbs the powerful waving chemicals much more quickly and in much greater quantities than normal hair. It is thus particularly susceptible to further harm. Pretreatments can decrease the amount of damage the waving products will do to the hair.

Pretreatments work in one of two ways. They can cause more bonds to form in the hair, thus increasing the number of bonds that the waving product must break. Or they can contain an "arresting agent" that will remain on the hair and limit the action of the thioglycolates by reacting directly with these bond-breaking agents. Ammonium persulphate is a common waving-action limiter. Of the bond-forming agents, glutaral and glyoxal are the most common. These chemicals can be sensitising to the scalp. If your scalp becomes sore or irritated after repeated use of a product which contains one of these ingredients, try switching to a pretreatment which uses an arresting agent.

Health and Safety

Permanent-waving components are extremely toxic and corrosive to the skin. These chemicals in combination may be more dangerous than the individual chemicals. They can be fatal in small quantities also. They can damage fabrics and metals, as well as skin and hair. Beauticians are warned to protect their hands against the corrosive action of thioglycolate waving lotions. Home users, too, are cautioned not to apply waving products if their scalp or skin is "sensitive, sore, scratched or tender," and they are advised to wear rubber gloves.

The use of surfactants (detergents) in the waving solutions may enhance the skin-penetrating abilities and toxic effects of the thioglycolates. All cold-waving lotions with a pH of 9.0 or higher may cause a "sensitisation after prolonged contact and may result in hair damage as well."

Ammonia and thioglycolate are corrosive and can cause blindness in case of contact with eye. However, surfactants (detergents)—particularly the non-ionics and amides—have an anesthetic effect on the eyes, which numbs them to sensations of burning. Thus, users are less likely to be aware of eye damage and to take appropriate remedies immediately.

Many of the active ingredients in permanent-waving products are caustic and can damage skin. Some alkalies have additional toxic effects. The pH adjusters triethanolamine (TEA), diethanolamine (DEA), and mono-ethanolamine (MEA) should be avoided. Nitrosamines, proven cancer-causing substances in animals, are sometimes found in TEA and may be found in the waving lotion penetrate the skin. The bromates found in neutralisers may also cause problems. They are extremely flammable in the dry form and are poisonous if swallowed or if they enter the skin through cuts or abrasions on the scalp or hands.

HAIR STRENGTHENING PREPARATIONS

Weakened hair, which may result from sensitisation due to hair treatment or over-processing or from environmental or internal causes, usually lacks bulk, tensile strength, lubricity, sheen, body and a barrier to penetration. It is often stringly, oversensitive to moisture, or even flimsy and, as a result, difficult to handle. The prevailing factor in the weakening of hair is the breaking of too many disulphide links, as may be the case with hair that is too frequently and excessively waved, bleached or dyed.

A process for improving the strength and elascity of weakened hair, which entailed treatment of the hair with solutions or dispersions of dimethylol urea or dimethylol thiourea for 15 minutes at a temperature of 30°– 40°C in the presence of a glycerophosphoric acid in an organic or aqueous medium at a pH between 1.5 and 6.0. The free amino groups of hair keratin, producing cross-links between the hair fibres, and polymerise in the presence of the acid catalyst to form a resin within the hair fibres. The resin strengthens the hair making it virtually insoluble in water and various organic solvents.

The commercial use of free formaldehyde and such compounds is banned in many countries. The release of free formaldehyde has subsequently been shown to be efficiently reduced by the inclusion of stabilising compounds such as urea, dicyandiamide, melamine or ethylene urea in the methylol compositions. The improved compositions for strengthening damaged or weakened hair contain, more specifically, monomethylol dicyandiamide and its methyl ether, methylolethylenethiourea, and methylolated melamines and 'methylols'.

Further compositions for strengthening degraded hair, which comprise at least one alkylated compound, to be so stable that virtually no formaldehyde was released during application. The methylol ether compounds proposed include mono- and dialkoxymethylureas or ethyleneureas and corresponding thioureas, tris(alkoxymethyl) melamines, N-alkoxymethylcarbamates or adipamides.

Other stable, easily obtainable yet more water-soluble compounds are condensation products of the former linear or cyclic methylolated substances with secondary amines such as N,N'-*bis*-(morpholinomethyl) urea or N,N'-*bis*-(ureidomethyl) piperazine and methylol derivatives of glyoxalurea or thiourea condensation products.

Lastly, addition of fairly small quantities of sulphites to any of the previous compounds prevents occurrence of free formaldehyde.

The 4–12 % of methylocated compound is used in strengthening preparation. The concentration depends on product used, water solubility, type of hair to be treated. Hydroxylated acids, acetic, phosphoric and hydrochloric acids (or salts thereof, such as acid phosphates) may be used as polymerisation catalyst. The reduction in solubility of hair in alkali indicate that remarkable strength is resulted by above composition.

Many new compounds have been proposed to enrich hair, enhance its protection against subsequent hair treatment, impart more body and bulk, eliminate a slimy feel on wetting and add lubricity, at the same time restoring it. Particularly interesting are, on the one hand, cationic resins with methylol groups and, on the other hand, methylolated compounds or derivatives with functional groups that may be transferred into the hair or onto its surface so as to modify its sensitivity to chemicals. The above-mentioned functional groups are disulphides—such as in N,N'-*bis*-(hydroxymethyl) dithiodiglycolamide and 2,2'-dithiodiethyl-*bis*-(morpholinomethylurea); tertiary amines—such as in diethylaminoethylurea and methylimino-*bis*-(3-propylurea)methylols; and quaternary ammonium compounds—such as in alkyldimethylammonium acetamide chloride and diethyl(ureido ethyl) ammonium acetamide chloride methylols. Dicarbonyl compounds such as glyoxal, glutaraldehyde, quinone, etc. and amino-dialdehydes have also been claimed to have strengthening properties.

HAIR STRAIGHTENERS

Many consumers need hair straighteners, particularly curly hair one. The idea held at one time that curly or kinky hair is eliptical in cross-section whereas straight hair is circular is now completely discarded. There are several types of hair straightening preparation : (i) Hot comb-pressing oil methods; (ii) Caustic emulsion and (iii) Chemical reducing agents.

(1) *Hot Comb Method.* In the hot comb method, hair is straightened by the use of petroleum jelly and a hot metal comb. This is called as 'hot pressing'. Alternatively a mixture of petrolatum and paraffin may be used. The petrolatum, major component acts as a heat transfer agent between the comb and hair, lubricating the latter to allow the comb to slide through it without drag. The hair is washed and dried before the pressing oil is applied and the heated metallic comb is used to straighten the hair.

(2) *Caustic Emulsions.* Caustic emulsions are in cream form. The use of caustic lye involves risks such as irritation of the scalp and even accidental eye damage.

A typical composition is given in formula 1. This paste is applied by combing and allowed to remain on the hair for 30 minutes at 40°C. The hair is then washed well with water to remove all the paste.

Formula 1

Constituents		*parts by weight*
A	Gum tragacanth	2
	Boric acid	1
	Water	40
B	Sodium carbonate	1
	Potassium hydroxide	1
	Glycerine	2
	Water	8

Formula 2

Constituents	*grams*
Hydroxyethylcellulose WP 4400	4.0
Lithium hydroxide	2.0
Sodium chloride	17.5
Water	to 100.0

The simple lye straighteners described above are mainly designed for people who want their hair straightened, not styled.

The multi-component lye straighteners conform to a general pattern of five components are also available. They have :

(1) A soft pomade based on mineral oil-jelly-wax applied to the scalp as a protective pretreatment.

(2) The 'relaxer', which consists of an oil-in-water emulsion containing about 3 per cent caustic soda and about 40 per cent fatty material.

(3) A cream shampoo which is used to follow the relaxer constitute 12 per cent alkyl sulphate, 2 per cent fatty acid diethanolamide and unspecified emulsified fatty material.

(4) A dilute oil-in-water emulsion containing about 2 per cent fatty esters and a cationic wetting agent referred to as the 'neutraliser', preferably slightly acid.

(5) A stiffer pomade containing petroleum jelly, fatty ester and lanolin to provide the final set and dressing.

It has also been recommended to follow the alkaline treatment. When calcium as in quicklime reacts to form chelates with keratin. This firmly set in the new hair shape. Keratin may then be released from the chelates by means of a complexing agent (EDTA) and a surfactant.

Straightening and colouring consist of alkali at pH 12–13.8, dyestuffs together with shading agents in a vehicle containing cetyl stearyl alcohol, sodium lauryl sulphate and a carboxyl vinylic polymer.

(3) *Chemical Hair Reducing Agents :* Hair straightening preparations contain a chemical keratin-reducing agent as the 'relaxer'. It effects the softening and straightening of the hair. The active agents are frequently thioglycollates at a slightly lower concentration.

In hair straightening the hair is free and kept in shape while combing by virtue of the high viscosity of the product. As a result, a cream-like formulation is no longer a disadvantage but becomes highly desirable, so that the majority of the preparations are oil-in-water emulsions, gels or thickened liquids.

This type of straightener is sold in two containers, one holding the 'relaxer', the other the 'neutraliser'.

Relaxer

The relaxer is an ammoniacal thioglycollate at pH 9.0–9.5. Organic bases such as monoethanolamine and an alkali carbonate of an amino acid may partly replace ammonia. Cream products are based on glycerol, glycol stearates, or on cetyl and cetyl stearylic alcohols emulsified in water by polyoxyethylated (usually 20 to 25 ethylene oxide) cetyl or oleyl alcohols. Gels or viscous liquids are obtained by means of carboxyvinyl polymer or copolymers. In formula 3, imidazoline is included with the aim of speeding up straightening.

Formula 3

Constituents	per cent
Emulsion base :	
Demineralised water	to 100.0
Cetyl alcohol emulsified by	
oxyethylated cetyl alcohol	22.0
Demineralised water	30.0
Sodium carbonate glycinate	5.0
Ammonium thioglycollate or	
thiolactate (50% aqueous soln)	12.0
EDTA (disodium salt)	0.3
Sodium *p*-hydroxybenzo	
ate methyl ester	0.05
Monoethanolamine	2.0
Imidazoline	0.2
Perfume	0.2

Neutraliser

Neutralisers are the usual materials, either sodium perborate or hydrogen peroxide. It is important to remember that hair which has already been damaged during hot combing treatment or by the use of lye straighteners should not be subjected to a thioglycollate straightener after several weeks have elapsed. The thioglycollate cream is applied liberally to the hair, which is then combed rapidly until the hair no longer has a tendency to curl. When the hair is sufficiently straight the cream is rinsed off and the neutraliser is applied to give a permanent set. Care should be taken to ensure complete neutralisation so as to avoid hair damage. Again a cream-like composition is preferentially used to weight down the hair and improve the maintenance of a straightened shape. The formulation is given in formula 4 :

Formula 4

Constituents	*per cent*
Ammonium carbonate	4.5
Sodium bisulphite	2.2
Sodium lauryl sulphate	4.2
Tallow soap	5.6
Oleic acid	1.0
Water	to make 100.0

The process of straightening is delicate hence require great care to avoid dryness, degradation and breaking of hair.

The addition of natural polymers promotes the setting of the straightened shape. They include soluble proteins, corn or rice starch, fruit galactomannan. More recently, the application of vinyl polymers used for wave sets has been claimed. An oxidation, chemical straightening process by alkaline monopersulphates changes the shape of natural hair.

Chapter 37

Oral Hygiene Preparations

INTRODUCTION

The substantial progress has been made by oral products in therapeutic or biological activity. It has strong influence on oral care products. The number of oral care products are growing in the market. Stain free teeth, prevention of diseases and freshness is achieved by oral hygiene products. The current dental science has limited success with maintenance of a healthy dentition. It is the driving force in dental research on over the counter (OTC) oral care products.

The two major arenas of preventive self-applied dental formulations have been reduction of dental caries and prevention of gum diseases. Dental caries affected as much as 95% of people in industrialised populations. Periodontal diseases are still rampant in all populations, but this must be put into perspective.

There are many types and stages of periodontal diseases. If the disease is categorised in terms of disease which threatens tooth loss or the need for major surgical measures, the incidence is relatively low in industrialised societies. If all conditions which cause mild inflammation or irritation are considered to be periodontal disease, the incidence is considerably higher.

The incidence and seventy of caries, periodontal and other dental diseases are reduced on proper oral hygiene. The application of OTC and practice of dental products for healthy mouth is discussed in this product. Certainly, a clean dentition is a first, necessary step towards a healthy dentition.

THE HUMAN DENTITION AND ITS ENVIRONMENT

Teeth and Associated Oral Structures

Oral hygiene practices are designed to maintain the crown enamel, dentin (if exposed) and gingiva in healthy states. Maintenance of these structures is expected to result in the general health of the dentition, including other hard and soft tissues and the supporting bony structure.

Humans have a complement of 20 teeth, which grow from 6 months of age (lower central incisors) to 20–24 months (second molars). Adults normally have a complement of 32 teeth, erupting from 6–7 years of age (first molars, central incisors) to 17–21 years (third molars). The teeth are supported by a bony structure. The principle diseases causing loss of teeth are dental caries and periodontal diseases. Good oral hygiene and dietary habits are effective preventive measures. If left undisturbed, the carious lesion will penetrate to the pulp and expand to destroy the crown. Similarly unless tended to, periodontal disease will result in inflammation and destruction of the soft tissue surrounding the tooth, and eventual loss of the tooth.

Saliva and Crevicular Fluid

Saliva is a mixture of the secretions of the parotid, submaxillary, sublingual and accessory salivary glands, occasionally along with the crevicular fluid.

Saliva performs two functions : (i) it is involved in protection of the oral cavity through several factors related to the bacterial population of the oral cavity; and (ii) it is involved in the initial process of food digestion. It is suggested that saliva could be anti-cariogenic by: (a) increasing the rate of carbohydrate clearance; (b) reducing acid production by fermentative pathways; (c) buffering the drop in pH caused by acid production in plaque; (d) increasing the rate of glycolysis in plaque, thereby reducing the time the plaque is below a critical pH for attack on enamel; (e) increasing enamel resistance to demineralisation by acid; (f) increasing the degree of saturation of plaque fluid with respect to hydroxyapatite or fluorapatite; (g) promoting remineralisation of initial subsurface carious lesions; (h) increasing bacterial clearance from the oral cavity; and (i) increasing the protection possibly afforded by the dental pellicle that selectively deposits from saliva immediately following cleaning of the teeth. Calcium ions present in saliva appear to play an important role in microbiological processes involved in both protection and destruction of the tooth by oral bacteria.

Crevicular fluid is introduced into the saliva through the gingival crevice. It is not noticeable in a healthy dentition or the absence of teeth, and can be an indicator of the extent of periodontal disease; more is exuded as the condition worsens. Crevicular fluid exhibits enzyme activity that changes with condition. Knowledge of crevicular fluid is still far from complete.

ORAL ACCRETIONS AND CONDITIONS

Dental Pellicle

Dental pellicle, a film from saliva, covers the tooth after cleaning the tooth. The bacterial mass (dental plaque) adheres and stains of exposed to chromagenic materials to this film. Dental plaque can be removed without removing pellicle. Substantive tooth stains, however, are removed only with removal of pellicle. Toothbrushing alone is inadequate to remove pellicle and must be used in conjunction with an abrasive. Chemical removal is possible, but only at the risk of damaging the underlying enamel.

Dental pellicle consists of glycoproteins selectively absorbed from saliva. The amount of pellicle formed following brushing with an abrasive (pumice) increases for about 1.5 hour, then levels off. Pellicle may hinder penetration of substances from plaque into enamel.

Dental Plaque

Dental plaque is primarily a bacterial accumulation. It occurs supragingivally (above the gum line) and subgingivally (below the gum line).

A relation between dental plaque, periodontal disease and caries known. Freedom from dental disease is usually a consequence of the absence of plaque.

A mass of oral bacteria, which initially accumulates around the teeth at the cervical margin and then grows apically is dental plaque. Dental plaque contains leucocytes, desquamated cells, and other oral debris. It undergoes maturation, which involves changes in the microbial flora, possibly including calcification and loss of viability of some of the bacteria. It can be removed only by mechanical means on maturation. This definition acknowledges that plaque is a dynamic system and that quantitative measurements must be reported only in the context of that dynamism. A corollary would be that only a totally plaque-free tooth can confidently be expected to be disease-free. Inadequate control of plaque can lead to either or both of two serious disease conditions, periodontal disease and dental caries.

The relationship between dental plaque and caries is clear. Dental plaque generates acid from carbohydrate, and this acid attacks the tooth enamel and initiates formation of the carious lesion.

Dental Calculus (tartar)

There are two types of dental calculus, supragingival and subgingival. The former is explained as a hard, mineralised dental plaque and/or materia alba permeated with crystals of various calcium phosphates covered with a layer of 'vital, non-mineralised plaque. Subgingival calculus is defined as an organic structure of micro-organisms and intermicrobial matrix, containing major amounts of crystalline calcium phosphates, in different amount and distribution from those of supragingival calculus. A major impetus for the study of dental calculus was the belief that this hard, rough accumulation, (or the plaque concentrated on it), in direct contact with the gum tissue, was an irritant leading to periodontal disease.

The nucleation of calcium phosphate occurs at salivary gland where saliva secretes which results into supragingival. The calcium phosphate crystals in both supra- and subgingival calculus consist of octacalcium phosphate, brushite, whitlockite and hydroxyapatite. Subgingival calculus has a major amount of whitlockite and amount of brushite. The contents of the periodontal pocket are apparently able to fulfil the conditions necessary for nucleation and subsequent crystal growth.

Periodontal Diseases

The marketers of oral hygiene products has more research interest in periodental disease. On the one hand, dental caries has to all intents and purposes been conquered with OTC fluoride toothpastes and rinses used in conjunction with water fluoridation and other fluoride treatments. On the other hand, OTC products offering the same degree of protection against tooth-threatening periodontal diseases as fluoride does against caries are not yet available.

The bacterial plaque formed spreads at the gingival margin. It generate toxins that inflame the soft gum tissue, which becomes soft, puffy and red, and bleeds easily. A periodontal pocket (a pathologically deepened gingival sulcus) may form in susceptible individuals, without destruction of tissue. Then the inflammatory process is aggravated: the pockets become enlarged and form the receptacles for subgingival plaque bacteria, debris and exudates. The surrounding connective tissue degenerates. The pockets continue to deepen, and the teeth start to loosen by the bacterial mass. Subsequently, the gums recede from the tooth crowns. Finally, microbial, plaque toxins and other materials in the pockets intensify the inflammatory process, supporting bone tissue is destroyed, and the teeth loosen and eventually fall out or are extracted. The relation between dental plaque and periodontal disease are :

(1) Removal of supragingival plaque once every 24–48 hours is adequate to preserve gingival health.

(2) There is a close relationship between the subgingival location of the advancing bacterial front and the level of connective tissue attachment. The bacteria produce an inflammatory reaction apically/laterally to the junctional epithelium.

(3) Periodontal disease may proceed by bursts of increased inflammatory activity. Dental plaque, gingival redness and bleeding on probing, and initial pocket depth probing do not relate to the occurrence of probing attachment loss. The incidence of periodontal disease sites in untreated patients is low.

(4) The level of oral hygiene correlates with the severity of periodontal disease.

Dental Stain

The tooth stains accumulated by a significant number of individuals are extrinsic in nature and result from discolouration of pellicle and/or plaque. The extrinsic staining has three steps : (i) Chromagenic bacterial produce colour substances in plaque; (ii) Dietary factors (berries, tea, coffee) or smoking retains colour and (iii) Chemical transformation of pellicle component produce colour products tea, wine, tobacco, pellicle reactive colour, oral use of antimicrobial agent produce stains.

Oral Malodour

Bacteria from virtually all sources in the oral cavity have been implicated. 'Morning breath' has been attributed to the overnight putrefaction of food deposits, salivary deposits, desquamated epithelial cells, and other types of oral accumulation and debris. Malodour of other than bacterial origin (for example due to the intake of odouriferous foods or to diseases which cause the exhalation of odouriferous metabolic products) cannot be eliminated by oral hygiene measures.

The main cause of oral malodour is probably the putrefaction of sulphur-containing protein substrates, predominantly by gram-negative bacteria. The process is not limited to specific bacterial populations. Micro-organisms indiginous to dental plaque, saliva, gingival crevice and tongue can participate, and mixed bacterial populations are most capable of odour production. The fermentation process releases hydrogen sulphide and methyl mercaptan (which together account for 90% of the volatile compounds generated), and dimethyl disulphide. The mouth air also contains aliphatic and aromatic alcohols, and indole. Mouth air and putrefying saliva from persons with periodontal disease show higher than normal levels of sulphur compounds.

The vigorous oral hygiene controls mouth malodour and can be considerably reduced by tongue brushing, toothbrushing, food ingestion and use of an oral rinse that reduces oral bacterial populations.

ORAL-CARE PRODUCTS

Product Categories

Oral-care products can be divided broadly into mechanical devices and chemical formulations.

Tooth brush

The tooth brush is a mechanical device. It is the most important part of the oral hygiene regimen. Toothbrushing alone, without toothpaste can maintain a reasonable state of oral health. Toothpaste enhances its value as it removes stain, freshen the breath and accomplish specific functions. Other mechanical devices include dental floss, water irrigators and pics of various designs.

Toothpaste

Toothpaste is a preferred vehicle to apply special active substances to the dentition. It enhances the role of toothbrushing in oral care. Toothpastes are now being marketed with special active agents to combat dental caries, reduce the formation of dental calculus, combat gum disease and reduce tooth hypersensitivity. Wide acceptance of toothpaste has made it the vehicle of choice for applying therapeutic or special cosmetic additives to the dentition. Toothpaste brands are proliferating as new health benefit additives are discovered, and older concepts of toothpaste formulation are no longer applicable. Oral-care products have been categorised as cosmetic or therapeutic, depending on whether or not they contain a health benefit agent.

Oral Hygiene Preparations

Toothpaste and tooth powder essentially contain abrasives, moisturisers, bonding and foaming agents, surfactants, sweeteners, preservatives, pigments, aromas, and special active ingredients. Tooth powders differ from toothpastes in their lack of moisturising agents such as glycerine, sorbitol and water, which the abrasive content can reach 90% of their total volume.

Abrasives

The abrasives found in tooth care preparations are usually insoluble, inorganic materials present in toothpastes in concentrations of between 15 and 60%. An average size of 15 μm (micrometers) is preferred. In general, the abrasives chosen for tooth care preparations must accomplish a maximum amount of cleaning and a minimum amount of damaging erosion.

The most widely used abrasives are aluminum hydroxide ($Al(OH)_3$), calcium carbonate ($CaCO_3$), calcium hydrogen phosphate ($CaHPO_4$), calcium hydrogen phosphate-dihydrate ($CaHPO_4.2H_2O$), silicic acid ($SiO_2.H_2O$), sodium aluminum silicate with, for example, zeolite structure ($Na_{12}(AlO_2)_{12}(SiO_2)_{12}.27H_2O$), insoluble sodium metaphosphate ($NaPO_3)_n$ and hydroxylapatite ($Ca_5(PO_4)_3OH$).

Some toothpastes also contain water-soluble sodium hydrogen carbonate ($NaHCO_3$) as an abrasive, usually in combination with calcium carbonate. The commonly coarse-grained, soft sodium hydrogen carbonate is only a temporary abrasive, as is slowly dissolves during brushing in the saliva. Polymethyl methacrylate is also used as an insoluble, organic abrasive in an extremely fine-grained form.

Chapter 38

Dentifrices

INTRODUCTION

The primary function of a dentifrice is to remove adherent soiling matter from a hard surface with minimal damage to that surface. This is done by using mild abrasive powder containing surface active agent. The foam produced also has a psychological effect in making tooth cleaning more pleasurable.

This cleaning function must be achieved in a short time—two minutes—and at body temperature. The paste, however, is convenient in packaging and use. The liquids haivng humectant properties is used to prevent the toothpaste drying out at the tube nozzle. The gelling agent is added to maintain a high solids suspension in a stable viscous form.

Finally it is necessary to add flavours and possibly preservatives, colours and active ingredients, and all these components must be non-toxic and non-irritant under the conditions of use.

The total product should maintain its consistency over a temperature range from 0°C to 37°C (that is, it should have a relatively flat viscosity-temperature curve). It should be stable when stored without physical or chemical change over the same temperature range. Most large manufacturers have international sales and may have to take into account local conditions in many countries.

BASIC REQUIREMENTS OF A DENTIFRICE

The minimum requirements are listed below :

(1) An efficient toothbrush should remove food debris, plaque and stains and clean if used properly.

593

(2) It should leave the mouth with a fresh, clean sensation.

(3) Its cost should be reasonable for frequent use by all.

(4) It should be harmless, pleasant and convenient to use.

(5) It should be stable in storage during its commercial shelf-life.

(6) It should confirm to accepted standards in terms of its abrasivity to enamel and dentine.

(7) If prophylactic claims are made, these should be substantiated by properly conducted clinical trials.

TOOTHPASTES

Ingredients

A balanced formula can only be achieved by considering all the ingredients together since many of them may have a dual function or may interact with one another. Cost and availability as well as local laws, regulations and even local habits may cause formulations to vary from country to country.

Abrasives

The most commonly used abrasives are precipitated calcium carbonate and dicalcium phosphate dihydrate. Other materials include tricalcium phosphate, calcium pyrophosphate, insoluble sodium metaphosphate, various types of alumina, silica and silicates. Particles of plastics may also be used.

The abrasive used in a toothpaste must always be a compromise between the ability to clean the surface and the necessity to avoid damage to the tooth surface.

Calcium carbonate

Chalk, or as it is normally purchased, precipitated calcium carbonate, is available in a number of grades varying in crystalline form, particle size and surface area.

By varying the conditions of precipitation, precipitated chalks of different densities and crystal habit may be obtained. The two common crystal types are aragonite (orthorhombic) and calcite (rhombohedral) and particle sizes in the range 2–20 μm are normally used.

Chalk is an efficient cleaner but does not produce good lustre on teeth. Probably the best compromise is to use a small proportion of chalk with a larger proportion of one of the less abrasive phosphates.

Waterworks chalk has been employed in toothpastes but quality is not always uniform and the higher level of water-soluble calcium may cause problems in some formulations.

All chalks give an alkaline reaction to toothpastes and it may be necessary to protect aluminium tubes from corrosion by adding sodium silicate.

Calcium phosphates

The varieties of calcium phosphates used in dentifrices are :

(1) Dicalcium phosphate $CaHPO_4.2H_2O$.
(2) Dicalcium phosphate anhydrous $CaHPO_4$.
(3) Tricalcium phosphate $Ca_3(PO_4)_2$.
(4) Calcium pyrophosphate $Ca_2P_4O_7$.

Dicalcium phosphate dihydrate

Dicalcium phosphate dihydrate (DCP) is most commonly used in dentifrices. The pH value of a toothpaste made with DCP is normally in the range 6–8. The taste of DCP-based toothpastes is normally better than that of chalk-based products and the flavour stability is improved.

DCP is in the metastable state and reverts to the anhydrous form with consequent hardening of the paste. This change is accelerated by the presence of fluoride ions. The DCP normally supplied is stabilised to delay or prevent this change. Trimagnesium phosphate, tetrasodium pyrophosphate and calcium sodium pyrophosphate are common stabilisers.

Anhydrous Dicalcium Phosphate (DCP)

It is more abrasive than the dihydrate it should be used in smaller quantities. It is less soluble than the dihydrate which is an advantage in fluoride-containing pastes.

Tricalcium phosphate

Tricalcium phosphate (TCP) is not used to a large extent. It too is less soluble than DCP.

Calcium pyrophosphate

Calcium pyrophosphate (CPP) was originally developed as the abrasive of choice for products containing sodium or stannous fluoride. It is claimed that the low availability of soluble calcium ions contributes to the stability of the fluoride.

Insoluble sodium metaphosphate

Insoluble sodium metaphosphate (IMP) is a particularly useful abrasive for fluoride-containing dentifrices since it contains no calcium ions.

Other abrasives

There have been a number of developments in abrasive systems. These have arisen because of particular demands :

The fluoride stability is increased by the use of hydrated alumina and synthetic plastics. The transparent dentrifices are obtained using silica in the form of hydrated xerogel and sodium aluminum silicates. Zirconium silicates are used to impart luster to the teeth.

Detergents

Tooth cleaning is essentially a detergent process surface-active agent is incorporated. Soap was the earliest detergent used, but the obvious disadvantages (high pH, taste and incompatibility with other components) has led to its replacement by synthetic detergents.

The detergent must of course be tasteless, non-toxic and non-irritant to the oral mucosa. The foaming qualities are important since they have a significant influence on the subjective assessment of toothpaste performance.

Some surface-active agents may have intrinsic prophylactic or therapeutic properties.

Sodium lauryl sulphate

Sodium lauryl sulphate (SLS) is probably the most widely used detergent for oral products and satisfies almost all requirements. In this context 'lauryl' denotes that the alkyl radical R in $ROSO_3Na$ is derived from a narrow cut alcohol predominantly C_{12} but with some C_{14}. The original sources are palm kernel or coconut oil fatty acids. Various grades are available from manufacturers and particular attention should be paid to taste. This is influenced by the free alcohol content. A low content of inorganic salts is also desirable. Recrystallised grades are excellent in quality but are expensive.

Sodium N-lauryl sarcosinate

Sodium N-lauryl Sarcosinate $(R.CO.N(Me)CH_2COONa)$ has been widely used in accordance with a Colgate patent which claims prophylactic effects as a consequence of its anti-enzyme properties. It is particularly useful in oral products because of its high solubility.

Sodium ricinoleate and sodium sulphoricinoleate

Sodium ricinoleate (castor oil soap) is highly soluble hence, used in dentifrices but is vulnerable (as are all soaps) to the presence of calcium ions. Sodium sulphoricinoleate (Turkey Red Oil) has also been used.

Other detergents

Sodium lauryl ether sulphate $(R(OC_2H_4)_nOSO_3Na)$, coco monoglyceride sulphate, alkane sulphonates and alkyl polyether carboxylates have all been proposed as surface-active agents for dentifrices.

Humectants

The humectant is required to prevent a dentifrice from drying out. This is most likely to happen if the cap is left off the tube.

More recently glycerine has been partly or wholly replaced by 70 per cent sorbitol syrup which has similar properties and is usually less expensive. It is available in crystallising and non-crystallising grades.

Propylene glycol has also been used as a third component of the humectant system.

Gelling agents

Gelling or binding agent is necessary to maintain a high-solids suspension in a stable form. It also modifies the dispersibility, foam character and 'feel' in the mouth. Gelling agents used in toothpastes are hydrophilic colloids which disperse in aqueous media. These include natural gums such as Irish moss and gum tragacanth, synthetic cellulosic products and silica.

Gum tragacanth

This gum is extensively used at one time and satisfactory pastes can be made with it. The final product may be variable because of the natural origin of the gum.

Cellulose derivatives

They are used as gelling agents for toothpaste. They can be tailored to suit any requirement in terms of solubility, gel strength, etc. They are non-coloured, non-toxic and relatively tasteless.

Carboxymethyl cellulose

Carboxymethyl cellulose (CMC) or, more strictly, sodium carboxymethyl cellulose (SCMC) is prepared by the action of sodium chloracetate on alkali cellulose. The physical properties may be controlled by adjusting the degree of breakdown of the cellulose before substitution and by the degree of substitution.

SCMC gels are anionic and sensitive to pH values outside the range 5.5–9.5. They are reasonably stable in the presence of electrolytes and calcium ions and in general are suitable for most toothpaste formulations. SCMC is used gelling agent for toothpastes.

Anionic SCMC is not suitable for toothpastes containing cationic agents such as certain antibacterials. Hence, a non-ionic cellulose derivative must be used.

One minor disadvantage of SCMC is its possible breakdown if toothpaste is infected with the organism *Penicillium citrinum* but this is a rare occurrence.

Cellulose ethers

Cellulose ethers are generally the methyl or hydroxyethyl ethers of cellulose. These ethers can be tailor-made to give prescribed properties by varying the degree of substitution. They are of course non-ionic and are stable over wide pH ranges and unaffected by metal cations. They are most valuable in formulations that contain antibacterials which are cationic.

Methylcellulose is more soluble in cold water which is advantage. Methylcellulose is somewhat incompatible with glycerine and this can be a drawback.

Hydroxyethylcellulose (HEC) has the general characteristics of the cellulose ethers. Toothpastes made with HEC are slower to disperse than those made with SCMC so that foam and flavour are slower to develop. Nevertheless HEC is probably nearest to the ideal binding agent for toothpastes, particularly products that contain cations.

Miscellaneous gelling agents

Starch ethers have been used in toothpastes and are satisfactory. Two synthetic resins, an ethylene oxide polymer and a carboxy-vinyl polymer have also been suggested for use in toothpastes.

Flavours

The flavour of a toothpaste is one of the most important characteristics influencing consumer acceptance. They have usually been based on the oils of spearmint and peppermint. These are often fortified with a trace of menthol to give a cooling effect. They are also modified with clove (or eugenol) wintergreen (or methyl salicylate), eucalyptus, aniseed, etc.

The nature of the foam and the dispersibility of the paste also affect the flavour impact in the mouth.

Other ingredients

Preservatives

The preservatives are added to a toothpaste to protect it from micro-organisms. Formalin and sodium benzoate and *p*-hydroxy benzoates were commonly used for this purpose.

Sodium benzoate is not effective at neutral and higher pH values. Flavour components themselves have some anti-bacterial action.

In case the product manufactured is sterile, preservatives are not added.

Corrosion Inhibitors

Sodium silicate is often added to high pH chalk-based toothpastes to prevent attack on aluminium tubes. Some phosphates also reduce the corrosion risk with alumina-based toothpastes.

Increasing the glycerine level in the water phase will often reduce the risk of this type of corrosion.

Colours

Colours are sometimes added to toothpastes. These must be chosen with care as colour fading, particularly at the nozzle, is not uncommon.

Bleaches

To enhance the whitening effect of toothpastes and powders and to assist in the removal of stains, oxidising agents are often added to the product. These included sodium perborate, magnesium peroxide, hydrogen peroxide - urea compounds, stabilised hydrogen peroxide compounds, etc.

Formulation of Toothpastes

A standard toothpaste constitute, gelling agent 1%, humectant 10–30% and abrasive, 15–50 %.The various formulations of toothpaste are given in formulae 1 to 8.

Formula 1 (Peppermint flavour)

Constituent	*per cent*
Peppermint oil	ca. 60
Menthol from peppermint oil	10-15
Anis oil, fennel oil	10-15
Cucalyptus oil	5-10

Spice oils	2-5
Nuances (flower and citrus oils)	0.5-10

Formula 2

Constituents	*per cent*
Spearmint oil	ca. 30-40
Peppermint oil	20-30
Menthol from peppermint oil	10-15
Anis oil, fennel oil	10-15
Spice oils	0.5-3
Nuances	0.1-5

Formula 3 Anticaries Toothpaste (base : calcium carbonate, sodium monofluorophosphate)

Constituents	*per cent*
Sodium carboxymethyl cellulose	1.30
Sorbitol (70%)	20.00
Water (demineralised)	37.54
Sodium saccharin	0.15
Sodium monofluorophosphate	0.76
Aroma	1.00
PHB methyl ester	0.20
PHB propyl ester	0.05
Calcium carbonate	35.00
Silicon dioxide (highly dispersed)	2.50
Sodium lauryl sulphate	1.50
	100.00

Formula 4 Anticaries Toothpaste (base : calcium hydrogen phosphate, sodium monofluorophosphate)

Constituents	*per cent*
Sodium carboxymethyl cellulose	1.10
Glycerine (86%)	10.00
Sorbitol (70%)	15.00
Water (demineralised)	35.76
Sodium saccharin	0.05
Sodium cyclamate	0.10
Sodium monofluorophosphaate	1.14
Benzoic acid	0.20
PHB methyl ester	0.15
Aroma	1.00
Calcium hydrogen phosphate (dihydrate)	25.00
Calcium hydrogen phosphate (non-aqueous)	7.00

Silicon dioxide (highly dispersed)	2.00
Sodium lauryl sarcosinate	1.50
	100.00

Formula 5 Anticaries Toothpaste (base : silicate acid/sodium fluoride)

Constituents	per cent
Hydroxyethyl cellulose	1.40
Sorbitol (70%)	28.00
Xylitol	5.00
Water (demineralised)	43.37
Acesulfam K	43.37
Sodium fluoride	0.25
PHB methyl ester	0.24
PHB propyl ester	0.18
Aroma	0.06
Abrasive silicic acid	1.00
Thickening silicic acid	11.00
Sodium lauryl sulphate	2.00
Fatty acid tauride	0.50
	100.00

Formula 6 Gel Toothpaste (base : silicic acid/sodium fluoride)

Constituents	per cent
Sodium carboxymethyl cellulose	0.40
Sorbitol (70%)	72.00
Polyethylene glycol 1500	3.00
Water (demineralised)	4.64
Sodium saccharin	0.07
Sodium fluoride	0.24
PHB ethyl ester	0.15
Aroma	1.10
Abrasive silicic acid	11.00
Thickening silicic acid	6.00
Sodium lauryl sulphate	1.40
	100.00

Formula 7 (base : calcium carbonate/carbamide)
Toothpaste for Gum Protection–I

Constituents	per cent
Methyl cellulose	1.10
Sorbitol (70%)	15.00
Water	31.00

Sodium saccharin	0.05
Carbamide	5.00
Sodium benzoate	0.20
α-bisabolol	1.10
Sage extract	0.50
Aroma	1.10
PHB methyl ester	0.15
Calcium carbonate	45.00
Soap (medical)	0.80
	100.00

Formula 8 (base : silicic acid/pyridyl carbinol/tartaric acid)
Toothpaste for Gum Protection–II

Constituents	*per cent*
Sodium carboxymethyl cellulose	1.00
Sorbitol (70%)	50.00
Water	25.72
Sodium saccharin	0.10
Sodium benzoate	0.15
Tartaric acid	0.12
β-pyridylcarbinol tartrate	0.15
Allantoin	0.30
Aroma	1.05
Silicic acid, xerogel	7.00
Silicic acid, aerogel	13.00
Red dye (C.I. 16255)	0.01
Sodium lauryl sulphate	1.40
	100.00

The addition of fluorides (sodium or stannous) or sodium monofluorophosphate to a toothpaste presents problems. For example, if free fluoride ions are present in a calcium carbonate formula they will quickly be precipitated as calcium fluoride and the cariostatic activity will be lost. In these cases the abrasive must be chosen with care to prevent or reduce this effect.

Insoluble sodium metaphosphate (IMP) and special grades of calcium pyrophosphate (CPP) are usually used with compounds releasing fluoride ions (for example SnF_2, NaF). Sodium monofluorophosphate is less of a problem since the ion is FPO_3^{2-} and not F^-. In this case DCP and even precipitated calcium carbonate may be used. Great care must be used in all fluoride formulations to ensure that the fluoride activity remains at a high level throughout the life of the product.

The level of fluoride ingredient used has conventionally been such that there is 1000 ppm of fluorine in the final product. This corresponds to 0.2 per cent of NaF, 0.4 per cent of SnF_2 and 0.76 per cent of Na_2FPO_3.

Manufacture of toothpastes

The processes involved in toothpaste manufacture are : the hydration of the gelling agent and the dispersion of the abrasive in the gel. The hydration of the gel is done by adding the solid gelling agent and water to the glycerine with vigorous agitation. It is not necessary to heat the mixture if CMC is used, but heating to 60°C is usual with Viscarin-type gelling agents. Over-stirring of CMC gels results in an irreversible diminution of viscosity and should be avoided. Toothpaste manufacturing process is outlined in fig. 38.1.

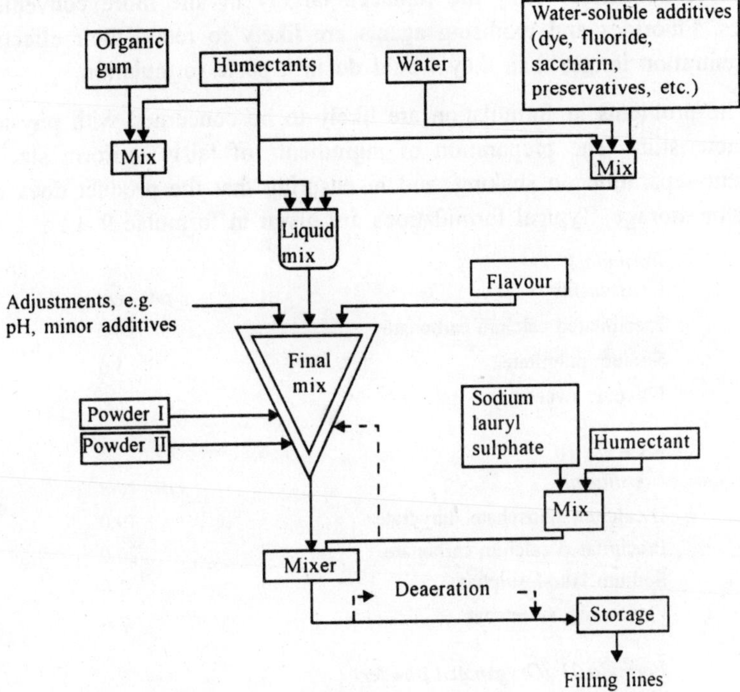

Fig 38.1. Schematic diagram of toothpaste manufacture.

Gel hydration can be continuous by means of an ejector (supplied by Hercules) in which the gel powder is introduced gradually into a stream of cold water which is then forced through a nozzle. The vigorous agitation produced gives a smooth uniform gel.

The heavy-duty mixing vesels like, Petzholdt, Fryma and Unimix are used for powder addition. The product is deaerated by mixing under vacuum. It is usual practice to add the active ingredient (if present) late in the mixing cycle and to add the surface-active agent and the flavour last of all. This is done to avoid excessive foaming and to reduce loss of flavour during evacuation.

The degree to which de-aeration is complete can be checked by density measurement. For a general formula such as that described above, a density of 1.55–1.60 would be expected. .

TOOTHPOWDERS

Toothpowders are the original, the simplest and the cheapest compounded forms of dentifrice. They are replaced largely by the more convenient pastes. Fluorides and oxidising agents are likely to retain their effective concentration longer than they would do in a paste formulation.

The problems in formulation are likely to be concerned with physical characteristics. The preparation of ingredients of fairly uniform size to prevent separation on shaking, and in ensuring that the product does not cake on storage. Typical formulations are given in formulae 9–12 :

Formula 9
Constituents	*per cent*
Precipitated calcium carbonate	95.0
Sodium palmitate	5.0
Flavour, sweetener	*q.s.*

Formula 10
Constituents	*per cent*
Dicalcium phosphate dihydrate	79.0
Precipitated calcium carbonate	20.0
Sodium lauryl sulphate	1.0
Flavouring, sweetener	*q.s.*

Formula 11 (Oxygenated powder)
Constituents	*per cent*
Precipitated calcium carbonate	96.0
Sodium lauryl sulphate	2.0
Magnesium peroxide	2.0
Flavouring, sweetener	*q.s.*

Formula 12 (Fluoride powder)
Constituents

	per cent
Dicalcium phosphate dihydrate	75.0
Precipitated calcium carbonate	23.0
Sodium lauryl sulphate	1.0
Sodium monofluorophosphate	0.8
Flavouring, sweetener	*q.s.*

Manufacture of Toothpowders

The manufacture of powders is very simple. The sweeteners, abrasive powder flavour, together with a little alcohol are mixed with the rest of the powders in a conventional powder mixer.

SOLID DENTIFRICE

Solid dentifrice is essentially a soap in which the abrasive powder is mixed. The proportion of soap varies from 10 to 30 per cent depending on the glycerine content of the finished product and the hardness desired. A higher proportion of flavour is added than in paste dentifrices. The product is usually coloured. Solid dentifrices, like toothpowders, have largely been replaced by toothpastes. A typical formula would be :

Formula 13
Constituents

	per cent
Dental soap	18.0
Precipitated calcium carbonate	79.0
Glycerine	3.0
Colour, flavour, sweetener	*q.s.*

A plastic mass is obtained by milling soap, abrasive material with glycerin and water. Then plodded by adding colour and flavour and extruded in a conventional soap plodder, cut into billets and stamped.

The abrasive nature of the product demands specially fabricated plodders and cutters.

PERFORMANCE TESTS

In recent years performance of dentrifices has significantly improved. The ready availability of human & animal teeth has made easy to carry out experiment on teeth *in situ* in the mouth to the performance of oral products. Chemists, physicists and dentists have all contributed to the performance of oral products.

Abrasive Action

The nature and quantity of abrasive influences the cleaning properties of dentifrice. The design of toothbrush and detergent play significant part. During cleaning, food debris, plaque, acquired pellicle, stains and calculus should be removed from the tooth surface, if possible without damage to the underlying enamel. Dentifrice abrasives are a compromise between the desire for perfect cleaning and the desire to avoid enamel wear.

The most obvious methods of measuring abrasion should be by weight loss, but this requires excessive abrasion and would lead to wear far in excess of that met with in a real life situation. The method for the measurement of the abrasive quality of toothpastes include the shadowgraph method, the surface profile method, interference microscopy replication techniques radio-tracer method.

In fact almost all conventional toothpastes fall comfortably within this standard; abrasivity ratings of 50–100 against dentine and 50–120 against enamel would probably cover most commercial toothpastes. There is no evidence to show that any conventional dentifrice, *when properly used*, has caused excessive wear of enamel or dentine. Cervical erosion, that is, erosion at the neck of tooth following gum recession, which occurs in older subjects, is due to bad brushing technique. The amount of material lost by abrasion with a relatively abrasive paste has been calculated as 1.2×10^{-8} g per brush stroke from enamel and 98×10^{-8} g per brush stroke from dentine.

It is possible to assess the relative wear rates of different particle sizes of different abrasives by radio-tracer technique. Within the fairly narrow limits of conventional toothpaste formulations there is an almost linear relationship between: (i) particle size and wear rate and (ii) percentage concentration of abrasive and wear rate.

Hardness, crystallinity of particles and particle shape all play some part in determining wear rate. It is now possible to pre-set the abrasivity of a toothpaste within fairly narrow limits. Probably the best compromise would be to have a relatively high proportion of a large soft particles (e.g., DCP 5–10 μm in size) and a small proportion of small hard particles (e.g., zirconium silicate or silica 1 μm in size). The large soft particles should remove most of the adherent soiling matter from teeth and the small particles should give some degree of polish without visible scratching.

DENTURE CLEANSERS

Denture cleansers are either in powder or tablet form or as liquids. They comprise essentially an oxidising agent, an electrolyte and an alkali.

The oxidising agent used is normally sodium perborate or sodium percarbonate, though hypochlorites, trichlorisocyanuric acid and its salts and persulphates.

Sodium percarbonate is more soluble in water than sodium perborate, but is not quiet so stable, though in the solid form its stability is adequate. Sodium perborate is available in two forms :

(1) *Sodium perborate tetrahydrate :* usually written as $NaBO_3.4H_2O$ but more properly $NaBO_2.H_2O_2.3H_2O$, has an active oxygen content of 10.38 per cent and is sold on the basis of 10 per cent active oxygen.

(2) *Sodium perborate monohydrate :* usually written as $NaBO_3.H_2O$ but more properly $NaBO_2.H_2O_2$, has an active oxygen content of 16 per cent and is sold on a basis of 15 per cent active oxygen.

The purpose of a denture cleanser is to loosen debris, which consists of saliva and food particles, to remove stains and to sterilise the denture.

Solid products are normally dissolved in water to form a solution in which the denture is immersed. Bubbles of oxygen formed helps mechanically to loosen food debris which is itself partly solubilised by the alkali present. Electrolytes have some solubilising action or mucous deposits. The combined effect is to loosen the debris so that, after a suitable period of soaking, it is easily brushed away.

In addition the denture is sterilised and stains are removed. Regular use of denture cleansers will also prevent the build-up of calculus on the denture surface. The amount of sodium perborate or percarbonate used is usually in the range 20–50 per cent.

The electrolyte used is invariably sodium chloride and the alkali is most commonly anhydrous trisodium phosphate, though sodium carbonate or bicarbonate and other alkalies may be used. Typical formulations are given in formulae 14 and 15 :

Formula 14

Constituents	*per cent*
Sodium perborate	40.0
Sodium chloride	30.0
Trisodium phosphate	30.0
Flavour, colour	q.s.

Formula 15
Constituents

	per cent
Sodium percarbonate	40.0
Sodium chloride	40.0
Sodium carbonate	20.0

Tables can also be prepared, either from conventional mixtures as in formulae 14 and 15, or as, for instance, in formula 16 :

Formula 16
Constituents

	per cent
Sodium percarbonate	88.0
Sodium chloride	10.0
Sodium silicate and/or other binders	2.0
Flavour, colour	q.s.

Liquid products are normally dilute solutions of sodium hypochlorite with sodium chloride. Such solutions of course lose some of their available chlorine over time.

The promotion of effervescence generated by carbomate acid mixture or copper salt helps in the break-up and dissolution of tablets and promotes the concept of activity. In the latter case great care should be exercised in checking the stability of the finished product. The inclusion of proteolytic and amylolytic enzymes in denture cleansers has been proposed. It is vital to keep moisture out of the product and to protect it from atmospheric humidity. The powder products must not cake on storage and solid products should dissolve quickly to give a clear solution.

The quality testing is performed on packed products. A performance of product under use should be monitored.

Chapter 39

Mouthwashes

INTRODUCTION

For medication of the mouth, gums or teeths the application in any form is referred as mouthwash. These mouthwashes are antibacterial. They are expected to give healthier, fresher mouth and good breath odour.

The use of antibiotics, for example, might destroy normal bacteria and thus permit the growth of undesirable organisms such as *Candida albicans*. It is possible to reduce and maintain the bacterial population due to adsorption of antibacterials. The Buccal Ephithelial test of Vinson and Bennet can serve as a standard technique. The total effect exerted is a combination of three factors: (1) the mechanical effect of rinsing food debris from the mouth; (2) the effect of the antibacterial agent on the oral flora; and (3) the effect of the flavour present. Since many flavouring substances have an antibacterial effect, there may be synergism between (2) and (3).

Mouthwashes used for general oral hygiene should nevertheless be rigorously tested for non-toxic and irritant properties. Prophylactic and therapeutic claims would require even more stringent clinical tests.

Choice of Antibacterial Agent

The antibacterial agents usually employed in mouthwashes include phenols, thymol, salol, tannic acid, chlorinated thymols, hexachlorophene and quaternary ammonium compounds.

Chlorinated phenols

*Para*chlormetacresol and *para*chlormetaxylenol are both suitable for use in mouthwashes due to their antibacterial properties and their flavour. They are not very soluble in water but may be solubilised with terpineol (or other suitable solubiliser) and soap to give a 1 per cent solution of active material. Such solutions, for example liquid chloroxylenol, are used at 10–20 per cent dilution.

Soap-based mouthwashes

For cleaning and freshening mouth soap based mouthwashes without antibacterial properties used is given in formula 1.

Formula 1

Constituents	*per cent*
Powdered soap	2.0
Glycerine	15.0
Alcohol	20.0
Water	63.0
Flavour	*q.s.*

Non-irritant synthetic detergents for extra foaming standard antibacterials may be added above basic formula.

Thymol (Isopropyl metacresol)

Thymol is soluble in suitable alcohols. An aqueous solution with borax is diluted to between 5 and 20 per cent concentration (Formula 2)

Formula 2

Constituents	*per cent*
Thymol	0.03
Alcohol	3.00
Borax	2.00
Sodium bicarbonate	1.00
Glycerine	10.00
Flavour	*q.s.*
Water	to 100.00

Hydrogen peroxide

Hydrogen peroxide is an excellent non-toxic antibacterial agent for use in mouthwashes. It can be used for cleansing ulcers and abscesses in the mouth, etc. A solution of one part hydrogen peroxide (10 vol.) diluted with 8 parts of water is useful as a mouthwash, or twice the strength can be used for septic cavities. Because of its instability it is not normally used in proprietary mouthwashes.

The dissolution of sodium perborate water gives an alkaline solution of hydrogen peroxide. Dissolve 17 g of sodium perborate and 6 g of citric acid in 80 ml of water which gives 10 volume strength. It is further diluted to 1:8 for use. Such products should be used only sparingly and to combat specific conditions, since the citrate present could lead to decalcification of the teeth.

Hexachlorophene

Hexachlorophene is substantive to the mucous membrane and is an effective antibacterial agent. The suggested concentration is 0.02 per cent in a 25 per cent alcohol–water mix.

Quaternaries

The use of quaternaries is now well established in mouthwashes. These compounds combine antibacterial and substantive properties. Many of them are non-toxic and non-irritant at the concentrations normally used. They are effective against plaque e.g. Benzethonium chloride. Probably the most effective antibacterials are of the chlorohexidine type. Unfortunately chlorohexidine, in common with most cationics, can produce a brown stain on teeth with continued use which is easily removed with good toothbrushing. A product of this type has been presented in the form of a gel, which, though intended for brushing onto the teeth, is really a mouthwash rather than a toothpaste since it contains no abrasive.

Other mouthwash components

Tannic acid, alum and zinc salts are used in mouthwashes because of their astringent properties. Which have antibleeding effects on the gums. A dilute solution of sodium hypochlorite is also used.

Flavouring of Mouthwashes

To generate freshness of the mouth after use flavour is used in mouthwash. Peppermint, menthol, eugenol, etc. are used flavours. Chlorinated phenols always have a characteristic flavour which, though not unpleasant, is difficult to cover. Cationics normally are slightly bitter and must be covered. For selecting unique flavour small manufacturers can use services of flavour house.

Aerosol mouth fresheners

Aerosol mouth fresheners are recommended for freshening the breath after eating, drinking or smoking. They contain only flavouring agents, though antibacterials could be added. Aerosol mouth fresheners are an alternative to chewing gum than as mouthwashes.

Chapter 40

Perfuming Cosmetic Preparations

INTRODUCTION

The terms 'fragrance' and 'perfume' are synonymous and are used interchangeably throughout the perfumery and cosmetics industries. Fragrance (or perfume) is a blend of two or more materials characterised by having olfactive properties, and is incorporated into a preparation with the intention of imparting specific odourous characteristics. Rarely, the fragrance may consist of one raw material only.

FUNCTIONS OF PERFUME

To impart a pleasant and suitable odour to the product is prime function of the perfume. The creative and evaluation team will find the suitability of the perfume.

Most cosmetic products have a mild but detectable odour. Although this 'base' odour is usually inoffensive, it can detract from the overall aesthetic character of the product 'masked'. Hence it is effectively using suitable fragrance. Very pungent bases, such as permanent-wave lotions, are almost impossible to mask, even with the strongest of perfumes. The role of the perfume in these products is therefore to produce as pleasant an odour as possible. This is attained by harmonising the fragrance with the base odour.

Fragrances can be designed to give subliminal support to a particular product. For example, a fragrance may appear to make a shampoo wash the hair more cleanly. A perfumed night cream may seem more nutritional to the skin, or a perfumed body spray may appear to impart all-over freshness, all-day. In other words, the perfume may have the role of apparently enhancing product performance.

Some fragrance materials have antimicrobial activity. Both bacteriostatic and bacteriocidal properties are evident. Consequently, fragrance blends often take on the antimicrobial activity of some of the individual components. Fragrances that have spicy and herbal notes usually exhibit some antibacterial activity owing to the presence of phenols such as eugenol (from clove oil), and thymol and carvacrol (from thyme oil). Consequently, the fragrance may also have the role of assisting the preservative system. Similarly, the fragrance may assist a bacteriocide/bacteriostat in its effect on the skin's microflora (reducing body odour) or may be required to carry this attribute of this finished product on its own.

Making of Unwanted Smells

Raw materials and active ingredients used to produce cosmetic preparations have distinctively strong odours of their own. Masking these odours presents serious difficulties. His work is made more difficult by the distinctive scent note. It is difficult to perfume the product in the desired scent direction. But sometimes the perfumer succeeds in binding a foreign odour of the preparation so successfully that a harmonious fragrance results.

For this reason it is necessary to have intensive cooperation between the cosmetic chemist and the perfumer, beginning as early as possible.

Aesthetics

The fragrance industry, is today able to develop technically perfected, dermatologically/toxicologically safe and harmonious perfumes. But the desire for a perfume that will become a trendsetter is comparable with the eternal search for the philosophers stone. This is the case when the preparation succeeds because of other characteristics or features, such as packaging, superior product characteristics or excellent marketing.

The fragrance sensibilities of the target market and differences in taste worldover should be considered while selecting perfume oil.

The luxury perfume is much costlier than perfumes used for soaps and shampoos. Fragrancing costs for many preparations can account for up to 25% of manufacturing cost. It is clear that perfuming costs need to be which often presents a difficult assignment for the perfumer. It is possible to keep the price limit for a good perfume oil from being too low by product engineers, marketing department and perfumers. In return, one should avoid the use of potentially expensive pseudo-active ingredients.

PRODUCT GROUPS

Soaps

Currently, soaps contain approximately 0.8 to 2% perfume oil. The past thirty years have seen an increasing tendency towards stronger perfuming. All the currently available aroma chemicals can not be used in perfuming soaps because of dermatological and toxicological considerations. About 35% of all aroma chemicals cannot be relied upon by perfumers to maintain their stability in soaps.

A further limitation of the number of aroma chemicals available for the perfuming of soaps results from the relationship between price and performance.

Creams and Lotions

The amount of perfume oils added with creams and lotions is 0.2 to 0.8%. Difficulties often arise when the demand for unified perfuming is placed on a range of body care preparations that consists of different emulsion types, each with their own unique smell. It is nearly impossible to achieve a perfect unity of fragrance, and there is no reason why preparations that serve different purposes should all have exactly the same aroma. While aroma chemicals in normal creams are seldom exposed to conditions that influence their stability, it is conversely more likely that perfume oils will influence emulsions (causing discolouring, decomposition). Special consideration must be given to the physiological compatibility of the aroma chemicals combined in perfuming oils for creams as emulsions remaining on skin is harmful.

Aqueous Preparations

Bath preparations

Bath preparations with respect to perfuming are more stable. The pH value of bubble baths and shower gels exhibit stability problems. The colouring with indicational dyes of weak acidic bubble baths enable visually perceptible colour changes during use. Even hydrolysis of esters can sometimes occur.

The quality of the surfactants, which today exhibit little smell of their own, makes it possible to apply more subtle fragrancings. This is, however, often rendered impossible by the addition of "active ingredients" that have characteristic odours, causing problems for perfumers, especially in bubble baths and shower gels.

A planned colouring is essential during the creation of a perfume oil for bubble baths or shower gels. The colouring is affected due to larger quantity of aroma chemicals of nitro-musk type or phenolic hydroxyl groups. When the packaging of the end product is to be light-permeable, the preparation should in any case be tested for photostability. Currently, scent notes whose prototypes are found in fine perfumes of both the feminine and masculine varieties are becoming increasingly popular.

Liquid soaps

The above descriptions of the perfuming of bubble baths and shower gels are largely valid also with respect to the perfuming of liquid soaps. Difficulties can emerge when, as a result of the additions of soaps, these preparations become so alkaline that discolouring and decreases in stability occur. These difficulties can be avoided if the perfumer falls back on aroma chemicals that are known from practical experiments to remain stable and non-discolouring when combined with fine soaps. The perfume oil concentration in liquid soaps lies between 0.5 and 2%.

Shampoos and hair care preparations

The proportion of perfume oil varies with shampoos. Baby shampoos generally contain up to 0.5% of a 'cosmetic' perfume oil. Normal shampoos are perfumed with 0.3 to 0.6%. Shampoos with specific effects must be more strongly perfumed 0.5 to 1% because of their own more intensive smell.

The shower shampoos which have become so fashionable recently contain around 0.8 to 2% perfume oil. The pH has no effect on stability of incorporated perfume oils. But especially in concentrations with high levels of perfume oils the viscosity can be often be influenced and discolouring can occur through exposure to light.

In case of water based hair care the long lasting scents are used as they remain in the hair for longer periods of time. These kinds of preparations are therefore perfumed from 0.3 to 0.5%. Perfume oils are generally not so strongly fixed and exhibit a more discreet scent note than shampoos, which are supposed to transmit a scent experience during application.

Deodorants

The type of perfume used in deodorants varies greatly. The concentration of perfume oil in deodorants and perfume deodorant is between 0.3 to 0.8%. It has become fashionable for scent notes to use successful masculine and feminine couturier perfumes as models. Because they will be in immediate contact with the skin, furthermore in the presence of alcohol and often without protective fatty substances, one must be especially careful in putting together deodorant perfume oils to use only aroma chemicals free from dermatological and toxicological side-effects.

Antiperspirants

Generally, what is true for deodorant sprays is also true for antiperspirants, with the restriction that most of the active ingredients used in antiperspirants negatively affect the stability of many aroma chemicals.

Hair sprays and hair setting preparations

In the perfuming of hair sprays and hair setting preparations, perfume oils used can exert a softening effect on the setting ingredients. The perfume oil component is kept to the lowest possible level (between 0.1 to 0.2%). Film-forming raw materials are good perfume oil fixatives and bring about a long lasting scenting of the hair. This can lead to a clash with any other perfume used at the same time. For this reason, perfume oils for hair sprays should not be fixed too strongly and should be compatible with as many other fragrances as possible.

Powders, Lip Care, Feminine Hygiene, Sunscreening and Hair Waving Preparations

Powders

The development of appropriate perfume oils for powders demands much more experience than that needed for perfume oils intended for other cosmetic preparations. The proportion of perfume oil in powders depends on application. Baby body and facial powders contain 0.3, 0.2, 0.1 and 3% perfume oil respectively.

In the perfuming of powders, one needs to consider the possibility of discolouration, loss of stability, evaporation and skin irritations. For example, many aroma chemicals discolouring the powder. Similarly, due to the large surface area of the majority of powders raw materials, they become more quickly oxidised than in other mediums due to more exposer to oxygen. This leads to change in the odour, reaction products and irritate the skin much more than the pure unaltered aroma chemicals. It also causes a greater evaporation of the aroma chemicals.

Lip care prepartions

While manufacturing the perfume oils not only smell but possess a pleasant taste as well. Hence, fruit note aromas like strawberry, raspberry, or cassis are often used, alone or in conjunction with certain flower notes. The fact that some aroma chemicals tend to change the colour of the lipstick. Lipsticks are perfumed to ca. 1%.

Feminine hygiene preparations

The factors applicable to deodorants are true for feminine hygiene preparations, except that with these preparations even more consideration has to be given to the dermatological/toxicological compatibility of the aroma chemicals used. The proportion in sprays (which are without exception dry sprays) lies between 0.2 and 0.3%.

Sun screening preparations

For sun protection all perfume oils can be used. Proportions vary from 0.2 to 0.6%. It is absolutely necessary to test the perfume oil and the perfumed preparation for eventual phototoxic characteristics.

Hair waving preparations

To date a satisfactory perfume oil for cold waving preparations does not exist. It is not the instability of the aroma chemicals but rather the unique smell of cold waving preparations that makes an adequate perfuming impossible. The acceptable solution to conduct a long series of experiments and examine all available aroma chemicals for their ability to balance out mercaptal ammoniac smells. Appropriate perfume oils can be found from bibliography (for example in Appell's work). The perfumer must attempt to create perfume oils that can provide the best possible disguise or balance out the particular smells of these preparations. The proportion of perfume oil in these instances is generally around 1 to 2%. It must be also taken into account that the high water content of most waving preparations makes the incorporation of emulsifiers for a homogeneous distribution of the perfume oils necessary.

APPLICATION TECHNIQUES

Tasks

Consistency or stability, disguise of the preparation's own smell, discolouration, fastness, solubility of emulsifying ability should be considered for suitability. The problems related to application are extraordinarily complex. Small changes in the formulation of the preparation often have a significant influence on fragrance aptitude. It is frequently advantageous to carry out one's own tests and not to rely on descriptions in the literature regarding the aptitude of aroma chemicals. In recent years, the quality of the raw materials used in these preparations has improved considerably, and allows for the incorporation of many aroma chemicals that earlier would have been incompatible. For example eugenol. leads to discolourants although it remains colourless in soaps produced by modern methods.

Also of major importance is the fact that the addition of small amounts of antioxidants to perfume oils can significantly improve the stability of many aroma chemicals. The surest way to determine the aptitude of an aroma chemical is evaluation by tests.

Testing Methods

The sample without perfume must be stored to find out which constituent is responsible for changes in odour, for evaluation following tests are conducted.

Year test

The year test is the one which is conducted on one year storage at 27°C. The perfumed and the unperfumed preparation are compared with fresh samples according to the testing criteria.

Short storage test

For quick information on the stability of an aroma chemical, the year test is often complemented with an accelerated storage test, the so-called short storage test. The conditions are generally storage at 40°C in glass bottles for two to three weeks.

Light test

Conclusions. regarding photocompatibility are drawn through storage in sunlight (up to six weeks) or accelerated in a xenotest apparatus (up to one week). This is important, for example, in the case of preparations such as bubble baths and shampoos in light-permeable containers.

Preparation test

For newly developed preparations, an additional test in the original packaging must be carried out before it is placed on the market. This test lasts up to six months at temperatures up to 50°C. If necessary, the preparations are subjected to the light test.

Skin test

It is a test in which scent of the skin is tested after washing the soap or application of cream at fixed interval of time. This test predicts preparations ability to remain on the skins.

FRAGRANCE SELECTION

The selection of the optimal fragrance presents a major problem. The selection should not be made by one person alone and absolutely not by the perfumer, because he would naturally give preference to his own recommendations. It is advisable to decide on a fragrance via a panel of experts and a panel of consumers.

A panel of experts is a group of 6 to 15 people schooled in questions of fragrance, and whose sense of smell has been tested. They judge the fragrances both in the preparation and in application under practical conditions, for exmple, soap on the skin or bubble baths in water. When determining the 'acceptability of the fragrance' due consideration should be given to uses, packaging, colour and target market.

The perfuming proposals preferred by the experts are then tested in a 'sniff test' (of around 100 different households), using the original package. Afterwards 100 households participate in a home use test to examine the proposed fragrance under conditions of actual use. If necessary, additional market research tests are carried out, such as TESI, a test-simulation model which evaluates the test preparation in relation to a buying situation with market competitors. For further security, the preparation can be tested on a regional level before it is finally brought on to the market.

EVALUATION OF SAFETY OF AROMA CHEMICALS

Naturally, all aroma chemicals must be tested in terms of their dermatological/toxicological safety. There are skin-damaging phototoxic substances, like furocoumarin in bergamot oil. Therefore only furocoumarin-free bergamot oil can be used in Eau de Cologne, creams and especially sunscreening preparations.

Similarly, many completely harmless musk-aroma chemicals, are available. A tetraline-musk body (1,1,4,4-tetramethyl-6-ethyl-acetyltetraline) cannot be used.

Chapter 41

Herbal Cosmetics

INTRODUCTION

As we prepare to enter the new millennium, the new-line of thinking is "nature is the best alternative." Beauty and health are used synonymously. This has promoted beauticians, cosmetologists and cosmetic companies to provide the best of 'herbal range to consumers'. Nature is not only a cosmetologist, but an expert chemist! An entire range of cosmetic products exists in nature! And, legends like Cleopatra have popularised herbal beauty care.

The herbal cosmetic market today has a share of almost Rs. 200 crores out of an estimated Rs. 2000 crores cosmetic industry in the country. The total cosmetic market is growing at the rate of 20–25 per cent per annum. Out of this growth about 60 per cent is that of the herbal cosmetic segment. Experts believe that with the kind of awareness spreading about the herbal products, the industry will not only flourish but it will be boom time for the herbal cosmetic industry in the turn of the century!

Due to growing competition with the liberalisation of the economy, many international brands have entered or are ready to take the plunge in the herbal sector. For those who want something different and are ready to pay a price for it, brands like Oriflame and Avon are the best alternative. Even these brands are keen to tap the 'Indian taste'.

In India we have specific products for the Indian customers and we call our herbal range as 'naturals'. These naturals are made from natural fruit extracts which are enhanced for amplified benefits through advanced science formulations. Basically it is beauty invented by nature and perfected through technology.

As for the Indian cosmetic market, there has been a distinct shift of stress on the quality of the product not only for the consumers but also for the manufacturers. The brand loyalty for cosmetic products is unfortunately very low and the propensity to switch is very high. To survive the producer is now forced to think more from a long-term and has to add more value to its existing products to keep the customer satisfied and motivated to buy the product.

HEALING WITH HERBS

Beauty and health are, two sides of the same coin. If nature has the qualities of a cosmetologist, its therapeutic treasure is equally vast and abundant. Unfortunately, till recently, ayurveda was never considered a mainstream method of healing. The reason, of course, is that the emphasis has been more on symptomatic healing than curing the disease from its roots.

A time when herbal cosmetics are taking over the cosmetic market, it seems that herbal medicines still have to make a noticeable impact. And that can happen only if you are prepared to see the same doctor again when he prescribes Dashmolarishta intead of Vitamin C for you.

HERBAL OR PLANT MATERIALS

Some of the herbal ingredients which are being commonly used in cosmetics are given below.

Almond

Fixed oil obtained from almonds is emollient in nature. It has extensive applications in cosmetics. A white milk used in cleansers and astringents is obtained by soaking ground almonds in milk or water and then by milling. Ground almonds are used in facial and body scrubs.

Apricot

Apricots contain highly nourishing oil and rich in vitamin A. It is believed to improve condition and elasticity of skin. The pulp of both fresh and dry fruit is used in face masks. Finely ground shell and kernel of the stone is used in facial and body scrubs.

Avocado

Oil of Avocado is rich in vitamin A. Its uses are similar to apricot oil.

Azadirachta (Neem)

Neem leaves are antiseptic. Neem fruits yield an oil called neem oil. Nimbidin is the chief bitter principle of neem oil. Neem oil is used in soaps. Extract of leaves can be used in toothpastes and skin care products.

Cabbage

Cabbage is chiefly valuable for its high mineral and vitamin contents. It has wonderful cleansing and reducing properties. Its juice can be used in skin care products.

Camomile

Camomile flower contains a dye hence used with henna in herbal hair dye. Infusion of flowers can also be used for astringency in cleansers, conditioners and hair rinses.

Carrots

Carrots are rich source of vitamin A. These are good for irritated or sunburnt skin. The raw grated vegetables and juice make excellent cleanser. Carrot juice can be used in masks and tonics.

Comfrey

Decoction of root and infusion of comfrey leaves have antiseptic and healing properties. These can be used in creams and lotions.

Common Basil (Tulsi)

Flowers and leaves of tulsi yield an essential oil which is considered antiseptic. Juice of its leaves can be used in skin creams and lotions.

Cornflower

Infusion of flowers can be used in skin tonics and freshners.

Cucumber

Cucumber juice is used in cosmetics. Cucumber juice is soothing, healing and slightly astringent. It can be used in masks, cleansers and toners.

Dandelion

Infusion of dandelion flowers can be used in masks, cleansers. Sap from the stem can be used in bleaching creams.

E. Alba (Bhringaraja)

It is a source of black stain and is used in hair preparations.

Elderflower

Infusion of flowers are mildly bleaching and therefore can be used in skin cleansers, conditioners and creams. Decoction of leaves of elderflower is healing and can be used for sun-burnt or wind-burnt skin.

Eucalyptus

Decoction of leaves of eucalyptus is antiseptic and invigorating. Decoction can be used in skin care products.

Fennel

Infusion of chopped leaves is soothing and stimulating. Infusion can be used in masks, cleansers and toners.

Funegreek

Fenugreek (Methi) is a traditional, component of hair care products. It is used for its cleansing and softening activity. It promotes scalp health. It can be used in hair care, skin care products.

Garlic

Garlic juice is considered antiseptic. Its juice can be used in antiseptic and smoothing creams.

Geranium

Geranium yields an essential oil which is used in perfumery. Geranium leaves and flowers can be used for healing or rejuvenating. Alternatively, geranium oil can be added in healing and cleansing creams.

Henna

Henna leaves contain a dye. Henna leaves are used in herbal hair dyes. Henna can be used with camomile flowers for dyeing the hair. Leaves and flowers yield an essential oil.

Indian Gooseberry (Amla)

Fruits, bark and leaves of Indian Gooseberry are rich in tannins. Fruits are used in herbal hair dyes, particularly with henna. Dried fruits are also used for washing the hair.

Indian Nard (Jatamansi)

Indian nard yields an oil, called spikenard oil. Oil is considered to improve hair growth and is also believed to darken hair. Spikenard oil can be used in hair preparations.

Jasmine

Flowers of jasmine yield an essential oil, called jasmine oil. Jasmine oil is used in traditional hair oils.

Lady's Mantle

Juice of leaves of lady's mantle has healing and mild bleaching properties. Its juice can be added to tonics for oily skin or to the creams for dry skin.

Lavender

Flowers and leaves of lavender yield an essential oil called lavender oil. Infusion of leaves and flower is considered antiseptic. Infusion can be used in skin creams and toners.

Lemon

Peel of lemon yields an essential oil which is used in perfumery and soaps. Infusion of leaves of lemon has soothing and mild astringent properties which can be used in skin tonics and cleansers.

Lime

Infusion of flowers and young leaves can be used in healing and moisturising creams and lotions.

Marigold

Infusion and decoction of flowers and leaves can be used in skin creams and lotions. These can also be used for hair lightening rinses.

Marsh Mallow

Infusion of flower of marsh mallow and decoction of its roots have antiseptic properties. These can be used in skin creams and cleansers.

Nutmeg

Nutmeg butter or fat obtained from nutmegs is used as external stimulant in hair lotions. Nutmeg oil is also used in spicy perfumes in soaps.

Orange blossom

Its essential oil is soothing and can be used in creams to soothe dry skin.

Parsley

Crushed leaves or juice of parsley can be used in poultices. Juice or infusion of leaves can be used in skin creams, lotions or cleansers. Also has anti-dandruff and deodorising properties. It can therefore be used in hair preparations also.

Peach

The pulp of fresh peach is used in face masks, cleansers and moisturisers. Peach nut oil is fragrant and rich and is used in nourishing creams.

Rose

Rose petals yield an essential oil, called rose oil. Infusion of petals can be used in skin toners, cleansers moisturisers and colognes.

Rosemary

Flowers and leaves of rosemary have antiseptic and anti-dandruff properties. Infusions of flowers and leaves can be used in skin creams and lotions. Herbal oil made from leaves can be used for the hair.

Safflower

Flower heads are a source of red and yellow dye called safflower. This dye is used in rouges.

Sage

Infusion of flowers and leaves of sage is healing and mild astringent. The intrusion can be used in skin tonics and cleansers.

Sandal

Sandalwood oil is extracted from heart-wood of sandal tree. Sandalwood oil is used in perfumes.

S. Racemosa (Lodhra)

Decoction of lodhra is used to prevent bleeding of gums.

Thyme

Leaves of thyme make astringent infusion. Leaves also make strongly perfumed herbal hair oil. Thyme also has deodorising anti-dandruff properties.

Turmeric

Turmeric has been long used in poultices and scrubs in India. Its water extract has antiseptic properties. It can be used in skin creams.

Wheatgerm

It is rich in vitamins and minerals. It is extremely beneficial in face masks and body scrubs. Wheatgerm oil is rich in vitamin E which is considered good for skin.

Witch hazel

Steam distillate of bark is astringent and antiseptic and can be used in skin creams, lotions and toners.

HOW TO USE HERBS

In earlier times, herbs were used for both medicinal purposes as well as for beautification. They have been used in fresh form and dried form. These can be used by mashing and directly applying to the body with or without using other ingredients. In fact, in earlier times these were used this way. But now-a-days, their extracts, decoctions, infusions, tinctures, steam distillates etc., are used rather than herbs themselves. Whenever these are prepared, preservatives should be added to them as these are perishable.

Infusions

Infusions are basically strong teas of herb(s) and can be prepared either in china clay pots or stainless steel vessels. Aluminium vessels should not be used as these can taint the infusions. The process of preparation is as follows :

Pour freshly boiled water in the vessel containing herb (coarse powder). Cover and leave for at least three hours. Strain and filter. The quantity of herb and solvent (water) may vary depending on whether herb is fresh or dried. Usually, for 100gm of fresh herb or 50 gm of dried herb 575 ml of water can be used.

Decoctions

Decoctions are prepared by boiling the herb with water. The following is the process :

Place the herb (coarse powder) in stainless steel or enamel pan, add sufficient water and cover the pan. Gently boil and continue boiling on gentle heat till the water is reduced to one quarter. This may require boiling for about 2-3 hours. Strain and filter. The same proportions of herb and water can be used as recommended in case of infusions.

Extracts and Tinctures

Extracts are generally prepared with hydro-alcoholic solvents and tinctures are prepared with either alcohol or hydro-alcoholic solvent with high percentage of alcohol. The following method can be used for preparing extracts and tinctures :

Place in a macerator crushed herbs or flowers. Add alcohol or alcohol-water mixture to cover the herb. Close the macerator tightly with lid and leave overnight. Add more solvent and further macerate for 2–3 days. Sometimes maceration may be as long as a week. Strain the mixture firmly with a strainer and filter the extract/tincture. Depending upon the constituents this extract/tincture could be used or it could be used as solvent for maceration of fresh herb. The recommended proportions are 100 gm of fresh herb or flowers and 575 ml. Extracts/tinctures should be stored in tightly closed containers.

Flower Waters

Flower waters are made in the same way as infusions. The same proportions of herbs and water can be used. However, the difference in infusion and flower waters is that solvent is allowed to remain in contact with flowers overnight in case of flower waters. Flower waters can also be prepared by using flower essence (essential oil) and purified water.

Oil Soluble Extracts

Oil soluble extracts are prepared by extracting herbs with petroleum ether. The herb (coarse powder) is left in contact with water overnight. More water is added next day. This herb-water mixture is placed with oil in a vessel and the vessel is heated till all the water has been removed. The oil is allowed to cool and is then filtered. In this way oil soluble principle of herb go into oil.

Some herbal extracts are available commercially. Balsara Herbal Products (Pvt.) Ltd. are making herbal extracts under the trade name Folicon. Some of the extracts are oil extracts and some are aqueous-glycol extracts. A list of these extracts and the products in which these can be used is given below in table 41.1.

Table 41.1. Folicon herbal extracts (cosmetic grade).

Application	Type	Code	
AMLA	Aqueoust extract	C 3101	Hair Care
AMLA	Oil Extract	C 3101	Hair care
ALOE	Aqueous Extract	C 3102	Skin & Hair care
BHRINGRAJ/MAKA	Aqueous Extract	C 3103	Hair care
BHRINGRAJ/MAKA	Oil Extract	C 3203	Hair Care
BRAHMI	Aqueous Extract	C 3104	Hair Care
BRAHMI	Oil Extract	C 3204	Hair Care
HENNA/MEHENDI	Aqueous Extract	C 3105	Hair Care
HALDI/TURMERIC	Aqueous Extract	C 3106	Skin Care
HIBISCUS	Aqueous Extract	C 3107	Skin & Hair Care
KHUS	Aqueous Extract	C 3108	Skin & Hair Care
LEMON PEEL	Aqueous Extract	C 3109	Skin & Hair Care
METHI/FENUGREEK	Aqueous Extract	C 3110	Skin & Hair Care
NEEM	Aqueous Extract	C 3111	Skin & Hair Care
ROSE PETAL	Aqueous Extract	C 3112	Skin & Hair Care
SHIKAKAI	Aqueoust Extract	C 3113	Hair Care
SANDALWOOD	Aqueous Extract	C 3114	Skin Care
TULSI	Aqueous Extract	C 3115	Skin Care

Recommended concentration in different preparations are shown in table 41.2.

Table 41.2. Folicon herbal extracts cosmetic products guidelines (use levels).

FOLICON Hair oil Herbal Extracts	Anti-Perspi--rants	Cleansers, Freshners, and Masks		Lotions, Creams	Hair-Condi--tioners	Medicated sun-screen	shampoo Creams
	%	%	%	%	%	%	%
Amla 5-10	–	5-10	5-10	5-10	–	1-5	–
Aloe	5-10	5-10	5-10	5-10	10-15	1-5	–
Bhringraj/Maka 5-10	–	–	–	–	–	–	–
Brahmi 5-10	–	–	–	–	10-15	1-5	–
Henna/Mehendi	–	–	–	5-10	–	1-5	–
Haldi/Turmeric	–	5-10	–	–	5-10	–	–
Hibiscus	–	–	–	–	–	1-5	–
Khus	5-10	1-5	–	–	–	–	1-5

(Cont'd)

FOLICON Hair oil Herbal Extracts	Anti-Perspi-rants %	Cleansers, Freshners, and Masks %	%	Lotions, Creams %	Hair-Condi-tioners %	Medicated shampoo sun-screen Creams %	%
Lemon peel	1-5	–	1-5	–	1-5	–	–
Methi/Fenugreek	–	–	–	1-5	–	1-5	–
Neem	5-10	–	–	–	5-10	–	–
Rose petal	–	1-5	–	–	–	–	–
Shikakai	–	–	–	5-10	–	–	–
Sandalwood	5	–	1-5	–	–	–	10
Tulsi	1-5	–	–	–	5-10	-m	5

HERBAL COSMETICS FOR THE SKIN

Creams and emulsions are all basically mixtures of water, oils, waxes and fragrances perfumery products in varying proportions. A slight change in the quantities in the preparations changes the character of the preparation and one should be careful when measuring the ingredients. When measuring the following rough equivalents should be kept in mind while formulating the product.

1 Teaspoon + 60 drops = 5 ml.

3 Teaspoons = 1 Tablespoon = 15 ml.

2 Tablespoons = 1/8 cup = 10 ml.

4 Tablespoons = 1/4 cup = 20 ml.

8 Tablespoons = 1/2 cup = 40 ml.

16 Tablespoons = 1 cup = 80 ml.

15 4 grains = gm.

28 3 gms = 1 oz.

16 oz = 1 lb. = 454 gms.

1 kg. = 2.202 lbs.

Equipment used for Herbal Preparations

(i) Enamel or pyrex bowls; (ii) One large frying pan; (iii) Measuring spoons; (iv) Measuring cups; (v) Electric heater or manual heater; (vi) A good strong sieve; (vii) A pipette or eye dropper; (viii) A wool of tissue paper for wiping jars, spoons, hands etc; (ix) Jars and Bottles and (x) Printed/plain labels.

Cleansing Creams

Cleansing creams are used to lift the dust. Also to remove stale make up. Soap can not remove waxy base of make-up thus, cleansing cream is used. The cleanser is generally applied with an upward movement and kept on the skin for about half a minute so as to dissolve the make up. Then it is removed with a tissue or damp cotton wool. The dirt and grime not only dogs the pores of the skin causing block heads but also coarsens the skin. A few formulations of cleansing creams are given below :

Formulation	
Bees Wax	1/2 Tablespoon
Emulsifying Wax (Petroleum)	1 Tablespoon
Baby oil	4 Tablespoons
Coconut oil	2 Tablespoons
Water	2 Tablespoons
Borax	1/4 Teaspoon
Witch hazel	1 Tablespoon
Perfume	3–5 drops

Note : Witch hazel is made from twigs of hamamelis or alderbush.

Process of manufacture

Melt the mixture of waxes and oil by transfering from enamel bowl to pan of boiling water. Prepare solution of borax in water in another bowl, add witch-hazel and heat it to dissolve. Mix solutions of both bowl in water with stirring cool to form, add perfume. It begins to thicken as it cools and beatens.

Rose Water Cleansing Cream

Formulation	
Beeswax or white Paraffin Wax	1 1/2 Tablespoons
Emulsifying Wax (Petrolatum)	1 Tablespoon
Mineral oil	4 Tablespoons
Rose water	6 Tablespoons
Borax	1/2 Teaspoon
Perfume	3–5 drops

Cucumber Cleansing Cream

Formulation	
Beeswax	3 Teaspoons
Coconut oil	4 Teaspoons

Mineral oil or Olive oil	4 Teaspoons
Cucumber Juice	4 Tablespoons
Glycerine	1 Teaspoon
Borax	1 Pinch
Green Colouring	1 Drop

Violet Cleansing Cream

Formulation

Lanolin	1/2 Tablespoon
Petroleum Jelly	1 Tablespoon
Mineral oil	4 Tablespoons
Water	10 Tablespoons
Violet extract	5 drops

Masks

Masks have been used for beauty treatment of the face. They are applied and kept on the face, left on the same for ten to twenty minutes and then washed off. They have nourishing, healing, cleansing and an astringent effect. These masks whip up the circulation and temporarily iron out wrinkles. They freshen the skin and leave it glowing. A few formulations of simple home made face masks are given as under:

Milk mask

This is one of the cheapest and simplest masks. Cotton wool is soaked in milk and rubbed on the face and left to dry for 10–15 minutes and then the face is washed.

Cleansing mask

Formulation

Powdered Brewers Yeast	1 Table spoon
Yogurt	1/2 Table spoon
Lemon Juice	1 Tea spoon
Orange Juice	1 Tea spoon
Carrot Juice	1 Tea spoon
Olive oil	1 Tea spoon

Procedure

Make a paste by mixing above ingredients, then apply and keep on face for 15 mins. If the skin is very dry more oil is added and discarded if the skin is oily. The brewer's yeast stimulates the flow of blood to the skin and gives the complexion a healthy glow. The yogurt cleanses the skin and the fruit juices provide vitamins and minerals.

Butter milk mask

A cup of butter milk is heated and three tablespoons of elderflower blossoms are put into it. The mixture is simmered on a very low heat for about 30 mins. and then allowed to cool. It helps in bleaching and cleansing large pores.

Meal mask

A paste of barley meal, pea flower and rose water is applied and left on the skin for several hours to make the skin soft and supple. One tablespoon each of finely ground oatmeal and milk along with a few drops of almond oil are allowed to form paste. The paste is applied and left on the skin for 15 mins. and then rinsed off.

Carrot and turnip mask

A paste is prepared by boiling carrot and turnip. It is applied for 10–15 minutes and then rinsed off with milk. This make leaves the skin feeling fresh and clean.

Oatmeal mask

Formulation of Oatmeal mask is given below :

Dried orange peel	2 Tablespoons
Oatmeal	2 Teaspoons
Cold cream	2 Tablespoons

The above ingredients are mixed into a paste and then applied generously all over the face and neck. After 10 to 15 minutes when it is dry the same is rubbed off using an upward circular motion. This leaves the skin smooth and shining.

Potato mask

Formulation of potato mask is given below :

Potato juice	1 Tablespoon
Fuller's earth	1 Tablespoon

The above ingredients are mixed to form a paste and applied on the face. Potatoes contain Vit-C and clear the skin of blemishes. Fuller's earth absorbs all the dirt and grease from the skin.

Paw paw mask

This is also known as Papaya Mask. Two tablespoons of ripe papaya flesh can be used as mask. Face is washed with water after applying mask.

Luxurious cleansing mask

Three strawberries are mashed into a paste and applied as a mask and then left on for 10 minutes and then washed off with rose water. Strawberries have slightly acidic properties and contain vitamin C, which cleanses the skin leaving it sparkingly clean. Another luxurious formulation consists of mashing and sieving the flesh of half a peach and then mixing it with a tablespoon of brandy.

Bath Salts

Two cups of ordinary washing soda are taken and two tablespoons of potassium carbonate is added to the same. A drop of an aromatic essential oil such as lavender or pine are added to the mixture and the mixture homogenised.

Bath Oils

Bath oil can be prepared by adding any vegetable oil such as almond, olive or sunflower to the bath water. Baby oil or mineral oil is not suitable as it cannot be absorbed by the skin and only remains on the skin surface. These basic oils can also be mixed with some of the herb oils or aromatic oil.

Bath Oil

Formulation	
Eggs	2 Nos.
Almond oil	1/4 Cup
Sunflower oil or Corn oil	1/4 Cup
Safflower oil or Olive oil	1/2 Cup
Honey	1 Teaspoon
Washing Detergent	2 Teaspoons
Vodka	1/4 Cup
Milk	1/4 Cup
Perfume	3-5 drops

Procedure

The thick emulsified mixture is prepared by adding eggs, honey, oil and detergent. Continue beating with addition of Vodka slowly, also add milk and perfume.

Massage Preparations

This is a special recipe for brides to be used daily at the time of bath for
ten days before marriage. The following ingredients are mixed to a paste
and rubbed vigorously. They cleanse and stimulate the skin leaving it soft.
supple and shiny.

Formulation

Dried, ground and sieved orange peel	1 Tablespoon
Dried, ground and sieved lemon peel	1 Tablespoon
Ground Almond	2 Tablespoons
Salt	1 pinch
Ground Thyme	1 Tablespoon
Ground all spice	1 pinch
Almond Oil	Sufficient to make a paste
Jasmine Oil (Perfume)	3–5 drops

Massage Paste

This consists of wheat flour or oatmeal mixed into a paste with double
cream. The paste has a deep cleansing effect on the skin and removes dead
cells and dirt.

Massage oil

The oils are heated and then the heated water is slowly added along with
a tablespoon of winter green or camphor oil to make a really invigorating
massage cream.

Formulation

Almond Oil	1/2 Cup
Castor Oil	1/2 Cup
Camphor Oil	1 Teaspoon

Body Lotions

Mix three tablespoons of rose water, one tablespoon of glycerine and two
tablespoons lime juice. This mixture is used after the bath if you suffer from
dry skin. The proportion of this lotion is to suit this type of skin. The lotion
no longer feels sticky and mixed into the skin easily. It has to be stored in
a refrigerator.

Blue Ratin Body Lotion

Formulation

Soap Flakes	3 Tablespoons
Glycerine	1 Teaspoon
Witch Hazel	1 Teaspoon
Olive Oil	4 Teaspoons
Water	5 Teaspoons
Blue Colouring	3–5 drops
Perfume	3–5 drops

Procedure

First soap flakes are dissolved in water with heating in a saucepan. Add glycerine, olive oil and witch hazel. Continue stirring until it cools, add blue colour and perfume. A lovely thin creamy blue mixture thus made is an ideal non-greasy, after-bath lotion which leaves the skin glowing like satin.

Lavender Body Lotion

Dissolve 1 teaspoon of borax in 1 cup of rose water and then add 2 tablespoons warmed olive oil with continuous beating. When all the oil has been added to the water and an emulsion has been formed the lavender water is added. The lavender water is made by infusing a handful of lavender flowers in 2 cups of boiling water. The above mixture is allowed to stand for an hour and then strained for use.

Almond Body Lotion

The formulation of almond body lotion is given below :

Formulation

Lanolin	3 Teaspoons
Almond Oil	1 Teaspoon
Petroleum Jelly	1 Teaspoon
Vegetable Lard	1 Teaspoon
Glycerine	2 Teaspoons
Water	10 Tablespoons
Soap Flakes	1 Teaspoon
Cornflower	1 Teaspoon
Perfume or Almond Essence	q.s.

Moisturising Creams

Almond oil moisturising cream

Melt two and one teaspoon of beeswax and emulsifying wax heating with water in a bowl and add five teaspoon almond oil. Now add hot water in melted mixture with stirring. Stop heating after sometime, cool and add lavender oil.

Herb Cream (Comfrey Cream)

A lanolin, white emulsifying wax and beeswax (1, 2 and 6 tablespoon) is melted. Add herb lotion or almond oil or infusion (6 tablespoon). This mixture is comfrey cream. Comfrey is a healing herb which is ideal for the sensitive skin or for skin having irritations or spots. A handful of comfrey is taken, liquidised and then strained. The comfrey residue need not be thrown away. A teaspoon honey may be added to tablespoon of this residue to form a paste and this can be used as a face mask.

Avocado Almond Cream

The formulation of avocado almond cream is given below :

Beeswax	3 Teaspoons
Emulsifying Wax (Petroleum Wax)	3 Teaspoons
Almond Oil	1/2 Cup
Avocado Oil	1/4 Cup
Rose Water	3 Tablespoons

Sun Flower Cream

The process to prepare sunflower cream is as follows. The waxes, oil and lanolin are heated over a water bath till they melt. The water, glycerine and borax are heated separately till the borax gets completely dissolved. While the melted oils and waxes are still on the heat, the water is added drop-by-drop with continuous beating with a wooden spoon till a whitish thick and creamy textured rich cream is formed. This is a good moisturising cream leaving the skin looking soft and moist.

Jasmine Moisturising Cream

The composition of jasmine moisturising cream is given below :

Emulsifying Wax (Petroleum)	2 Tablespoons
Almond or Sunflower Oil	2 Tablespoons
Lanolin	1 Teaspoon
Borax	1/2 Teaspoon
Witch Hazel	1 Teaspoon
Glycerine	1 1/2 Teaspoons

| Water | 8 Tablespoons |
| Jasmine Oil | 3–5 drops |

Nourishing Creams

Preparations of this type include all skin nutrients whether they are oil based or of the moisturising type. They may be light and immediately absorbed or heavy and sticky requiring to be managed into the skin. The nourishing creams comprise of Hormone creams and lotions, vitamin products, screen ampoules, antiwrinkle creams and lotions, and biologically active preparations. A few formulations of nourishing creams are given as under:

Cocoa Butter	2 Table spoons
Emulsifying Wax (Petroleum Wax)	2 Table spoons
Beeswax	1 Table spoons
Sesame Oil	4 Table spoons
Almond Oil	1 Table spoons

Elderflower Cream

The process to prepare elderflower cream include imxing of almond oil, elder flower blossoms and lanolin (3, 4 and 1 tablespoon) in a bowl over a water bath and allowed to simmer together for an hour. It is then strained and then warm water is added slowly with constant stirring.

Wheat Germ Cream

The composition of wheat germ cream is given below :

Beeswax	1 Teaspoon
Emulsifying Wax (Petrolatum)	1 Teaspoon
Lanolin	5 Tablespoons
Sesamc Oil	3 Tablespoons
Wheat Germ Oil	2 Tablespoons
Water	5 Tablespoons
Witch Hazel	1 Tablespoon
Borax	1/2 Teaspoon
Perfume	3–5 drops

Honey Cream

Honey is a very useful ingredient creams used for dry, coarse and sensitive skin. For a nourishing cream for sallow skin melt lanolin, honey, lectithin (3, 0.5, 1, tablespoon) in a bowl with heating. Add slowly warm water (4 tablespoon) with beating and cool it.

Vitamin Cream

The composition of vitamin cream is given below :

Beeswax	1 Tablespoon
Emulsifying Wax	1 Tablespoon
Lanolin	1 Tablespoon
Wheat Germ Oil (or vitamin-E Capsules 2 No)	3 Tablespoons
Carrot Oil (or Vit-A Capsules -2Nos)	3 Tablespoons
Borax	1/2 Teaspoon
Distilled Water	6 Tablespoons
Tincture of Benzoin	2 drops
Orange Flower Oil	3–5 drops

Mixed Vegetable Cream

Beeswax	2 Tablespoons
Emulsifying Wax (Petrolatum)	4 Tablespoons
Lanolin	3 Teaspoons
Almond Oil	4 Tablespoons
Sesame Oil	4 Tablespoons
Avocado Oil	2 Tablespoons
Safflower Oil	2 Tablespoons
Sun flower Oil	2 Tablespoons
Water	5 Tablespoons
Borax	1/2 Teaspoon
Amber Oil	3–5 drops

Water-in-Oil Emulsions

These are grooming preparations hence more popular. The emulsifiers like polyvalent soaps, beeswax and borax haven been used. The other emulsifiers for the water-in-oil emulsions which are used are calcium oleate, calcium stearate, zinc stearate, aluminium stearate. The composition of hair grooming preparation is given below, 1 to 5.

No. 1	*per cent*
Mineral oil	65.55
Beeswax white	2.5
Water	34.8
Borax	0.15
Perfume	q.s.

Formulations (Emulsified Hair Grooming Preparation)

Ingredients	per cent			
	2	3	4	5
Oleic acid	12	20	1.0	-
Beeswax	2	1	1.5	2.5
Mineral oil	45	-	41.7	48
Lanolin	2	0.5	-	-
Lime water	19	33.5	53.8	-
Saccharated lime water	20	5	-	-
Olive oil	-	40	-	-
Magnesium sulphate (25%)	-	-	2.0	-
Stearic acid	-	-	-	1.5
Absorption base	-	-	-	5.0
Petrolatum	-	-	-	12.5
Magnesium oleate	-	-	-	2.5
Water	-	-	-	28
Perfume and Preservative	q.s	q.s.	q.s.	q.s.

HERBAL COSMETICS FOR THE HAIR

The cosmetics having ayurvedic base with some medicinal compound yielding desired properties are implied as herbal cosmetics. The preparations like herbal hair oil, herbal shampoos are best examples. Hair Care products and cures comprising of herbal Hair Rinses, Hair tonics, Shampoos, Henna powders, are designed to restore beauty and health to the hair. They contain extracts known for their qualities of stimulating hair growth, improving hair texture and colour. The range of Market Products are :

Herbal Hair Conditioners

This is a powerful combination of invaluable herbal ingredients like Amla, Sandalwood, Brahmi, Lichens and other precious herbs. It makes the hair strong, healthy and lustrous. Prevents hair loss and acts as a scalp deep cleanser while maintaining the natural moisture balance. It provides body to dull lifeless hair.

Herbal Hair Oil

This is an effective combination of arnica, Henna, Shikakai and other herbal extracts, especially created to prevent hair loss and promote luxurious hair growth.

Herbal Henna

The preparation of herbal henna include combination of pure henna leaves and various herbal ingredients. They make the hair lustrous, manageable and silky. It also works as a powerful treatment for scalp disorders and gives a mild natural colour and sheen to the hair.

Herbal Hair Tonic

This preparation is for controlling hair loss/dandruff/greying hair. It stimulates hair growth and thickness. It is a specialised tonic treatment for revitalising the hair and generating a natural lustroics glow. It serves well for texturising rough brittle hair and split ends.

Herbal Henna Shampoo (for normal to oily hair)

This preparation is to cleanse the scalp and controls excessive secretion of scalp oils and imparts sheen to dull hair making it tangle free.

Herbal Amla Shampoo (for normal to dry hair)

This preparation is made from extracts of Amla, Dater and Arnica and Other rare herbs. This shampoo makes the hair lustrous, healthier and manageable and cleanses the scalp while retaining and stimulating its natural oil.

Hair Treatment Cream

This preparation is an effective, medicated antiseptic ointment for the treatment of dandruff scalp disorders, seborrhoea and sealy skin conditions and is used for the treatment of dandruff and falling hair.

Herbal Mint/Hair Gloss Conditioner

It is used as a hair sheen and can be used just before shampoo for providing texture and conditioning to the hair.

Herbal Hair Rinse

This is a combination of mint, brahmi and other herbal extracts and is an effective control for dandruff and scalp disorders and promoters hair growth.

Herbal Hair Oils (Medicated)

Medicated herbal oils are also used as scenting material for the hair. The base oil may be a mineral oil such as coconut oil, groundnut oil, castor oil, neem oil etc. To make an oil ayurvedic or a herbal product certain essential oils are added which provide certain specific characteristics of the individual used in the formulation.

Base formulation for herbal hair oil

Til Oil	91 ml
Mollha Kankola	0.3 ml
Whees Charrila	0.3 ml
Sailag Nagar	0.03 ml
Chander	0.04 ml
Long Dalchini	0.03 ml
Ratanjot	0.02 ml
Kapur Kachuri	0.05 ml
Satabar	0.02 ml
Tejpal	0.03 gms
Howber	0.03 gms
Balhere	0.06 gms
Amla	0.06 gms
Harral	0.03 gms
Chanpa	0.03 gms
Keora (keya)	0.06 gms
Gulabphul	0.03 gms
Alaichi (Big)	0.02 gms
Pudhina	0.4 gms
Perfume	0.3-15 gms
Quatah water	1.5 kg
Menthol	0.2

There is no unique formulation for herbal hair oil. The formulations of the manufacturers are trade secrets. However, the above formulation can be taken as a standard base and it can be changed according to requirements.

Herbal Shampoos

These products are more suitable in hard-water areas. *Quillaja saponania* is one of a whole class of Saponias widely distributed in nature. Commercially saponin is extracted with water and alcohol from the Quillaja bark or from the soap root. Preparations of saponin are generally intended to cleanse the scalp and to reduce scaliness. These are generally compounded with rosemary or celandine and in addition to its use in shampoos saponin has found some use in bubble baths. Some formulations for Herbal Shampoos are given as under :

	Ingredients	*per cent*
(i)	Quillaja bark, powdered	5.0
	Ammonium carbonate	1.0
	Borax	1.0
	Bay leaf oil	0.1
	Water	92.9
(ii)	Sodium carbonate anhydrous	19.4
	Powdered soap	79.0
	Saponin	1.6
	Perfume	q.s.
(iii)	Borax	20.0
	Sodium, sesquicarbonate	40.0
	Powdered soap	35.0
	Saponin	5.0
	Perfume	q.s.

Hair Dye

Henna is the only important vegetable dye. The presence of 2-hydroxy-1, 4-naphthaquinone gives henna a dyeing properties. It consists of the dried powdered leaves of *Lawsonia alba*, *Lawsonia spinosa* and *Lawsonia inemis*, which are removed from the plants prior to flowering. It is soluble in hot water and is, in acid solution, a substantive dye for Keratine. The powdered henna is mixed with hot water. Citric or adipic or other suitable acid is added to adjust pH 5-8. Then hair is dyed using this paste and kept for five to sixty minutes, then shampooed, rinsed and dried. The "henna Pack" is kept in place by means of a towel and is allowed to remain on the head for the required time varying from five to sixty minutes after which the hair is thoroughly shampooed, rinsed and dried.

Henna

The henna plant is shrub, *Lawsonia alba lans* of bears small, fragrant, greenish-white flowers, but as the only plant utilised for dye are the leaves and stems, the plants are frequently at back. For infusions, used as raises the leaves are usually left whole, but is the more common commercial form of henna, the leaves and stems are ground to a powder, about the colour of dark unstand and with characteristic earthy odour.

The principal advantage offered by henna as a modern hair dye is that it is harmless to the system and causes no irritations on the skin. A few applications impart a slight amount of colour to the entire hair shift and depending on the original colour of the hair, a whole range of reddish pure vegetable shades - tomato, beet, egg plant - can be produced. If it is plied often enough, therefore, hair of any and all original shades can be brought to the same characteristics orangered.

Applications of Henna

Despite the intense dark red shade of solutions of henna, the active dyeing principle, is only about 1% of it. This dye is used mostly in beauty shops, as a rinse or as a pack.

Henna Rinse

For slight change of colour, bring out highlights on dark hair, henna levels are steeped in boiling water until all the dye is extracted. This solution is then poured several times over freshly shampooed hair. Commercial preparations of a henna rinse should be acidified with 1 to 2% of some mild organic acid viz adipic acid, citric acid, tartaric.

Henna Pack

For a more decided, or more lasting, change in shade, powdered henna leaves are made into a paste with boiling water. This is applied to all the hair, which is allowed to remain covered and undisturbed until test on a single strand shows that the desired shade has been attained. The subsequent treatment, the colour can be kept uniform by applying the paste to only the new growth of hair.

To prepare henna powder, the fresh leaves are ground extremely fine. Fresh powder is dark brownish-green, as it ages bearing yellowish red. It is usually packed in waxed paper in cans or boxes that can be tightly sealed.

Henna mixtures

To modify the often unpleasing reddish shades that are the only ones possible with henna alone, this plant was long ago mixed with indigo. The dried and powdered leaves of indigo origin are either mixed with ground henna or applied alternatively with henna, to produce a good black. Shades from light brings to black can be produced by varying the proportions of henna and indigo.

Camomile

Camomile is another plants used for colouring the hair. Several varieties of related plants are found in U.K., Western Europe, and the U.S.A., of which only two - camomile and German or Hungarian camomile - are used to any extent. The essential oil distilled from camomile flowers in pale blue, due to the presence of some hydrocarbon azuline. Like henna, camomile can be used as a rinse, a pack, or a shampoo.

Camomile pack

Camomile pack is prepared by mixing 2 parts of its powder with Kaolin or 1 part of fullers. Boiler water is then added to make paste. The shade produced depends on the original shade of the hair and the time of contact, and the change in colour lasts considerably longer than the glints produced by a rinse. As with the henna pack the paste must be kept hot during the entire application, and the operation should be completed as quickly as possible.

Camomile shampoo

German camomile is usually prepared because it produces a slightly better colour and because it is considerably cheaper than Roman variety. Camomile shampoos are marketed as powders and as liquids. It constitutes: Camomile German Powdered (10%); Mild organic acid (5%); Sodium sulphate (85%); Oil of camomile (q.s. (for perfume)).

A liquid shampoo is most effective when freshly prepared. Hence it may be made in concentrated form. It constitutes : Camomile infusion (10%); Sodium laurge sulphate (30–40%); Water (50–60%).

Camomile-henna mixtures

Pack of this type may be prepared to very good effect to produce a fair range of colours on hair of different basic shades as shown below.

Formulations for Camomile-Henna Mixture

Ingredients	1	2	3
Camomile	75	50	25
Henna	25	50	75
Original Colour	Shade Produced		
Blond	Golden-blond		
	Red - Gold		
	Golden - Red		
Light Brown	Light auburn		
	Auburn		
	Bright auburn		
	Chest Nut		
Dark Brown	Reddish - Brown		
	Dark auburn		

Henna Reng Dye

It is possible to produce other shades instead auburn. The blue black shade shade is obtained by the mixture of powdered indigo and henna. It is known as henna renges.

Camomile Herbal Dye

Of the various species of camomile, only 'ic' *Anthemis nobilis*' (Roma Camomile) and *Matricaria camomile* (German camomile) have been used in cosmetics and are useful in tinting the hair. The active ingredient of these flowers in 1, 3, 4 -trihydronyflavore, known as apigenin. Either an aqueous extract or a paste of the ground flower heads are generally used. To lighten the hair a paste consisting of 2 parts camomile and 1-2 parts kaolin mixed to a thin cream with hot water is applied to the head for a period varying from 15 minutes to 60 minutes depending upon the shade required. While making a formulation it is essential that at least 5% camomile or its equivalent in extract, must be present to produce any effect at all and the azulene also present in camomile contributes to the brightening effect.

Fresh Soap-wort Shampoo

When water is added to soap-wort foam is produced due to saponina. All parts of the plant produce a gentle cleansing lather that does not sting the eyes or make the hair brittle. It can be combined with other herbs that suit particular types of hair and colour.

Formulation No. 1

Leafy Soapwort stems 6-8 long	10 Nos.
Water	500 ml.
Strong Herb or flower infusion	3 Tablespoon

For making an infusion use the following amounts. To each 1/2 litre of boiling bottled spring water add.

Dried Herbs :	14 gms for weak infusion
	28 gms for normal strength
	56 gms for strong infusion
Fresh Herbs :	1 1/2 handfuls for weak infusion
	3 handfuls for normal strength
	6 handfuls for strong infusion

Dried herbs of any of the flowers given below can be used.

– Camomile, Lime flower, yarrow.

- Rosemary, sage, comfrey, henna (powder).
- Marigold flowers.
- Lavender, peppermint, white, dead nettle.
- Marsh mallow, comfrey
- Southerriwood, stinging nettle.
- Elder flower, Uiyme, rosemary.

Procedure

The stems are cut into short lengths and put into an enamel saucepan. Then bruised lightly with a wooden spoon and water is added. The mixture is boiled and then covered and allowed to simmer for 15 minutes with continuous stirring, till cool. Then the contents are strained and bottled for future use. A flower infusion is prepared and 3 tablespoon of strong herb or flower is added. The shampoo is covered and then allowed to cool and then strained. The whole of the above liquid is used for one hair wash.

Dried Soap-Wort

The process of preparation of dried soap wort root shampoo include soaking of dried soap-wort in a boiling water for 12 hours or overnight. Transfer it to enamel and, boil and simmered for 15 minutes. The dried herbs or flowers are then added, stirred, covered and left to cool. The mixture is then stirred into a fug and used for one hair wash.

Nettle Rinse and Conditioner

Nettle is an excellent hair conditioner. It is an acidic plant and promotes a healthy gloss. It is prepared by first cutting nettles, then put it saucepan and add cold water and boil. Simmer for 15 minutes by covering it. Strain the mixture into a jug and allow it to cool.

Lotion for Unruly Hair

Formulation No. 2

– Herb or flower infusion of normal strength	– 8 Tablespoons
– Eau de cologne	– 1 Tablespoon
– Glycerine	– 1 Tablespoon

Horsetail Hair Rinse and Tonic

Herbal hair tonics and conditioners help restore the scalps natural acidity and strengthen the circulation giving a healthy shine to the hair. Horsetail contains saponins which create lather, flavour-glycosides which act on the scalps tiny blood vessels to stimulate the circulation and silica which provides body to the hair. The horsetail stems (long) are bruised with a spoon before adding the boiling water to make an infusion. They are then covered and left until lukewarm and then strained. After shampooing and rinsing, the infusion is poured on the hair and massaged on the scalp. Blot the mixture with towel and the hair is combed. Before dry hair cover the head with warm towel and wait for 10 mins.

Tonic Shampoo

Put 135 gms each of dried reetha, shikakai and amla into one litre water. Keep it for 24 hours. Boil and then allow to cool. Filter the mixture and use as a shampoo.

Herb Shampoo

Soak overnight in an iron pot 210 gms of dried olives and shikakai in cold water. Then boil for 10-15 minutes and strain liquidige the flesh that remains and use the paste to wash your hair. This shampoo leaves your hair soft and shining.

HERBAL INDUSTRY: PROBLEMS AND PROSPECTS

Besides our rich heritage of health care systems, it was only for last 100-150 years that these systems slowly but steadily have been taken up at the industrial level. Prior to this, these systems were confined to individual Vaidyas, Hakeems, traditional Dais, Bone setters, Massagers etc. But for last one century, various small, medium and big companies have been manufacturing traditional formulations at various levels of operations. The scale of the operations are many fold as compared to the individual hakeems or vaidyas. Though the traditional healers continue to play a very important role in the health care scenario of the country, especially in remote villages where amenities of the modern world are not available, industries play a key role in popularising these traditional formulations among the masses by making them available at every nook and corner of the country. Inspite of facing a number of threats, industry is playing a significant role in popularising the traditional heritage overseas as well. Though, TSM based industry is well established today, there are a number of problems surrounding it which need to be addressed. Some of these problems have been summarised below.

Lack of awareness

The herbal industry has to face enormous problems in educating the consumers about systems and the benefits of their products. There is a high cost involved in using the print or electronic media. TSM industry does not make enough money to invest adequately in educating the consumers using these effective media. Furthermore, out of the vast number of generic products of these systems, only a few can be advertised as an OTC products to educate the people.

In such a situation, the limited information provided by the manufacturers, chemists and vaidyas is the only thing a consumer can rely on. At times due to inadequate knowledge and lack of updating the knowledge of these personnel, there flows wrong or incomplete information about the products. These lacunae are hindering the growth of these systems.

Language

As per the existing laws, the herbal industry which manufactures the TSM based medicines have to comply with the Drugs & Cosmetics Act and Rules. The experts and policy makers who framed this law with their wisdom have appreciated the philosophy of these traditional systems to be naturally different from the philosophies and approaches of the modern systems. If we look into the history of these systems, we find that, these systems have been generated by our ancient Rishis, Vaidyas, Hakeems etc., after spending several centuries on research and validation. Whatever formulation of these systems we are having today have already been validated on the lines of Anuman (Presumption), Pratyaksha (Direct Experimentation) and Praman (Proof). The agony of the situation, however, is that the policy makers who decide on the matters related to TSM are not always the people who have been trained in these systems. They are the people who have been trained in the (so called) modern sciences like, pharmacy, chemistry, toxicology etc., whose language is quite different from what a TSM has. While these disciplines are equally important for understanding TSM also, they should not be made the exclusive basis for understanding, evaluating and approving TSM products. Apples and oranges cannot be compared using the same parameters, nor are they substitutes for each other.

A number of multinational pharmaceutical companies acquiring herbal companies world over is increased. India also did not remain untouched, where more than 20 Indian pharmaceutical companies have diversified to develop and market herbal products (Table 41.3).

Table 41.3. Indian companies diversified into herbal products.

A. Pharma Companies

Albert David Ltd.
Alembic Chemical Works Ltd.
Biological E Ltd.
Cadila Healthcare Pvt. Ltd.
Cadila Pharmaceuticals Ltd.
Cipla Ltd.
Concept Pharmaceuticals Ltd.
Dabur Pharmaceuticals Ltd.
Deepharma Ltd.
Dey's Medical Stores (Mfg.) Ltd.
East India Pharmaceutical Works Ltd.
Franco-Indian Pharmaceuticals Ltd.
Lupin Laboratories Ltd.
Lyka Labs. Ltd.
Merind Ltd.
Parke-Davis (India) Ltd.
Rallis India Ltd.
Ranbaxy Labs. Ltd.
TTK Pharma Pvt. Ltd.
Wockhardt Ltd.
Elder Pharma
Croslands
Themis (Phytomedica)
Dr. Reddy's Labs.

B. Non-pharma Companies

Bajaj
Velvette
Dalmia
Godrej
and many more

Environment

The third, but important, problem ISMs are facing today is the availability
of raw material and the concern for the environment by some
conservationists. Government of India has banned 29 plants and their value-
added products from being exported from the country on the grounds that
these herbs are getting endangered. This list is likely to increase in future.
Though industry itself is also in favour of safeguarding the environment, the
word 'conservation' should be viewed in its totality. Merely banning the use
of herbs is not a solution to the problem. One has to make sure that our
traditional heritage is conserved along with the environment. Banning the

herbs will slowly and slowly kill our rich legacy of traditional systems of healthcare which our forefathers have given to us and it becomes our moral duty to pass it down to generations to come. There is a need to work together for protecting our environment as well as our traditional heritages of healthcare.

Several herbal companies have put in efforts to generate quality control methodology, adopted improved technological methods to handle herbs on large scale without wastage and have generated efficacy data based on experimental and clinical studies. Industry has also adopted modern packaging technology to prolong the shelf-life of the products and to safeguard their potency while being transported or stored in the shops or homes. All these efforts were made with a view to compete with the nations like China in the world herbal market. But the conservationists' efforts and resultant government policies in an era of liberal economy has put these efforts of the industry into the doldrums.

Prospects

The ever-increasing fascination worldwide for the herbal products has opened new vistas in the herbal industry. There are tremendous market opportunities in the world today. Today the world herbal products market is worth Rs. 51,500 crores. Even if India tries to grab a minimum of 5% share of the world trade, it comes roughly Rs 2,600 crores as againsts a current total exports Rs. 308 crores (Table 41.4). Unfortunately the TSM based products in India are lagging far behind the modern pharma products in our country. While the turnover of former is merely Rs. 2,492 crores, the latter is doing business worth Rs. 11,228 crores. This shows our apathy towards the TSM based medicines inspite of the fact that they are in high demand across the globe.

Table 41.4. Indian health care market — 1995-96.

Category	Value (million US $)	Value (Rs. crore*)
Natural Products, medicinal plants based and others	639.00	2492.10
Western medicine based	2879.00	11228.10
Total annual production	3518.00	13720.20
Export of natural products	79.20	308.88

* US $ = Rs. 39.00

Countries other than India, have developed practical strategies to promote their traditional systems, while we have not gathered significant mass in these lines. More open, coordinated, progressive policies are to be framed to develop the herbal industry by co-operation of various ministries like Ministry of Commerce, Ministry of Industry, Ministry of Health, Ministry of Agriculture and Ministry of Environment and Forests, apart from representatives from industry. Strategies involving the development of information in today's language, evolving GMP standards with adoption of enhanced quality control parameters, and practical approach towards usage of natural resources would be required to encash the prospects of the herbals.

Chapter 42

Skin Lighteners or Bleaches

INTRODUCTION

Some knowledge of the physiology and biochemistry of skin colour and particularly of the pigmentation processes is essential for an appreciation of the mode of action of skin lightening agents.

SKIN LIGHTENING AGENTS AND FORMULATIONS

Opaque Covering Agents

In a wider sense, normal make-up with a higher than average content of white or pale pigments such as titanium dioxide, zinc oxide, talc, kaolin and bismuth pigments may be used.

Oxidising Agents

The use of hydrogen peroxide has limited use as a skin bleaches.

Mercury Compounds

Red mercuric oxide and mercurous chloride have been used, but mercuric chloride and ammoniated mercury give the most effective preparations.

When mercuric chloride used in lotion reacts with skin generates hydrochloric acid. This causes sloughing of the epidermis. Ammoniated mercury — $NH_2.HgCl$ — does not cause the same degree of exfoliation of the skin.

In a modern context, ammoniated mercury can scarcely be considered to be a suitable material for use in a cosmetic product which will be applied regularly, day-by-day, despite its existence in pharmacopoeial formulae.

Hydroquinone

The most favoured material for use in skin bleach preparations is hydroquinone which is to be effective at 1½–2 per cent in a vanishing cream in producing a temporary lightening of skin colour.

Hydroquinone has been 'found to be safe and effective (for skin lightening) by a Food and Drug Administration.

Hydroquinone is a white crystalline material, melting point 170–171°C, soluble in water (1:14), freely soluble in alcohol and ether. Solutions become brown on exposure to air as a result of oxidation and must be stabilised. The oxidation is very rapid if the solution is alkaline.

In formulation, the usual stabilisers are sodium sulphite or meta-bisulphite, often with the addition of a little ascorbic acid. Compositions should preferably be slightly acid (pH 4–6). Manufacturing equipment must be stainless steel or glass-lined to avoid discolouration, and contact with air minimised. Formulation is given in formulae 1 and 2 :

Formula 1 (Hydroquinone bleach cream)

Constituents	*per cent*
Lexemul	15.0
Lexate	6.0
Stearyl alcohol	3.0
Silicone oil	1.0
Propyl paraben	0.1
Water, deionised	67.8
Propylene glycol	5.0
Hydroquinone	2.0
Sodium metabisulphite	0.05
Ascorbic acid	0.05

Formula 2 (Hydroquinone lotion)

Constituents	*per cent*
Lexemul	5.0
Mineral oil light	2.5
Isopropyl myristate	2.5
Cetyl alcohol	1.0
Brij	0.9
Deionised water	79.85
Propylene glycol	5.0
Sodium lauryl sulphate 30%	0.9

Sodium metabisulphite	0.15
Ascorbic acid	0.1
Hydroquinone	2.0
Citric acid	0.1

The process involves addition of the oil phase to the water phase at 75°–80°C with high shear mixing. When a smooth emulsion has been formed, transfer to low shear mixing. Cool with stirring and at 55°–50°C add, in order, sodium metabisulphite, ascorbic acid, hydroquinone and citric acid perfume at 40°C.

The following formula for a medium viscosity lotion which will function as a sunscreen lotion as well as lightening and evening out skin tone is given in formula 3 :

Formula 3 (Oil-in-water toning lotion for black skin)

Constituents	per cent
Veegum	1.5
Titanium dioxide	0.2
Cetyl alcohol	1.8
Polychol	0.6
Glycerine	5.0
Uvinul	1.0
Sodium myristyl sulphate	1.0
Hydroquinone	3.0
Perfume	1.0
Sodium metabisulphite	0.2
Aluminium chlorhydroxyallantoinate	0.2
Water	to 100.0

A soft cream with a dry emollience may be formulated as given in formula 4 :

Formula 4 (Skin lightening cream)

Constituents		per cent
A	Arlacel 165	8.0
	Crodamol CSP	8.0
	Liquid base CB.3929	4.0
	Crodalan IPL	2.0
	Polychol 15	3.0
	Cetyl alcohol	1.0
	Nipasol M	0.05

B	Propylene glycol	5.0
	Kelzan	2.0
	Nipagin M	0.15
	Deionised water	64.4
C	Sodium metabisulphite	0.1
	Ascorbic acid	0.1
	Hydroquinone	2.0
	Perfume	0.2 or *q.s.*

The preparation include heating A to 72°C; disperse the Kelzan in B and heat to 70°C. Add A to B with stirring until emulsified. Cool with stirring, adding the sodium metabisulphite, ascorbic acid and then the hydroquinone at 55°–50°C. Add the perfume at about 40°C. Adjust the pH value if necessary to 5.0–5.5 with citric acid. The clinical experiences with the cream shown in formulae 5 and 6 :

Formula 5

Constituents	per cent
Tretinoin	0.1
Hydroquinone	5.0
Dexamethasone	0.1
Hydrophilic ointment	to 100.0

Formula 6

Constituents	per cent
Hydroquinone	5.0
Hydrocortisone I.P	1.0
Retinoic acid	0.1
Butylated hydroxytoluene	0.05
Polyethylene glycol 300	47.0
Methylated Spirit 74 o.p.	to 100.0

Ascorbic Acid and its Derivatives

Ascorbic acid is an active constituent of skin bleaching preparations. Scientists have reported on the successful use of creams containing 3 and 5 per cent ascorbyl oleate for bleaching freckles in human skin, resulting also in a marked improvement in the condition of the skin, which became softer and more supple.

By reviewing the chemistry of a number of esters of ascorbic acid and the stability of a few when incorporated into cosmetic creams. One in particular, the magnesium salt of ascorbic acid-3-phosphate, had shown to have a bleaching effect on pigmentation in human skin and to be useful in clinical practice.

The Natural Way

Limited skin lightening effects can be achieved with natural materials which have been used for centuries and which are useful for bleaching out fading suntan and freckles. Among the materials used are cucumber juice, lemon and lime juice, buttermilk, crushed strawberries and fresh horseradish.

Chapter 43

Quality Control of Cosmetics

INTRODUCTION

Quality is the totality of features and characteristics of a product or service that bear on its ability to satisfy a given need! Thus, quality is seen not as a single factor, but as a concept made up of several criteria.

Quality is the sum of all factors which contribute directly or indirectly to the safety, effectiveness and acceptability of the product!

Quality means conformance to requirements. Quality department personnel cannot be concerned with the various mystical aspects of quality — beauty, elegance, value for money, or even fitness for purpose (unless this means conformance to a defined purpose). The emphasis on conformance throws the responsibility on the marketing and technical departments to define the requirement and the product, such that a product specification is produced. This specification should define the environmental and reliability characteristics that the product must have, the workmanship standards that are applicable, and so on. The quality department are responsible for checking conformance to this specification.

The traditional view of quality requires modification, and that a new philosophy of quality is needed. The terms quality and productivity are not only to be related to the production of goods but also to process. For this reason, these concepts of quality must encompass every level of a business organisation — from administration, sales, to financing and marketing. The new concept of quality "built in" instead of 'tested for' emphasises prevention in all phases of the production cycle and beyond that in every work process within a business, rather than verification or post-improvement procedures. In this chapter, we discuss this new conception of quality.

657

The Analytical Laboratory — Its Significance for Raw Materials/ Active Ingredients and the Finished Preparation

All the raw materials and ingredients needed for cosmetic preparations are tested for usability in an analytical laboratory. The finished cosmetic preparations are also tested here.

The prerequisites for optimal chemical control (optimal is used here to signify the lowest expenditure that provides the highest degree of protection) are the most modern technical aids, such as IR, Uv, AAS spectrometers as well as gas chromatography and personal computers. Qualified and experienced technical personnel are also indispensable.

The Microbiological Laboratory — Assurance of Purity and Hygiene

It is the task of a microbiological laboratory to control mould or bacteria in cosmetic preparations. During production of a cosmetic preparation it must already be known which micro-organisms the preparation is exposed to; the germs discovered must be analysed and tested to ensure that a precisely fixed germ count is not exceeded.

Sterile working conditions are not required for cosmetic preparations, but hygiene regulations must be observed during the production process. The Good Manufacturing Practice (GMP) and Good Laboratory Practice (GLP) guidelines are generally regarded as the applicable regulations. Cleaning and disinfecting are just as essential as the provision of hygiene education to personnel. Only informed employees can guarantee that all regulations are adhered to. Disinfecting machines and machine components is normally cost- and time-intensive. The "clean in place" system is an example of innovative attempts in this area, which no business can ignore.

"Clean in place" refers to washing, cleaning and disinfecting machines by an installed system of pipes, without having to take apart machine components or pumps.

Raw Material Inspection

The inspection of cosmetic ingredients is based on specifications. It takes into consideration the specific requirements of the company as well as legal guidelines. Each detail of the specification must be coordinated with the producer of the raw materials. An inspection certificate which is given to the user with each delivery makes it easier for laboratory administrators to decide which properties of the specification need to be subjected to chemical analysis. It is not necessary to analyse every property of each delivery. The decision regarding how often to test is dependent upon the frequency of delivery, coordination of all specification properties between supplier and buyer, the inspection certificate accompanying each delivery, the history of the raw material in question, the degree of trust between buyer, supplier, and the complete documentation of all data. A sensory test of smell, colour and/or appearance, together with an IR spectrum is often sufficient to identify a delivery accurately. The decision as to whether a full analysis of all specific values needs to be carried out on every third, fourth or fifth delivery is made by the laboratory administration. It is also absolutely necessary to store all analytical data in a personal computer. An evaluation of the raw material or supplier can be made at any time with the help of this data.

Inspection of the Finished Product

Specific manufacturing instructions are essential for the production of the finished preparation. Instructions regarding temperature, mixing speed, instruments or the order in which ingredients are incorporated make it easier to keep to the specification. Here too an in-process control which registers the actual values (temperature, mixing speed, etc.), in relation to the specified values should be adopted.

A particular risk can be bound up with the weighing of the individual raw materials. However great the inspection expenditures are (for example, the additional presence of a specialist from the analytical laboratory while weighing in active ingredients), mistakes by personnel cannot be excluded. For example, too much of an active ingredient (such as a preservative) could cause fatal effects. For this reason active ingredients are often inspected again by the analytical laboratory as to their proportion in the final preparation. This doubled or tripled safeguarding is cost-intensive and hence not justifiable according to modern conceptions of quality. In the search for preventative test characteristics, the use of computerised weighing controls has therefore become prominent. A preparation manufactured under these conditions will generally meet the prescribed standards.

Inspection of Containers and Packaging Materials

It would not be possible to inspect each of bottles delivered for packaging purposes. Yet it must be guaranteed that they are all in flawless condition. Breakage and structural faults which would lead to leakage or glass splinters must be totally excluded. It is impossible to compromise on this issue. The problem can be resolved with the help of statistics. Probability statements regarding the specific quality of a delivery can be made through the use of spot tests and fault evaluation methods that practically eliminate error. The results of tests on a small representative sample of a total delivery are used to decide whether an entire delivery of an item should be accepted or rejected. In terms of cost-benefit analysis, this is the only practicable system for inspecting large lot quantities, from the viewpoint of both buyers and manufacturers of packaging materials.

Skip Lot Inspection

The cosmetics industry also makes use of packaging materials which are relatively simple to produce. They have therefore retained consistently good quality over the years, but each delivery must nevertheless be thoroughly inspected. If the history of the packaging component has remained unchanged, it is sufficient to inspect every third, fourth, or fifth delivery (hence the term skip lot). Precise knowledge of the packaging components and the supplier are however essential prerequisites for using this method of inspection. For deliveries which are not inspected, an identity check (appearance, potential transport damage, code number and so on) will suffice.

Source Inspection

Packaging materials that are produced in large quantities and ordered in partial lots also do not require comprehensive inspection of each delivery. This is especially important when using the just-in-time inventory method, where small quantities of the material are ordered with greater frequency. Source inspection requires that the total amount be statistically checked once by the producer. An identity check is then sufficient for all deliveries of partial lots. In this case too, cooperation between supplier and buyer is necessary, especially regarding the lot size per order.

Vendor Certification

There are suppliers whose preparations have proved themselves over the years to be of exceptional quality. Packaging components supplied by them do not need to be routinely checked with every delivery, since the risk remains small. A control can be made via skip lot or source inspection methods at any time.

Computer-Assisted Delivery Testing

An exact record and evaluation of all available data is an essential prerequisite of the above mentioned tests employed in less frequent quality inspection methods. A computerised system of measurement at the packaging delivery point makes this possible. Such a system is organised so that it contains all the basic data regarding a supplier and a component, and the data from every further test procedure, regardless of whether it is a skip lot or routine test method, is fed into its memory. The results of attributive tests (of appearance, colour and so on), are entered manually. The analysis of a number of these tests shows immediately whether there is a tendency towards deviation from the norm, or whether the values remain constant within the specification. The computer is "queried" every time goods arrive. It provides the decision on whether to test that particular delivery. This system has proved its reliability in practice and can therefore be recommended for use in every business.

In-Process Control

Prevention is the only possible means of recognising and correcting errors, and cutting costs. In-process controls are preventative technical-operational procedures which monitor the work process by means of measurement devices integrated into that process. An example of this is the computerised weighing control of ingredients. In this system, the quantity of an ingredient which is given in the formulation instructions is stored in memory and automatically compared with the actual amount weighed. Each deviation is communicated immediately by a signal or a print-out. The sum of the individual weight readings must correspond exactly to the quantity of the formulation. Projections of the actual amount of a basic ingredient used in relation to the target amount with reference to the amount in stock provides additional security.

Integrated Inspection Processes in the Manufacture of the Finished Product

Different companies offer memory-programmed management and inspection systems. Their purpose is to automate a mixing process with computer assistance according to predetermined specifications. This optimises effective time management of a facility and consistent reproducibility of the finished preparation, as well as the storage of formulation data. Control systems continually monitor temperature, mixing speed and process duration.

These systems also help in prevention, or advance inspection (in-process control procedures). The check weigher belongs to this group of systems, for example. This is a weighing system which is integrated into the production line and automatically checks the weight of each individual preparation. A computer evaluates the measurement results. In production facilities which cannot implement check weigher, weight inspection takes place during the production process by means of spot checks (computer assisted inspection of fill quantities). The evaluation and projection of individual results at a central point provides certainty that the proper fill quantities are always contained. Torque measurement, an ongoing check for container leakage, and inspection of the finished product with automatic removal of faulty items—those with missing labels or lot codes, empty containers, containers without lids, or insufficiently filled containers — are also possible in this way.

Finished Goods Control

The systematic prevention of error encompasses errors related to machines on the one hand and to human error on the other. In production of the finished product, this is especially important because it is the last opportunity to spot and rectify errors before the product is released. Specifications features for the packaged preparation are prescribed by the Cosmetics Decree, the Finished Package Decree, and the Guidelines for Aerosols, as well as by the specifications. The "in-process controls" are employed as a rule in this inspection. The manufacturer must establish the extent of inspection activities in such a way that every preparation which reaches the consumer fully meets all legal requirements.

Good Laboratory Practice (GLP) and Good Manufacturing Practice (GMP)

Good Laboratory Practice (GLP) and Good Manufacturing Practice (GMP) are used widely for the production of cosmetic preparations. GLP offers guidelines for laboratory investigations of chemical substances in terms of their characteristics and/or safety. The guidelines cover the direction of laboratory tests, the responsibilities of the laboratory administration, and quality assurance. Similarly, GMP provides fundamental advice concerning hygienic production practices in the cosmetics industry, including recommendations concerning raw materials and packaging materials, disinfection, microbiological quality control, personnel and preservation.

Environmental Protection

In accordance with regulations regarding waste products, tariffs are levied on companies in proportion to the burden they place on the waste disposal system.

The determination of disposable substances and of the chemical oxygen demand (COD) should be part of the routine of a laboratory concerned with environmental protection. Expenditure on such controls is justifiable and inspection can be restricted to 2 or 3 times per year once values have been standardised. The biological oxygen demand (BOD) must as a rule be determined by an independent body. The quality of the air should also be known, at least for the production facility. With the help of this kind of data, a company can have an influence on air and wastewater quality and thus keep these costs under control.

QUALITY PROMOTION

Improvement of quality is an attitude which every level of a company needs to adopt, because the goal of improvement demands that every work procedure, every service and product meets the norms of quality *every time*. The process of improving quality and productivity is an ongoing process which has in view nothing less than the approach and the engagement of all co-workers. This approach requires a philosophy of quality whose goal is to encourage employees in the direction of improving quality within the framework of the company's strategy.

The following plan is based on bringing the classic works proposal system together with an innovative quality team.

Works Proposal System

The works proposal system, which is usually regulated by agreements between management and the works committee, is an instrument of the employees for voluntarily improving current conditions.

A transitional stage from the works proposal system to quality team is offered by the group proposal. The regulation stipulates that "suggestions for improvement may be prepared and submitted by individual workers (individual proposals), or by a group of workers (group proposals)." Employees are thus motivated to develop new ideas and initiatives that lie outside of their immediate assignment areas.

It may seem strange to integrate the classic works proposal system into the general category of "quality assurance." However, a modern conception of quality must concern itself with systems of improvement that ultimately secure success through implementation of quality standards.

Quality Teams

One theory and productivity has proved itself especially useful in practice. The four principles are :

(1) *The definition of quality* : maintaining the norm.

(2) *The way to assure quality* : prevention.

(3) *Performance standard of quality* : zero error.

(4) *The yardstick of quality* : the price of deviating from the norm.

Improved quality and service, higher customer satisfaction, increased employee enthusiasm and pride in production, as well as lowered costs, are the primary goals. An essential prerequisite for successfully carrying out this concept of quality is a good understanding of the four principles and a positive relationship to them on the part of workers at all levels of the company. All employees should be educated, trained and made familiar with the problem areas by means of quality seminars.

The practical application of the theory is achieved through the help of specially formed quality teams whose task is to eliminate deviations from the norm. They establish, in accordance with given models, which norms should be applied within the process, how these norms can be maintained in the future by means of preventative measures, thereby eliminating deviations from the norm in the short-term (researching causes), and what measures must be undertaken to finally rectify existing problems. It is surprising how quickly solutions to problems are found when the given logically constructed procedural model is strictly adhered to.

Chapter 44

Testing of Cosmetics

INTRODUCTION

Cosmetic analysis is both specialised and diversified. The range of tests covered in this chapter reveals that it utilises the full gamut of modern analytical chemistry to identify and determine inorganic and organic components in extremely complex mixtures. However, as is often the case, techniques developed by cosmetic chemists to solve their particular problems also have great utilitarian value in other areas of analytical chemistry.

A cosmetic is considered to be adulterated if it contains a poisonous or deleterious substance which may cause it to be injurious to users under customary conditions of use or if it is contaminated in a manner which may render it harmful to health. It is misbranded if its labeling is false or misleading, if it does not bear the required labeling information, or if it is not truthfully packaged.

The composition of cosmetic raw materials themselves requires increasing scrutiny by the analytical chemist for the support of safety testing programmes for ingredients and for the establishment of precise standards and specifications for such ingredients.

This chapter provides both specific procedures and general methods for approaching the analysis of cosmetics, toiletries, in a logical and systematic manner. The analyst must supplement this information with laboratory experience gained in the actual analysis of products. The chemist should have a good understanding of spectrophotometry and chromatography. It is hoped that the experienced cosmetic chemist will find this text a source of new ideas from which to increase the range and scope of his investigations.

SPECTROSCOPY IN COSMETIC ANALYSIS

Ultraviolet and visible, fluorometric, infra-red, mass, and nuclear magnetic resonance spectroscopy plays vital role in cosmetic analysis.

Ultraviolet and Visible Spectroscopy

Ultraviolet (uv) spectrophotometry plays an important role in cosmetic analysis for identifying and determining compounds which, through electronic transitions, absorb radiation in the region from 200 to 380 nm. Primarily this technique applies to aromatic compounds, although uv spectra are useful for characterising some aliphatic substances such as compounds with conjugated double bonds, ketones, and mercaptans.

Constituents that absorb in the ultra-violet

The following partial listing of cosmetic raw materials which absorb in the ultra-violet :

Alkyl aryl polyether alcohols, Alkyl aryl sulphonates, Aminobenzoates, Aminophenols, Arylamines, Aryl sulphonamide-formaldehyde resin, Benzoyl peroxide, Camphor, Cinnamates, Coumarins, Dichlorophene, Furocoumarins, Hexachlorophene, *p*-Hydroxybenzoates, Nitroarylamines, Phenosulphonates, Phthalates, Quaternaries containing aryl groups, Salicylates, Silicons containing phenyl groups, Tricresyl phosphates.

For several cosmetic ingredients, e.g., sunscreens, preservatives, and arylamine oxidation hair dye intermediates, uv spectrophotometry is an invaluable analytical technique.

Visible spectrophotometry (colourimetry) is based on the same principles as uv spectrophotometry, except that the absorption range of interest extends from 380 to 800 nm. The substances to be determined appear coloured to the human eye. Many coloured inorganic and organic cosmetic materials have been investigated. For example, ethanol can be measured through its reaction with potassium dichromate; a coloured chromic complex is determined. The coloured Schiff's derivatives of citral and citronellal have been evaluated. Polyvinyl pyrrolidone and its copolymers form a coloured complex with a red dye which enables their determination down to 0.1 ppm. As a final example, the amount of tartar emetic in certain toilet waters can be measured through its coloured iodoantimonous acid complex.

Infrared Spectroscopy

The infrared (IR) region of the spectrum most frequently studied by the analytical chemist lies between 4000 and 660 cm^{-1} (2.5 and 15 μm).

Organic molecules selectively absorb energy at specific wavelengths within that region. A plot of the wavelengths where absorption occurs vs. their intensities provides an IR spectrum which can be useful for characterising or identifying a cosmetic ingredient.

Sample preparation

Analysis of a cosmetic may be initiated by obtaining an IR spectrum of the non-volatile portion of the sample. Almost all preparations containing water (which absorbs IR radiation), such as creams, lotions, antiperspirants, depilatories, shampoos, and hair dyes, can be dehydrated by heating them on a steam bath or at 105°C in an oven. The IR spectrum of the liquid or solid residue may be obtained by placing the sample between salt plates either as a neat film or in mulls made with mineral and/or halocarbon oil. A neat film is deposited on a plate by evaporation of a volatile solvent, such as alcohol or acetone, containing the dispersed or soluble residue. Solids with low melting points may be placed on the plate, heated to the liquid state, and resolidified by cooling into thin glassy films. Alternatively, about 2 mg of a solid sample can be mixed with approximately 200 mg of KBr powder and pressed into a pellet in a die at 20,000–50,000 psi pressure.

IR spectra of solutions and gases are generated with the aid of sealed cells of various pathlengths. Spectra of volatile solvents in nail lacquers are obtained by dissolving a sample aliquot in carbon disulphide, a solvent which is transparent to most IR radiation, and adding the solution to a selected liquid cell. Gases, such as propellant mixtures in aerosol products, are allowed to flow into a previously evacuated gas cell. (If the concentration of gas in the cell is too great, it can conveniently be reduced by using a pipet to which a rubber bulb is attached to withdraw some of the gas.)

Characteristic absorbances

A cosmetic sample IR spectrum should first be examined in the 2–9 μm region, where absorption peaks at specific wavelengths can be associated with given functional groups. Such a study can frequently provide a quick indication of the presence or absence of alcohols, amines, acids, esters, ethers, amides, soaps, nitriles, mercapto compounds, nitro compounds, alikyl sulphates, sulphonates, and sulphonamides. Frequently, closely related ingredients can be differentiated by comparison of the "fingerprint" region (9–15 μm) of the spectrum where secondary information may be provided. The relationship among the observed absorbances at the various wavelengths is studied for an estimation of the relative amount of each ingredient.

Structural elucidation

Infrared spectrophotometry is a helpful too for structural elucidation of previously uncharacterised compounds which may be encoutered in many of the complex raw materials used in cosmetic formulation. When IR spectra of pure substances differ to only a minor degree, the substances are very similar chemically. On occasion, the absence of a key absorption peak can provide important in formation which may be used to reduce the number of suspected compounds.

Fluorometry

When selected excitation wavelengths of uv light impinge upon certain types of chemical substances, characteristic light emissions occur at longer ultraviolet or visible wavelengths. Usually, an excitation wavelength which elicits a maximum emission response is chosen for a given compound in order to achieve the greatest sensitivity. The most appropriate emission wavelength is the one with the greatest intensity whose maximum is well removed from that of the excitation band. Fluorometric sensitivity is very good; it is common to measure microgram levels of material with ease.

The application of the fluorometry technique is greatly expanded by conversion of non-fluorescing compounds into fluorescent derivatives. For example, formaldehyde, a cosmetic preservative, is determined through quantitative conversion to its fluorescent lutidine derivative. Aluminum can be determined qualitatively and quantitatively in deodorants and antiperspirants through the fluorescent aluminum salt of 8-hydroxy-quinoline.

Mass Spectrometry

Submicrogram samples of pure compounds may be identified readily by mass spectrometry. A sample is introduced into a highly evacuated chamber (ca 10^{-10} atmospheres) and bombarded with a beam of electrons powerful enough to strip away one or more electrons from its molecular orbitals. These collisions produce a number of positively charged molecular fragments along with a frequently occurring positively charged species of the same molecular weight as the sample molecule. These species are separated electromagnetically by various instrumental techniques into discrete particle types represented by different relative mass-to-charge ratios. A plot of particle mass vs. relative abundance is obtained which represents the mass spectrum of the sample. Some samples may be identified by comparison with standard reference mass spectra but the structural elucidation of unknowns requires skilled interpretation. The positively charged molecular fragments form spectral peak patterns which are characteristic of certain structural features. The molecular weight of the unknown compound may be determined if its corresponding parent ion peak is present in the spectrum.

Mass spectral analysis has been used for the verification of suspected structures of derivatives of cosmetic compounds. Various alkanolamines in hair sprays must be acetylated prior to their determination by gas chromatography. Some contain several —OH groups as well as a —NH group. Mass spectral analysis determines whether or not the —NH group was acetylated and the number of —OH groups converted to acetates.

Nuclear Magnetic Resonance

Many spinning atomic nuclei, such as those of hydrogen (protons), boron, fluorine, and phosphorus, possess angular momentum and a magnetic moment. Because of the most widely studied of the nuclei are protons, the description of their nuclear magnetic resonance (NMR) behaviour is emphasised. When organic molecules containing protons are placed in a large magnetic field of known fixed strength and this field is swept with small known magnetic fields at right angles to the sample, certain of the applied frequencies (or small magnetic fields) interact with the magnetic moments of the protons, causing the protons to "flip" from their orientation with the large magnetic field to an orientation against it. Energy absorbed during this "flip" is considered to be "nuclear magnetic resonance" energy. The amount of absorbed energy required to cause a proton to flip depends upon its structural location in the molecule, because the electronic envelope surrounding a proton has a significant influence on its magnetic properties. An NMR plot of the radio frequency energy absorbed by a molecule vs. the changing magnetic field strength can yield valuable information about the various types and numbers of hydrogen atoms in the molecule. Thus nuclear magnetic resonance is an important tool for the structural elucidation of many complex molecules.

Nuclear magnetic resonance reference spectra are useful for verifying the identities of many compounds; however, it should be emphasised that no single spectral technique should be relied upon for proof of structure. Confirmation is best obtained by a number of mutually supportive instrumental and chemical approaches. As a case in point, the structure of β_1-cadinene (now known as α-cadinene), a sesquiterpene hydrocarbon isolated from olibanum oil, could not be verified by comparison with any available standard reference spectra (infrared or nuclear magnetic resonance). Through a process of elimination with a combination of chemical and spectral techniques, the number of possibilities was narrowed down to two alternative cadinene isomers, δ_1 and β_1 (now known as α).

The NMR spectrum provided confirmatory evidence that the unknown is the β_1 isomer, based on the number of olefinic protons.

CHROMATOGRAPHIC TECHNIQUES IN COSMETIC ANALYSIS

Adsorption Column Chromatography

Adsorption column chromatography is a useful technique for separating mixtures of organic compounds possessing varying degrees of polarities. In principle, the more polar the compound the more strongly it is held by the adsorbent. In applying this technique, a solution or slurry of the sample in a solvent of relatively low polarity is introduced into a chromatography column containing a porous bed of finely divided adsorbent such as silica, alumina, Florisil (magnesium silicate), or magnesia. The components of a mixture are separated through the successive use of eluting solvents of increasing polarity. Even though a practical knowledge of the relative polar strengths of various solvents or mixtures may be gained from experience, some information on this subject has been published and may be used by the analyst in establishing an experimental design for his intended purpose.

Column preparation, sample application and elution

The preferred method for preparing the column is to prepare a wet slurry of the adsorbent using the least polar solvent of the proposed solvent system. This slurry is then added in increments to the column containing some of this least polar solvent, and each increment is "packed" into the column with the aid of 5–6 psig of air pressure. To prevent channeling, the liquid level should not be allowed to fall below the top of the adsorbent bed within the column, either during its preparation or during use.

The sample is dissolved or dispersed in a small amount of the initial eluting solvent with gentle warming if necessary to increase the solubility of the sample. This sample solution is then added to the column and a constant flow rate is maintained by gravity or by the application of air pressure. The column is then eluted successively with solvents of increasing polarity. Uniform aliquots are collected and the chromatographic procedure is continued until all components of interest have been collected.

Qualitative and quantitative analysis

Components in the eluted aliquots may be determined by gravimetric or volumetric analysis or such instrumental techniques as infrared (IR), ultra-violet uv, nuclear magnetic resonance, and mass spectrometry. Other helpful techniques are thin layer chromatography (TLC), refractometry, fluorometry, and polarimetry.

Variables and efficiency

Reproducibility and efficiency of separation are dependent on certain variables, most of which can be controlled by the analyst. For example, the selection of a smaller mesh size adsorbent offers a correspondingly larger surface area for adsorption which favours improved separations. Other factors affecting accuracy, reproducibility, and efficiency of adsorption chromatographic procedures include proper selection of solvent polarity, viscosity, and compatibility of successive solvents; packing procedure; method of sample application; and flow rate. Although ambient temperature is satisfactory in most cases, the temperature of the column can affect the separation process.

Application to cosmetic analysis

The cosmetic analyst is frequently confronted with the problem of separating complex mixtures of cosmetic ingredients with polarities ranging from non-polar aliphatic hydrocarbons to such highly polar compounds as polyols. The selective elution of components of increasing polarities can be demonstrated by the analysis of a commercial grade of glyceryl monostearate. The order of elution is aliphatic hydrocarbons, triglycerides, fatty acids, diglycerides, monoglycerides, and finally glycerol.

Partition Column Chromatography

The partition column chromatography is generally a glass tube packed with a solid support which has been coated with a stationary or immobile phase, one of two immiscible, equilibrated liquids. The sample is introduced at the top of the column and the column is then developed by introduction of a moving or mobile phase, the other liquid of this immiscible pair of liquids. As the mobile phase moves down the column, the sample component of interest (solute) will move in the system at a rate which depends on its relative solubilities in the two phases. This is expressed as the partition coefficient and is defined as the concentration of the solute in the mobile phase divided by its concentration in the immobile phase under equilibrium conditions. This value is constant for any given solute in a specific system. Usually, development of the column is continued until all solutes of interest have been eluted from the column. These eluates are collected in uniform aliquots and reserved for analysis by chemical or instrumental techniques.

In conventional practice, the more polar liquid of the solvent pair is used as the immobile phase. This is referred to as normal partition chromatography. In reverse phase partition chromatography the less polar solvent is the immobile phase.

Partition chromatography is generally used for the separation of chemically similar compounds that have some solubility in both liquid phases and different partition coefficients. Adsorption chromatography is primarily useful for the separation of compounds into different chemical classes.

Solid support

The ideal solid support consists of small, uniform inert particles with high permeability and large surface areas. Of the materials commonly used as solid supports, diatomaceous earths such as Celite or Filter-Cel most closely fulfill these requirements. Other materials that have been used as solid supports include cellulose powder, potato starch, modified cellulose, polyamides, and silica gel.

Commercial grades of diatomaceous earth require sieving to the desired mesh range followed by washing with HCl, distilled water, and finally methanol or ethanol to remove impurities. After drying at 120°C, the support is suitable for preparing columns. Analytical grades of diatomaceous earth do not require further treatment. If the support is to be used for reverse phase partition chromatography, it must be silanised with a preparation such as dimethyldichlorosilane by exposing the support to vapours of the silanising reagent in a closed container or by treating the support with a toluene solution of the silanising reagent. If non-silanised support is used to prepare a reverse phase partition column, the polar mobile phase will wet the support, displace the immobile phase, and render the column useless.

Selection of the liquid phases

The first step in the design of an analytical partition chromatography columns is to determine the composition of the mobile and immobile liquid phases. The analyst first selects two partially immiscible solvents in which the component to be separated is soluble. Typically, this may involve a solvent pair such as hexane and methanol. Usually a third and possibly a fourth solvent is introduced into the system to modify the partitioning of the solute between the two phases. Chloroform, for example, may be introduced to increase solubility in the non-polar phase. Water can be used to increase immiscibility of the two phases as well as to change the solute solubility in the polar phase. Equal volumes of the two liquids selected as the mobile and immobile phases are added to a separatory funnel along with several milligrams of the compound to be determined. The system is then mixed thoroughly by shaking to establish equilibrium. After the liquid phases have separated, the amount of solute in each phase is determined. The partition coefficient should be approximately 0.01. If it is not, other solvents are introduced to adjust the partition coefficient to the desired value. The phase in which the solute has the highest concentration is always used as the immobile phase.

Preparation of column

After a suitable solvent system has been selected, the appropriate amount of each component of the solvent system is added to a separatory funnel and mixed thoroughly. The equilibrated phases are then separated and stored in stoppered flasks until used. Silanised support is used for reverse phase and non-silanised support for normal partition chromatography.

For analytical columns a glass chromatographic tube approximately 2 cm id is usually used. There are several methods for packing the column. For example, the support is coated by mixing with a calculated amount of immobile phase until uniform, and then is added to the column in increments, with tamping, until the column bed is the desired height. Alternatively, the uncoated support is packed into the empty tube and then washed with immobile phase. Mobile phase is then added and allowed to flow through the column until no immobile phase is observed in the eluate. Other methods of column packing are variations of the above techniques. After the column has been prepared, approximately 10 mg of the sample to be analysed is added to the column in one of the following ways :

(1) The sample is dissolved in the immobile phase, mixed with a small portion of the support and packed on the top of the prepared column.

(2) The sample is dissolved in a minimum amount of mobile phase and added to the column.

(3) When neither of the above methods is applicable, the sample is dissolved in a volatile solvent and mixed with a small amount of support. The solvent is then removed and the dried mixture is added to the column.

After the column has been prepared and the sample added, the column is developed with the mobile phase. During development, the level of mobile phase should not be allowed to fall below the level of the coated support. Ideally, the flow rate should be about 8 ml/hr/cm^2 obtained either by gravity or applied air pressure. Development is continued until the components of interest have been eluted. If the components are coloured or can be rendered visible under uv light, each resolved band is collected separately. If components are not visible, a series of uniform aliquots are collected for analysis to determine composition.

Application to cosmetic analysis

Many active ingredients in sunscreen preparations, such as *p*-hydroxybenzoates, cinnamates, and aminobenzoates, are successfully separated by partition chromatography and determined by uv spectrometry. Sixteen cosmetic preservatives have also been separated and determined in different cosmetic products.

Ion Exchange Chromatography

Organic or inorganic compounds containing ionic functional groups can be separated from other ionic or non-ionic compounds by ion exchange chromatography. The ion exchanger is normally a solid, finely divided, insoluble material containing ionic sites which are available for the attraction of oppositely charged ions in solution. Although naturally occurring materials such as the alkaline aluminosilicates (zeolites) have served in this capacity, synthetic organic polymers (resins) containing various functional groups are nearly always used for analytical purposes. The separation is usually carried out in a glass column containing the ion exchange resin selected to perform a specific separation. The sample is introduced onto the column in the appropriate solvent and the column is then usually washed with a solvent to elute all non-ionic compounds.

After sample introduction and removal of non-ionic components, the column is developed by displacement chromatography or elution chromatography. Displacement chromatography is carried out by forcing the sample component of interest (solute) from the resin by adding a solution (eluant) containing an excess of an ionic solute that is more strongly attracted to the ionic sites on the resin than is the sample solute. In elution chromatography, the eluant contains an ionic solute that is less strongly attracted to the ionic sites on the resin. The sample solute is equilibrated between the resin and the eluant. As the eluant flows, each solute moves down the column at a rate dependent upon pH and ionic concentration of the eluant and the ionic attraction between the solutes and ionic sites on the resin.

Column preparation, sample addition and elution

For most analytical purposes, a 1×20–25 cm chromatographic tube is filled with 10–15 ml of pretreated resin. In order to ensure column uniformity, the resin is added in increments as a thick slurry to a tube already one-third filled with liquid, which is gradually drained as the addition proceeds. Each successive addition should be made before the resin has completely settled. Ion exchange resins in acid or base form are packed in columns containing distilled water or a non-aqeuous solvent, while those in the "salt" form are added to the appropriate salt or buffer solution. After the column is packed, the resin in the column is washed with the same liquid used for packing until the pH of the effluent is the same as that of the liquid added at the top. The resin in the column should always have liquid above it.

After the level of liquid in the column is adjusted to just cover the top of the ion exchange resin, the sample is added as a solution in the appropriate solvent, i.e., distilled water or organic solvent if the resin is in acid or base form, or buffer if the resin is in "salt" form. The sample solution is added slowly to the top of the column and allowed to percolate into the resin bed at the same flow rate that will be used during the chromatography step. The walls should be rinsed several times with the selected solvent and the rinse solution allowed to settle into the resin bed. Uniform aliquots are collected continuously until development is complete.

The amount of sample added to the column depends on the total ion exchange capacity of the resin and the type of chromatography used. The total "sample load" in the elution procedure should not exceed 1% of the total ion exchange capacity of the resin in the column for analytical purposes or 5–10% for preparative work. Sample load in displacement chromatography may range from 25 to 50% of the total ion exchange capacity of the resin.

A flow rate of 0.005–0.05 ml of solution/ml of resin/minute is suitable in analytical determinations for particle sizes of 0.15–0.5 mm but in any case it should not exceed 0.05 ml/minute.

After eluates have been collected they can be analysed by IR or uv spectrometry or any other applicable chemical or instrumental technique.

Applications to cosmetic analysis

Alkanolamine salts of various acid polymers are used in some hair spray formulations. The alkanolamine can be separated from the acid polymer by a strongly acid ion exchange resin, Type I. The polymer is eluted with 75% ethanol. The amine is displaced from the resin with HCl (1+1) and determined by gas-liquid chromatography (GLC). Triethanolamine lauryl sulphate is separated into its acidic and basic components by chromatography on the strongly basic ion exchange resin, Amberlite CG-400. The triethanolamine is eluted with water and determined by GLC.

Amberlite CG-400 is also used to remove interfering surfactants from solution prior to TLC identification of hexachlorophene. These surfactants, if not removed, frequently alter the R_f of hexachlorophene, making its identification more difficult.

Alkanolamines may interfere with the GLC determination of glycerol in cosmetic creams. Alkanolamines and glycerol remain in the aqueous phase when the sample is extracted from an acidified solution of methanol-water with $CHCl_3$ or petroleum ether. Glycerol may be separated and recovered by elution with water in a column which contains pretreated, strongly acidic resin. The alkanolamine is retained by the resin.

Boric acid in certain deodorant preparations can be accurately determined after removal of interfering zinc or aluminum ions with the cationic exchange resin Amberlite.

Alkanolamides and/or soap can be separated from shampoos containing alkyl sulphates and/or sulphonates. A weakly basic anion exchange resin, Amberlite, is suitable for this separation. It must, however, be thoroughly pretreated by being washed with all of the same solvents to be used during the separation. Acidified ethanol elutes the alkanolamide and/or soap from the column, and the sulphate-sulphonate mixture retained by the resin is then isolated by elution with a basic (ammonium carbonate) methanol solution.

Thin Layer Chromatography

Over the past decade TLC has become an increasingly popular analytical tool for the analysis of cosmetic raw materials and products. The principles of adsorption TLC are based on those previously discussed in the section on adsorption chromatography. The principal differences are: A smaller mesh size adsorbent is supported as a thin layer with the aid of a binder, such as calcium sulphate, on a plastic sheet or a glass plate, and the mobile liquid phase (developer) separates but does not remove the sample components from the TLC adsorbent layer. Silica Gel and aluminum oxide are the adsorbents most frequently used. Although adsorption TLC is the most common type, TLC separations may also be achieved with partition or ion exchange coating materials according to the principles of ion exchange and partition chromatography discussed previously.

General techniques

The sample is spotted on the plate near one edge (origin) and the plate is placed in an almost vertical position in an enclosed chamber containing developer previously added to a height of 1.0 cm. The origin should be above the developer after the plate is in place. As the developer rises through the adsorbent layer by capillary action, the components of the sample are separated. After the developer reaches a predesignated height (solvent front), the plate is removed from the chamber and allowed to dry.

Each component may appear as a circular or elliptical spot somewhere between the origin and the solvent front. If the spot is not visible, chemical or other pretreatment may be required for its detection.

The use of several developments (multiple development) is often helpful in achieving a separation. After an initial development, the plate is removed from the chamber and allowed to dry, and one of its edges is immersed in the same or a different developer. If the same edge is immersed, the second development occurs in the same direction, but if the plate is rotated 90° it is developed perpendicular to the original direction. This technique is referred to as two-dimensional thin layer chromatography.

Each separated component can be isolated by preparative TLC, performed with the use of thicker adsorbent layers (500 μm) than those used for analytical purposes (250 μm). The adsorbent is lightly scored around the area containing the component. All of the material within the scored area is removed from the plate, and the component is separated from the adsorbent by extraction with a suitable solvent. Any chemical or instrumental procedure previously described may then be applied to its identification and/or determination. Components may also be determined without their removal from the plate (*in situ*) by visible, uv, or fluorometric spectral scanning with a densitometer.

Variables

Many factors influence the reproducibility of R_f values as well as separation of components. In adsorption TLC, separations are influenced by variation in the composition, purity, particle size, uniformity, layer thickness, and moisture content (activity) of the adsorbent, as well as the composition of the developing solvent, degree of vapour saturation and shape of the developing chamber, temperature, development technique, and sample composition.

Application to cosmetic analysis

Citral and citronellal are identified and determined in various essential oils. After their separation on Silica Gel, the plate is sprayed with a modified Schiff's reagent to produce distinctively coloured derivatives which are removed from the plate and determined spectrophotometrically.

Mixtures of naturally occurring coumarins and psoralens (furanocoumarins) in certain citrus oils are separated on Silica Gel and each compound is tentatively identified through its R_f and its fluorescent colour when viewed under uv radiation. Their isolation by the preparative TLC technique provides sufficient material for further investigations. Bergapten, a psoralen with the potential for causing a phototoxic reaction on human skin, occurs naturally in certain essential oils. It is identified in fragrance products through comparison of its fluorescent colour and spatial relationship with those of a standard. It is also determined at levels as low as 0.001% by weight by comparison with various amounts of standard. Separation is achieved by using two-dimensional multiple development with two different solvent systems.

Natural resins can sometimes be differentiated by the respective patterns of their acidic constituents on TLC plate. The components are visualised with the acid indicator spray, bromocresol green.

Thirty different hair dye intermediates used in commercial products are detected through two-dimensional development of the sample with accompanying standards in two different solvent systems. Although some of the compounds or their oxidation products are naturally coloured, others are identified only after the application of various chemical sprays. Their spatial relationship and colours are compared to those of the standards. Some of these intermediates can be detected in commercial hair dyes at levels as low as 0.02%. TLC is helpful in the identification and determination of the products formed by the oxidation of one of the above intermediates, *p*-phenylenediamine in alkaline solution.

The identification of preservatives in cosmetic products is facilitated by TLC on plates coated with Silica Gel impregnated with fluorescent additives. The uv-absorbing compounds show up as reddish brown spots under short wave uv light, which is useful as a preliminary identification step.

High Pressure Liquid Chromatography

High pressure liquid chromatography (HPLC) is merely an extension of adsorption, partition, and ion exchange column chromatography. All theoretical and practical considerations that apply to these forms of chromatography apply to HPLC as well. Although efficiency and resolution can be significantly increased by reducing column diameter, support or adsorbent particle size, and sample size, practical considerations limit the steps that can be taken in this direction. Instrumentation is now available which overcomes these limitations and significantly shortens analysis time. The high inlet pressures available in HPLC systems allow the use of narrow diameter columns and small uniform particles of packing materials. High efficiency chromatographic separations also require systems of low dead volume, a problem effectively handled by the use of micro detectors and capillary tubing.

HPLC systems usually employ: (*a*) inlet pressures of 500–5000 psig; (*b*) column lengths of 10–100 cm; (*c*) column diameters of 1–6 mm id (average, 2 mm); and (*d*) eluant flow rates of 10 to several hundred ml/hour. Many HPLC systems are also equipped with gradient elution devices which can change the composition of the eluant continuously during the course of an analysis. The detectors most commonly used are uv/visible of fixed or variable wavelength. Refractive index (RI) detectors are also used. The uv detectors are sensitive but are limited to the detection of aromatic and other compounds that are strong uv-absorbers. The RI detector is universal but is relatively insensitive and difficult to use with gradient elution.

GAS CHROMATOGRAPHY IN COSMETIC ANALYSIS

Gas chromatography has become an increasingly important analytical tool for separating, identifying, quantifying, and isolating ingredients from a wide variety of complex cosmetic formulations. Individual components may be separated rapidly by this technique (usually in less than one hour) and in many cases identified and determined at very low levels.

Gas chromatography actually comprises two alternative separation modes, one based on adsorption (gas-solid chromatography, GSC) and the other based on partition (gas-liquid chromatography, GLC) interactions.

Gas Liquid Chromatography

In GLC, a light inert carrier gas such as helium or nitrogen serves as the "mobile phase." This mobile phase flows continuously through a copper, , glass, or stainless steel column packed with a relatively non-volatile film of "stationary liquid," which is coated either on a large number of small, inert, solid particles ("solid support") contained therein or directly on the inside column walls. Such "packed columns" will be emphasised because of their high frequency of use in current cosmetic analysis and their dual utility for analytical and preparative functions. Narrow bore capillary columns (0.04–0.05" id), which have the stationary liquid phase coated directly on their inner walls, are used almost exclusively for analytical purposes because of the very limited amounts of sample they can handle successfully. Either column is supported in a thermostatically controlled oven in the gas chromatograph so that is may be maintained at a preselected temperature or range of temperatures.

A sample in the solid, liquid, or gaseous state is introduced into the chromatograph via an "injection port," which is usually maintained at an elevated temperature to aid in volatilisation. (Lower injection port temperatures are sufficient for sample constituents with higher vapour pressures, such as gases and low boiling liquids.) In some cases, a chemical derivative of the sample may have to be prepared to form a product volatile enough so that it can be separated by GLC. No decomposition or destruction of components should occur at the temperatures required for volatilisation. The solid support coated with stationary phase should be readily permeable to the carrier gas flow. Vapour pressure of the stationary liquid phase should be minimal or non-existent at the operating temperature of the column.

The separation of volatilised components is influenced by a number of factors as the mixture is swept through the column by a flow of carrier gas. At the head of the column, each component of the vapourised mixture partitions itself in an equilibrium between the carrier gas and the stationary liquid phase. These multiple equilibria are continually established as the carrier gas sweeps the solutes through the column. The degree of interaction of each of the solutes with the liquid phase depends upon combinations of various electronic forces which selectively retard each component until it exists from the column as a separate entity. Separation is therefore based on the continuing competition between the liquid and gas phases for the volatilised solutes. The velocities of the solutes differ at a rate reflected mathematically by the partition coefficient, K, which is a measure of the degree that each component is retained by the liquid phase. For a given solute A,

K = Concentration of solute A in stationary phase/concentration of solute A in mobile phase.

Preparation of Packed Columns

Selection and preparation of tubing

Stainless steel, copper, aluminum, and glass are all common materials from which tubing for columns may be manufactured. Some chemical entities in the sample or the stationary phase may be catalytically converted by the composition of these metallic tubes. Glass is the least likely material to contribute to such adverse activity, and it has the added bonus of furnishing a clear view of the progress made during the column packing procedure. However, glass tubes are fragile and expensive, and they usually must be ordered specially to fit a given gas chromatograph. Packed columns may serve in an analytical (1/8–1/4" od) or preparative (3/8–1/2" od) capacity. In order to fit in the oven, tubes must be either U-shaped or, more often, coiled. After the tubes are coiled, they are cleaned with solvents, filled with packing material (solid support coated with stationary phase), and connected to the chromatograph with Swagelok fittings.

The tubes are cleaned with solvents prior to packing to remove industrial contaminants used in their manufacture. Usually a vacuum is applied to one end and a detergent solution and/or a series of organic solvents is introduced at the opposite end. Each successive organise solvent should have a higher polarity, the final one being the same as the solvent used for coating the liquid phase onto the solid support—e.g., pentane, chloroform, acetone, and the liquid phase solvent.

Preparation of solid support

The support should be sieved in order to select the appropriate narrow range of particle (mesh) sizes required for efficient column performance, as previously discussed. Most of the fines should be removed; if they are created by sieving an easily fractured support, the support should instead be added to water and the fines eliminated by decantation. Many solid supports are commercially available in the required mesh range and are adequate for use as is.

Packing the column

A well packed column should have a uniform distribution of closely packed solid particles (not crushed or fractured) to keep the pressure drop of carrier gas between the column's inlet and outlet at a minimum. One end of the clean, previously shaped tube is plugged with a gas-permeable material such as silanised glass wool or a thin metal fritted disk so that the packing material will remain in the tube. As the packing material is poured through a funnel temporarily attached to the open end, a full vacuum is applied to the plugged end while the column is tapped lightly and/or vibrated to facilitate proper settling. When no more material can be transferred into the column, it is assumed to be full and properly packed. Alternatively, 40–50 psig of inert gas or air can be applied to a specially designed reservoir (commercially available) containing the packing material. The reservoir is attached to the open end of the column, and the column is tapped or vibrated mildly until no more material is transferred into the column. Care should be taken not to disconnect any part of the apparatus until after the pressure is turned off and enough time has elapsed for the pressure to equilibrate to ambient conditions.

After the column has been packed, the open end is plugged with silanised glass wool or a fritted metal disk, and nuts and ferrules are attached to both ends (if they have not been attached previously) to provide for connection of the column to the instrument. (Special plasticised ferrules are available for glass columns).

Specialised details must be observed in packing certain types of supports. For example, only gentle tapping or minimal vibration should be used for easily fractured solids; packed porous polymer beads, because of their shrinkage, should be conditioned in an oven for at least two hours at 25–50°C below the maximum recommended operating temperature with a slow carrier gas flow, followed by column removal and further packing.

Column conditioning

After packing, all freshly prepared columns should be conditioned, primarily to eliminate various volatile impurities introduced by the coating and packing procedure or previously present in the liquid phase or solid support. Common impurities include water, unpolymerised monomeric species from which the liquid phase was synthesised, and solvents. Conditioning may also aid in achieving a more even distribution of the liquid phase throughout the packing material.

One end of the column, usually the one to which the packing was added, is attached to the inlet portion of the chromatograph while the other end is left unconnected in order not to contaminate or damage the detector.

Generally, the column, under low carrier gas flow, is heated gradually to 10–20°C above the expected operating temperature. The column is conditioned overnight at this temperature.

Certain liquid phases, require specialised treatment; for example, they may be conditioned initially for various time periods at specified temperatures with no carrier gas flow. To obtain accurate quantitative results, it is frequently necessary that certain other highly specialised and unique steps be taken. Several examples are presented below under determination.

Sample Preparation

One or more physical and/or chemical steps frequently must be taken in order to separate and isolate the components to be measured in the sample from the rest of the ingredients that may interfere with the analysis. It is also useful to concentrate the sample, especially when handling small quantities.

Reduced or ambient pressure distillation is helpful for the isolation or removal of volatile materials. The higher boiling sesquiterpene hydrocarbon portions, frequently present in small amounts in essential oils used in fragrance compositions, may be isolated along with other higher boiling compounds as the residue remaining after distillation of the more abundant, volatile terpene hydrocarbons. Such a residue is then separated further on alumina into sesquiterpene hydrocarbon fractions by liquid column chromatography. Azeotropic distillation is also useful. In this process, a liquid immiscible with the sample is added and the temperature is raised to the boiling point. The components of interest are quantitatively distilled over along with the immiscible liquid. This procedure is useful for the separation and isolation of propylene glycol from a variety of cosmetic products. In this case, iso-octane is the immiscible liquid of choice since one of the prerequisites for its selection is that only a minimum amount is required for quantitative transfer of the component of interest, propylene glycol.

Extraction with immiscible solvent systems is effective when selectivity is achieved for the ingredients of interest. Hexachlorophene is extracted from whole blood with an ether-alcohol (18+7) solution in a study of the extent of its penetration through mammalian skin. Glycerol is separated from most other ingredients in cosmetic creams through the addition of water (in which it is soluble) followed by extraction (and separation) of most other organic ingredients in the cream with chloroform or petroleum ether. Most cosmetic ingredients, except for highly polar water-soluble compounds such as glycerol, are soluble in these organic solvents. Glycerol is isolated from toothpastes or gums by extraction with 2-methoxyethanol, in which it is soluble. Any insoluble materials are separated by filtration of the extract.

The preparative liquid column chromatographic methods are also useful tools for separation and isolation. As indicated above, an alumina column is useful for fractionating various sesquiterpene hydrocarbon mixtures; a silica gel column is helpful for removing interfering hydrocarbons from fatty acid esters and alcohols determined in lipsticks.

The chemical reactivity of any functional group of a cosmetic ingredient may be useful for its retrieval from a mixture. The acidic (phenolic) nature of hexachlorophene, for example, facilitates its alkaline extraction from whole blood.

In headspace analysis, vapours of solutions of sufficiently volatile ingredients may be injected directly into the gas chromatograph without any need for elaborate sample clean-up or separation procedures. The sample is placed in an enclosed system so that an equilibrium of the component(s) is established between the gaseous and liquid phases. An aliquot of the vapour is then withdrawn through a septum or membrane covering and injected into the chromatograph. Cosmetics containing methanol, ethanol, or isopropanol, as well as chloroform in toothpastes, are easily handled by this technique. These ingredients can be identified and also determined by headspace analysis. A number of propellants and solvents found in aerosol hair spray formulations may also be identified.

In practice, the solute molecules not only partition between the two phases but may also be adsorbed by the so called "inert" solid support. Such an adverse effect may be significant even after the support has been "deactivated". The tendency toward adsorption on the active sites of the support is directly proportional to solute polarity, i.e., acids are adsorbed most readily, followed by phenols and amines. Adsorption, manifested in peak tailing, leads to incomplete separations and, in turn, to inaccurate qualitative and quantitative results. If such highly polar compounds are quantitatively converted to less polar derivatives not able to form hydrogen bonds with the stationary liquid phase and/or support, this problem can be minimised or eliminated. For example, phenols or alcohols are treated with acid anhydrides or azomethanes to form ester or ether derivatives, respectively. In addition, alkanolamines, used as neutralisers for acidic polymer resins added to various aerosol hair spray formulations, are converted to their corresponding difunctional amide and acetate derivatives with acetic anhydride-pyridine (2+1) reagent (if the amine is tertiary, only an acetate is formed).

Because high molecular weight compounds may have inordinately long retention times, it may be necessary to convert them, by chemical means, to lower molecular weight products which are more volatile and therefore more amenable to GLC. In some cases, these products may also have the potential, because of their high polarity, for being easily adsorbed on the support. Both problems are eliminated in an analysis of castor oil in lipsticks by the following technique: Castor oil, the relatively non-volatile triglyceride of ricinoleic acid, is converted by transesterification with a methanol-benzene-sulphuric acid solution (300+100+3.7) to glycerol and the methyl ester of ricinoleic acid. The amount of castor oil present is then determined by a GLC analysis for methyl ricinoleate. Multiple products from reactions such as this one also afford further confirmatory evidence for identification by the accumulation of additional retention time data. Mixtures of certain higher molecular weight isomeric fatty acid esters of fatty alcohols are difficult to separate; however, after transesterification with methanol, they become readily separable mixtures of fatty alcohols and methyl esters of fatty acids. The alcohols can be verified by acetylation. The alcohol peaks then disappear and retention times for the corresponding acetate peaks are obtained.

Detectors

In selecting a detector, points to consider are (1) sensitivity, (2) signal-to-noise ratio, (3) linearity, (4) stability, (5) specificity (range of application), and (6) destructive effect on the effluent (reactivity with the sample molecules). Thermal conductivity, flame ionisation, and electron capture detectors are used more frequently than other types in the GLC analysis of cosmetics.

Most common is the thermal conductivity detector in which a metallic filament, heated by the passage of a constant electrical potential, exhibits a given resistance which is subject to change when a corresponding change is induced in its environmental temperature. An inert carrier gas is diverted over two such heated filaments that are part of a Wheatstone bridge assembly. The heat loss from both filaments, maintained through a block set at constant temperature, is equal until a component is swept over one of them. At this point, the heat conductance away from this filament is affected, along with its temperature and resistance. The bridge circuit measures the difference in the electrical signal generated between the two filaments because of their difference in resistance and emits a signal which restores balance and indicates the presence of the solute. In order to achieve optimum sensitivity and wide linearity of response, the carrier gas should have a low molecular weight, high thermal conductivity, and low viscosity. Helium is ideal in all these respects and is usually employed as the carrier gas when using a thermal conductivity detector. The thermal conductivity detector performs best with a high bridge current and low block temperature, and its performance may be adversely affected by carrier gas flow fluctuations or ambient temperature effects. This detector is less sensitive (about 1×10^{-9} g/second) than those discussed below.

In the flame ionisation detector, the organic vapour of the solute enters a flame supported by hydrogen and air, and it is converted to cationic species and free electrons. A gap between two electrodes with a difference in applied potential is oriented so as to monitor the current increase over that generated from background. This detector is destructive to the compounds measured and relatively insensitive to inorganic compounds such as water, hydrogen sulphide, carbon disulphide, and carbon dioxide; however, it is more sensitive to organic compounds (about 3×10^{-12} g/second) than the thermal conductivity detector. The presence of inorganic compounds may decrease the sensitivity of the flame ionisation detector to organic compounds. This detector has a wide linear response. Either nitrogen or helium may be used with it as a carrier gas because both (i) have small background conductivity—necessary for the detection of proportionately large currents produced by sample components; (ii) have minimal viscosity, and (iii) are inert. Optimum sensitivity is dependent on carrier gas and combustion gas (hydrogen and air) flow rates.

The electron capture detector is also dependent upon ionisation of the sample but, unlike the flame ionisation detector, it does not destroy the compounds measured. Certain types of vapour molecules, when bombarded with electrons, accept or "capture" them and become negatively charged. These highly energised electrons are emitted from a radioactive source, usually ^{63}Ni. The negatively charged molecules produced enter a gap between oppositely charged electrodes where a current has previously been established by ionisation of the carrier gas. When the negatively charged, slower moving components enter this gap, the current is diminished and the sample is detected accordingly. This detector is much more selective than those discussed above, since it detects only electron-accepting molecules containing halogen, sulphur, organometallic, phosphorus, or polyaromatic substituents. It is also the most sensitive of the three at about 3×10^{-14} g/second. Although nitrogen has been used as a carrier gas with electron capture detectors, it may foster other ionic mechanisms that may erroneously be attributed to electron capture. In order to avoid such effects, argon containing 5–10% (v/v) methane (sometimes butane) is used as carrier gas, and pulsed electrical waves are applied across the electrode gap.

Table 44.1 lists the range of operating temperatures chosen for each type of detector with corresponding injector and column temperature ranges selected from a cross-section dealing with the gas chromatographic analysis of cosmetic products and raw materials.

Table 44.1. Operating temperature ranges (°C) for various types of detectors with accompanying injector and column temperatures[a].

	Thermal Conductivity	Flame Ionisation	Electron Capture
Detector[b]	240–320	190–300	300
Injector[b]	240–320	200–310	270–275
Column[c]	65–250	70–270	125–250

[a] Based on methods developed for the analysis of cosmetics and cosmetic raw materials.

[b] A selected isothermal temperature is chosen for any given analysis.

[c] Most temperatures are selected for isothermal runs within this range; some are a programmed span of temperatures from this range for a given analysis.

Identification

The time between injection of a sample onto the column and elution of a component represented by its peak on a chromatogram is known as the "retention time" of that component. Its reproduction is difficult because of many inherent variables associated with GLC. However, reproducible relative retention times may be calculated by adding a known compound, called an internal standard, to the sample and comparing the time it takes for emergence of this compound to that required for the unknown. Relative retention times are frequently expressed in terms of Kovats indices or McReynolds constants.

Many of the raw materials used in cosmetic products are complex mixtures of ingredients. A distinctive GLC peak pattern (fingerprint) may emerge based on the relative retention times of the various components. Comparison of these relative retention times with those of standard materials can be very helpful in identifying unknown constituents. For example, the sesquiterpene hydrocarbon fractions of many essential oils used in fragrance compositions contain characteristic identifiable combinations of specific sesquiterpene hydrocarbons which generate peaks that are representative of or unique for a given oil. Comparison is made of their peaks with those peaks from a similar fraction obtained from a standard oil.

Mere knowledge of the retention time of an ingredient does not provide confirmatory evidence of its identity. The pure unknown compound may be isolated by trapping as it elutes from the column and then subjected to various "wet" chemical or instrumental techniques to confirm its identity or elucidate its structure. Instrumental techniques frequently used for this purpose include ultraviolet and infrared spectrophotometry, fluorometry, nuclear magnetic resonance spectroscopy, mass spectrometry, refractometry, and optical rotation (polarimetry and circular dichroism). Melting point and boiling point measurements are also useful. With mass spectrometry and, less often, infrared spectrophotometry, it is possible, and indeed common, to connect the instrument directly in series to the exit port of the GLC column via an interface which obviates trapping or handling of the compound.

Determination

The internal standard method is the preferred approach for GLC analysis of cosmetics because :

(1) Errors which could arise from non-linear detector response are minimised.

(2) Identical aliquots of the standard and sample solutions are not required for injection.

(3) Identical concentrations of the sample and standard are not required. The standard is a volumetrically prepared solution consisting of a known concentration of a purified compound that is identical in structure to that to be determined in the sample. The internal standard is a compound different from the one being determined and should meet the following criteria: absent from the sample, miscible with the component to be determined in the chosen solvent system, pure, non-reactive with any chemical in the sample or solvent, and chemically and thermodynamically stable. The internal standard should also generate a symmetrical peak on the recorder with a retention time close to but not identical with that of the compound to be determined or any other constituent of the sample solution.

In applying the internal standard technique, a suitable volume of a known concentration of the internal standard is added to the sample solution containing all of the unknown constituent such that the peak heights generated by both in the resulting chromatogram are similar (within 10–25%). Increasing accuracy is usually assured as the ratio of peak heights approaches unity. Although each peak should be no less than 50% full scale on the recorder, 75–95% is preferred.

An identical (or similar) amount of internal standard is then added to a separate vessel into which an experimentally determined amount of the compound of interest from a standard solution containing only that compound has been added such that the peak height ratio of unknown to internal standard is similar to that in the sample solution. The total solution volume is adjusted so that it is similar to that of the sample solution containing the internal standard. The above criteria for peak heights and ratios also apply here.

Aliquots of each solution should be injected alternately at least twice so that all peak heights for the unknown in all chromatograms vary by less than ± 10%. An average value is obtained for the peak height ratios for each solution. Values which vary by more than 10% from this average should be rejected.

The percentage by weight of unknown in the sample may be calculated as follows :

% Unknown = $[(R_u R_s) \times$ weight unknown in standard $\times$ (weight internal standard in sample/weight internal standard in standard)/weight of sample] $\times$ 100

where R_s = average peak height ratio of sample solution (peak height unknown in sample/peak height internal standard in sample of several injections), and R_s = average peak height ratio of standard solution (peak height unknown in standard/peak height internal standard in standard of several injections).

If possible, all determinations should be made on the same day in order to minimise instrumental errors resulting from carrier gas flow rate change, temperature deviations, detector response variables, etc. It is also best to work at the same attenuation and range settings of the instrument for both sample and standard solutions throughout the determination.

Two examples of potential errors that may arise in internal standard analyses are illustrated: In the headspace determination of chloroform in toothpaste, the difference in absorption by the rubber dam and smaller concentration (10%) of the 1,1,1-trichloroethane internal standard compared to the chloroform necessitates that all measurements be made over a short time span before peak ratios change drastically. Second, the peak response for ethanol may vary because of the dependence of this response on the time between emergence of the internal standard from the previous injection and the injection of the current ethanol sample. This phenomenon, caused by support adsorption, requires injection of all ethanol samples at a preset time differential.

ANALYSIS OF CREAMS AND LOTIONS

Three types of ingredients are essential in the formulation of cosmetic emulsions. One ingredient is water. Another is "fatty" or water-insoluble material such as beeswax, spermaceti, hydrocarbons, lanolin, fatty acids, alcohols of high molecular weight, glycerides, isopropyl myristate, etc. The third essential ingredient is a surface active agent that emulsifies the water and "fatty" material. Some common emulsifiers are soaps, glyceryl monostearate, alkyl sulphates, lanolin, quaternary ammonium salts, and polyoxyethylene compounds.

Because these products are either water-in-oil or oil-in-water types of emulsions, they can, in most cases, be analysed in a similar manner.

General Analysis

Net Contents

(1) If the product is a cream, weigh the intact container at the beginning of the analysis. After the analysis is completed, remove any remaining sample and weigh the empty container. Calculate the weight of the product by difference.

(2) If the product is a lotion, mark the outside of the bottle at the surface level of the liquid before beginning the analysis. Upon completing the analysis, empty the bottle and note the volume of water required to fill it to the mark.

(3) For aerosol products in metallic containers, weigh the intact container, chill in a dry ice chest for 2 hours, and punch out the top of the chilled container with a cold chisel and hammer.

>   (a) If contents are liquid, pour out the contents and weigh the parts of the container together. Calculate the contents by difference.

>   (b) If contents are solid, place the opened can with the frozen material in a 4 L beaker and let it warm to room temperature. Decant the liquefied contents and proceed as in above.

Description of Product

Note colour, odour, and other physical characteristics of the product.

Type of Emulsion

(1) Smear a thin film of the product, ca 1" square by about 1/16" thick, on a watch glass. Sprinkle small amounts of a finely ground oil-soluble dye and a water-soluble dye on separate areas of the film. Spreading of the oil-soluble dye indicates a water-in-oil emulsion; spreading of the water-soluble dye indicates an oil-in-water emulsion.

(2) As an alternative procedure, determine the type of emulsion by noting whether a portion of the product readily mixes with mineral oil (w/o emulsion) or water (o/w emulsion).

(3) Another procedure involves electrical conductivity, as follows :

"Equipment for this may be easily constructed by wiring in series a 30,000 ohm 1/2-watt resistor, electric contacts for the test sample, a resistorless neon lamp (1/4-watt 104 to 120 v.), and a push-button switch. The sample is placed across the test contacts and the circuit is closed. If the neon lamp glows, the emulsion is o/w; if it does not, the emulsion is w/o. There are occasions when the lamp will glow dimly or will start to glow upon continued application of the electric current. This usually indicates either a dual emulsion or a gradual inversion of the emulsion. The presence of electrolytes, particularly at high levels, may result in conductivity even though the emulsion is essentially w/o."

pH of Emulsion

(1) For o/w cream emulsions, mix 1 g of cream with 9 ml of water and determine the pH of the resulting mixture with a glass electrode instrument. Report the pH (1 part cream plus 9 parts water).

(2) For o/w lotion emulsions, determine the pH directly on the liquid.

(3) Alternatively, determine the approximate pH of (a) and/or (b) with short range pH test paper.

Ashing at 600°C

Weigh about 5 g of the product in a flat-bottom platinum dish and heat on the steam bath under a jet of air for 1 hour. Remove the dish and add 1 g of ashless cellulose powder, then mix with a glass stirring rod. Wipe any adhering material from the stirring rod with a piece of filter paper and add the paper to the dish. Heat the dish under an infrared (IR) heating lamp until the sample is charred. Complete ashing at 600°C in a muffle oven.

Examination of Ash

(1) *Borates*—Mix a portion of ash with a few drops of H_2SO_4 in a platinum dish, add several ml of methanol, stir well, place in a darkened hood; and ignite. A green flame indicates the presence of boron.

(2) *Carbonates*—Mix a portion of ash with a drop of HCl. An odourless effervescence indicates the presence of carbonates.

(3) *Other water-soluble salts*—Dissolve the remainder of the ash in water or dilute HNO_3, filter off any insoluble material, and test aliquots of the filtrate for chlorides with $AgNO_3$, for sulphates with $BaCl_2$, and for phosphates with ammonium molybdate. Also check for Na and K by platinum wire flame tests.

(4) *Water-insoluble ash* (commonly TiO_2, ZnO, or talc)—Ignite the ash in a platinum crucible. The appearance of a yellow colour that fades on cooling indicates the possible presence of ZnO. Dissolve the ZnO in acid and precipitate as ZnS for confirmatory test.

Solubilise TiO_2 by heating in a covered porcelain crucible with 8 ml of H_2SO_4 and 4 g of Na_2SO_4. Cool and carefully pour the solution into 50 ml of ice-cold water. Test for titanium colorimetrically with H_2O_2.

Fuse talc with a 1 + 10 mixture of Na_2CO_3 and K_2CO_3 and test for magnesium and silica.

(5) *Infrared spectrum of ash*—Prepare a mineral oil mull or KBr disk of the ash and obtain its infrared spectrum. This spectrum is helpful in identifying the nature of the ash.

(6) *X-ray fluorescence spectroscopy*—Elements with atomic numbers equal to or greater than that of aluminum can readily be identified by X-ray fluorescence spectroscopy.

Non-volatile Matter at 105°C

Weigh about 1 g of the product into a large weighing bottle and heat on a steam bath under a jet of air for 30 minutes. Continue heating at 105°C in an oven for 2 hours. Cool in a desiccator, weigh, and report as non-volatile matter (2 hours at 105°C).

Infrared Examinations of Non-volatile Matter

Smear a film of the non-volatile matter on a salt crystal and obtain the IR spectrum of the film. Examine this spectrum carefully for clues to the composition of the sample. The positions of the absorbance peaks indicate the identity of the ingredients present; the absorbance values indicate the relative amounts of the various components present.

In particular, look for the presence of esters, hydrocarbons, alcohol polyhydroxy compounds, polyoxyethylene compounds, amides, soaps, fatty acids, and alkanolamines.

Chloroform-Extractable Matter

This extraction is applicable to soap emulsions such as the beeswax-borax, alkanolamine, glyceryl monostearate, and alkali metal types. The emulsification of the $CHCl_3$ may preclude its use with creams emulsified by other surface active agents.

(1) Transfer about 3 g of sample to a separatory funnel with 50 ml of water, strongly acidify with HCl, and extract with four 35 ml portions of $CHCl_3$. Combine the $CHCl_3$ extracts, wash with 10 ml of water, and add the washing to the reserved extracted aqueous solution. Filter the $CHCl_3$ extracts through a cotton plug into a tared 250 ml beaker, evaporate the solvent on a steam bath under a jet of air, and dry the residue at 105°C in an oven for 15 minutes. Cool in a desiccator and weigh as $CHCl_3$-extractable matter.

(2) Test $CHCl_3$-extractable matter for lanolin or sterols, or both, by the Liebermann-Burchard reaction :

   Dissolve a small portion of the material in 10 ml of $CHCl_3$, add 5 ml of acetic anhydride and follow with 5–10 drops of H_2SO_4. Stir well. The appearance of a characteristic green colour indicates the presence of lanolin or sterols.

(3) Obtain an IR spectrum of a film of the $CHCl_3$-extractable matter and examine the spectrum to ascertain what material may be present.

Material Not Extractable by Chloroform from Acid Aqueous Solution

(1) *Total non-extractable matter*—Dilute the reserved extracted aqueous solution from above to 100 ml with water in a volumetric flask. Pipet out a 40 ml aliquot into a tared 100 ml beaker and evaporate the water on a steam bath under a jet of air. Dry the residue at 105°C in an oven for 10 minutes, cool in a desiccator, and weigh as non-extractable matter.

   (a) Obtain an IR spectrum of a film of the material and note whether a polyhydroxy compound or an alkanolamine hydrochloride (usually triethanolamine) or both may be present.

   (b) As an additional test for alkanolamine, add excess *n*-propyl alcohol to the residue in a beaker, boil to a small volume, chill, scratch the sides of the beaker with a glass rod to induce crystallisation, filter off the crystals, and dry. Determine the melting point of the crystals of alkanolamine hydrochloride or obtain an IR film spectrum of a mineral oil mull of the crystals.

(2) *Chemical test for glycerol*—Neutralise a 10 ml aliquot of the extracted aqueous solution, (a), to methyl red with CO_2-free $0.1N$ NaOH, making the final adjustment with $0.02N$ NaOH and leaving the solution just barely yellow. Add 30 ml of $0.02M$ KIO_4 and stir well. If the red colour of methyl red appears immediately, glycerol, is probably present.

(c) *Additional chemical test for glycerol*—Place 3 ml of the aqueous solution containing glycerol in a large test tube. Add 3 ml of freshly prepared 10% catechol solution, then 6 ml of H_2SO_4. Heat in boiling water for 30 minutes. An orange-red colour indicates the presence of glycerol.

Saponification of Chloroform-Extractable Matter

(1) Reflux the $CHCl_3$-extractable matter for 2 hours with 25 ml of 95% alcohol, 1 g of KOH, and 50 ml of benzene. Transfer the saponified mixture to a separatory funnel (with Teflon or ungreased glass stopcock), add 50 ml of hot water, shake well, and draw off the aqueous layer. Continue the extraction with two additional 50 ml portions of hot benzene. Reserve the extracted aqueous solution. Combine the benzene extracts, and wash with three 30 ml portions of 30% alcohol, shaking very gently with the first washing and more vigorously with the other two. Add the washings to the reserved extracted aqueous solution. Filter the washed benzene extract through a cotton plug into a 250 ml tared beaker, evaporate the benzene on the steam bath, and dry the residue at 105°C in an oven for 15 minutes.

Cool and weigh as unsaponifiable matter. Obtain its IR Spectrum.

(2) Acidify the reserved aqueous solution (a) with HCl and extract with three 30 ml of portions of $CHCl_3$. Reserve the extracted aqueous solution and use it to test for combined glycerol and a polyxyethylene compound. Wash the combined $CHCl_3$ extracts with water. Filter the washed $CHCl_3$ extract through a cotton plug into a tared 250 ml beaker, evaporate the $CHCl_3$ on the steam bath under a jet of air, dry the residue at 105°C in an oven for 15 minutes, cool, and weigh as fatty acids.

Examination of the Saponifiable Matter

(1) *Equivalent weight of fatty acids*—Dissolve saponifiable matter in 50 ml of alcohol that has been neutralised to phenolphthalein with $0.1N$ alkali. Titrate with $0.1N$ NaOH to a phenolphthalein end point. Calculate the equivalent weight of fatty acids.

(2) *Hanus iodine absorption number.*

Hydrocarbons and Alcohols in Unsaponifiable Matter

Dissolve the unsaponifiable matter in 50 ml of boiling heptane and cool the solution to room temperature with stirring.

(1) If there is any precipitated matter, filter the resulting mixture through filter paper, wash the residue well with petroleum ether, and reserve the filtrate. Dissolve the residue by pouring hot $CHCl_3$ through the filter paper, and evaporate the $CHCl_3$ on the steam bath. Dry at 105°C in an oven for 10 minutes, cool in a desiccator, let stand in air 10 minutes, and weigh as alcohols insoluble in heptane-petroleum ether.

Evaporate the reserved heptane-petroleum ether solution on the steam bath and take up the residue in 50 ml of warm petroleum ether (b.p., 30–75°C). Transfer the solution, with the aid of 25 ml of petroleum ether, to a glass chromatographic tube holding a 9 × 3/4" activated alumina column (Alcoa grade F-20, 80-200 mesh), and let the solution flow through the column at a rate of 3–5 ml/minute, collecting the petroleum ether eluates in a 400 ml beaker. Follow this solution with 175 ml of petroleum ether. Reserve the petroleum ether eluate.

Remove the petroleum ether from the column by passing 50 ml of 95% alcohol through the column under air pressure. Then, again using air pressure, force 125 ml of boiling alcohol through the column at a rapid rate (4–7 minutes for 125 ml). Collect these two eluates in a tared 250 ml beaker labelled *First Hot Alcohol Fraction*. Pass another 50 ml of boiling alcohol through the column and collect it in a tared 250 ml beaker labelled *Second*

Hot Alcohol Fraction.

Evaporate each of the hot alcohol fractions to dryness on the steam bath under a jet of air, redissolve the residue in 10 ml of $CHCl_3$, and evaporate to dryness again. Dry the residue at 105°C in an oven for 10 minutes, cool, and weigh as alcohols soluble in heptane-petroleum ether.

Evaporate the reserved petroleum ether eluate to about 50 ml on the steam bath under a jet of air, transfer to a tared 250 ml of beaker with 50 ml of petroleum ether, and evaporate the solvent on the steam bath. The residue will be hydrocarbons. Dry at 105°C in an oven for 10 minutes, cool, and weigh. Obtain an IR film spectrum of the hydrocarbons.

(b) If there is no precipitated matter, evaporate the heptane solution on the steam bath and proceed as in (a), beginning "take up the residue in 50 ml of warm petroleum ether. . .".

Examination of Alcohols in Unsaponifiable Matter

(a) Obtain the IR spectra of the alcohols.

(b) Combine all alcohol fractions and determine the solubility of the combined fractions in cold methanol to aid in identifying the wax.

Weigh a sample not exceeding 0.5 g into a 150 ml beaker, add 80 ml of methanol, cover the beaker with a watch glass, and heat on a hot plate until solution is complete. Transfer the beaker to an ice bath and stir the solution vigorously with a thermometer until the temperature is 5°C. Pour the resulting mixture through a 4" Buchner funnel containing an 11 cm No. 595 S&S filter paper, with 1/2" of the edge turned up, wetted with methanol. Filter by gravity or gentle suction. Do not allow the material on the filter to dry or cake. Wash the residue with 20 ml of ice-chilled methanol and drain it dry with suction. Transfer the filtrate to a tared beaker with the aid of $CHCl_3$, evaporate to dryness on the steam bath, heat at 100°C in an oven for 10 minutes, cool in a desiccator, and weigh. Repeat the drying in the oven until the weight is constant to 1–2 mg. Table 44.2. gives the results obtained with beeswax alcohols.

Table 44.2. Solubility of beeswax alcohols in cold methanol[a]

Sample, g	Soluble Alcohols, g	Insoluble Alcohols, g (By Difference)
0.1	0.027	0.073
0.2	0.040	0.160
0.3	0.045	0.255
0.4	0.054	0.346
0.5	0.062	0.438

[a] A 0.500 g sample of spermaceti alcohols contains a maximum of 0.008 g of insoluble alcohols (10).

Composition of Chloroform-Extractable Matter

(1) *Hydrocarbons*—For direct and easy identification of hydrocarbons, determine their IR spectra.

(2) *Fatty acids*—If the IR spectrum of the $CHCl_3$-extractable material indicates only fatty acids, no further analysis is necessary and the IR spectrum of the non-volatile matter will probably have the characteristic absorption maximum of soap at about 6.5 μm.

The presence of saponifiable matter but no unsaponifiable matter suggests several possibilities. The fatty acids may have originally been present as a triglyceride, a monoglyceride, the ester of some other polyhydroxy compound, or the ester of a water-soluble alcohol such as isopropyl alcohol.

(3) *Waxes*—Some of the more common waxes such as spermaceti, beeswax, and lanolin contain roughly equal amounts of saponifiable and unsaponifiable matter. If the IR spectrum of the $CHCl_3$-extractable matter indicates that only esters and possibly hydrocarbons are present, and if the amounts of fatty alcohols and acids are approximately equal, it may be assumed that the waxes are equal to the sum of the alcohols and fatty acids.

Some indications as to the types of waxes may be obtained from the IR spectra, the equivalent weight of the fatty acids, the solubility of the alcohols in methanol, and a test for lanolin.

(4) *Waxes plus other materials*—If both saponifiable fatty acids and unsaponifiable alcohols are present in unequal amounts, there are several possibilities. If the alcohols greatly exceed the acids, multiply the weight of the acids by two and calculate the resulting value as waxes. Subtract the weight of acids from the weight of alcohols and calculate the remaining value as free alcohols. (It is assumed that the IR spectrum of the material before saponification indicates esters and alcohols.) Conversely, if the fatty acids exceed the alcohols, multiply the weight of alcohols by two to obtain a value for waxes. Calculate the unassigned fatty acids as free fatty acids, soap, or some fatty acid compound which has no unsaponifiable matter. (Again, reliance is placed on the IR spectra and chemical tests).

(5) *Charts*—A diagrammatic summary of the analysis of mixtures of hydrocarbons, beeswax, and spermaceti is presented in Figs. 44.1, 44.2 and 44.3. With suitable modifications this scheme is applicable to the analysis of most "fatty materials" or their mixtures.

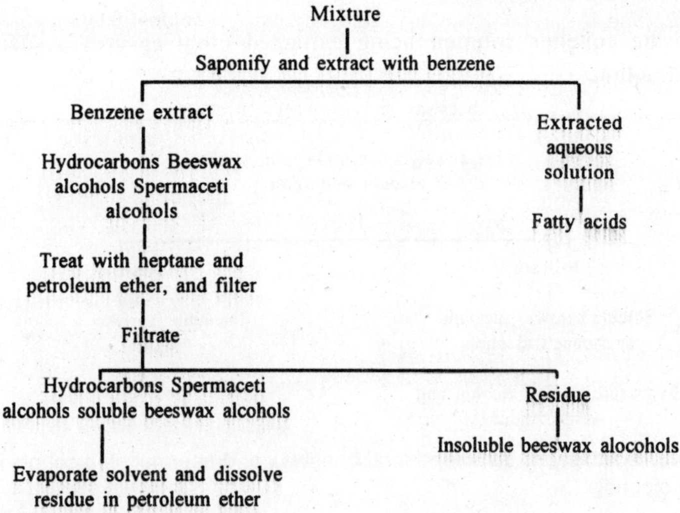

Fig. 44.1. Preparation of a mixture of hydrocarbons, beeswax, and spermaceti for chromatography.

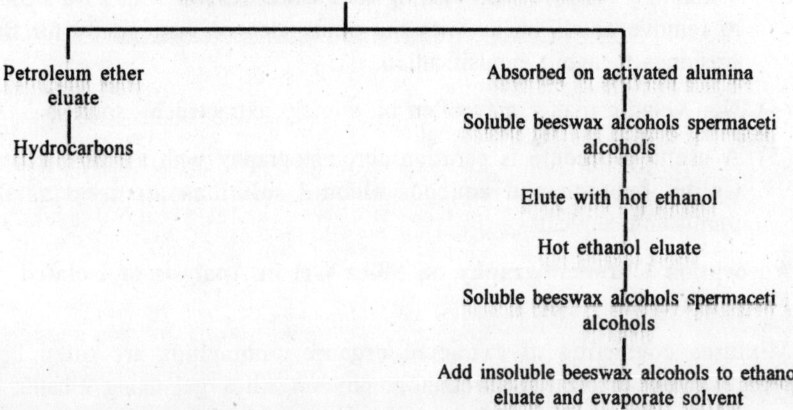

Fig. 44.2. Chromatography of hydrocarbons, spermaceti alcohols, and soluble beeswax alcohols.

Other Solvent Extraction Procedures

(1) In attempting to isolate materials from products by immiscible solvent extraction, the successive use of solvents other than $CHCl_3$ is very helpful. In general, the selectivity of solvents for organic materials decreases in the following order: petroleum ether, carbon tetrachloride, benzene, ether, ethyl acetate, and $CHCl_3$. The correct adjustment of pH

of the aqueous solution being extracted often ensures a successful extraction.

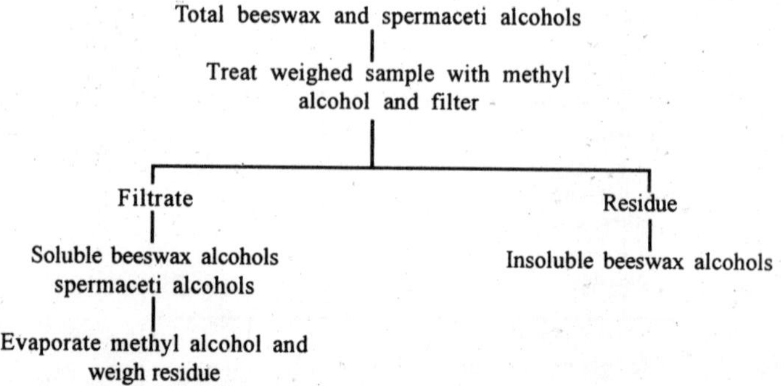

Fig. 44.3. Solubility of combined total beeswax and spermaceti alcohols in cold methyl alcohol.

(2) The system petroleum ether-80% methanol separates organic compounds on the basis of polarity.

(3) Ethyl ether does not extract fatty acids from a 25% alcohol solution containing 1% KOH. In washing such ether extracts with 25% alcohol to remove alkali, shake with the dilute alcohol very gently for first washings to avoid emulsification.

(4) Non-volatile matter may often be directly extracted by solvents.

(5) A useful procedure is partition chromatography with a silicon-coated Celite; heptane and aqueous alcohol solutions are used as the immiscible solvents.

Adsorption Chromatography on Silica Gel in Analysis of Isolated Fatty Materials

Mixtures consisting of extracted organic compounds are often best separated by adsorption chromatography on silica gel using eluants of increasing polarity. This technique, employing solvents ranging in polarity from iso-octane to methanol, is usually applicable to mixtures of materials increasing in polarity from aliphatic hydrocarbons to highly ethoxylated compounds or polyols.

(1) *Preparation of silica gel column*—Insert a small glass wool plug into the bottom of a 24 × 1/2" id glass chromatographic tube, stopper the bottom of the tube with a rubber policeman, and fill with 20–30 ml of iso-octane. Slurry 10 g of silica gel with 30 ml of iso-octane, pour ca half of the slurry into the chromatographic tube, remove the rubber policeman, and pack the column by applying 5–6 psi pressure to the

tube. Do not permit the solvent to drain below the surface of the silica gel at any time during the analysis. Add the remainder of the slurry to the tube and pack as before.

(2) *Sample addition*—Accurately weigh ca 200 mg of material, dissolve or disperse in 20 ml of hot iso-octane, transfer the hot solution to the chromatographic tube, and elute at the rate of 1/2 to 1 ml/minute. Regulate all flow rates by gravity or application of air pressure to the tube. Collect all eluates in 20 ml fractions in consecutively numbered, tared 50 ml beakers. Rinse sample beaker with two 10 ml portions of hot iso-octane, allowing each rinsing to elute separately at a flow rate of 2 ml/minute.

(3) *Column elution*—Increase the flow rate to 3 ml/minute and continue the elution with the following solvents in the order indicated: 80 ml of iso-octane; 100 ml of ether-iso-octane (1+19); 100 ml of ether-iso-octane (1+9); 100 ml of ether-iso-octane (1+3); 200 ml of ether-iso-octane (1+1); 200 ml of ether; and 100 ml of methanol.

(4) *Analysis of eluates*—Evaporate the eluates on the steam bath under a jet of air and heat the residues at 105°C for 10 minutes in a drying oven. Cool and weigh the residues.

Plot as a bar graph the weights of the residues as the ordinate against the corresponding eluate numbers as abscissa. Indicate on the bar graph the eluant for each residue.

The plot will consist of $\geq$ 1 peak. Choose the largest residue in each peak and obtain its infrared film spectrum. Use the IR spectra to characterise the residues.

It is worth noting that the interpretation of the IR spectra by visual inspection could occasionally be in error. For this reason it may sometimes be necessary to compare the eluting pattern of the isolated component to that of the suspected raw material.

Determination of Water by Toluene Distillation

Transfer a 10–20 g sample to a 250 ml Erlenmeyer flask and add 50 ml of toluene and a few glass beads. Connect the flask to a Dean & Stark distilling tube receiver, and distill until no more water collects in the receiver. Cool, read the volume of water under the toluene at room temperature, and from this volume calculate the percentage of water.

Examination of Aqueous Fraction from Toluene Distillation

(1) *Specific gravity or density*—Determine with a small specific gravity bottle, or pipet 2 ml into a weighing bottle and weigh.

(2) *Propylene glycol*

    (a) Transfer 3 ml of an aqueous solution of propylene glycol to a large test tube, and add 3 ml of freshly prepared 10% catechol solution and then 6 ml of H_2SO_4. Heat in boiling water for 30 minutes. If the solution turns red and a precipitate forms, propylene glycol is present.

    (b) To several ml of solution add 5 ml of $0.02M$ KIO_4. After 1 minute add an excess of 5% alkali followed by $0.1N$ I_2 solution. If a precipitate of iodoform forms immediately, propylene glycol is present.

Silicons in Creams

The IR spectra of silicons are very characteristic, but often the silicons are present in such small percentages that they are easily overlooked in the IR spectra of the product. Some selective separation is necessary. The extraction procedures used in the course of the analysis may concentrate the silicon enough so that it can be recognised. One of a number of special procedures for silicons is as follows :

Shake the emulsion with twice its volume of (1+1) dioxane-toluene mixture, centrifuge, and draw off the top layer, which is mostly toluene containing the silicon plus a small amount of organic material. Evaporate the toluene. Examine the IR film spectrum of the residue for the presence of a silicon.

Determination of Esters of *p*-Hydroxybenzoic Acid

Esters of *p*-hydroxybenzoic acid can be isolated from emulsion products by partition chromatography and determined by ultraviolet uv spectrophotometry. The *p*-hydroxybenzoates are quantitatively eluted by the acidified 10% alcohol eluate and are determined as combined esters of *p*-hydroxybenzoic acid.

Emulsifiers

(1) *Beeswax-borax w/o cream* gives positive tests for boron and sodium in ash; it contains less than 35% water.

(2) *Beeswax-borax o/w cream* gives positive tests for boron and sodium in ash; the pH of a mixture of 1 part cream plus 9 parts water is usually between 8 and 9.

(3) *Soap o/w cream* has the following characteristics: the pH of a mixture of 1 part cream plus 9 parts water is usually between 8 and 9; a Na_2CO_3 ash is obtained; the water content is over 70%; the IR

absorption maximum of the soap may be discerned at about 6.5 μm in the spectrum of the non-volatile matter; and glycerol or other polyhydroxy compounds are probably present.

(4) *Triethanolamine-soap o/w cream* yields no ash; the pH of a mixture of 1 part cream plus 9 parts water is usually between 7 and 8; the water content is over 70%; the IR absorption maximum of soap may be observed at about 6.5 μm in the spectrum of the non-volatile matter; glycerol or other polyhydroxy compounds are likely to be present; and triethanolamine will be found in the acid aqueous solution after extraction by $CHCl_3$.

(5) *Glyceryl monostearate-soap o/w cream* gives a positive test for combined glycerol; more than 60% water is present; and glyceryl monostearate may be detected in the IR spectrum of the $CHCl_3$-extractable matter.

It may be possible to isolate some of the glyceryl monostearate by dissolving the $CHCl_3$-extractable matter in hot heptane and chilling the solution in an ice bath. Remove any precipitated glyceryl monostearate and identify by an IR film spectrum.

Aerosol Products

(1) Analyse the product that emerges from the pressurised can. The propellant dissipates almost immediately and is not available for analysis.

(2) To identify a halogenated propellant, spray a small amount of product into a suction flask and immediately stopper the top of the flask. Allow the gases emerging from the side arm of the flask to flow into a cell suitable for determination of spectra of gases. Obtain the IR spectrum of propellant, and identify. If the concentration of propellant in the gas cell is too great, remove the excess with a pipet fitted with a rubber bulb.

(If too much of the product is released into the suction flask, the flask becomes chilled, and only the more volatile constituent of the propellant vapourises).

ANALYSIS OF LIPSTICKS

A lipstick generally consists of a moulded, solid fatty base containing dissolved coal-tar dyes and suspended pigments. The base is usually composed of waxes, oils, and fatty material; typical ingredients might include carnauba wax, candelilla wax, beeswax, hydrocarbons, castor oil, oleyl alcohol, butyl stearate, polyethylene glycols, propylene glycol, lanolin, and cocoa butter.

The characteristics and quantity of the waxes present in the lipstick determine the gloss and hardness of the product. The oils are selected primarily on the basis of their solvent effects on the fluorescein dyes. Oils may account for more than half the total weight of the lipstick. The fatty material gives the lipstick film more body, softens the skin of the lips, and promotes the dispersion of insoluble pigments.

Dyes, including the fluoresceins, are used predominantly to stain the lips; lakes and other pigments are used for their brightening and covering effects. A typical product may contain as much as 10% lakes and other pigments and only 2–3% pure dyes in a fatty base. The use of colour additives in lipsticks is regulated by the Food and Drug Administration.

Because of the tendency of all oil-fat-wax mixtures to decompose rapidly, antioxidants and preservatives are added to lipsticks to prevent rancidity and deterioration. Flavours are also added to camouflage any unpleasant fatty taste and odours of the base.

General Analysis

Determine net content, non-volatile matter, ash and infrared spectra as described earlier. Note: colour, odour and other physical characteristics of the lipstick.

Lakes and Fillers

Accurately weigh 0.25–0.30 g of lipstick into a 100 ml beaker. Add 10 ml of trichloroethylene, cover with a watch glass, and heat on the steam bath to dissolve hydrocarbons, waxes etc. Filter the warm mixture under vacuum into a 250 ml suction flask through a tared. Gooch crucible containing double glass fibre filter disks for extra fine precipitates. To prevent contamination of the solvent by rubber fittings, wrap the Gooch crucible with a piece of Al foil so that the foil projects slightly below the bottom of the crucible. Wash the beaker and Gooch crucible with five 5 ml portions of trichloroethylene, followed by 5 ml portions of acetone until the washings are colourless or nearly colourless. Reserve the filtrate for determination by 5.7. Dry the crucible in an oven at 105°C for 2 hours, cool in a desiccator, let stand in air for 10 minutes, and weigh as lakes and fillers.

Trichloroethylene-Acetone Solubles

Transfer the reserved filtrate in the suction flask from above into a tared 100 ml beaker, using two 5 ml portions of trichloroethylene and two 5 ml portions of acetone. Evaporate the solvents on the steam bath under a gentle jet of air. Add 10–15 ml of $CHCl_3$ to the beaker, cover with a watch glass, and reflux on the steam bath until the sides of the beaker are free of any adhering material. Remove the watch glass and evaporate to dryness on the steam bath under a gentle jet of air. Dry the beaker in an oven at 105°C for 15 minutes, cool in a desiccator, let stand in air for 10 minutes, and weigh as trichloroethylene-acetone solubles. This fraction contains all of the ingredients of the lipstick except lakes, fillers, and volatile materials. Reserve for determination of trichloro ethylene - acetene soluble and lenolin or sterols.

Chromatographic Analysis of Trichloroethylene-Acetone Solubles

Prepare silica gel column, introduce the sample and carryout elution as per the process described earlier.

Treatment of eluates—Evaporate the eluates on the steam bath under a jet of air and heat the residues in an oven at 105°C for 10 minutes. Cool in a desiccator and weigh the residues. Obtain an IR spectrum of each residue, using the smallest amount possible to obtain the spectrum. Use the spectra to characterise the residues as containing aliphatic hydrocarbons, liquid or solid alcohols, castor oil, and esters, including isopropyl esters, acetates, etc. Reserve the residues.

Gas-Liquid Chromatographic (GLC) Determination of Castor Oil

(1) *Apparatus*

    (a) *Gas chromatograph*—with hydrogen flame detector. Operating conditions: temperatures (°C)—detector 235, injection port 300, column 235; helium carrier gas flow rate, 85 ml/minute.

    (b) *GLC column*—6'× 4 mm id glass tube packed with 100–120 mesh Chromosorb W HP coated with 5% SP-2100. Coated the support with the liquid phase by the funnel coating procedure. Condition the column overnight at 250°C with a helium flow rate of ca 10 ml/minute.

    (c) *Refluxing apparatus*—All glass, with T24/40 joints, consisting of 100 ml round-bottom flask and water-cooled refluxing condenser, heated with an electric heating mantle.

(2) *Reagents*

    (a) *Transesterification mixture*—Menthol-benzene-sulphuric acid (3+1+0.037). Prepare fresh every 2 or 3 days.

(b) *Dicyclohexyl phthalate internal standard solution*—20 mg/ml. Accurately weigh ca 1 g of dicyclohexyl phthalate in a 50 ml volumetric flask, and dilute to volume with toluene.

(c) *Methyl ricinoleate standard solution*—209 mg/ml. Purify ether-extractable methyl ricinoleate by adsorption chromatography as follows: Tamp a small glass wool plug into the bottom of a 24 × 1/4" id glass chromatographic tube. Fill the tube with ca 100 ml of $CHCl_3$. Slurry 40 g of silica gel in 100 ml of $CHCl_3$, and pack half of the slurry in the tube under 5–6 psi of air pressure. In this and subsequent steps, do not let the solvent drain below the surface of the silica gel. Pack the remainder of the slurry as before. Dissolve the ether extract of the methyl ricinoleate in 40 ml of hot $CHCl_3$ and add to the chromatographic column. Rinse the beaker with two 20 ml portions of hot $CHCl_3$ and add to the column after eluting the initial solution. Continue eluting the column with 120 ml of $CHCl_3$ and 200 ml of $CHCl_3$-ether (9+1). Discard the $CHCl_3$ eluate and collect the $CHCl_3$-ether (9+1) eluate in 25 ml portions in tared 50 ml beakers. Evaporate the eluates on the steam bath under a gentle jet of air, and dry the residues in an oven at 105°C for 5 minutes. Cool and weigh the residues. Combine the 4 largest residues and rechromatography as before. Again combine the 4 largest residues and reserve. Accurately weigh ca 1 g of residue in a 50 ml volumetric flask, and dilute to volume with toluene.

(3) *Transesterification*—Accurately weigh a 200–300 mg lipstick sample into a tared 100 ml round-bottom boiling flask. Add 20 ml of transesterification mixture and 3 glass beads to the flask. Connect the flask to a water-jacketed reflux condenser with a 24/40 male joint. Attach a heating mantle and adjust the voltage on the power source so that the solution in the flask boils gently. Reflux the solution for 2 1/2 hours. Cool the condenser and rinse with 10 ml of water. Transfer the transesterification mixture and rinse into a 125 ml separatory funnel. Rinse the condenser and boiling flask with an additional 10 ml of water and add to the separatory funnel. Rinse the condenser and flask with 25 ml of ether, transfer the rinsing to the funnel, and shake well. Draw off the lower aqueous layer into a second 125 ml separatory funnel and extract with 20 ml of ether. Discard the water layer. Combine the two ether extracts, rinse the second 125 ml separatory funnel with a small amount of ether, and add to the ether extracts. Wash the combined ether extracts with three 5 ml portions of water and discard the wash each time. Filter the ether extracts through

a cotton plug washed with 20 ml of ether into a 125 ml Erlenmeyer flask. Rinse the separator with two 5 ml portions of ether, and filter the rinsings through the cotton plug. Wash the plug with 20 ml of ether. Reduce the volume of ether to 30–40 ml by evaporating under a gentle jet of air, and reserve the remainder for (4).

(4) *Determination*—Add 1 ml of dicyclohexyl phthalate internal standard solution to the reserved ether extract from (c). Inject 2–5 μl of sample solution into the gas chromatograph. Adjust the operating conditions to elute the methyl ricinoleate peak in ca 8–10 minutes. Compare the peak heights of the sample and internal standard, and, if necessary, add 1.0 ml aliquots of dicyclohexyl phthalate internal standard solution until the ratio of sample peak height most nearly approaches 1.

Prepare a standard solution containing 6.0 ml of dicyclohexyl phthalate internal standard solution and 4.0 ml of methyl ricinoleate standard solution, and inject 2–5 μl into the gas chromatograph. Compare the peak heights and, if necessary, add 1.0 ml aliquots of standard solution or internal standard solution so that the ratio of peak heights most nearly approaches that obtained before for the sample (ca 1).

Inject a 2–5 μl aliquot of standard solution, and adjust the range and attenuation to keep the peaks from 60 to 95% of full scale. Under the conditions determined previously, alternately inject 2–5 μl aliquots of the standard and sample solutions, making a minimum of 2 injections of each.

Calculate the peak height ratios as follows :

R_u = peak height of sample methyl ricinoleate/peak height of dicyclohexyl phthalate

R_s = peak height of standard methyl ricinoleate/peak height of dicyclohexyl phthalate

Neither R_u nor R_s values from each set of injections should differ by > 5%. Determine average values of R_u and R_s, and calculate the amount of methyl ricinoleate in the sample as follows :

Methyl ricinoleate, mg

$$= (R_u/R_s) \times M \times (I_s/I'),$$

where M = mg methyl ricinoleate in standard, I_s = mg internal standard in sample, and I' = mg internal standard in standard.

The castor oil content of the lipstick can be estimated by dividing the methyl ricinoleate content by a factor of 0.9.

ANALYSIS OF SHAMPOOS

Every shampoo contains one or more surface active agents. In commercial preparations the two most widely used surfactants are soaps and alkyl sulphates. They may be salts of either alkali metals or alkanolamines. The vehicle for the surface active agent is ordinarily water. Many other materials are used in conjunction with the surfactant to sequester cations, control the viscosity of a liquid product, and alter the characteristics of the foam; or to serve as hair conditioners, bacteriostats, opacifiers, etc. Some of these materials are lanolin, the Versenes, water-soluble gums, hexachlorophene, fatty acid-alkanolamine condensates, polyoxyethylene compounds, alcohols of high molecular weight, hydrocarbons, and inorganic salts.

General Analysis

Net Contents

(1) At the beginning of the examination, mark the outside of bottle at the surface level of the liquid. At the end of the examination, empty the bottle and note the volume of water required to fill it to the mark.

(2) Determine net content, pH, ash at 600°C, distillation examination of ash, non-volatile matter and water by toluene using the procedure described earlier.

Description of Shampoo

Infrared Examination of Non-volatile Matter

(1) Obtain an infrared (IR) film spectrum of the non-volatile matter on a salt crystal.

It may be necessary to prepare the sample by slurrying the non-volatile matter with alcohol, placing a film of the mixture on a salt crystal, and drying at 105°C in an oven for 5 minutes.

(2) Examine the spectrum for the possible presence of soap, alkyl sulphates, alkanolamines, fatty acid-alkanolamine condensates, polyoxyethylene compounds, polyhydroxy compounds, and quaternary ammonium compounds.

Test for Ammonia

(1) Make a portion of the shampoo strongly alkaline with 30% NaOH, and note whether the odour of ammonia can be detected.

(2) Alternatively, hold a piece of moistened red litmus paper over the shampoo which has been made strongly alkaline. If the paper turns blue, ammonia is present.

Test for Basic Nitrogen Compounds Including Ammonia

Mix about 1 g of shampoo with 8 g of anhydrous Na_2CO_3 in a large Pyrex test tube, cover with another 2 g of Na_2CO_3, and heat the mixture strongly over a gas flame. If a moistened red litmus paper turns blue when it is held in the vapours, ammonia or some other basic nitrogen compound is present.

Lanolin and/or Sterols

The extract from any method of extraction that removes the fatty material will contain the lanolin and/or sterols. Test a portion of such an extract by the Liebermann-Burchard colorimetric test.

Water-soluble Gums

(1) *Preparation of water-repellent glass plate on which film is formed—*
Wash a 4" square of window pane glass thoroughly with soap and water, and dry it with a towel. Dip a solid glass stirring rod in a bottle of Desicote and streak the adhering liquid across the top of the plate. Repeat the streaking process several times. Rub the plate with lens paper to distribute the Desicote evenly, and then rub it with clean lens paper to remove any excess Desicote. The plate is now ready for use.

The coated plate may be used for two or three films provided it is washed with cold water and dried with a towel after each use. The plate can be recoated by repeating the described procedure.

(2) *Preparation of gum solution*—Precipitate any water-soluble gums by adding alcohol to the shampoo. Centrifuge to concentrate the gum, decant, and discard the supernatant liquid. Redisperse the gum in a minimum amount of water, reprecipitate with alcohol, centrifuge, and decant as before. Dissolve the residue in 10–25 ml of water.

(3) *Preparation and identification of gum film*—Place the water-repellent glass plate over an open 2" aperture in the steam bath. Pour the aqueous gum solution on the plate so that a circle of liquid about 2" in diameter is formed (about 10 ml is required). Heat the plate on the steam bath until all the liquid has evaporated. Remove the film with forceps (if the film is stuck to the plate it can usually be removed by scraping with a razor blade). Transfer the film to a beaker and dry at 105°C for an hour. Place a piece of dried film between two salt plates and obtain the IR spectrum. To identify the gum, compare the spectrum with spectra of known gums.

ANALYSIS OF NAIL LACQUERS

Most nail lacquers are solutions of nitrocellulose, plasticisers, and resins in mixed volatile organic solvents. Colour and opacity are produced by pigments suspended in the lacquer.

In nail lacquers, *n*-butyl phthalate, camphor, and tricresyl phosphate are common plasticisers; an aryl sulphonamide-formaldehyde polymer is a frequently used resin, and the pigments are likely to be organic lakes, guanine, and titanium dioxide. The solvents are usually mixtures of toluene, butyl acetate, ethyl acetate, ethanol, isopropanol, and butanol, and occasionally include acetone, xylenes, or methyl ethyl ketone.

A typical nail lacquer might contain 12% nitrocellulose, 5% *n*-butyl phthalate, 5% aryl sulphonamides-formaldehyde resin, 1–3% camphor, and 1–2% pigment. The solvent may approximate 35% toluene, 40% butyl acetate, 15% ethyl acetate, and 10% ethanol. Ethanol is sometimes present as the packing solvent for the nitrocellulose.

General Analysis

Net contents

Remove the brush and cap from the bottle and mark the height of the liquid on the outside of the bottle. When analysis has been completed, empty the bottle, rinse it with acetone, and fill it with water to the previously inscribed mark; then empty the water into a graduated cylinder and record the volume of liquid.

Description of Nail Lacquer

Note colour, odour, opacity, and other physical characteristics of the nail lacquer.

Infrared Film Spectrum of the Nail Lacquer

(1) Prepare a sample by coating a salt crystal with a thin film of the lacquer and drying at 105°C in an oven for 5 minutes.

(2) Observe whether the infrared (IR) spectrum can be interpreted as a mixture of nitrocellulose, *n*-butyl phthalate, and aryl sulphonamide-formaldehyde resin.

Non-volatile Matter at 105°C

After discarding the brush from the bottle top, weigh the closed bottle of nail lacquer. Pour ca 1.0–1.2 g into a tared weighing bottle 65 mm high and 45 mm in diameter, and weigh the nail lacquer bottle again. The difference in weight is the sample weight. With the top removed, manipulate the weighing bottle so that the lacquer covers the entire inside surface of the bottle as a thin film. Then heat the lacquer at 105°C in an oven for 2 hours. Cool and weigh the residue as non-volatile matter.

Determination of Non-volatile Constituents

The overwhelming majority of commercial nail lacquers contain these non-volatile components: nitrocellulose, organic lakes, inorganic pigments such as TiO_2, *n*-butyl phthalate, and an aryl sulphonamide-formaldehyde resin. The following is a method of analysis (Fig. 44.4) for these nail lacquer ingredients :

(a) Dilute a 4–6 g sample with 5 ml of acetone, add 20 ml of benzene, and pour slowly with stirring into a 400 ml beaker containing 150 ml of hot benzene. Rinse the sample container with 5 ml of acetone and add the rinsings to the benzene. Evaporate the benzene on the steam bath under a gentle jet of air to about 80 ml, dilute with 90 ml of benzene, and cool to room temperature. Pour the mixture into a 250 ml centrifuge tube, rinse the precipitation beaker with 10 ml of benzene, and add the rinsings to the centrifuge tube. Reserve the precipitation beaker and any precipitate that adhered to the sides. After centrifuging, decant the supernatant liquid into a 250 ml beaker labelled no. 1 and set it aside.

(b) Dissolve the residue in the precipitation beaker and the centrifuge tube from (a) with 15, 10, and 10 ml portions of acetone, and pour the combined acetone solutions into a 100 ml beaker. Evaporate the acetone on the steam bath, redissolve the residue in 10 ml of acetone, add 20 ml of benzene, and repeat the precipitation and centrifuging procedure described in (a). Decant the supernatant liquid into a beaker labelled no. 2. Reserve the precipitation beaker and the centrifuge tube containing the residues.

(c) Filter the decanted liquids in beakers No. 1, (a), and No. 2 (b), through the same 12.5. cm filter paper into two tared beakers labelled I and II. Reserve the filter paper and beakers Nos. 1 and 2.

(d) Evaporate the filtrates in beakers I and II, (c), on the steam bath under jets of air, and dry the residues in an oven at 105°C for 10 minutes. Cool, and weigh I and II as combined resin and plasticiser.

(e) Rinse the reserved beakers Nos. 1 and 2, (c), with two 30 ml portions of hot methyl ethyl ketone, and pour the rinsings through the reserved filter paper, (c), into a tared beaker. Discard the filter paper, evaporate the filtrate on the steam bath under a jet of air, dry the residue in an oven at 105°C for 20 minutes, cool, and weigh as a mixture consisting essentially of nitrocellulose and pigments.

Obtain an IR film spectrum of the residue from a film prepared in the following manner: Dissolve a little of the material in acetone, pour some of the solution on a salt crystal, let the acetone evaporate in air, dry the film on the crystal in an oven at 105°C for 5 minutes, and cool to room temperature.

(f) Dissolve the residues in the reserved precipitation beaker and the centrifuge tube, (b), in acetone, and transfer the acetone solutions to a tared 250 ml beaker. Evaporate the acetone on the steam bath, redissolve the residue in 5 ml of acetone, and add 75 ml of alcohol-ether solution (1+2) followed by 10 ml of water. Evaporate the solvent on the steam bath under a gentle jet of air and dry the residue in an oven at 105°C for 1 1/2 hours. Cool, and weigh as nitrocellulose plus pigments.

Obtain an IR film spectrum of the residue as described in (e)

(g) Use 32 ml of methanol to dissolve and transfer the resin and plasticiser residues in beakers I and II, (d), into a 250 ml separatory funnel. Add 8 ml of water and extract with four 40 ml portions of petroleum ether. Reserve the extracted methanol solution. Set the separatory funnels aside for rinsing later.

(h) Filter the combined petroleum ether extracts, (g), through a 12.5 cm filter paper into a tared 250 ml beaker. Follow with an additional 40 ml of petroleum ether wash through the filter paper. Reserve the filter paper. Evaporate the filtrate on the steam bath under a jet of air, dry the residue in an oven at 105°C for 10 minutes, cool, and weigh as butyl phthalate.

Obtain an IR film spectrum of the material from a liquid film spread on a salt crystal. Also obtain the ultraviolet uv spectrum of an alcohol solution of the phthalate.

(i) Transfer the reserved extracted methanol solution, (g), to a 500 ml separatory funnel, add 50 ml of $CHCl_3$, dilute with 200 ml of water, and acidify with a little HCl. Shake the mixture well and draw off the $CHCl_3$ layer. Continue the extraction with additional 50, 50, and 30 ml portions of $CHCl_3$. Reserve the extracted aqueous solution. Set the separatory funnel aside for rinsing later.

(j) Filter the combined $CHCl_3$ extracts, (i), through the reserved filter paper, (h), and follow with an additional 40 ml wash with $CHCl_3$.

Reserve the filter paper. Evaporate the filtrate on the steam bath under a jet of air, dry the residue in an oven at 105°C for 10 minutes, cool, and weigh as aryl sulphonamide-formaldehyde resin.

Obtain an IR film spectrum of the resin as described in (e). Also obtain a uv spectrum of an alcohol solution of the material.

(k) Filter the extracted aqueous solution, (i), through the reserved filter paper, (j). Discard the filtrate. Rinse all the separatory funnels used in the extractions, (g) and (i), with two 30 ml portions of acetone, and filter the acetone solutions through the filter paper into a tared 250 ml beaker. Discard the filter paper. Evaporate the filtrate on the steam bath, dry the residue in an oven at 105°C for 10 minutes, cool, and weigh as a mixture of resin and nitrocellulose.

Obtain an IR film spectrum of the material as in (e).

Separation of Nitrocellulose from Pigments

(1) *Guanine*—The mother-of-pearl appearance of some nail lacquers arises from the presence of guanine. To analyse such a product, dilute a weighed sample with acetone, centrifuge, and decant the acetone. Identify the guanine residue from the infrared spectrum of a mineral oil mull or from a film prepared by rubbing the guanine between two salt crystals with a drop of acetone. Separate the crystals and evaporate the acetone at 105°C in an oven.

(2) *Opaque coloured nail lacquers*—The opaque coloured effect is achieved through the dispersion of finely divided organic lakes and inorganic pigments such as TiO_2. The nitrocellulose may be separated from the colouring agents by the following procedure (3) :

Dissolve, disperse, and transfer the residues of precipitated nitrocellulose and pigments, from steps of the non-volatile estimation process into a 50 ml heavy-duty centrifuge tube with the aid of 25 ml of hot methyl ethyl ketone. Add 4 drops of water and 50 mg of silicic acid (Mallinckrodt's chromatographic grade) to the solution. Centrifuge at high speed for one hour and decant the supernatant liquid into a 250 ml separator. Treat the residue in the centrifuge tube with successive 15 and 10 ml portions of methyl ethyl ketone, centrifuging 15 minutes each time and decanting as before into the separator. Discard the residue in the centrifuge tube.

Add 1 ml of HCl to the separator, shake 2–3 minutes, and extract with 50 ml of water saturated with methyl ethyl ketone. Discard the extract. Make the methyl ethyl ketone solution slightly basic with ammonia, and extract with 50 ml of an aqueous solution which is weakly ammonical and saturated with the ketone, and which contains 0.5% ammonium chloride. Continue the alkaline extracts until no further colour is extracted (three extractions usually suffice). Discard the extracts, re-acidify the ketone solution with HCl, and wash with two 25 ml portions of water that is weakly acid with HCl and saturated with the ketone. Discard the washings.

Filter the extracted methyl ethyl ketone solution through an 11 cm filter paper into a 250 ml tared beaker. Wash the separator and the filter paper with two 25 ml portions of hot methyl ethyl ketone, collecting the washings in the tared beaker.

Evaporate the volatile solvent on the steam bath under a gentle jet of air. Redissolve the residue in 5 ml of acetone and add 75 ml of alcohol-ether solution (1+2) followed by 10 ml of water. Again evaporate the solvent on the steam bath under a gentle jet of air and dry at 105°C in an oven for 1 1/2 hours. Cool, and weigh as nitrocellulose.

In separating the nitrocellulose and the pigments from the nail lacquers, traces of pigment may adhere to the sides of the glassware or on the filter paper. These traces can be dissolved by alcohol strongly acidified with HCl.

Tricresyl Phosphate in Nail Lacquers

When tricresyl phosphate is used with *n*-butyl phthalate as the plasticiser, the two substances are isolated together. The IR spectra of these two materials differ sufficiently so that each can be identified in the presence of the other. If it is necessary to determine the amount of each component, the uv spectra differ enough so that the methods for the analysis of two-component mixtures can be applied.

Acrylonitrile-Butadiene Polymer

An acrylonitrile-butadiene polymer may be detected in the infrared spectrum of the non-volatile matter. The CN absorbance peak is at 4.45 μm.

When the polymer is present as an additional ingredient in a nail lacquer, it is isolated by precipitation with methanol. The methanol solution may be used for the usual nail lacquer analysis.

Base Coats

(1) Base coats are preparations which are applied prior to the nail lacquer. They are usually solutions of polymeric substances. The polymers are most readily recognised by examination of the IR spectrum of the non-volatile matter. Ultra-violet spectra, sometimes of films on a quartz surface, are also helpful if the polymers have aromatic constituents. When a base coat contains two or more polymers, it is good practice to separate them, e.g., by extraction.

(2) Acrylonitrile-butadiene and phenol-formaldehyde polymers are among those used in base coats.

Infrared Spectrophotometric Analysis

Description and Interpretation of Spectra

A commercial colourless nail lacquer, known to contain nitrocellulose, *n*-butyl phthalate, and aryl sulphonamide-formaldehyde resin as non-volatile ingredients, was analysed. The IR spectra of the following fractions were examined.

(1) *Non-volatile matter*

    (a) The presence of nitrocellulose is immediately apparent from the absorption maximum just beyond 6 μm and the one at 12 μm.

    (b) Although there is no positive identification for *n*-butyl phthalate, the indication of carbonyl absorption at 5.85 μm suggests that it might very well be a constituent of the product. Likewise the absorption at 8.6 μm may arise from an aryl sulphonamide-formaldehyde resin.

(2) *Nitrocellulose*

    (a) The principal absorption maxima at 6.05 and 7.8 μm are characteristic of the covalent nitrate groups. The strong absorption at 11.9 μm is also associated with the nitrate group and that at 9.4 μm with the cellulose ring. Note the broad but weak OH bond absorption just before 3 μm and the weak C–H absorption just before 3.5 μm. The minor absorption maxima also aid in identifying nitrocellulose.

    (b) In the analysis of coloured opaque nail lacquers containing suspended pigments, the precipitation of the nitrocellulose entrains the pigments. However, it is difficult to differentiate the infrared film spectrum of the contaminated nitrocellulose from that of pure nitrocellulose. In part this is due to the small amount of pigment, 1–2%, ordinarily present in nail lacquers.

(3) *n-Butyl phthalate*

    (a) The characteristic strong absorption maxima of phthalates are the carbonyl ester absorption at 5.8 μm and the maxima at 7.8, 8.9, and 9.3 μm. The weak doublet centred around 6.3 μm always appears in phthalates. The other absorption peaks are also helpful for identification.

(2) The intensity of the C–H absorption in the 3.5 μm region is indicative of the length of the alkyl carbon chain of the ester.

(4) *Aryl sulphonamide-formaldehyde resin*—is the spectrum of a commercial sample of the resin.

(a) The major absorption maxima attributable to sulphonamides appear at 7.5 and 8.6 μm. The weak C–H bond absorption in the 3.5 μm region as well as the relatively strong absorption between 12 and 15 μm indicates the aromatic nature of the material. The weak but sharp absorption peak just before 3 μm suggests N–H linkages. The absorption at other wavelengths aids in uniquely identifying the resin.

(b) There is some difference in absorption between the commercial and isolated resins in the 13.5 μm region. The resin may be slightly altered chemically, in the course of the analysis, or else some component of the resin may be lost in the extraction procedure.

Infrared Spectra of Guanine and Tricresyl Phosphate

(1) *Guanine from commercial nail lacquer*—Spectra of a known sample of guanine obtained by heating the hydrochloride to 200°C.

Guanine has many identifying absorption maxima. Two readily recognisable linkages are the N–H absorption between 2.9 and 3.5 μm and carbonyl absorption below 6 μm.

(2) *Tricresyl phosphate*—Technical grade (80% *para* and 20% *meta* isomers). The spectrum of an *n*-butyl phthalate-tricresyl phosphate mixture spectrum can easily be interpreted as a composite of the widely differing individual spectra of the phthalate and phosphate.

The weak C–H absorption maximum at 3.5 μm, the absorption between 6 and 7 μm, and that beyond 12 μm strongly suggest the aromatic character of tricresyl phosphate. The strongest absorption maximum, however, occurs between 10 and 11 μm and is attributed to the phosphate radical.

Gas-Liquid Chromatography

Gas-Liquid Chromatographic (GLC) Determination of Solvent and Camphor (Plasticiser)

(1) *Apparatus and reagents*—(1) *Gas chromatograph*—F&M Model 810, or equivalent, equipped with a thermal conductivity detector capable of operating in a rapid program mode after an initial isothermal period. Operating temperatures (°C): injection port 260; detector 250; column

initially 118 until solvents elute (ca 20 minutes), then programmed to 240 at 60/minute, held isothermal until camphor elutes (ca 7 minutes), and returned to the starting temperature, Helium carrier gas flow rate: 80 ml/minute.

(a) *GLC column*—Copper tubing, 20'×1/4" outer diameter packed with 70–80 mesh Gas-Chrom R coated with 10% PEG 20M.

(b) *Stock solutions*—Accurately weigh ca 5 g each of isopropanol, ethyl acetate, toluene, butyl acetate, butanol, camphor (dl), and propyl acetate (internal standard) into separate 25 ml volumetric flasks and dilute each to volume with iso-octane (distilled in glass).

(2) *Preparations of samples*—Accurately weigh ca 1 g of nail lacquer into a tared 25 ml glass-stoppered Erlenmeyer flask. Slowly add, with stirring, 10–15 ml of iso-octane. Pulverise the resulting precipitate with a stirring rod. Add 1.0 ml of *n*-propyl acetate internal standard solution. Stir the mixture and let it settle for 4–5 minutes before analysis.

(3) *Preparation of standard solution*—Prepare a standard solution containing the following amounts of the stock solutions from (a): 2.0 ml of ethyl acetate (ca 400 mg), 3.0 ml of propyl acetate internal standard (ca 600 mg), 4.0 ml of toluene (ca 800 mg), 4.0 ml of butyl acetate (ca 800 mg), 6.0 ml of butanol (ca 1200 mg), and 2.0 ml of camphor (ca 400 mg).

(4) *GLC determination*—Inject 10–15 μl of sample onto the column, using the operating temperatures given in (a). Adjust the sample volume and/ or attenuation to bring each peak to ca 50–90% of full scale deflection (do not adjust the iso-octane peak). The sample volume should not exceed 15 μl. In a similar manner, inject the standard solution and a determine the correct injection volume and attenuations. Identify the solvents in the sample by comparing their retention times with those of the components of the standard solution. After determining the correct operating conditions, alternatively inject the sample and standard ≥ 3 times each.

(5) *Calculations*—Calculate the amount of each solvent or plasticiser in the sample as follows :

% Solvent or plasticiser = $(R_u/R_s) \times K_s \times (IS_u/IS_s) \times (100/W)$,

where R_u = peak height of solvent in sample × attenuation/peak height of internal standard in sample × attenuation, R_s = peak height of standard solvent in standard solution × attenuation/peak height of internal standard in standard solution × attenuation, K_s = mg of standard solvent in standard solution, IS_u = mg of internal standard in sample solution, IS_s = mg of internal standard in standard solution, and W = mg of sample.

Gas-Liquid Chromatographic Determination of Acetone, Xylene, and Methyl Ethyl Ketone

Acetone, xylene, and methyl ethyl ketone may also be determined by GLC with a minor temperature adjustment (1–2°C) for methyl ethyl ketone and a longer initial isothermal operating period for xylene. If either ethanol or isopropanol is present by itself, the alcohol can be identified by its retention time. Mixtures of these alcohols cannot be resolved under the conditions described. The shape of the curve indicates the presence of a mixture and some estimate can be made of the relative amounts of each. It should be noted that mixtures of alcohols may also be determined by headspace analysis.

ANALYSIS OF SUNSCREEN PRODUCTS

Sunscreens are incorporated into cosmetic preparations for the selective absorption of potentially harmful burning radiation between 290 and 320 nm while permitting the transmission of longer wavelengths that permit tanning. A variety of chemical compounds, mostly aromatic, are available for incorporation into creams, lotions, and solutions as sunscreens. These preparations can be roughly classified according to the vehicle or base in which the sunscreen is incorporated, e.g., hydrocarbons, vegetable oils, alcohols, and oil-in-water or water-in-oil emulsions.

Determine net contents, description of product, non-volatile matter at 105°C, ash content and pH as per the procedure described earlier.

Infrared Examination of Non-volatile Matter

Examine the infrared (IR) spectrum for possible identification of sunscreens as well as other materials.

Determination of Sunscreen

(1) *In an alcohol base*—Dilute a known amount of the product to a definite volume with alcohol. Identify and determine the sunscreen from the ultra-violet uv spectrum of the resulting solution. Make dilutions as necessary.

(2) *In an emulsion base*—It is generally necessary to isolate the sunscreen from the emulsion before its determination. The method for the isolation and determination of amyl *p*-dimethylaminobenzoate described in Official Methods of Analysis, (AOAC), with some modifications, is also applicable to the determination of certain other sunscreens. For

example, elute 2-ethoxyethyl *p*-methoxycinnamate with acidified 50% alcohol and homomenthylsalicylate with acidified 70% alcohol. Before eluting with eluants of high alcohol concentration, wash the eluant with 2–3 ml of the immobile solvent, *n*-heptane-CCl_4 (1+1). Determine these sunscreens in acid solution by uv spectrophotometry.

(3) *In a hydrocarbon or vegetable oil base*

    (a) Dilute the preparation with spectrophotometric grade iso-octane. Identify the determine the sunscreen from the uv spectrum of the isooctane solution.

    (b) Alternatively, it may be necessary to isolate the sunscreen from the hydrocarbon or vegetable oil base by partition chromatography before the determinative step. If so, chromatograph the preparation directly without prior solvent extraction, and then proceed as in (2).

Analysis of Sunscreen Vehicle

(1) *Alcholol base*—See determination of sunscreen.

(2) *Hydrocarbon or vegetable oil base*—An examination of the IR spectrum of the non-volatile matter should be sufficient for the characterisation of these base ingredients.

(3) *Emulsions*—These preparations are normally mixtures of several ingredients. The base ingredients can generally be determined by methods used for creams and lotions. (Fig. 44.4).

SKIN SENSITISATION AND SENSITIVITY TESTING

A variety of substances are used in the manufacture of cosmetics. Therefore, finished cosmetics when used on human body have potential for several type of adverse reactions. The adverse effects that may be caused include skin irritation and allergic sensitisation, contact urticaria, stinging, phototoxicity and photoallergy. Skin irritation and allergic sensitisation is the most frequently encountered adverse effect. With growing consumer awareness and enforcement of consumer protection act, it is necessary for cosmetic manufacturer to assess the potential of adverse effects of his product. Safety considerations are not only relevant to consumers but also to factory workers who handle cosmetic raw materials in large quantities.

The substances that induce inflammation are known as irritants. Inflammation may be caused by use either immediately or through prolonged use or on repeated use. Irritants may be classified as primary

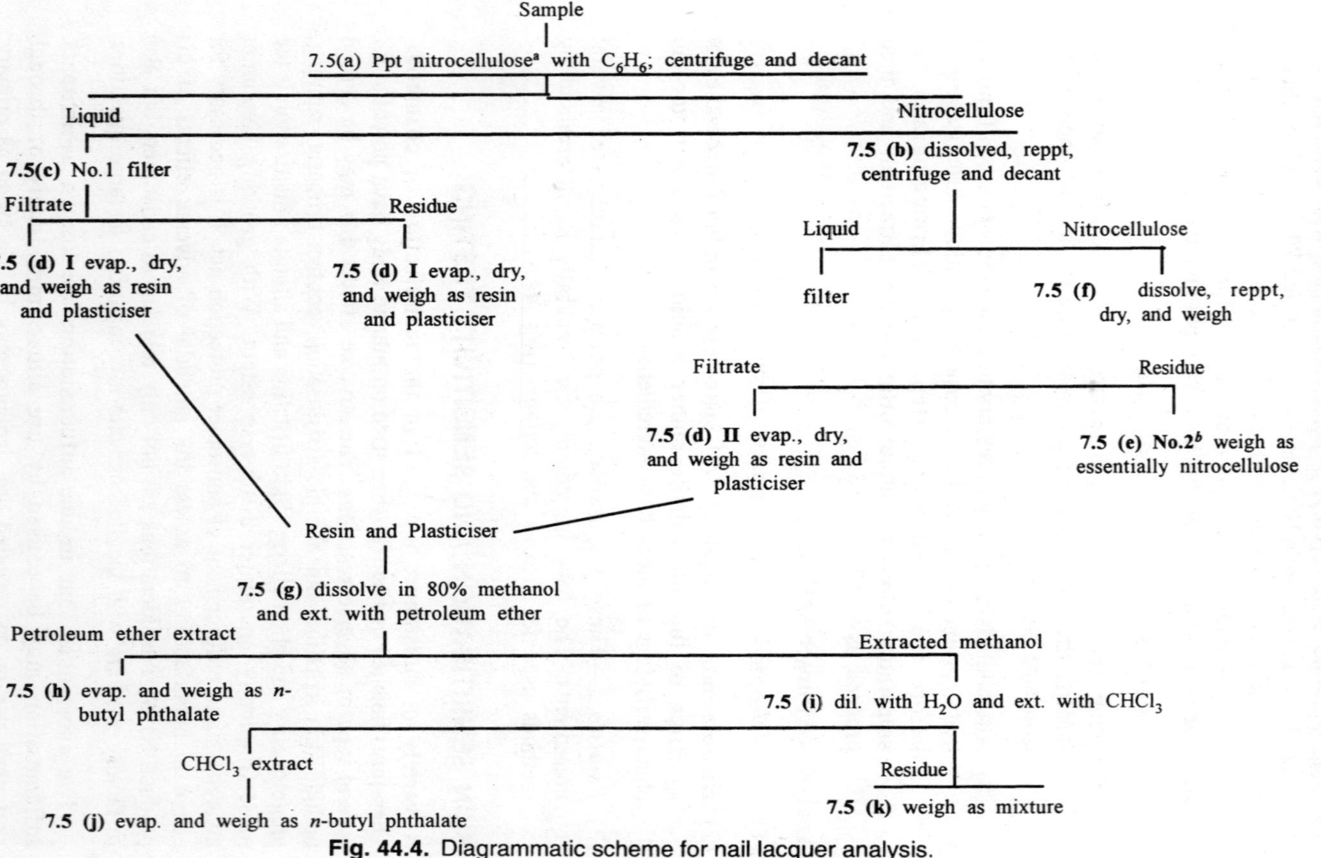

Fig. 44.4. Diagrammatic scheme for nail lacquer analysis.

irritants and secondary irritants. Primary irritants cause inflammation on the first contact (the duration of contact may be of several hours). Secondary irritants are harmless on first contact but cause inflammation on repeated contact and inflammation becomes more severe with progressive contacts.

In view of the potential of the cosmetic for harmful effects mentioned above, some tests have been designed to predict the potential of substances to induce irritation or sensitisation.

Sensitivity Testing

For detection of potential primary irritation, Draize test or its slight modification is used. In this test, Albino rabbits are clipped and the substance to be tested is applied to : intact skin, abraded skin, or lightly scarified skin.

All of them are covered with a patch for 24 hours. The sites of applications are examined at intervals and changes are assessed and recorded. The skin of rabbit is more susceptible than man. However, this method of testing can lead to false positive and false negative. It is advisable to compare results of test substance with the results of known harmless substance. Bureau of Indian Standards 4011 - 1982 (6.1 - Test for Irritant Potential) recommends that if there is no reaction in any of the animals, the same test should be performed on 10 human volunteers applying the substance on the skin of the forearm.

Patch Test

On humans, sensitivity testing of cosmetics may be performed either as a diagnostic or as a prophetic test. By diagnostic test if is intended to discover whether the cosmetic used has caused dermitis and if cosmetic is known, the ingredient which has caused it. It is known as diagnostic patch test. Prophetic test is done to assess whether a new cosmetic should be placed on the market or not. This test is known as prophetic patch test. General procedure of patch test is given below.

Place about 0.1–0.3 gm of cosmetic to be tested on a piece of cotton fabric or flannel (2–3 sq.cm. in size) and apply this to the skin of arms, thighs or back. This patch is covered with a patch of cellophane (about 5 sq.cm.) and sealed with adhesive plaster (about 40 sq.cm.). Apply several patches at one time. Out of these several patches some of them should be of similar cosmetics of other brand available in the market and known not to cause any harm to the skin. Other similar cosmetics act as controls. These patches are allowed to remain on the skin for 24–72 hours. If there are no reactions, the same patch may be reapplied to the same place or a fresh patch of the same material may be made and applied. This may be continued till :

(1) Either a reaction is produced under one or more patches, or

(2) Investigator is confirmed that no reaction will occur.

Reading and grading of patches is best done by an experienced dermatologist. Sites of patches should be examined after 30 minutes of removal of patch. Observation can be made a little earlier but not earlier than 15 minutes. Usually the skin under adhesive tape gets inflammed, the skin under cellophane tape remains clear, the skin under test patch may not have reaction or may have reactions like erythema, erythema with papules, papulo-vascular reaction or ulceration or neurosis. Patch test reactions, usually, are graded as under :

Description of Observation	Symbol
No reaction	−
Erythema only	+
Erythema with papules	++
Papulovascular	+++
Ulceration of Neurosis	++++

Open patch test

In case of cosmetics containing higher percentage of potential irritants like hair dyes, shampoos, hair tonics, patches should not be sealed. These should be used as open patches. Open patch test is performed on sensitive part of skin, e.g., bend of elbow, popliteal space of the skin behind the ears. The site of patch is inspected after 24 hours. If there is no reaction, the test is repeated once more on the same site. If still no reaction is there, the test is repeated a third time. If no reaction is observed on third application, the person may be taken as not hypersensitive.

Prophetic patch test

Before a programme of prophetic skin test is undertaken, the investigator should perform test on himself to ascertain that the material is not primary irritant. After satisfying himself, the investigator can select ten subjects. He should get signed from each of the subjects a waiver for any damage. The amount of money to be paid should also be got agreed upon. If the results of the test are favourable, the investigator may use more subjects. For full scale prophetic patch test, 200 normal subjects are used. The cosmetic to be used is placed on the skin of the subjects for one to five days depending upon the nature of the cosmetic and judgement of investigator. The purpose of this application is inducing or establishing sensitivity and also detecting the presence of any primary irritant the patch sites are examined and observations are made. Subjects are observed for three more days for development of any late reactions. After 7–10 days, patches are again applied to the same area of those subjects who did not show reaction to the test. One or more similar cosmetics known to have been used for years without complaints are used as controls. If there are no reactions or one or two reactions out of 200, new product can be placed on trial sale in small community of about 10,000 persons.

Prophetic patch test may be covered or open. Many a time, question is asked what is percentage of hypersensitive persons that is allowable. Schwartz suggests that incidence more than one in 10,000 may not result in safe cosmetics.

Repeated Insult Test

The prophetic patch test has certain shortcomings, to name :

(1) Quick absorption of potential skin irritant through skin.

(2) Rapid evaporation of volatile skin irritant from the patch.

(3) Small amount of cosmetic in patch in comparison to large amount in actual use.

(4) Small area of skin used in comparison with skin surface area used in actual use of cosmetic.

(5) Short exposure of skin to cosmetic than the exposure actual in use.

To overcome some of the shortcomings of prophetic patch test, its modification in the form of repeated insult technique has been suggested. Draize described this technique as under.

The repeated insult technique consists in applying the finished cosmetic, or if the test substance is an ingredient of the cosmetic, it is applied in the same concentration as is found in the finished formulation incorporated in a bland base. The bland base must necessarily vary according to the nature and solubility of the test substance. Therefore, the diluting bland base may vary from physiologic saline to such nonirritating, nonsensitising organic solvent as dimethyl phthalate. Ordinarily 0.5 gram or 0.5 ml. of the test sample is applied by usual patch test procedure. The skin of the back or upper arm is used, such areas as are ordinarily covered by wearing apparel. The test material is maintained *in situ* for 24 hours. Allowing 15 to 20 minutes after the removal of the patches, readings are made and the reactions recorded. The test subject is allowed 24 hours of rest and the second test patch is applied to a different test site, otherwise the procedure is identical to that of the first application. Each individual is thus subjected to a series of 10 such consecutive exposures, exclusive of Sundays and nonwork days. Following the 10 individual exposures the subject is given 10 to 14 days rest after which time a "retest application" is made similar to one of the original 10 applications. A comparison of the reaction following "retest" with the average reading of the 10 original applications permits an appraisal of the sensitising propensities of the test substance. Preferably the test subjects should be equally divided as to sex and cover as wide an age range as practicable.

To be comprehensive, a panel of 200 test subjects should be employed. Two hundred subjects represent an auspicious number and smaller groups possibly should be used at first, gradually building up to 200 subjects. It would be unwise to use a large number if it were possible to demonstrate on a smaller panel, 25 or 50 individuals that the substance is a potent sensitiser and unfit for its intended use.

Another modification of patch test for volatile substances has been described in BIS : 4011 in section 5.3.1 reader may refer to the same.

In BIS : 4011 - 1982 it has been suggested that this test should first be performed on 10 laboratory animals and if there is no reaction in any of the animals, it should be performed on 10 human volunteers.

Photopatch Test

Certain substances are not harmful by themselves but they become harmful when exposed to sunlight. It has also been mentioned earlier that substances that absorb light wavelengths between 300–800 nm have potential of phototoxicity. In case, a substance is considered phototoxic, photopatch test may be performed. To perform this test, the substance to be tested is applied in duplicate patches in the same manner as for standard patch test. After 24 hours, one of the patches in the pair is exposed to sunlight for 30 minutes. Alternatively the patch can also be exposed to ultraviolet light. The light exposed patch is covered again. One additional site in the adjoining area of the skin is exposed to sunlight or ultraviolet light as has been done in case of one patch of the pair. The time of exposure is also same as for the exposed patch. This site acts as control. After further 24 hours the patches are opened and examined. If the patch not exposed to light and the skin area exposed to light do not show reaction but the patch site which has been exposed to light shows reaction, the test indicates that substance is phototoxic. If no reactions are observed on patch sites and control skin site, the substance may be taken as non-phototoxic.

Test for Sensitising Potential

An individual is not likely to be allergic to a substance, if he is exposed to it for the first time. In immunological terms, it can be said that lymphocytes of that individual acquire capability to recognise and elicit immunological response. Thus next exposure of that individual to the substance has chance of sensitising him. With the number of exposure his chances of developing hyper-sensitivity increase.

Test for sensitising potential has been described in BIS : 4011. It has been recommended that standard patch tests with same chemical or cosmetic are repeated in the same 200 volunteers after an interval of 10–14 days. The number of persons who will show positive reaction will represent the sensitising potential of the substance or cosmetic which has been tested.

Provocative Patch Test

Weak sensitising agents, generally, do not sensitise a person with first exposure. Repeated exposures, however, to even weak sensitising agent may make a person hyper-sensitive. This can be indicated by provocative patch test. In this test, 10–15 applications of a substance or cosmetic are made on alternative days on the same spot of skin in 10 volunteers. Standard patch test is performed after 10 days of the last application. The test will show, how many persons have been sensitised. It has also been recommended in the Indian Standard that in case of unknown chemicals, the test should be first performed on animals like guineapigs or rabbits.

Use Test

In use test, the cosmetic to be tested is actually used and its adverse effects, if any, are observed. BIS : 4011 suggests that 15 volunteers should be asked to use the cosmetic and they should make 15 applications of it. If there is no adverse reaction, cosmetic can be released for trial.

Skin Testing with some Specific Cosmetics

Creams

Dermatitis, usually with cleansing and emollient type creams, is not frequent. Dermatitis may occur with creams containing substances like mercurials, salicylic acid, deodorant creams containing salts of aluminium, bleaching creams containing oxidising agents and vanishing creams (because of alkalinity). Another important ingredients in creams is perfume which may cause dermatitis. Allergy has also been reported because of lanolin. Hydroxy fatty acids in lanolin are said to be allergic in nature.

To carry out diagnostic test, rub cream into an unaffected suitable part of the body (e.g., forearm) for five days. However, if positive reaction occurs before five days, rubbing of cream may not be continued. While carrying out prophetic patch test, the patch may be covered. A similar cream available in market for a long time without any adverse effects may be chosen as control. The best test remains the user test.

Deodorants and Antiperspirants

Dermatitis from deodorants is infrequent. However, some phenolic antimicrobials can cause dermatitis. Perfumes of deodorants can cause dermatitis in hyper-sensitive persons.

Deodorants in powder and cream form can be tested using diagnostic patch test. In these forms the patches can be covered. But perfumes in solution should not be sealed in patch test. Perfume should be sprayed on the skin and the skin should be kept open.

On the other hand, dermatitis from antiperspirant is not uncommon. Aluminium chloride in antiperspirant is more irritant than aluminium sulphate. Often modern antiperspirant preparations are buffered to minimise irritation. Both diagnostic and prophetic patch tests can be employed for these cosmetics.

Depilatories

Depilatories (chemical) contain sulphides of alkali (e.g., NaS), alkaline earth (BaS) or thio-glycolic acid. These substances are reducing agents and disrupt cystine linkages of hair protein and thereby act as depilatories. These are capable of acting as primary irritants. Often some degree of erytherma is associated with depilatories. Intensity will depend upon concentration, pH and time of exposure. Therefore losed patch tests for diagnostic purposes are not recommended.

In case of testing of new depilatory, actual user test should be employed with well established depilatory in market as control. However, before undertaking actual user test, irritant properties of new depilatory may be assessed by carrying out prophetic test on 10 subjects with well established depilatory as control. The following procedure could be followed for the prophetic patch test.

Prepare inactive base of depilatory. Prepare dilutions given below of the well established depilatory and depilatory to be tested :

one part depilatory + one part base

one part depilatory + two parts base

one part depilatory + three parts base

one part depilatory + four parts base

+

One part depilatory + n parts base

Place patches of well established depilatory dilutions on one thigh and patches of depilatory dilutions to be tested on other thigh of the subject. Examine the patches on two thighs after appropriate time. This time could be 24–48 hours.

Hair dyes

If prophetic patch test is to be carried out on new hair dye, its composition should be ascertained. Because it may contain primary irritant, e.g., ammonia (more than 1% ammonia is primary irritant). Prophetic patch should not be closed, if the hair dye contains primary irritant.

Normally hair dye is applied to grey hair and is allowed to remain in contact with hair for about 1–1 1/2 hour. Excess cf unconsumed hair dye is washed with shampoo. Hair dye is applied every four to six weeks to colour emerging roots of hair. In view of this open prophetic patch test should be used in case of hair dyes. The test could be carried out as follows :

Prepare hair dye for application and paint one square inch skin on the forearm or behind the ear. Repeat it every day for three days (application should not be washed off during this period). On the other forearm run a control with well established hair dye available in market. Repeat the procedure after seven to fourteen days. Reaction of the dye to be tested should not be more than the reaction with control dye.

If the new hair dye passes this test, it should be tried on 1000 subjects twice. When found satisfactory on trial, it may be placed on the market.

Cold wave lotion

Thioglycollates are used in permanent waving lotions. Thioglycollates in higher pH on long exposure can cause dermatitis. However, dermatitis from cold wave lotions is not frequent. Closed prophetic patch test can be performed with long established cold wave lotion. The following procedure could be followed :

Place 0.2 ml of new preparation on a piece of flannel (3 sq.cm. in area). Place this patch on forearm, cover it with uncoated cellophane (about 9 sq. Cm. in area) and cover this with adhesive plaster (about 18 sq. cm. in area). A similar patch of control waving lotion is placed on the other forearm. Remove the patches after 24 hours and examine the patch sites after 15–30 minutes of removal of patches. Reaction should not be more than the reaction with control. The test is repeated after 7–10 days for finding out sensitisation potential. This test should be performed on 100 female subjects.

This test is more stringent than actual user test as in practice cold wave lotion does not come in contact with scalp and even if it comes in contact, the contact time is not more than 3 hours. Another factor for stringency is that in practice, cold wave lotion treatment is given once in a few months while the second patch test is followed within 7 – 10 days.

Lipsticks

Before introduction of indelible type lipstick cheilitis was rare. Cheilitis occurred frequently with the use of bromofluorescein type dyes in lipstick formula. This adverse effect was caused by photosensitising effect of fluorescent dyes. However, the fluorescent dyes being used in modern cosmetics are less photosensitive than earlier ones. Addition of pigment lakes which act opacifiers reduce photo-sensitisation.

Closed prophetic test can be performed on lipsticks. After removing the patch, the patch site should be exposed to bright sunlight for several hours to observe the tendency of the lipstick to photosensitisation. A well-estbalished lipstick in market should be used as control. At least 200 subjects should be taken for this test. Along with patch test, they should also use the lipstick under test and should continue to use for at least one month. If results of patch test and actual user test are favourable, an extensive user test on additional 800 subjects should be carried out before the lipstick is put on the market.

Nail polish

With use of nail polish fingers have been rarely affected. However, dermatitis may be caused at those places which are frequently and habitually touched, stroked or scratched by finger nails. Solvents of nail polish evaporate quickly. As such, causative factors could be resins, colour and pigments.

Earlier actual loss of nail occurred due to phenolic type of resin and synthetic rubber. Their use has been discontinued since then. Splitting of nails on long usage of nail polish has been observed.

In case of nail polishes, open prophetic patch test is carried out. The following procedure could be used :

Apply new nail polish about 1 sq. inch in area on forearm. A well-known and established nail polish should similarly be applied to the other forearm. Do not cover the patch. Apply test nail polish on nails of hand which has patch of test nail polish. Similarly apply control nail polish on the nails of forearm which has patch of control nail polish. Remove patches after 48 hours and after 7 – 10 days repeat the test on same subjects. Continue the use of nail polishes for two months. In performing this test take 200 subjects. Compare the results of test nail polish and control nail polish. Reactions with test nail polish should not be greater than the control nail polish. If new nail polish passes this test, user test on additional 800 subjects should be carried out before nail polish is put on the market.

Testing of hair & bath preparations for eye irritation properties

Eyes can be accidentally exposed to shampoos, other hair preparations and bath preparations. Draize and Kelly designed test on eye mucosa of albino rabbit. Such tests can be conducted as follows :

Instil 0.1 ml of test substance in conjunctival sac of one eye of nine albino rabbits (the other eye works as control). Divide rabbits in three groups comprising of three rabbits each.

Group – 1   Leave the eyes of this group unwashed.

Group – 2   Wash the treated eyes with 20 ml of lukewarm water after 2 seconds of instillation of test substance.

Group – 3   Wash the treated eyes after 4 seconds of instillation of test substance with 20 ml of lukewarm water.

Read the occular reactions with hand slit lamp for seven days or till any residual injury persists.

Any preparation, which leaves corneal or iris lesions for more than 7 days, is considered severe eye irritant. Eye irritant effect of combination of synthetic detergents is greater than produced singly by them. All synthetic detergents preparations used on hair or during bath should be assessed for their eye irritating effect.

Efficacy Testing

The majority of skin-care products are sold to improve the appearance and feel of the skin, and are broadly classified as moisturisers. The condition and appearance of the skin is a function of its softeness and flexibility, which is adversely affected by loss of water. Tests for moisture content and moisture transmission, as well as the viscoelastic response, ultrasound techniques, and electrical properties of skin have been the subject of much study, with the development of increasingly sophisticated instrumentation to show the condition of the skin and its water content. Most of these new methods have evolved to help evaluate the potential performance of products, and are very useful in screening formulations, levels of ingredients and other stepwise formula variables.

Cell turnover testing

A very useful test to determine cell turnover rate utilises the dansyl chrloride test, in which dansyl chloride, which fluoresces under a long-wave uv or woods lamp, is applied to the skin. This material penetrates the stratum corneum to a uniform depth. The cell turnover rate, or transit time, is determined by disappearance of the stain compared to a non-treatment site.

Instrumental tests

The instrumental test is proving to be of great value is profilometry, which is especially useful in the assessment of the changes to the skin's surface caused by the anti-ageing effect of retinoid therapy on sun damaged skin. Two primary methods are used inprofilometry. One method employs replicas and the other uses photographs. Fully hydrated skin has altered surface characteristics, as measured by both lower peaks and greater distance between peaks. This is also the case with skin that has fewer lines and wrinkles due to effects other than hydration. In evaluation with a replica, an instrument with a stylus produces a tracing of the topography of the skin, based on a replica produced by dental impression material. With two-dimensional photographs, computerised image analysis techniques are used to review the image.

The gas-bearing electrodynamometer (GBE), has been widely used to evaluate the viscoelastic properties of the stratum corneum, both in *vivo* and *in vitro*. More recently computer handling of data generated by this device has improved its utility. The Twistometre device measures resistance to torsion of the skin, as applied by a rotating disc inside an external ring, which acts as a stator. This device is used to evaluate the skin's softness and suppleness. There has been a long history of the measurement of not only the water content of skin, but also the flow of moisture from the skin, which is known as the moisture vapour transmission rate (MVTR) or the moisture/water loss or transepidermal moisture loss (TEML). MVTR is a technique that has shown continual refinement. Progress has continued to the more recent development of a well-accepted instrument, the Servo-Med evaporimeter. Most of the work on the direct measurement of water content in skin has utilised an infrared (IR) spectrophotometer.

Testing for moisturisers

The tests consist of the flow of moisture through a barrier, which is often a synthetic film with a transmission rate similar to that of skin. This technique (MVTR) combined with a water-holding technique to measure hydration or the ability of the formula or test material to maintain humectancy is very useful in evaluating or predicting potential *in vivo* results.

Test methods other than MVTR and water-holding capacity include scanning colourimetric techniques, scanning electron microscopy (SEM) and biomechanical properties of skin under various *in vitro* conditions. *In vitro* test methods can be instructive when their correlation to *in vivo* performance has been established. In the author's experience, a battery of tests and an excellent history of correlation are necessary for the tests to be of predictive clinical value. The problem is that, when *in vitro* tests are used with excised skin or a synthetic membrane, unlike the human skin, there is a lack of the biological underpinnings of the epidermis, which is anything but an inert, static or dead layer.

Test for sebum

This device (Sebutape) is a film that visualises and traps sebum. The tape can then be analysed, and total relative sebum amount computed using image analysis. If quantification is not the only requirement, the sebum can be solvent-extracted, gravimetrically analysed and broken down into component parts. This can be a very useful technique for assessing the instant removal of sebum by cleansers, or the effect on sebum production of various treatments.

Test for cleanser mildness

A wide variety of cleansers are available on the market and will be covered within this chapter. Tests for cleansers have focused primarily on testing soap or detergent formulations for mildness. In this test, dilute solutions of soaps or test solutions are maintained in a cup held against the skin of human volunteers. The skin is then evaluated for redness, scaling, and fissuring, often against high irritancy (non-fatted soap bar) and low irritancy (fatted sodium cocoisethionate bar) controls. A more recent modification of this method calls for a shorter test, and measures the increase in MVTR by a Servo-Med evaporimeter. This has shown close correlation with the longer duration visual assessment technique. These methods differ from the

Soap Chamber Test in that the Forearm Test adjusts frequency for humidity, while the Flex Test is reported to be unaffected by humidity. For evaluation of facial cleansres of all types for mildness and drying potential, the method outlined by Frosch utilises half-face testing with realistic use conditions. In order to obtain suitable dryness scores, this test is best run during periods of low relative humidity. Cleanser tests for efficacy, evaluating soil and make-up removal, as well as esthetic properties and after-cleansing skin-feel, must also be done to obtain a complete picture.

A combination of all these test methods, together with adequate clinical and consumer trial and observation, should enable one to have a great insight not only into the performance of products but, more importantly, as to how and why products work, and how variations in formula affect performance. The is necessary for the enlightened formulator.

Stability Testing

Stability testing is done to ensure that a developed product will be fit for use during its expected life. Stability testing should be done early in the development cycle to remedy any problems before final testing. A well-run stability trial can provide much information about a product in a relatively short period of time. Prototype products in packaging, representing the ultimate trade package material, can be placed at ambient, and elevated (e.g., 37°C and 45°C) temperatures, refrigerated and cycled through freeze/thaw cycles, and placed in high humidity chambers. During these trials, testing should focus on attributes most important to the products performance, or on the integrity of any active ingredients. In addition, there should be physical testing, for example pH and viscosity, chemical content of active ingredients, water content, etc. The formulator must not overlook attributes relating to consumer perception, such as fragrance, colour, rub-in characteristics and appearance. Two or three months of successful elevated temperature testing and three of four freeze/thaw cycles will usually indicate that products will have an adequate shelf-life. It is important, however, to continue testing for longer periods at ambient temperature, to obtain an understanding of the product's ultimate shelf-life. Further testing whenever changes are made to the supply of raw materials or to the formulation is essential.

Microbiological Testing

During the development phase, close attention should be paid to the microbiological integrity of the formulation. Professional advice should be sought from a well-trained industrial microbiologist. If this is not immediately available, input and guidance on preservatives from the suppliers' staff and consultants should be sought. Suppliers will often have more than one preservative and can make recommendations about particular types of formulation. Another method of finding appropriate preservatives is to review ingredient labeling on similar types of products that are available from reputable manufacturers and which have similar constituents, pH and package types. This, coupled with a microbiologist's or a preservative vendor's input can often give several trial systems. Finished products are usually tested using a challenge technique. If this essential testing is not available in-house, a qualified contract lab can assist. The test involves adding large numbers of a variety of organisms to the product and checking the ability of the product to reduce them to an acceptable level. To ensure that the product retains its ability to remain preserved, adequacy of preservation should also be checked during stability testing.

Safety Testing

The product must be safe under all conditions of use and potential misuse. Safe products arise from a combination of careful formulating, good background data and adequate testing. A review of materials used in products for similar purposes is an excellent place to start. Most successful commercial products have been adequately tested and have a history of safe usage. Safety data for each raw material should be available from its manufacturer and should be reviewed. After a thorough background check, finished products should be tested through a competent third party. Many testing companies not only do safety testing, but also recommend appropriate testing that should be done for the class of product that has been developed. It is often useful to solicit testing protocols from two testing companies, asking them to explain and justify each test they recommend. It must be remembered, however, that they can only make enlightened recommendations based on input. Divulging as much information as possible at this stage, will give better, more appropriate testing input, and insight into possible untoward reaction.

After adequate clinical testing has been completed, and during all stages of consumer trial, it is important to check for unexpected reactions or patterns of misuse, while carefully extending exposure. Careful attention should be paid to reported reactions from usage or misusage situations. After launch, there should be a mechanism to check on reported reactions to determine if, on a broad exposure, these levels are higher than one would expect for the category.

Chapter 45

Environmental Aspects

INTRODUCTION

The environment refers to those factors, other than process equipment and people, which directly influence the microbiological quality of a finished product. These include the design of buildings and processes, the choice of construction materials and surface finishes, the provision of ancillary services such as lighting, ventilation, power, etc. and the exclusion of vermin.

Exposures to toxic substances are linked to a variety of health problems. Corporate managements have now reoriented their environmental strategies towards development of eco-friendly products. These products are conceived to reduce the burden on environment both during use and disposal while ensuring the desired product performance with environmental compatibility.

Provision of Services

When providing services to any manufacturing area, the cardinal rule is to avoid dust traps. The usual routes for electrical suppliers, air and steam lines and product transfer lines are the overhead route or via in-wall trunking. As a general principle, surface wall mountings should be avoided. They trap dirt and make cleaning difficult. Overhead services should be run in the ceiling void above a suspended ceiling. Services can be kept conveniently out of the manufacturing area and structural roof-members covered up whilst still maintaining easy access.

Ventilation is probably the most difficult service to get right. Air flows through manufacturing areas should be kept to an absolute minimum. This almost always entails the provision of local supply and extraction. Air supplies should be filtered to remove dust and screened to prevent the entry of insects. Air-conditioning systems can be a particular problem. Evaporation coolers and humidifiers can be very efficient breeding grounds for bacteria. If they must be used, then a bacterial filter should be placed downstream and a regular cleaning and sanitation routine implemented and recorded.

If parts of the manufacturing process generate large volumes of steam or dust, then the ventilation system can become overloaded. In this case, local extraction must be provided to prevent cross-contamination with dust or condensation. The filtered air supply must be able to provide the make-up air volume, or contaminated air will be drawn directly into the building. For this reason it is better to have continuous, rather than intermittent, local extraction.

Hygiene policy

Cleaning regimes for the manufacturing area should be designed to produce as little dust as possible. It is better to have major clean-downs when the plant is not in production, at night or between shifts. Floors should be washed daily. During production any spillage of products or accumulation of moisture should be cleaned up as soon as possible and, preferably, immediately. A schedule should be developed for the periodic washing of walls and ceilings and a responsible person should be nominated to see that the programme is carried out.

Cleaning equipment should not, itself, be allowed to become a source of contamination. Mops, buckets, squeegees and brushes should be kept in a sanitary condition, which means clean, dry and stored off the floor. Floor-washing machines, particularly those with integral tanks for detergents, can be dangerous. Bacteria can breed in the solutions or the delivery system and the machines then become disseminators of contamination. Such machines should themselves the designed for easy cleaning and sanitation. Cleaning equipment should be stored in a designated wash area, which should itself be kept clean and tidy.

Toilets and hand-washing facilities should be provided outside the production area. Return to the production area should only be possible by passing a hand basin.

People can be one of the greatest sources of contamination in the plant. All employees should be provided with clean, protective clothing, including head coverings, as often as necessary. All personnel should be encouraged to report minor ailments that may serve as a potential source of contamination. They, and anyone having exposed sores, cuts or abrasions, should be assigned to duties that do not involve exposure to products. The wearing of jewellery, badge, hair clips, etc., should not be allowed.

Control of vermin

A vermin control policy should be in place. This should include physical systems for the exclusion of insects, rodents and birds. A pest control officer should be appointed. He or she may be a suitably qualified employee, or a consultant from an outside contractor.

Some simple rules can help to reduce the chances of infestation. Doors should be kept closed at all times. Loading bay doors should be provided with seals so that loading operations are carried out under cover. Eating and drinking, except in the designated place, should be strictly forbidden. Food attracts vermin. Rubbish and excessive quantities of packaging materials or finished stock should not be allowed to accumulate in the production area.

Environmental monitoring

Environmental conditions may directly or indirectly affect the quality of the finished product.

An environmental monitoring programme has to be properly constituted to be of any practical use. It is left to the discretion of the manufacturer to determine what type of programme would be most suitable for his or her own facility, since each plant is unique. External standards for the microbiological quality of air and surfaces do not exist for non-sterile production. Most manufacturers, therefore, develop their own 'in-house' standards. Such standards must be based on knowledge of the normal background levels of contamination. These must be determined at different times of the day over an extended period, preferably to include all the seasons, before a realistic picture can be determined and standards sets. Regardless of the approach used, a manufacturer should establish guideline and action levels for all the areas tested, once a normal range of recovery has been established. If any results exceed the action levels, the ground responsible for cleaning and sanitising should be immediately informed and corrective action should be taken. Exceeding the manufacturer's guidelines should precipitate further testing, including the finished product.

The quality of the manufacturing environment will be influenced by many factors. Understanding these influences is an essential part of an environmental control programmes. Influences on the environment can be divided into four basic categories: buildings, equipment, personnel, and sanitisation and housekeeping. These are discussed elsewhere in this chapter.

The microbiological quality of physical surfaces within a manufacturing environment can be defined as either

direct contact surfaces : e.g.,	processing equipment
	filling equipment
	pumps
	valves
or non-direct contact surfaces : e.g.,	walls, floors, ceilings
	overhead lighting or walkways
	support beams
	pallets, forklift trucks.

The type and frequency of microbiological monitoring of physical surfaces will depend on the susceptibility of the product to microbial contamination, the scale of manufacturing and the type of plant and process. Those physical surfaces which come in direct contact with the product would normally be more frequently monitored than those not in direct contact with the product. Direct contact surfaces should be examined for the presence of micro-organisms that are known to cause harm to the consumer or product spoilage. Indirect surfaces should be monitored to determine background levels of microorganisms which are intrinsic to the manufacturing environment. Any increase from these predetermined levels may be considered an indication of a potential microbiological problem that may affect product quality and may necessitate a revision of the frequency of testing and sanitisation.

The most commonly used methods include the use of swabs, contacts plates and the collection of rinse water. Personnel involved in environmental monitoring should be properly trained according to a written procedure and should be establish a properly documented programme. In general, all processing and filling equipment, valves, working surfaces etc., should be monitored on a defined and periodic basis. Transfer lines should be taken apart, if necessary, and rinsed thoroughly with an appropriate rinse solution of known microbiological quality.

A schedule should be prepared and followed for the routine monitoring of the air in each designated area. The frequency of monitoring should be determined by the type of activities in the area. Again, frequency and standards should be established by 'in-house' needs. The most commonly used methods, many of which are available commercially, include :

Sedimentation	:	setting agar plates
Filtration	:	membrane filter
Impingement	:	all glass impinger
Impaction	:	Andersen-sieve, slit-to-agar centrifugal air sampler

Water treatment systems

Water systems are particularly susceptible to microbial contamination and must be designed to prevent microbes from proliferation and hence being introduced into products. There should be systems in place to remove suspended matter, silicates, organic materials, chlorine, and colloidal material form the raw water before it enters the treatment system.

Modern water treatment systems have a multi-media filter specifically designed according to the quality of the incoming water. Such filters combine depth filtration with activated charcoal and coarse and coarse ion-exchange resins. The routine microbial monitoring programme should provide information for the operation to determine whether chemical disinfection of the demineralisation equipment is required. The suitability of any such disinfection agent should be confirmed with the resin supplier. Where required, anti-microbial treatment systems must be incorporated.

Ozonisation, at 0.1 ppm maximum, is an affective disinfectant system, although with relatively high maintenance and operation costs. Appropriate ultraviolet uv lights must be installed at a points of use to remove the ozone. Water must be recirculated at an adequate velocity (minimum of 1 m/s) to ensure even distribution of the ozone.

uv irradiation at 254 nm is effective in clear, colourless water. uv transmission can be reduced by particulated matter, organic material and some pigments. uv lamps generate large amounts of heat in use and can only operated when water is flowing. They take up to 30 seconds to reach operating wavelength after being switched on and any water flowing through in that time will not be fully treated. It takes a set time for uv to disable or kill microbes and, in a moving system, there may not be sufficient dwell time in the lamp chamber to ensure effectiveness. These problems can be resolved by moving the water throught the uv lamps in a continuous recirculation system.

Heat can be used to prevent microbial growth in water systems. The water must be maintained above 85°C at all times throughout the system.

In order to ensure even distribution of the heat, recirculation is required. Heated systems have high operating charges and high installation costs for insulation. Cooling systems would be needed for processes requiring water at less than 85°C.

Bacterial filters should not be used as the sole mechanism for microbial control due to the potential for bacterial to grow on the filter. Filters used at point of use should either be capable of resterilisation or be thrown away after such use. Filters can become blocked after a period of usage and may restrict water flow. It is therefore important to monitor the pressure differential across the filter.

In order to maintain the microbial integrity of the treated water, it must continually flow through the anti-microbial systems. This can be achieved by keeping the treated water moving in a continuous loop. Water should be distributed in a pipework loop separate from that of the supply system. When there is no water usage, water flow in both loops can then be maintained. This is frequently referred as a 'figure-of-eight' system. Fig 45.1. Illustrates a simple 'figure-of-eight' water system.

Water flowing through a pipe creates a 'viscous' boundary layer at the pipe surface due to the effect of drag caused by water surface tension. Any microbes falling into this layer will be in a relatively calm zone and will proliferated to form a film of cells across the inner surface. This biofilm will eventually grow into the flowing water and clumps will be broken away. If these clumps are large, anti-microbial systems may not affect those cells in the centre of the clump.

Flow rates should be high enough to ensure that the viscous layer is not more then 5 µm. This should limit any biofilm to no more than five cells in depth and any clumps broken off will be small enough for the antimicrobial treatment to handle. The minimum flow rate to achieve this condition depends on the diameter of the pipe.

ENVIRONMENT ISSUES

Environment regulation is widespread and virtually all embracing throughout the world.

By the large the more draconian measures being put into effect in Europe and the UK under the Integrated Pollution Prevention and Control (IPPC) and the Integrated Pollution Control (IPC) bodies do not apply directly to the formulation or the manufacture of cosmetics and toiletries. The prescribed processed under these regulations do not include this industry and are more concerned for the moment at any rate with the heavier industry polluters of air, water and the land. Of much more concern to the formulator and manufacturer of cosmetic and toiletry products are the regulations and proposed tax levies associated with waste and with packaging.

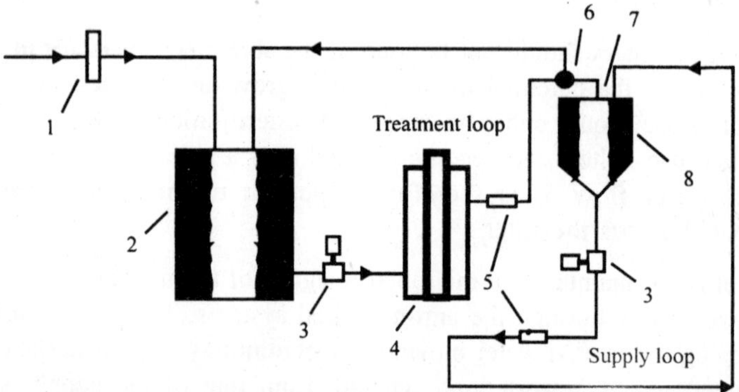

Fig. 45.1. Figure-of-eight water system: 1, multi-media filter; 2, city water storage tant; 3, pumps; 4, demineraliser; 5, antimicrobial treatment; 6, return valve; 7, sanitary level switch; 8, treated water storage tank.

Waste

As far as the cosmetic and toiletries industry is concerned the significant waste factors can be separated into two major areas; that produced as a result of the formulation and manufacture of the product and that produced due to the packaging of the product. Waste produced in the factory has of course been a traditional area of control from the cost point of view but it has taken on a greater significance in recent years due to environmental factors. Waste produce from rejected batches of products must be disposed under strict control. Hazardous materials may be involved. Rejected raw materials fall into the same category. In Europe disposal must be carried out by registered waste carriers and of course paid for. The same principles apply to reject packaging materials and faulty components on line. It is important to ensure good quality componentry to avoid rejects and down time on the lines but now the disposal factor assumes greater significance in preventing cost increases. Preventative quality assurance measures and built in fail safe considerations at the design stage thus become even more important.

Packaging Waste

It will be necessary as time passes to pay great attention to the way cosmetics are packaged for sale due to the regulations now being discussed in the EU and other countries. These regulations will in fact impose a levy on the packaging industry and allied industries designed specifically to discourage of use of excessive packaging. The imposition of tax on waste destined for landfill will also affect the design of future packs. Point of sale and merchandising units will have to be looked at with environmental considerations in mind. Packaging waste provides a challenge and an opportunity particularly in the product development area. The prime target of most governments concerned with environmental problems is to minimise waste. The retail packs of consumer products thererore provide an avenue of approach towards achieving this aim offering the consumer an environmentally friendly product.

Design of packs that will use of minimum of material whilst still adequately protecting the product will become increasingly more important. Reusable packs and packs that can be recycled will offer further opportunities and failing all else packs that can be disposed of with no environmentally deleterious effect or indeed which may provide an environmental benefit must be considered. Use of materials from sustainable sources and those which cause little or no environmental problems or do not use large amounts of energy in their preparation will increase in significance.

As ever the development of a product must ensure that the product and the pack are designed together and that all members of the team are aware of the environmental consequences.

Eco labelling

Consumer demand for information on the environmental acceptability of the products they buy has led to the spread of 'Eco labelling' schemes throughout the world. Individual countries and the EU are devising their own definitions and specifications required for the award of an 'Eco' label, an important factor in the development of products for the future.

Volatile Organic Compounds (VOC)

The present situation on the use of VOCs is far more stringent in the USA where state legislation limits considerably the use of volatile materials in cosmetic and toiletry formulations. This drastically affects the formulation of hairsprays and fixatives, mouses and conditioners as well as nail polish remover, antiperspirants, deodourants and fragrances. These limitations will lead to new technology since at the moment solutions offered to date have not gained great consumer support. Elimination of the aerosol as we know it today seems to be the main target of the regulatory bodies.

In Europe at the moment the situation is still developing and is not yet clear. No overt regulations are aimed specifically at cosmetic and toiletry products. There is more concern over the sale and storage of petroleum, the coating, printing and paint industries where great changes have already taken place in formulations and processes. The forward looking cosmetic and toiletry formulator will be well advised to watch developments and use innovation to lessen the dependence on the use of volatiles.

Environmental Management Systems

It is worthy of note that public concern about the environmental performance of manufacturers of consumer products is growing. This concern may well influence the choice of products the consumer may make in the highly competitive marketplace. Any company that is able to convince the consumer of its concern and care about the effects of its activities on the environment may therefore well steal an edge on its competitors. It is also a distinct possibility that the retailers who place a great deal of emphasis on their environmental image may well force the pace and insist that their suppliers show evidence of their commitment to protecting the environment.

Throughout the world therefore a number of schemes are being developed that will enable companies to demonstrate such commitment through standards and certified environmental management schemes. The British Standards Institute have introduced BS 7750 and on the international front an ISO Scheme (ISO 14000) is being finalised. These schemes set out the requirements for a management system and structure which is adopted and accepted by independent accredited assessors enable companies to claim certified status. It is essential therefore that workers in the field of consumer product maintain awareness of the developments in this important area.

ENVIRONMENTAL PROTECTION ASPECTS

Questions regarding the disposal and the environmental compatibility of consumer items such as detergent and cleaning products, lacquers, hobby materials or body care preparations have in recent years come under the scrutiny of public concern.

Cosmetics and toiletries do not belong to those consumer products which are of particular environmental concern. This can be attributed to their comparatively small application amounts on the one hand, and their composition on the other hand – these preparations are after all characterised by their compatibility with the human body. Even if soaps, shampoos and shower gels, like other domestic cleaning products, go directly into the sewage system after appropriate use, in their case the principal consideration is understandably directed at human compatibility aspects.

In discussing aspects of environmental protection, we are in fact addressing themes which have a more general ecological significance. Because some cosmetic preparations fulfil a cleaning function (of the body) in a sense of the law concerning the environmental compatibility of detergents and cleaning preparations and, "as experience indicates, can end up in the water system after use," then the question of protection of the water supply will be considered. In addition, we will discuss environmental aspects of body are preparations which are produced in aerosol form and which earlier contained chlorofluorocarbons (CFCs) as propellants.

Ecological aspects of Household Preparations

The fact that cleansing and care products are used in the household on a regular basis and are a part of everyday life has still further consequences. For example, the consumer's level of knowledge concerning the correct application of a preparation for his particular needs – a particular shampoos, bathing preparation, or detergent – is naturally very high. The optimal dosage of a preparation or product is therefore determined by the individual needs of the consumer. Corresponding to degree of consumer familiarity with these products, manufacturer information on packaging is therefore restricted to essential consumer information such as recommendations for specific dosage and effectiveness. The demand raised occasionally on grounds of environmental protection for more extension package information (for example dosage directions instead of dosage recommendations) assumes that consumers are unnecessarily high dosage and therefore "unnecessarily burden the environment". This raises the question, however, of whether an unnecessary conflict is not thereby created between the consumer's desire for individual choice and an intended improvement in environmental protection.

From a scientific viewpoint, a much more important consequence of regular use of household preparations that pass into the sewage system after use is the fact that the resulting environmental exposure can be determined relatively easily and is accessible to reliable analytical investigations. The indispensable knowledge needed to assess the environmental compatibility of cleaning and care products, namely the concentration of their ingredients in the water supply, is therefore easily obtained. This relative homogeneity furthermore makes it easier to arrange methods for testing environmental compatibility as practically as possible, so that secure conclusions regarding actual ecological effects can be drawn. In determining the ecological behaviour of a substance, it is necessary to draw in, along with its expected or observed distribution routes within the environment, all the factors that decisively influence its concentration in the environment: its biological, hydrolytic or photolytic degradability in water, abiotic decomposition processes in the atmosphere, adsorption or accumulation in surface areas, as well as its integration into natural physiological cycles.

Knowledge of these factors in different ecological systems is supplemented by the investigation of possible or actual effects. Depending on the characteristics of the substance in question and the production quantities, a variety of investigations ranging from simple laboratory experiments to extensive field studies may be in order. With household products, which enter the environment overwhelmingly via the sewage system, investigations focus primarily on water ecology. Investigations regarding possible effects on water treatment (for example, bacterial toxicity) and the determination of acute and chronic toxicity on the different trophic levels of aquatic organisms are of essential importance here. If sludge is one of the distribution routes of the substance, studies of its behaviour here and its further ramifications (for example, to agriculture) are also undertaken. Also drawn into the evaluation are the effects of a substance on other substances, such as its influence on the sedimentation behaviour of heavy metals, as well as aspects of specific accumulation (biological accumulation, biological magnification, geological accumulation) or the influence on the nutrition of water organisms are also explored.

The purpose of these investigations, which are conducted on all ecologically relevant ingredients of household products, is to provide a scientifically based evaluation of their ecological compatibility. Active ingredients can only be incorporated into a household product intended for daily use when it has been established that it is not expected to cause irreversible damage or harm to the environment.

Cosmetic Preparations as Washing and Cleaning agents – Water Pollution Protection

The environmental compatibility of cleaning agents washing and cleaning agents were defined in legal terms as, among other things, products "intended for cleaning or in assisting cleaning which can end up in the sewage system after use." Although cosmetic preparations which fall under this definition are exempted in some cases from special legal requirements such as content declarations or notification of the formulation it is nevertheless important to note that the legal principle and the regulations regarding specific substances also apply fully to relevant body care preparations.

Cosmetics and toiletries which fall under the jurisdiction of this law are, for example, hair-washing preparations, toothpastes, bath preparations or hand-washing preparations. They are also subject to the requirement in ("Principle"), which stipulates that such preparations only be brought onto the market in a form "in every avoidable impairment of the character of water, especially in respect to the natural environment and to drinking water, and every impairment of the operation of waste water plants is avoided.

Principle of this law (concerning the degradability of organic substances) also applies to cosmetics and toiletries. Until now, the biological degradation of anionic and non-ionic surfactants was regulated on the basis of this paragraph in two statutory orders. After a prescribed analysis (methylene blue active substance for anionic surfactants and bismuth active substance for nonionic surfactants), as well as a two-step test system (selection and confirmation tests), a minimum degradability of 80% must be established for the surfactant within the testing period.

The fast and complete biodegradability of the surfactants incorporated into body cleansing and bath preparations has been thoroughly investigated and verified. Exact knowledge concerning the ecological characteristics of these surfactants is not primarily derived from their use in cosmetic preparations, but rather from their use in considerably largely quantities in other learning products).

Although this law was certainly aimed at products which are used in significantly larger quantities outside the field of cosmetics we may proceed from the assumption that the principle of environmental and water compatibility applies to cosmetic preparations in the same way.

Chlorofluorocarbons as Propellants

The chlorofluorocarbons are used as the propellant in nearly all aerosols, including those in the area of body care, because of their advantages in application, which are related to their chemical inertness and excellent spray quality. The technolically most important compounds in this group are trichlorofluoromethane and dichlorodifluoromethane.

It has been found, that chlorofluorocarbons break down the ozone layer in the stratosphere.

The physical and chemical processes that influence the ozone layer in the atmosphere have proved to be much more complex than originally thought when the thesis concerning of ozone in the stratosphere is the result of a balance between processes of formation and decomposition. It is dependent on sunlight and largely independent on human activities. Both natural and anthropogenic processes of decomposition are influenced not only by CFCs, but als by an array of other trace gases such as methane, nitrogen monoxide and carbon dioxide.

All changes in the emission rates of these traces gases can exert an influence on the ozone layer. In order to predict possible changes in concentrations of ozone, computer-assised mathematic models are needed which can keep track of all of the complex processes involved. Current model predictions, however, rely on a number of assumptions about future emissions of different trace gases, part of which is of natural origin. Of particular significance here is methane, because it acts to reduce the amount of active chlorine which emerges with the decomposition of CFCs in the ozone layer from uv radiation. The extent to which CMCs can cause a breakdown of the ozone layer is therefore among other factors directly related to methane emissions.

It remains to note that because of their advantages to the consumer, packaging in aerosol form has also become an established element for many cosmetic preparations in recent decades. Industry is now, as a result of several years of innovative efforts, in a position to get by largely without CFC propellants and hence to continue to provide the consumer with reliable product packaging.

Environmental impact

Laws covering the pollution of air, water and land exist both nationally and internationally, and are constantly under review. The consumer interest in environmental issues, together with trend towards green politics and consumerism has developed public opinion to a degree where products may be accepted or rejected according to their environmental performance. Controversy concerning the ozone layer and the use of aerosol products is a particular example. Marketing edge may be achieved by developing products that can be shown to be harmless or beneficial to the environment. In this respect, ingredients must be chosen with due care. Packaging, labelling, natural materials and use of energy all require careful consideration and, without doubt, new issues will arise in the future. Vigilance and current awareness is essential.

Animal protection and rights

The problems encountered in this area are associated not only with the use of animals for laboratory testing, but also with the use of materials and ingredients derived from animal sources. The use of animals for cosmetic testing has decreased substantially over the past decade and continues to decline worldwide. However, international regulations still require safety data based on animal studies, therefore new materials not used for other industries such as plastics, food, chemicals or pharmaceuticals, would require animal testing. Fortunately, consumer pressure is now aimed at the regulators of international law (e.g., the EC Commission) and ultimately these requirements will be modified to make the situation much clearer and easier to cope with.

Due to the growing pressure from vegetarians, religious groups and animal rights groups, this area of concern is constantly changing. The labelling of products as 'non-animal tested,' 'contains no animal ingredients,' 'kosher', etc., requires careful consideration.

Evaluation of Safety of Aroma Chemicals

Naturally, all aroma chemicals must be tested in terms of their dermatological/toxicological safety. There are for example skin-damaging phototoxic substances, such as furocoumarin in bergamot oil, that are not appropriate for preparations which come in contact with the skin and cold be exposed to the sun. For this reason, only furocoumarin-free bergamot oil can be used for example in Eau de Cologne, cream and especially sunscreening preparations.

Similarly, alongside many completely harmless musk-aroma chemicals, there is a tetraline-musk body (1,1,4,4-tetramethyl-6-ethyl-acetyltetraline) that cannot be used because of its toxicological characteristics.

All toxicological data on aroma chemicals have for years been determined by the RIFM/USA (Research Institute for Fragrance Materials) and published on an ongoing basis. Recommendations on the use of aroma chemicals are issued by the IFRA (International Fragrance Association. These recommendations are adopted by the whole research industry. Of course, chemical regulations also apply to aroma chemicals.

Future Ecological and Economic Challenges

Consumer, trade, manufacturer and state interests that have previously shaped the catalogue of requirements for packaging will continue to influence the future development of packaging materials for cosmetic preparations.

Producers will attempt to increase unit sizes in order to be able to manufacture economically. This trend will certainly be at the expense of individual style. The consumer will recognise this and differentiate between preparations for everyday needs and luxury items. High-quality cosmetic preparations must stand out from the field by their packaging in order to signal their luxury contents to the consumer. Product and package developers will support the introduction of new preparations onto the market with the help of innovative packaging. The increasingly stronger contributions made by the packagers of pastes and creams are an example of this.

Ecological considerations will exert the greatest influence on packaging in coming years. Even the state will intervene occasionally by issuing ordinances in order to protect resources on the one hand while on the other preventing the mountains of garbage from growing any further. This will influence the ongoing trend towards leaner packaging, and new, more easily workable materials will also play a role. The desire for simple and effective disposal will only increase. Plastic materials have recently been developed in laboratory settings which decompose in compost because of their sugar and starch bases. If such raw materials can be manufactured economically, the future of packaging will belong to them.

Glossary

Perfumery Materials

Absolutes. An ethanolic extract of a concrete or a resinoid which contains the maximum concentration of odouriferous components and is free from natural waxes and/or any solvent used in the processing.

Acid Value. It is numerically equivalent to the number of milligrams of potassium hydroxide required to neutralise the free acids present in 1 g of the material.

Alcohol Perfumery Grade, Denatured. Rectified ethyl alcohol, specially denautred for the addition of denaturants. It does not add any undesirable by-odours to it.

Aldehydic Blend. Blend deriving their unique character from the predominance of aldehyde notes.

Amber Note. A heavy full-bodied warm ambergris like note.

Animal Note. Odours or notes with a sensuous character.

Aromatic Chemicals/Aroma Chemicals. Organic chemicals derived by organic synthesis or as isolate from natural essential oils possessing distinct aroma. Used as raw material for the preparation of perfumery blends or flavours.

Aromatic Plant. Plant bearing a characteristic aroma.

Aromatic Water. Aqueous odouriferous condensate of hydro-distilled and/ or steam-distilled material of vegetable origin containing fully dispersed essential oil.

Attar (Indian). A perfume concentrate characteristic of single flower or a mixture of flowers and/or other materials of plant or animal origin with oil of sandalwood as the base.

Balsam. An odouriferous exudate from plants/ trees which flows naturally or is artificially induced by incision.

Blend, Cologne. Any harmonious combination of fragrances, the main characteristics of which are derived from citrus oils.

Blend, Oriental. A blend with heavy, full-bodied sweet balsamic and animal note.

Blend, Spicy. Any fragrance combination having spicy overtone.

Blend, Woody. Any fragrance dominated by a woody character.

Body. Main fragrance theme.

Boiling Range. It is the range of temperature within which a specified percentage of the material distils.

Bouquet. Generally a harmonious combination of two or more floral notes.

By-Note. A temporary or permanent odour effect additional to the main pattern of odour associated with the material.

Carbonyl Value. It is numerically equivalent to the number of milligrams of potassium hydroxide, that is, equivalent to the amount of hydroxylamine required to oximate the carbonyl compounds present in 1 g of the material.

Cell. A unit of plant tissue.

Cellular. Composed of cells.

Chypre. A mossy-woody fragrance, complex with a characteristic sweet citrus top note, frequently encompassing some floral tones.

Citrus. Odours reminiscent of citrus fruits, such as orange, lemon, bergamot, grape-fruit, etc.

Cologne. Name used traditionally for solution of citrus perfume blends in aqueous ethanol.

Concrete. A material derived from a single source of vegetable or animal origin by extraction with a suitable solvent. It generally contains non-odouriferous constituents, such as waxes, colouring matter, etc., in addition to odouriferous components and is free from any solvent used in the process.

Condensate. Vapours that have been condensed.

Condenser. Part of distillation apparatus where the hot vapours are cooled and condensed for recovery.

Congealing Point. It is the maximum constant temperature at which a liquefied solid resolidifies.

Deterpenised Oil. Natural essential oils which are free from terpenes and/or sesquiterpenes.

Diffusion. The ability of a fragrance to radiate and permeate the environment.

Distillation. A process of evaporation and recondensation used for purifying liquids.

Distillation, Dry. Distillation of semi-solid and solid materials in the absence of steam, water, or any other solvent.

Distillation, Hydro. Distillation of a substance carried out by indirect contact with boiling water.

Distillation, Steam. Distillation of a substance by passing steam through it.

Distillation, Vacuum. Distillation of a substance under reduced pressure.

Dry out. Final phase of the main fragrance after the main volatile constituents have evaporated.

Enfluerage. Process of extracting fragrance of fresh flowers by intimate contact with mixtures of purified fats preferably at low temperatures.

Essential Oil. It is a volatile perfumery material derived from a single source of vegetable or animal origin by a process, such as hydrodistillation, steam distillation, dry distillation or expression.

Essential Oil, Synthetic. It is a composition generally consisting of natural essential oils, aromatic chemicals, resinoids, concretes, absolutes, etc, but excluding animal or vegetable non-essential oils and not having a nonvolatile residue in excess of 10 per cent by mass. It is so composed that it bears a close resemblance primarily in odour to a naturally occurring essential oil.

Ester Value. It is numerically equivalent to the number of milligrams of potassium hydroxide required to neutralise the acids liberated by the hydrolysis of the esters present in 1 g of the material. It represents the difference between the saponification value and the acid value of the material.

Ester Value after Acetylation. It is numerically equivalent to the number of milligrams of potassium hydroxide required to neutralise the acids liberated by the hydrolysis of 1 g of the acetylated material.

Evaporation Residue. Represents the percentage of perfumery material which is not volatile when heated on a steam-bath under specified conditions.

Expression. The process of extracting essential oil from the plant cells by application of mechanical pressure.

Extract. A concentrated product obtained by treating a natural perfumery material with a solvent which is subsequently evaporated.

Extraction. The process of isolating essential oil with the help of a volatile solvent.

Extract, Alcoholic. A French word, now universally used in perfumery, meaning an alcoholic extract of odourous part of a pomade. It is generally used to mean alcoholic solution of a perfume concentrate.

Fixative. A substance which is compatible with and provides body and substantivity and rounds off a perfume composition by regulating the rate of evaporation of its volatile constituents.

Flavour. A combined organoleptic sensation of aroma and taste in a flavouring material is also called a flavour.

Floral. The fragrance characteristic of an existing known flower type.

Fore Runnings. Initial fractions of the distillate obtained during a distillation process.

Fougere. Perfume composition having a citrus/lavender top note with sweet powder rosaceous body with mossy/woody background.

Fractionation. The process of distillation by which an essential oil is separated into various fractions.

Fruit Flavour/Essence. Suitably blended mixtures of flavouring materials, permitted chemicals and food colours, in a solvent medium of either ethanol or the permitted non-alcoholic solvents.

Fruity Note. The impression of fruit odours within the fragrance theme.

Full Bodied. A well-rounded-out fragrance that possess depth and substantivity.

Green Note. Notes that recall fresh-cut grass, leaves and stems or other parts of plants.

Gum. A natural water soluble anionic material, often of glycoside-like structure and of high molecular mass which collects in or exudes from certain plants. It forms neutral or slightly acidic solution or a sol with water and has a typical mild odour.

Gum Resin. Natural exudation from plants and trees consisting of gums and resin with very small amounts of essential oils.

Harmonious. Order, accord and symphony in a fragrance.

Heavy. Oriental balsamic as against floral/green.

Infusion. A process of treating a substance with water or organic solvent, with or without heating.

Isolate. Either a single constituent or a multi-component fraction or a composited fraction, rich in desired odouriferous components and derived from a natural perfumery material.

Lasting Qualities. The ability of a fragrance to retain its character over a given period of time.

Leathery Note. Any fragrance conveying the dominant characteristic of tanned leather.

Melting Point. The temperature at which the material melts and becomes liquid throughout as shown by the formation of a definite meniscus.

Melting Range. The range between temperatures at which the material begins to form droplets and at which it becomes liquid throughout.

Middle Note. The main overall odour effect experienced by olfactory nerves on smelling a strip impregnated with a material and exposed to the atmosphere for some time.

Mossy Note. The notes that recall to mind moist dark forest having moss on the trees.

Natural Perfumery Materials. Perfumery materials of natural origin.

Odour. That property of a substance which stimulates and is perceived by the olfactory sense.

Oleoresin. Exudations from tree trunks or barks of trees and are characterised by the fact that these consist of entirely or mainly resin accompanied with an essential oil in varying percentages, soluble in organic solvents.

Oleoresin, Gum. An exudation from plants mainly consisting of essential oil, resin and gum.

Oleoresin, Spice. Extractibles of spice having resin and essential oil obtained by solvent extraction.

Perfume. A solution of perfumery compound/compounds in ethanol or other suitable solvents meant for use as a personal adornment. Here ethanol or other suitable odourless solvents are used as carriers for the fragrances.

Perfume Concentrate. A non-alcoholic concentrated perfume blend.

Perfumery Compound. A concentrated base which is further diluted with or without toning and further modifications to suit various end-uses.

Perfumery Material. A naturally occurring substance, or a derived material, or a preparation obtained by physical and/or chemical means, which diffuses or imparts an odour or a flavour.

Perfumery Materials, Synthetic. Man-made single perfumery materials, by chemical processes.

Pomade. Refined and deodorised animal fat(s) saturated with volatile oils present in and exhaled from the flowers especially the rose and the jasmine.

Rectification. Method of separation of undesirable substance to improve the quality of the materials.

Relative Density. The ratio of density of material at 27°C to that of distilled water at 27°C or 4°C when all masses are made in air is called relative density at 27°C or 4°C. Originally, it was known an specific gravity.

Residual Note (Dry Out Note). An odour effect experienced by olfactory nerves on smelling a strip impregnated with a material and exposed to the atmosphere for a period of time when the top and the middle notes have disappeared.

Resin. Solid or semi-solid translucent exudation from trees of plants. These are soluble in organic solvents.

Resinoid. A semi-fluid or a solid material obtained from a single resinous source of vegetable or animal origin by extraction with a suitable solvent and is free from solvent used in the process.

Saponification Value. It is numerically equivalent to the number of milligrams of potassium hydroxide required to neutralise the free acids liberated by hydrolysis of the esters present in 1 g of the material. It represents the sum of acid value and ester value.

Saponification Value after Acetylation. It is numerically equivalent to the number of milligrams of potassium hydroxide required to neutralise the free acid and the acids liberated by hydrolysis of the esters present in 1 g of the acetylated product.

Sesquiterpene. Term denoting a hydrocarbon composed of one-and-a-half terpene units, a single terpene unit being equal to two isoprene units.

Sesquiterpeneless Oil. An isolate obtained by suitably removing the sesquiterpenes (O_{15} H_{24}) from an essential oil.

Tail Running. The last fraction of distillate obtained in a distillation process.

Terpeneless Oil. An isolate obtained by removing almost all monoterpenes (C_{10} H_{16}) from an essential oil.·

Thin. The lack of body, richness and substantivity.

Tincture. A cold alcoholic extract of the soluble part of a natural fragrant material of vegetable or animal origin, the solvent being left in the extraction as a diluent.

Tissue. Plant structure composed of cells.

Top Note. The first odour effect experienced by olfactory nerves on smelling a strip freshly impregnated with a perfumery material.

Vacuum Distillation Residue. It is the percentage of material left behind undistilled when a known quantity of the material is distilled in vacuum at specified temperature and pressure.

Volatile. A material is said to be volatile when it has the property of evaporating at room temperature when exposed to atmopshere.

Woody Note. The impression of wood or woody odours within the fragrance theme.

Cosmetics

Absorbencis. The substances which impart this property in face powders include colloidal kaolin, starch, precipitated chalk & magnesium carbonate.

Adhesion. It is the property that is required in talc powders. This property in face powder can be imparted by talc, some metallic soaps of stearic acid like magnesium stearate and zinc stearate.

Aerosol. A suspension of liquid or solid particles in a gas, the particles often being in colloidal size range. Fog and smoke are common examples of natural aerosol. Fine sprays (Perfumes, insecticides, inhalants) are man-made aerosols.

Anti-foaming agent (anti-foamer). A substance which prevents the formation of a foam or considerably reduces foam persistence.

Anti-Perspirants. It is a product which reduces the sweating rate.

Antioxidants. An organic compound, added to cosmetics, food, fats and oils, etc., to retard oxidation. An antioxidant prevents the oxygen from beginning this reaction.

Antiseptic. A substance applied to humans or animals that retards or stop the growth of micro-organisms without necessary destroying them.

Brightening agent. Product intended to improve or restore the purity of the colour; it may also play a part in giving the textile articles the desired finish from the point of view of feel and sheen. *Note* – These products are generally those mentioned under the heading of processing aid.

Cake Eye Shadow. It is usually a combination of waxes such as castor oil, mineral oil, and cetyl alcohol.

Cake Mascaras. Cake mascaras are carefully blended mixtures of fatty wetting agents, which make the product spread smoothly on the lash when applied with a wet brush, and waxes like beeswax, carnauba or paraffin wax.

Chelating Agent. A substance having a molecular structure embodying several electron-donor groups which render it capable of combining with metallic ions by chelation.

Cold Creams. Cold creams are generally emollient emulsions from which water evaporates on application, producing a cooling effect on the skin.

Depilatories. Depilatories are normally waxy mixtures based on beeswax, resin, paraffin wax and petroleum jelly.

Deodorants. A deodorant is a product which contains germicides to prevent the bacteriological breakdown of sweat into unpleasant smelling substances.

Denture Cleansers. Denture cleansers are in powder or tablet form and may effervesee when put into water. These are prepared from mixed alkalies such as calcium and sodium phosphates with an oxidising agent like perborate.

Disperse Phase. It is the discontinuous phase of a dispersion.

Dispersion. Mixing a substance system consists of two or more phases, in which one substance (dispersed phase) is distributed in the finest form (dispersed) into another substance (dispersion agent). The particle of dispersed phase as well as those of the dispersion agent can be solid, liquid or gaseous.

Emolliency. Emolliencies or Emollients are terms often used to describe the action or component of cosmetic creams. An emollient is a substance which, because of its chemical or physical nature will help the skin retain water.

Emulsions. A heterogeneous system made by dispensing small globules of one liquid in another liquid which forms a continuous phase creams and milks are all emulsions; that is, they are products prepared by mixing oils and water to form an opaque fluid, uniform in appearance.

Eye Make-up. The basic products of eye make-up are *mascara* in cake, cream and liquid forms; *Eye liner* in stick, cake, or brush–applied fluid forms, eyebrow make-up; and eye shadow.

Eye Shadows. Eye shadows, in liquefying cream form contain water in lower quantities and waxes such as beeswax, ozokerite, and spermaceti to give body to the product.

Foaming Agent (foamer). A substance which, when introduced into a liquid, confers on it an ability to form foam.

Homogenisation. Changing the condition of distribution and the particle size of the inner phase of emulsion and suspension so that, from a microscopic perspective, a homogeneous system is produced and the dispersed phase neither settles at the bottom nor rises to the surface (creaming) without the influence of external forces.

Lipsticks. Lipstick the shiny stick of colour applied from a propelling type holder, is by far the most important lip colouring sold.

Lash-building Mascaras. Lash-building mascaras are similar to the ordinary products but contain finely chopped fibres, frequently nylon, which adhere to the natural lash and add length and thickness.

Liquefying Creams. Liquefying creams are normally translucent products of the correct consistency for smoothing on the skin without drag.

Liquid Eye Shadow. Liquid eyeshadow may be suspension of colour ingredients in oil bases such as isopropyl myristate, palmitate or stearate with mineral or vegetable oils.

Liquid Mascaras. Liquid mascaras may be based on water or water mixed with ethyl alcohol.

Manicure Products. Manicure products are nail lacquers providing colour and polish to the nails; nail lacquer remover; nail creams and lotions, to combat dryness and brittleness of the nails; and cuticle removers, used to define the shape of the nail by removing the overlapping cuticle.

Mixing. The basic operation of chemical technology that serves to create the most complete homogenisation of substances. The substances should be so extensively combined that every portion of the mixture contain the most uniform compound of the individual components possible.

Mouthwashers. Mouthwashers are mainly water/ethyl alcohol solutions, and are available in many forms for many purposes.

Oral Hygiene Products. The important products sold for oral hygiene purposes are: Toothpastes; Toothpowders and solid dentifrices; mouth washes and denture cleaners.

Preservatives. Preservatives inhibit or show down damaging changes brought out by micro-organisms.

Rouge. Rouge or blushers as they are now often called, are obtainable in five main forms; powder; pressed powder, liquid, cream, and stick.

Sequestering Agents. Certain chemicals incorporated in small quantities into cosmetic preparation have the ability to lock up these traces of metals within their structure, and render them inactive. Such compounds are called sequestering agents.

Skin Toners. Skin Toners are aqueous or aqueous alcoholic lotions & these have become part of the facial skin treatment.

Slip. Slip is the quality of powder which makes it easy to spread & imparts a characteristic smooth feeling on the skin.

Sunscreen Creams and Lotions. Sunscreen preparations are products which prevent harmful radiation from reaching the skin and causing the well known symptoms of sunburn.

Suspension. The distribution of very small but not molecular particles of a solid substance within a liquid. Suspensions, like emulsions, are generally cloudy to the eye and tend to settle under the influence of gravity.

Thickening Agents. These are employed to adjust viscosity.

Toothpaste and Toothpowder. A tooth paste is composed of four basic components other than water, i.e., abrasives, humectants, binders and detergents. Toothpowders, normally contains only the abrasive and detergent, whereas the solid dentifrice has in addition a binder or merely a humectant.

Ultra-Violet Absorbers. Certain chemicals when added to the products in small quantities have the ability to absorb the sun's damaging ultra-violet radiation and prevent the spoiling of the products.

Vanishing Cream. Vanishing creams are emulsions which rub into the skin and leave a dry feel. Creams of this type are used in general skin care and form the basis of many hand, foundation, and moisturising creams.

Wetting Agent. A substance which, when introduced into a liquid, increases its wetting tendency.

BIS Specifications

BIS SPECIFICATIONS FOR COSMETIC RAW MATERIALS

Name of Cosmetic Raw Materials	:	BIS Number
Acetic Acid	:	695-1986
Acetic Anhydride	:	1235-1988
Acetone	:	170-1986
Ammonium chloride for cosmetic industry	:	11015-1984
Beeswax, bleached for cosmetic industry	:	4028-1992
Benzyl Acetate	:	3858-1993
Benzyl Alcohol	:	3924-1980
Benzyl Chloride	:	6412-1971
Borax for cosmetic industry	:	11023-1984
Boric Acid for cosmetic industry	:	263-1990
Butyl p-hydroxy benzoate for cosmetic industry	:	6334-1987
Calcium Carbonate, precipitated for cosmetic industry	:	918-1985
Calcium stearate for cosmetic industry	:	2519-1983
Castor oil for cosmetic industry	:	11486-1985
Coconut diethanolamide	:	7101-1983
Di-calcium phosphate for dentifrice	:	1767-1980
Ethyl acetate for cosmetic industry	:	10283-1982
Ethyl p-hydroxy benzoate for cosmetic industry	:	4652-1987
Glycerine for cosmetic industry	:	12590-1988
Glyceryl monostearate for cosmetic industry	:	4236-1985
Groundnut oil for cosmetic industry	:	11375-1985

Isopropyl alcohol for cosmetic industry	:	10301-1982
Isopropyl myristate for cosmetic industry	:	5356-1983
Kaolin for cosmetic industry	:	1463-1983
Lanolin, anhydrous for cosmetic industry	:	5340-1981
Magnesium carbonate for cosmetic industry	:	2528-1984
Magnesium oxide for cosmetic industry	:	2529-1983
Magnesium stearate for cosmetic industry	:	2521-1984
Methyl p-hydroxybenzoate for cosmetic industry	:	4653-1985
Mineral oil for cosmetic industry	:	7299-1974
P-phenylene diamine	:	10377-1982
Petroleum Jelly for cosmetic industry	:	4887-1980
Perfumery Material		
Olfactory Assessment	:	2284-1988
Sampling & Tests	:	326-1984
Terminology	:	6597-1988
Propyl p-hydroxy benzoate for cosmetic industry	:	6333-1985
Sesame oil for cosmetic industry	:	11376-1985
Sodium carboxyl methyl cellulose	:	9831-1987
Sodium lauryl ether sulphate	:	11487-1985
Sodium lauryl sulphate for cosmetic industry	:	3986-1988
Sodium silicate for cosmetic industry	:	9601-1980
Sorbitol solution (70%)	:	3987-1983
Stearic acid for cosmetic industry	:	9681-1980
Talc for cosmetic industry	:	1462-1985
Titanium dioxide for cosmetic industry	:	2851-1983
Water soluble sodium carboxymethyl cellulose for cosmetic industry	:	9830-1988
Zinc oxide for cosmetic industry	:	2850-1983
Zinc stearate for cosmetic industry	:	2520-1984

BIS SPECIFICATIONS FOR COSMETICS

After-shave Lotion	:	9255-1979
Bindi (Liquid)	:	10998-1984
Cologne	:	8482-1977
Cosmetic pencils	:	9832-1981

Depilatories, chemicals	:	9336-1988
Hair Cream	:	7679-1978
Hair dye Powder	:	10350-1993
Hair oil	:	7123-1993
Henna powder	:	11142-1964
Kum Kum powder	:	10999-1964
Lip salve	:	10284-1982
Lipstick	:	9875-1990
Nail Polish (Nail enamel)	:	9245-1975
Oxidation, hair dyes, liquid	:	8481-1977
Pomades & brilliantines	:	9339-1988
Powder hair dyes	:	10350-1982
Shampoo, synthetic detergent based	:	7884-1992
Shampoo, soap based	:	7669-1990
Shaving cream	:	9740-1981
Skin cream	:	6608-1978
Skin powder for infants	:	5339-1978
Skin powder	:	3959-1978
Toothpaste	:	6356-1993
Tooth powder	:	5383-1976
Toilet soap	:	2888-1983
Antibacterial Toilet soap	:	11479-1985
Shaving soap	:	5784-1970
Transparent Toilet soap	:	11303-1985

Appendices

APPENDIX -I

MANUFACTURERS OF RAW MATERIALS

CALCIUM CARBONATE

Citurgia Biochemicals Ltd., Neville House, J. N. Heredia Marg, Ballard Estate, Mumbai - 400001.

Sturdia Chemicals Ltd., Neville House, Graham Road, Ballard Estate, Mumbai - 400001.

Radha Chemicals Ltd., 3-Synagogue Street, Calcutta - 700001.

Searole Chemicals Ltd., 6-Church Lane, Calcutta.

Carbonate (I) Ltd., Hong Kong House, 9-Dalhousie Square, Calcutta.

Thirani Chemicals Ltd., 1439-40, Loni Road, Delhi - 110032.

Gulshan Sugar & Chemicals Ltd., 9th km. Janpath Road, Muzaffar Nagar (U.P.).

Aditya Chemicals Ltd., 407 Red Rose, Nehru Place, New Delhi 110019.

Lime Chemicals Ltd., 161-Bara Iman Road, Null Bazar, Mumbai.

PERFUMES & AROMATIC CHEMICALS

S H Kelkar & Co. Ltd., L. B. Shastri Marg, Mulund, Mumbai.

Industrial Perfumes Ltd., Hay Bundar Road, Tank Road, P.O. Sweri, Mumbai.

Naarden (I) Ltd., Plot No. 6, Saki Vihar Road, Saki Naka, Mumbai - 400072.

Calcutta Chemical Co. Ltd., 35 - Panditia Road, Calcutta.

Bhavna Chemicals Ltd., 53-57, Laxmi Insurance Building, Sir Phirozshah Mehta Road, Mumbai.

Camphor & Allied Products Ltd, Jahangir Building, 133-Mahatma Gandhi Road, Mumbai - 400001.

Dr. Paul Lohmann (I) Ltd., 145/1-Jessore Road, Calcutta.

Richardson Hindustan Ltd., Tieckon House, Dr. Moses Road, Mumbai - 400057.

Dabur (I) Ltd., 145 - Rash Bihari Avenue, Calcutta - 700029.

Maschineiyer Aromatics (I) Pvt. Ltd., Grand Southern Trunk Road, Chennai.

Hindustan Lever Ltd., Hindustan Lever House, 165-166, Backbay Reclamation, Mumbai - 400020.

Organoma (I) Ltd., Toli Chowki, Hyderabad - 560008.

Sunanda Aromatic Industries, KKS Road, Metagalli, P.O., Mysore.

Karnataka Soaps & Detergents Ltd., 5-Crescent Road, High Ground, Bangalore - 560001.

Vanillin Fine Chemicals Ltd., 20 - Begumpat, Hyderabad (Andhra Pradesh).

Bush Boake Allen (I) Ltd., 1-5-7 – Sells Street, St. Thomas Mount, Chennai - 600016.

Asian Chemical Works, 124-126, Shamal Das Gandhi, Princess Street, Mumbai - 400002.

ACETIC ACID

Andhra Sugar Ltd., Venkataraya Puram, Tanuku - 534211 (Andhra Pradesh).

Gujchem Distilleries Ltd., National Chambers, Near Depali Theatres, Ashram Road, Ahmedabad - 9

Indian Organic Chemical Industries Ltd., Near Excelsior Building, Wallace Street, Mumbai - 400001.

Sirsilk Ltd., Lingapur House, 3-6–237, Himayat Nagar Road, Hyderabad - 29.

Somaiya Organic Chemicals Ltd., Fazalboy Building, Mahatma Gandhi Road, Fort, Mumbai - 400023.

Vam Organic Chemicals Ltd., Skyline House, 85 - Nehru Place, New Delhi - 110019.

ACETONE

Sirsilk Ltd., Lingapur House, 3-6–237, Himayat Nagar Road, Hyderabad - 29.

National Organic Chemicals Ltd., Mafatlal Centre, Nariman Point, Mumbai - 400021.

Herdillia Chemicals Ltd., Air India Building, Nariman Point, Mumbai - 400021.

Shri Chalthan Vibhag Khan Udyog, Sahkari Mandli Ltd., District Surat, Gujarat.

ACETANILIDE & ANILINE

Hindustan Organic Chemicals Ltd., Rasayani, Kulaba, Maharashtra.

BENZENE

National Organic Chemicals Ltd., Mafatlal Centre, Nariman Point, Mumbai - 400021.

Indian Petrochemicals Corp. Ltd., P.O. Petrochemicals, Vadodara (Gujarat).

Indian Oil Corporation Ltd., 254-C, Dr. Annie Besant Road, Prabhadevi, Mumbai - 400025.

Steel Authority of India, Ispat Bhawan, Lodhi Road, New Delhi - 110003.

BUTANOL (BUTYL ALCOHOL)

Kolhapur Sugar Mills Ltd., Kasava Bhavada, Kolhapur - 416006.

National Organic Chemicals Ltd., Mafatlal Centre, Nariman Point,
Mumbai - 400021.

Somaiya Organic Industries Ltd., Narang House, 34-Shivaji Marg,
Mumbai - 400029.

BUTYL ACRYLATE

Colour Chem Ltd., Dinshaw Vachha Road, 194-Churchgate Reclamation,
Mumbai - 400020.

Indian Petrochemicals Ltd., P.O. Petrochemicals, Vadodara (Gujarat).

Rajprakash Chemicals Ltd., 114-Mittal Chambers, Nariman Point,
Mumbai - 400020.

BENZYL CHLORIDE

Indian Organic Chemicals Industries Ltd., New Excelsior Building, Wallace
Street, Mumbai - 400001.

Organoma (I) Pvt. Ltd., 4-5-146 Sultan Bazar, Hashmatganj,
Hyderabad - 500001.

CHLOROFORM

Mettur Chemicals & Industrial Corporation Ltd., Mettur Dam, Salem,
Tamil Nadu.

Standard Alkali Chemicals, Mafatlal Centre, Nariman Point, Mumbai - 400021.

CARBON TETRACHLORIDE

Mettur Chemicals & Industrial Corporation Ltd., Mettur Dam, Salem,
Tamil Nadu.

National Rayon Corporation Ltd., Eros Theatre Building, 1-Jamshedji Tata
Road, Mumbai - 400020.

Standard Alkali Chemicals, Mafatlal Centre, Nariman Point, Mumbai - 400021.

DIACETONE ALCOHOL

National Organic Chemicals Industries Ltd., Mafatlal Centre, Nariman Point,
Mumbai - 400021.

Herdillia Chemicals Ltd., Air India Building, Nariman Point, Mumbai - 400021

DIMETHYL TETRA PHTHALATE

Indian Petrochemicals Ltd., P.O. Petrochemicals, Vadodara (Gujarat).

Bombay Dyeing & Mfg. Co. Ltd., Patalganga, District Raigarh, Maharashtra.

Bongaigaon Refinery & Petrochemicals Ltd., 112-Surya Kiran Building, 19-K.G. Marg, New Delhi - 110001.

ETHYL ACETATE

Andhra Sugar Ltd., Venkataraya Puram, Tanuku -534211.

Sirsilk Ltd., Lingapur House, 3-6–237 Himayat Nagar Road, Hyderabad - 29.

Indian Organic Chemicals Ltd., Near Excelsior Building, Wallace Street, Mumbai - 400001.

Somaiya Organics (I) Ltd., Narang house, 34 Shivaji Marg, Mumbai - 400 029.

Mysore Acetate & Chemicals Ltd., P.O. Mandya, Karnataka - 571404.

HEXAMINE

Aegis Chemicals Industries Ltd., Baldora Bhawan, 117 Maharishi Karve Marg, Opp. Churchgate Station, Mumbai - 400020.

Allied Resins & Chemicals Ltd., 13 Camac Street, Calcutta - 700017.

Cibatul Ltd., P.O. Atul - 396020, District Valsad, Gujarat.

ISOPROPYL ALCOHOL (ISOPROPANOL)

National Organic Chemical Industries Ltd., Mafatlal Centre, Nariman Point, Mumbai - 400021.

METHYL ISOBUTYL KETONES

National Organic Chemical Industries Ltd., Mafatlal Centre, Nariman Point, Mumbai - 400021.

XYLENE

Indian Petrochemicals Ltd., P.O. Petrochemicals, Vadodara (Gujarat).

TOLUENE

Indian Oil Corporation Ltd., 254-C, Dr. Annie Besant Road, Prabhadevi, Mumbai - 400025.

Indian Petrochemicals Ltd., P.O. Petrochemicals, Vadodara (Gujarat).

Steel Authority of India, Ispat Bhawan, Lodhi Road, New Delhi - 110003.

FACE CREAM AND POWDER

Bengal Chemical & Pharmaceutical Works Ltd., 164 - Maniktala, Main Road, Calcutta - 700013.

Burrough Welcome & Co. Ltd., 38-C, Old Prabhadevi Road, Mumbai.

Ciba Geigy Ltd., Royal Insurance Building, 14 - J. Tata Road, Mumbai

Colgate Palmolive (I) Ltd., Steel Creta House, 6-Dinshaw Vacha Road, Churchgate Reclamation, Mumbai.

Ponds India Ltd., 26 - Commander-in-Chief Road, Chennai.

J. K. Helen & Curtis, Kharan Road, Jekegram, Thane - 400646.

Ranbaxy Laboratories, 19 - Nehru Place, New Delhi - 110019.

Dabur (I) Ltd., 142 - Rash Behari Avenue, Calcutta.

Godrej Soaps Ltd., Eastern Express Highway, Vikroli, Mumbai.

Hindustan Lever Ltd., Hindustan Lever House, 165/166 - Backbay Reclamation, Mumbai 400020.

Johnson & Johnson (I) Ltd., 39 - Forget Street, Mumbai.

Lakme Ltd., Govind Station Road, Troubay Road, Deonar, Mumbai.

HAIR OIL

Bengal Chemical & Pharmaceuticals Works Ltd., 164-Maniktala, Main Road, Calcutta - 700013.

Dabur (I) Ltd., 142-Rash Behari Avenue, Calcutta.

Calcutta Chemical Co. Ltd., 35 - Panditia Road, Calcutta.

Godrej Soaps Ltd., Eastern Express Highway, Vikroli, Mumbai.

Swastik Household & Industrial Products Ltd., 13/5 Walchand Hira Chand Marg, Ballard Estate, Mumbai.

TOOTHPASTE

Bengal Chemical & Pharmaceuticals Ltd., 164-Maniktala, Main Road, Calcutta - 700013.

Colgate Palmolive (I) Ltd., Steel Creta House, 6 - Dinshaw Vacha Road, Churchgate Reclamation, Mumbai.

Ciba Giegy Ltd., Royal Insurance Building, 14 - J. Tata Road, Mumbai.

Geoffrey Manners & Co. Ltd., Bombay Agra Road, Ghatkopar, Mumbai.

Hindustan Lever Ltd., Hindustan Lever House, 165/166-Backbay Reclamation, Mumbai 400020.

J. L. Marison & Jones (I) Ltd., P.O. Box No. 2214, Yashwantpur, 9th Mile Tumkar Road, Bangalore.

APPENDIX II

MANUFACTURERS OF PLANT AND MACHINERY

AGITATORS & MIXERS

A.P.V. Equipment Co. Ltd., Kalwa, Hamilton House, 3, Graham Road, Ballard Estate, Mumbai, 400038.

A.P.V. Engg. Co. Ltd., Jessore Road, Dum Dum, Calcutta - 700028.

Larsen & Tourbo Ltd., L & T House, Ballard Estate, Mumbai - 400038.

The National Rayon Corp. Ltd., Ballard Estate, Mumbai - 400038.

The New Standard Engg. Co., NSE Estate, Goregaon East, Mumbai - 400002.

Rajinder Mechanical Industries Pvt. Ltd., 325, Kalabadevi, Mumbai - 400002.

Shreno Ltd., 4/22-23, Industrial Estate, Gorwa Road, Baroda - 390003.

Sohal Engg. Works, Sohal Industrial Estate, L.B. Shastri Marg, Bhandup, Mumbai - 400078.

Sudershan Chemical Ltd., 162, Wellesley Road, Pune - 411001.

DRYERS

Anup Engg. Ltd., Anil Starch Premises, Anil Road, P.B. No. 1164, Ahmedabad - 380025.

A. Stock & Company Ltd., 8, Garagacha Road, Calcutta - 700027.

Blanden Cole, Div. of Chemaux Ltd., Range Udyan Sitladevi, Temple Road, P.B. No. 6498, Mahim, Mumbai - 400016.

Bucka-Wolf New India Engg. Works Ltd., P.B. No. 22, Pimpri, Pune - 444918.

Hindustan Development Corpn. Ltd., Hindu Family Building, 27, Sir R.N. Mukerjee Road, Calcutta - 700039.

Indian Sugar & General Engg. Corp., P.O. Yamuna Nagar - 135001.

Larsen & Toubro Ltd., L & T House, Ballard Estate, Mumbai - 400038.

Utkal Machinery Ltd., Kansbahal - 770034 (Orissa).

EVAPORATOR & CONDENSER

Anup Engg. Ltd., Anil Starch Premises, Anil Road, P.B. No. 1164, Ahmedabad - 380025.

The APV Equipment Co. Ltd., Kalwa, Hamilton House, 3, Graham Road, Ballard Estate, Mumbai - 400038.

APV Engg. Co. Ltd., 2, Jessore Road, Dum Dum, Calcutta - 700028.

Bharat Heavy Plate & Vessels Ltd., Visakhapatnam - 530012

Larsen & Toubro Ltd., L & T House, Ballard Estate, Mumbai - 400038

Saurashtra Engg. Corpn. Pvt. Ltd., Khatani Indl. Estate, Mumbai - 400070.

Testeels Limited, Navdeep Opp. Akashwani, Ashram Road, P.B. No. 5 Navjivan, Ahmedabad.

Anup Engg. Ltd.. Anil Starch Premises. Anil Road. P.B. No. 1164.
Ahmedabad - 380025.

PRESSURE VESELS, REACTIONS VESSELS, AUTOCLAVE CHEMICALS STORAGE TANKS, CHLORINE TANK CONTAINERS

Bharat Heavy Electricals Limited, 18-20 Kasturba Gandhi Marg, H.T. House,
New Delhi - 110001.

Bharat Heavy Plate & Vessels Ltd., Vishakhapatnam - 530012.

Hindustan Development Corp. Ltd., Hindu Family Building, 27, Sir R.N.
Mukherjee Road, Calcutta - 700039.

Indian Sugar & General Engg. Corp., P.O. Yamuna Nagar - 135001.

The K.C.P. Limited, Rama Krishna Building, 38, Mount Road,
Chennai - 600006.

Larsen & Toubro Ltd., L & T House, Ballard Estate, Mumbai - 400038.

Testeels Limited, Navdeep, Opp. Akashwani, Ashram road, P.B. No. 5,
Navjivan, Ahmedabad.

Texmaco Limited, Birla Building, 9/1, R.N. Mukherjee Road, Calcutta-700001.

Vijay Tanks & Vessels Pvt. Ltd., 93. L.B. Shastri Marg, Mulund (East),
Mumbai - 400080.

WEIGHING MACHINERY

Avery India Ltd., Avery House, 28/2, Waterloo Street, Calcutta - 730001.

India Machinery Co. Ltd., Dass Nagar, Howrah - 711105.

Gilanders Arbuthanot & Co. Ltd., A-1, Gillander House, Netaji Subhash Road,
Calcutta - 730001.

George Salter India Ltd., Chartered Bank Bldgs., Calcutta - 700001.

Power Build Ltd., P.B. No. 28, Vallabh Vidyanagar - 388120 (Gujarat).

Rollatainer Ltd., 13/6 Mathura Road, Faridabad - 121003 (Haryana)

Jyoti Weighing Systems Ltd., 14, Joy Builders Colony, Indore - 520006.

INDUSTRIAL FILTERATION EQUIPMENT

Purolator India Ltd., Hauz Khas, P.O. Yusuf Sarai, New Delhi - 110049.

Tak Machinery Ltd., Janmabhoomi Bhavan, Ghoga St. Fort, Mumbai.

Vulcan Laval Ltd., Mustafa Bldg., 7-A, Sir P.M. Road, Mumbai - 400001.

Vacuum Plant & Instruments Mfg. Co. Pvt. Ltd., 48-A, Mundhawa, District
Pune, Maharashtra.

SPRAY DRYING EQUIPMENTS

Vulcan Laval Ltd., Mustafa Bldg., Sir P. M. Road, Mumbai - 400001.

Larsen & Toubro Ltd., L & T House, Ballard Estate, Mumbai - 400038.

Nestler Boilers Pvt. Ltd., Dalal Chambers, Sir V. Thackersey Road,
Mumbai - 400020.

Macneil & Mager Ltd., Mackinnon Mackenzie Bldg., Ballard Estate,
   Mumbai - 400038.

ALUMINIUM COLLAPSIBLE TUBES

Bharat Aluminium Industry, Plot No. D-95, M.I.D.C., Industrial Area, Satpur,
   Nasik.

Bharat Containers Pvt. Ltd., Civil Court Lansdowne Road, Appolo Bundur,
   Mumbai - 400039.

Universal Cans & Containers Ltd., Century Bhawan, 771, Dr. Annie Besant
   Road, P.B. No. 19116, Worli, Mumbai.

Zenith Tin Works Ltd., Clerk Road, Opp. Race Course, Mahalaxmi,
   Mumbai - 400034.

Oriental Containers Ltd., Plot A-1, M.I.D.C., Murhad, Distt. Thane
   (Maharashtra).

CRUSHING & GRINDING MILLS

Alfred Herbert (I) Ltd., 13/3 - Strand Road, Calcutta - 700001.

KCP Ltd., 38-Mount Road, Chennai.

Heavy Engineering Corporation Ltd., P.O. Dhurwa, Ranchi, Bihar.

Tata Robin Fraser Ltd., 11- Station Road, Burna Mines, Jamshedpur - 831007.

ACC Vickers Babcock Ltd., Express Tower, Nariman Point, Mumbai - 430021.

Elecon Engineering Co. Ltd., Vallabh Vidyanagar, Gujarat.

Kusum Engineering Co. Ltd., 25 - Swallow Lane, Calcutta - 700001.

PACKAGING MACHINERY

Hindustan Lever Ltd., Hindustan Lever House, 169/166 - Backbay Reclamation,
   Mumbai - 400020.

Rollatainers Ltd., 13/6 - Mathura Road, Faridabad - 121003.

Malins of India Ltd., 12 - Mohali Biren Roy Road, Calcutta - 700034.

Wimco Ltd., Indian Mercantile Chambers, Ramjidas Komanti Marg, Mumbai.

Ravalgaon Sugar Farms Ltd., Construction House, Walchand Hira Chand Marg,
   Ballard Estate, Mumbai - 430038.

Reed Medway Packaging Co. Ltd., 179 - Jorbagh, New Delhi - 110003.

Printpack Machinery Ltd., Link Road, Faridabad, Haryana.

FILTER PRESS

Punjab Oil Expeller Co., Patel Marg, Ghaziabad, (U.P).

Lyallpur Engg. Co., G.T. Road, Ghaziabad, (U.P).

S. P. Engg. Corporation, 7/97 L. Road, Kanpur, (U.P).

Pioneer Iron & Steel Corporation Ltd., 6 - Tarachand Ganguli, Belur, Howrah,
   Calcutta.

CAN SEALING MACHINE

Metal Box Co. of India Ltd., Barlous House, Chowringhee, Calcutta - 700020.

PULVERISING MACHINERY

International Combustion (I) Ltd., 101 - Park Street, Calcutta.

References

Yoshiro Masada, *Analysis of Essential Oils by Gas Chromatography and Mass Spectrometry*, John Wiley and Sons, Inc., New York, London.

E. Sagarin, *The Science and Art of Perfumery*, McGraw Hill Books Co. Inc., New York.

Urdang George, *Essential Oils*, Van Nostrand, New York.

W.A. Poucher, *Perfumes, Cosmetics and Soaps*, Chapman and Hall Ltd.

G. Fenaroli, *Handbook of Flavour Ingredients*, Chemical Publishing Co., New York.

Ernest J. Perry, *The Chemistry of Essential Oils*, Reinhold Publishing Corporation, USA.

M. Gutcho, *Synthetic Perfumery Material*, Noyes Data Corporation, USA.

R.W. Moncrieff, *The Chemical Senses*, Leonard Hill London.

Y.R. Naves and G. Mazuyer, *Natural Perfume Materials*, Reinhold, New York.

S. Arctander, *Perfumes and Flavour Materials of Natural Origin*, Chemical Publishing, USA.

J. Pickthall, *Blending and Fixation in Perfumery*, Pergamon, USA.

S.F. Wells, *The Essential of Perfume Compounding*, Davis Brothers, London.

L. Ruzicka, *Olfaction and Odours*, Interscience, USA.

F.W. Brown, *Chemistry of Aromatic Chemicals*, Wiley Interscience, USA.

B.H. Kingston, *Perfumes and Essential Oils*, Noyes Data Corporation, USA.

Simon, Mendelsohn, *The Behavior of Perfumery Ingredients in Products*, Chemical Publishing Co. Inc., New York.

A.G. Arend, *Spray Dried Perfumes*, John Wiley and Sons Inc., New York.

G. Ohloff, *Analytical and Instrumental Techniques in Perfumes*, Symposium Held in Tokyo, 1989.

S. Vodoz, *Perfume and Flavor Chemicals*, Noyes Data Corporation, USA.

H. Fayol, *Packaging Perfumed Products*, Reinhold Publishing Corporation, USA.

F.V. Wells, *Perfumes and Consumers Acceptance*, United Trade Press, London.

E.S. Maurer, *Imitation in Perfumery*, Ellis Horwood Limited, New York, London.

J.M. Juran, *Quality Control Handbook*, Tata McGraw Hill, New York.

A. Scherm, *Active Ingredients in Cosmetics and Toiletries*, Chemical Publishing, USA.

E. Charlet, *Cosmetic Preservatives*, Noyes Data Corporation, USA.

D.F. Williams and W. H. Schmitt, *Chemistry and Technology of Cosmetics and Toiletries Industry*, Blackie Academic and Professional, London, New York, Tokyo.

G.A. Nowak, *Process Technology of Cosmetics*, Noyes Data Corporation, USA.

A.L.L. Hunting, *The Encyclopaedia of Shampoo Ingredients*, London Micelle Press.

CTFA, *International Cosmetics Ingredients Dictionary*, CTFA, Inc., Washington D.C.

H. Talsma and D.J.A Crommelin, *Cosmetics Preparation*, Mack Publishing Co. Easton, USA.

H.J. Wing, *Chemistry and Manufacture of Cosmetics*, Chemical Publishing Co., New York.

P.M. Love, *Cosmetics, Toiletries and Perfumes*, Reinhold Publishing Corporation USA.